Distributive Lattices

Raymond Balbes
and
Philip Dwinger

University of Missouri Press

ISBN 0–8262–0163–6

Library of Congress Catalog Card Number 73–94309
Printed and bound in the United States of America
University of Missouri Press, Columbia, Missouri 65201

Contents

II. Lattices

III. The Prime Ideal Theorem

IV. Topological Representations

V. Extension Properties

VI. Subdirect Products of Chains

VII. Coproducts and Colimits

VIII. Pseudocomplemented Distributive Lattices

IX. Heyting Algebras

X. Post Algebras

Preface

In the last four decades lattice theory has become a well-established branch of modern mathematics, an important role being played by the theory of distributive lattices. Although the systematic study of lattices started in the thirties with the pioneering work of G. Birkhoff and his school, the origin of lattice theory goes back much further. In particular, George Boole's studies in classical logic around 1850 led to the investigation of algebraic structures, which eventually became known as Boolean algebras. A systematic study of these algebras—which form a special class of distributive lattices—was not carried out until much later, again in the thirties, by M. H. Stone. The modern concept of lattice was introduced in the last decade of the 19th century by Charles S. Peirce and Ernst Schröder, while during that same period Richard Dedekind introduced the notion of modular lattice. Thus, the theory of distributive lattices can be considered one of the oldest branches of lattice theory, and although many other parts of lattice theory have meanwhile been developed extensively, it remains an area of considerable interest to researchers in the field. To a great extent this interest is due to the intimate connections between distributive lattices and other fields, notably topology, algebra, and logic.

The authors have attempted to present a fundamental treatment of the theory of distributive lattices which is designed to bring the reader to the frontiers of current research. Indeed, considerable attention has been paid to this treatment in the later chapters of the book, where the basic theory is applied to various special classes of distributive lattices. There are numerous exercises scattered throughout the book which the reader is urged to solve. An extensive bibliography is included to enable the reader to pursue the various topics more fully. The book is addressed to graduate students and to those mathematicians who work in the field or want to become acquainted with it.

Besides G. Birkhoff's three editions of Lattice Theory, *there are many other books which survey the whole area. Our purpose is rather to provide a thorough treatment of only a portion of lattice theory, namely distributive lattices. G. Grätzer's book* Lattice Theory: First Concepts and Distributive Lattices *also deals exclusively with distributive lattices and there is, of course, overlap in some areas; however, our treatment is essentially different, as are most of the special topics.*

Distributive lattices can be defined as a special class of partially ordered sets but also as a class of algebras with certain operations which satisfy a given set of identities. In recent years the development of universal algebra has emphasized the algebraic approach and has provided lattice theory with new tools and methods and has opened up new avenues of investigation as well. However, both the algebraic and order theoretic approaches are important, and they are therefore employed simultaneously throughout the book. Moreover, where relevant, the methods and terminology of category theory are applied. Category theory, which has also been developed to a great extent in recent years, has provided various branches of modern mathematics with a very convenient apparatus by means of which certain kinds of results can be expressed. In particular, the language of category theory makes it possible to exhibit clearly the relationship which exists between different classes of distributive lattices. These results can often be formulated in a very succinct and more elegant form within the framework of category theory. It should be emphasized, however, that only the most basic notions and tools of category theory will be used in this book and that the authors have tried to avoid the use of any of the more complicated methods and results of category theory. For example, the concept of reflective subcategory is defined and employed without making use of the notion of adjoint functor. In this way, the approach to the theory of distributive lattices in this book is essentially threefold: order theoretic, algebraic, and categorical. These viewpoints are often used in conjunction, and there is an emphasis on one or more of them whenever the topic under discussion makes this desirable.

In Chapter I all those elements of universal algebra and category theory which the reader will need—and in addition, some notions of set theory—are presented. The reader who has a working knowledge of category theory can proceed directly from Part 1 of Chapter I to Chapter II, referring back to the rest of Chapter I when necessary. It is also assumed that the reader has some knowledge of the elements of (point set) topology.

The fundamental theory of distributive lattices is developed in Chapters II–VII. Some highlights in these chapters are the prime ideal theory, the representation theory, free algebras, coproducts, and extension theorems. Chapter VI, which deals with a more special problem, may be omitted upon first reading. Chapters VIII–XII are all special topics in which the results obtained in the earlier chapters are applied. In the choice of these topics the authors have been led, to a large extent, by personal taste and their own interest. There are quite a few important classes of distributive lattices which have been extensively investigated but which have not been included in this book, the main reason being the limitation of space. Also, several of these omitted topics have been presented to the mathematical community in separate treatises. These areas include: Boolean algebras (which, however, are discussed in this book in some detail because of their essential role), partially ordered algebraic systems, cylindric algebras, and applications to logic and topology. The reader who is interested in one or more of these areas is referred to the existing literature. We mention the following references, which include some extensive bibliographic sources: J. C. Abbott [2]; *G. Birkhoff* [6]; *H. B. Curry* [1]; *Ph. Dwinger* [2]; *L. Fuchs* [1]; *G. Grätzer* [2], [4]; *P. R. Halmos* [2], [3]; *L. Henkin, J. D. Monk and A. Tarski* [1]; *H. Rasiowa* [3]; *H. Rasiowa and R. Sikorski* [1]; *J. B. Rosser* [1]; *R. Sikorski* [8]; *G. Szász* [1]; *W. Thron* [1].

The special classes of distributive lattices which are discussed in this book are pseudocomplemented distributive lattices (*Chapter VIII*), *Heyting algebras* (*Chapter IX*), *Post algebras* (*Chapter X*), *de Morgan algebras and Lukasiewicz algebras* (*Chapter XI*). *Finally, Chapter XII is entirely devoted to complete and* α*-complete distributive lattices, which may satisfy a higher degree of distributivity. The authors have made a serious attempt to make Chapters VII–XII as independent of one another* *as possible so the reader who has read the first chapters can devote himself to one or more topics which are of special interest to him. There are occasional references between these chapters which the reader should be able to use even if he has not read the other chapters. A natural relationship exists, however, between Chapters VIII and IX and between Chapters X and XI, whereas Chapter XII is in fact entirely independent of the other chapters. The reader should be able to read most of Chapter XII with no more knowledge than that of Chapter II.*

The numbering system for references is exemplified as follows: A theorem (corollary, definition, etc.) which is the second one in section 5 of Chapter VI will be referred to as Theorem (Corollary, Definition, etc.) VI.5.2. The chapter or section number is omitted if the reference occurs in that chapter or section; thus, Theorem 5.2 refers to theorem 2 in section 5 of the same chapter and Theorem 2 refers to Theorem 2 in that same section.

Raymond Balbes
University of Missouri—St. Louis
St. Louis, Missouri

Philip Dwinger
University of Illinois at Chicago Circle
Chicago, Illinois

March, 1974

Acknowledgments

The writing of this book began during the academic year 1969–1970 when Raymond Balbes was visiting at the University of Illinois at Chicago Circle. While continuing work on the manuscript, Philip Dwinger spent the academic year 1971–1972 at the California Institute of Technology and the University of Amsterdam. We are grateful to these institutions for their hospitality.

We are greatly indebted to many persons who have given us valuable assistance, advice, and support. J. Berman has made numerous contributions and generously provided us with new results and improvements in existing proofs of theorems. The second author is grateful for the many stimulating discussions he had during his stay at the California Institute of Technology with W. A. J. Luxemburg. Much of the material was presented in graduate courses at the University of Illinois at Chicago Circle and in a seminar at the University of Amsterdam; the remarks and comments of the students and participants have led to quite a few improvements in the text. We also received many valuable comments and suggestions from M. E. Adams, W. J. Blok, and A. L. Burger. Our special thanks are due to D. C. Feinstein, who read the entire manuscript and who provided a great number of corrections and improvements. We are deeply aware that these persons have greatly benefited our work.

We express our thanks to Cheryl Anderson for the actual typing of the manuscript. Finally, we would like to express our gratitude to our institutions, the University of Missouri—St. Louis and the University of Illinois at Chicago Circle, for the support we have received during the preparation of the manuscript.

Distributive Lattices

Note that $f: \varnothing \to B$ is a function if and only if $f = \varnothing$. The term *mapping* or *map* is often used instead of function. Also, concerning the function (A, B, f), we call A the *domain* of f and B the *codomain* of f. We write

$$\operatorname{Im} f = f[A] = \{f(a) : a \in A\}$$

and call this set the *image* of f. If the domain and codomain are clear from the context, we refer to $f: A \to B$ simply as f. For functions $f: A \to B$ and $g: B \to C$, the *composition* $g \circ f: A \to C$ is defined to be the function such that $(g \circ f)(x) = g(f(x))$ for each $x \in A$. A function $f: A \to B$ is *one-one* if $f(x) = f(y)$ implies $x = y$ and is *onto* if for each $b \in B$ there exists an $a \in A$ such that $f(a) = b$. If both conditions hold then f is called a *one-one correspondence*. In this case the function $f^{-1}: B \to A$ defined by $f^{-1}(b) = a$ where $f(a) = b$ is called the *inverse* of f. Now let $A \subseteq B$. The function $1_{A,B}: A \to B$ defined by $1_{A,B}(x) = x$ is called an *inclusion* and if $A = B$, then $1_{A,A}: A \to A$ is the *identity* on A. We write 1_A for $1_{A,A}$. Clearly if $f: A \to B$ is one-one and onto then $f \circ f^{-1} = 1_B$ and $f^{-1} \circ f = 1_A$. Also for $A \subseteq B$ and a function $g: B \to C$ we obtain the *restriction* of $g: B \to C$ to A by defining the function $g|A: A \to C$ by $(g|A)(x) = g(x)$ for each $x \in A$. If $B \subseteq A'$ then a function $f: A' \to C$ for which $f|B = g$ is called an *extension* of g.

An *equivalence relation* on a set A is a relation R on A which satisfies

(i) $(x, x) \in R$ for each $x \in A$ (*reflexive law*).

(ii) $(x, y), (y, z) \in R \Rightarrow (x, z) \in R$ (*transitive law*).

(iii) $(x, y) \in R \Rightarrow (y, x) \in R$ (*symmetric law*).

For each $x \in A$, we write

$$[x]_R = \{y \in A : (x, y) \in R\}$$

and we call the sets, $[x]_R$, *equivalence classes*. The set $\{[x]_R : x \in A\}$ is denoted by A/R. There exists a *natural map* $\nu_R: A \to A/R$ defined by $\nu_R(x) = [x]_R$. If only one relation R is under consideration then we will write $[x]$ instead of $[x]_R$.

A *family* $(a_s)_{s \in S}$ of elements of a set A is a function f from S to A such that $f(s) = a_s$. Note that we may have $a_s = a_t$ with $s \neq t$. In particular $(a_s)_{s \in \varnothing}$ is the function $(\varnothing, A, \varnothing)$. A *non-empty family* will be one in which $S \neq \varnothing$. If we write $(a_s)_{s \in S} \subseteq A$ then this will always mean that $(a_s)_{s \in S}$ is a family of elements of A. Moreover when we write $A = (a_s)_{s \in S}$ we mean that the image of $(a_s)_{s \in S}$ is exactly A. The notation $(A_i)_{i=1}^n$ is used for $(A_i)_{i \in \{1, \cdots, n\}}$. Now let $(A_s)_{s \in S}$ be a family of sets. We define the *cartesian product* $\bigtimes_{s \in S} A_s$ of the family $(A_s)_{s \in S}$ to be the set of all functions $f: S \to \bigcup_{s \in S} A_s$ where $f(s) \in A_s$ for each $s \in S$. This is a natural generalization of the former definition of cartesian product, for if S is finite then there is an obvious one-one correspondence between the n-tuples of the first definition and the functions of the second. In the sequel we will use whichever terminology is most convenient. Note that if $S = \varnothing$ then $\bigtimes_{s \in S} A_s = \{(\varnothing, \bigcup_{s \in S} A_s, \varnothing)\}$ so that, in this case, the cartesian product has exactly one member. Again, let $(A_s)_{s \in S}$ be a family of sets. For each $t \in S$ the map $p_t: \bigtimes_{s \in S} A_s \to A_t$ such that $p_t(f) = f(t)$ is called a *projection*; projections (in the n-tuple notation) are maps

$$p_i: A_1 \times \cdots \times A_n \to A_i$$

such that

$$p_i((a_1, \cdots, a_n)) = a_i.$$

Finally, if $A_s = A$ for each $s \in S$ we write A^S for $\times_{s \in S} A_s$.

Let n be a non-negative integer. An *n-ary operation* on a set A is a function $f: A^n \to A$. For $n = 0$, the term *nullary operation* is used. In this case, f maps from a one-element set into A and can therefore be thought of as a specific element in A. If f is 1-ary (called *unary*) we often write $f(a) = a'$ or $f(a) = \bar{a}$ etc. In case of a 2-ary (*binary*) operation $a + b$ or ab usually denotes $f(a, b)$.

2. Partially ordered sets

The notion of a partially ordered set, although basic to lattice theory, is also so important to the study of universal algebra that we make this our first definition. (See also Section II.1.)

Definition 1. A *partially ordered set* (or briefly, a *poset*) is a pair $(P, \leq)$ where P is a non-empty set and $\leq$ is a binary relation on P satisfying for $x, y, z \in P$:

(i) $x \leq x$ (reflexive law).
(ii) $x \leq y, y \leq z \Rightarrow x \leq z$ (transitive law).
(iii) $x \leq y, y \leq x \Rightarrow x = y$ (anti-symmetric law).

Where no confusion is likely to arise, it is customary to also use the symbol P to denote $(P, \leq)$. Also, if $x \leq y$ but $x \neq y$ then we write $x < y$.

Example 2. The set Z of integers determines a poset $(Z, \leq)$ where $\leq$ has the usual meaning.

Example 3. For the positive integers N, the divisibility relation $|$ makes $(N, |)$ a poset and is distinct from $(N, \leq)$ where $\leq$ has its usual meaning.

Example 4. Again let N be the positive integers and assume $\leq$ has its usual meaning. Then $(N, \leq)$ is a poset. We will only use the symbol N to denote the partially ordered set $(N, \leq)$.

Example 5. Let P be any non-empty set of subsets of a set X. Clearly $(P, \subseteq)$ is a poset.

The poset of all subsets of a set (partially ordered by inclusion) is called the *power set* of X.

We make the convention that when the partial ordering of a set of subsets is not explicitly stated, then it is to be taken as inclusion (see Example 5). Two members a, b of a poset are *comparable* if $a \leq b$ or $b \leq a$. If P is a poset then a non-empty subset C of P is called a *chain* in P if any two elements in C are comparable. If $C = P$ then P is called a *chain* (chains are also referred to as *linearly* or *totally ordered posets*). A non-empty subset of a chain, endowed with the same partial ordering, is called a *subchain* of the chain. For $n \geq 1$, denote by $\mathbf{n}$ the subchain $\{0, 1, \cdots, n-1\}$ of Z. In particular, $\mathbf{1}$ is a one-element chain. If no two distinct elements of a poset P are comparable, then P is said to be a *totally unordered set*.

Let P be a poset. An element $a \in P$ is called *maximal* if $x \in P$, $a \leq x$ implies $a = x$. There is at most one element b in P with the property that $x \leq b$ for all $x \in P$. This element, if it exists, is called the *greatest element* or the *unit* of P. Dually, a *minimal element* a is defined by the property that $x \leq a \Rightarrow x = a$ and *a least element* b (or *zero*) by the property that $b \leq x$ for all $x \in P$. The zero and unit of P are denoted, when they exist, by 0_P and 1_P (or simply by 0, 1). For a subset $S \subseteq P$, an element $u \in P$ is an *upper bound* for S if $s \leq u$ for all $s \in S$ and $v \in P$ is called a *lower bound* for S if $v \leq s$ for each $s \in S$.

Definition 6. A poset P is *well ordered* provided that if $\varnothing \neq S \subseteq P$ then S contains a least element.

In particular every well ordered poset is a chain: $\{x, y\}$ contains a least element. $(N, \leq)$ provides an example of a well ordered set.

We will make use of the axiom of choice which states that if $(A_s)_{s \in S}$ is a non-empty family of non-empty sets, then $\times_{s \in S} A_s \neq \varnothing$. It is well known (cf. H. Rubin and J. Rubin [1]) that this axiom is logically equivalent to each of the following:

(1) If every chain in a poset P has an upper bound then P has a maximal element. (Zorn's lemma).

(2) For every non-empty set S there is a relation $\leq$ on S such that $(S, \leq)$ is well ordered.

3. Ordinals and cardinals

It will be advantageous for us to consider ordinals and cardinals as posets.

Definition 1. An *ordinal* is a set α such that

(1) If $\beta \in \alpha$ then $\beta \subset \alpha$.

(2) If $\alpha \neq \varnothing$ then $(\alpha, \in')$ is a well ordered poset (where $\beta \in' \gamma$ provided $\beta \in \gamma$ or $\beta = \gamma$).

We name the finite ordinals:

$$\mathbf{0} = \varnothing$$

$$\mathbf{1} = \{\mathbf{0}\} = \{\varnothing\}$$

$$\mathbf{2} = \{\mathbf{0}, \mathbf{1}\} = \{\varnothing, \{\varnothing\}\}$$

$$\mathbf{3} = \{\mathbf{0}, \mathbf{1}, \mathbf{2}\} = \{\varnothing, \{\varnothing\}, \{\varnothing, \{\varnothing\}\}\}$$

and in general, for an integer $n \geq 1$,

$$\mathbf{n} = \{\mathbf{0}, \mathbf{1}, \cdots, \mathbf{n-1}\}$$

and is partially ordered by

$$\mathbf{0} \subseteq \mathbf{1} \subseteq \cdots \subseteq \mathbf{n-1}.$$

Recall from page 4 that $\mathbf{n}$ was also defined by

$$\mathbf{n} = \{0, 1, \cdots, n-1\}.$$

It is convenient to use the same symbol to represent both sets. The meaning of $\mathbf{n}$ in each case is then determined by context.

The class $\mathfrak{O}$ of all ordinals* is "well ordered" in the sense that if $\mathfrak{S}$ is a non-empty subclass of $\mathfrak{O}$ then there exists an $\alpha \in \mathfrak{S}$ such that $\alpha \subseteq \beta$ for each $\beta \in \mathfrak{S}$. This justifies the adoption of poset terminology when dealing with $\mathfrak{O}$. In particular, we write

$$\alpha \leq \beta \text{ whenever } \alpha \subseteq \beta, \alpha, \beta \in \mathfrak{O}.$$

We note that for each ordinal α we have $\alpha = \{\beta \in \mathfrak{O} : \beta < \alpha\}$. The smallest infinite ordinal is:

$$\omega = \{\mathbf{0}, \mathbf{1}, \mathbf{2}, \mathbf{3}, \cdots\}$$

Since there is a one-one correspondence between the non-negative integers and the finite ordinals we will sometimes identify $\mathbf{n}$ with n. For any ordinal $\alpha \neq \mathbf{0}$ we define the dual ordinal $\breve{\alpha}$ by simply reversing the partial ordering on α. That is, the elements of α and $\breve{\alpha}$ are the same and $\beta_1 \leq \beta_2$ in α if and only if $\beta_2 \leq \beta_1$ in $\breve{\alpha}$.

A *limit ordinal* is an ordinal α which has the property that

$$\alpha = \cup\{\beta \in \mathfrak{O} : \beta < \alpha\}.$$

Equivalently, α is a limit ordinal if for each ordinal $\beta < \alpha$ there exists an ordinal γ such that $\beta < \gamma < \alpha$. For example, $\mathbf{0}$ and ω are limit ordinals.

Definition 2. A *cardinal* is an ordinal α with the property that there does not exist a one-one correspondence between α and any ordinal β such that $\beta < \alpha$.

Thus the finite ordinals are all cardinals, as is ω, but the ordinal $\omega \cup \{\omega\}$ is not a cardinal.

In the theory of sets it is shown that for each set S there is a unique cardinal α for which there exists a one-one correspondence with S. We denote this cardinal by $|S|$ and say that α is the *cardinality* of S. Suppose A, B are disjoint sets such that $|A| = \alpha$, $|B| = \beta$. We define the cardinals:

$$\alpha + \beta = |A \cup B|, \alpha\beta = |A \times B| \text{ and } \alpha^\beta = |A^B|.$$

It is to be noted that these notions are, in fact, well defined. There is a one-one correspondence $\alpha \to \aleph_\alpha$ between the ordinals and the infinite cardinals such that

* In the von Neumann approach to set theory the term "class" is undefined and "set" is defined as a class that is a member of some class. In particular this approach avoids the contradiction implied by considering "the set of all sets". However, the ordinals then form a class and not a set so that the definition of poset is not applicable to $\mathfrak{O}$.

$\alpha \leq \beta \Leftrightarrow \aleph_\alpha \leq \aleph_\beta$; thus $\aleph_0 = \omega$. A set S is *countable* if $|S| \leq \aleph_0$, otherwise S is said to be *uncountable*.

A brief—but for our purposes complete—development of ordinals and cardinals can be found in Chapter 1 of R. S. Pierce [3]. In particular it is shown that every well ordered set can be identified with an ordinal.

Part 2. Universal algebra

Although the field of universal algebra is itself extensive, we present here only those parts of it which suit our specific needs. The reader who is interested in pursuing the topic should consult one of the standard works (P. M. Cohn [1]; G. Grätzer [2]; R. S. Pierce [3]).

4. Algebras and homomorphisms

Definition 1. A *similarity type* τ is an m-tuple $(n_1, \cdots, n_m)$ of non-negative integers. The *order* of τ is defined to be m; in symbols we write $o(\tau) = m$.

Definition 2. An *algebra* of type $\tau = (n_1, \cdots, n_{o(\tau)})$ is a pair (A, F) where A is a non-empty set and F is an $o(\tau)$-tuple $(f_1, \cdots, f_{o(\tau)})$ such that for each $1 \leq i \leq o(\tau)$, f_i is an n_i-ary operation on A.

In considering a class of algebras of a given similarity type τ we will, in general, use the same symbol f_i to represent the ith operation for each algebra in the class. Thus, F will denote the $o(\tau)$-tuple of operations for each of these algebras. For example, when considering two arbitrary groups, we usually denote the binary operation by $+$ (or by $\cdot$) in both groups. Where confusion is not likely to arise, we will also use the symbol A to denote the algebra (A, F). If A and B are algebras of the same similarity type then they are called *similar*. Finally, when denoting a binary operation by the symbol $\cdot$ it is customary to omit this symbol whenever possible.

Definition 3. Let A, B be algebras of type τ. A function $f: A \to B$ is a *homomorphism* if for each $1 \leq i \leq o(\tau)$,

$$f_i(f(a_1), \cdots, f(a_{n_i})) = f(f_i(a_1, \cdots, a_{n_i}))$$

whenever

$$(a_1, \cdots, a_{n_i}) \in A^{n_i}.$$

If f is one-one it is called an *embedding*. f is an *isomorphism* if f is one-one and onto; and f is called an *automorphism* of A if, in addition, $A = B$. An automorphism f of A is *proper* if $f \neq 1_A$.

Note that if f_i is a nullary operation which selects the element a from A and b from B and $f: A \to B$ is a homomorphism then $b = f(a)$.

If $f: A \to B$ is an isomorphism then f^{-1} is also an isomorphism and we write $A \cong B$. Indeed, it suffices to prove that f^{-1} is a homomorphism. For each $i \le o(\tau)$ and

$$(b_1, \cdots, b_{n_i}) \in B^{n_i},$$

$$f(f_i(f^{-1}(b_1), \cdots, f^{-1}(b_{n_i}))) = f_i(f(f^{-1}(b_1)), \cdots, f(f^{-1}(b_{n_i}))) = f_i(b_1, \cdots, b_{n_i}).$$

So

$$f_i(f^{-1}(b_1), \cdots, f^{-1}(b_{n_i})) = f^{-1}(f_i(b_1, \cdots, b_{n_i})).$$

Definition 4. Suppose (A, F) is an algebra of type τ and B is a non-empty subset of A. If for each $1 \le i \le o(\tau)$,

$$(b_1, \cdots, b_{n_i}) \in B^{n_i}$$

implies

$$f_i(b_1, \cdots, b_{n_i}) \in B,$$

then

$$(B, (f_1 | B^{n_1}, \cdots, f_{o(\tau)} | B^{n_{o(\tau)}}))$$

is called a *subalgebra* of (A, F).

Thus, the subalgebras of A are themselves algebras of type τ. The assertion that B is a subalgebra of A means: B is closed under the operations of A and that the operations f_i on B are those of A, restricted to B. Note that if f_i is nullary and selects $a \in A$ then its restriction selects a in the subalgebra B.

Lemma 5. *If $f: A \to B$ is a homomorphism then $f[A]$ is a subalgebra of B. In particular, if $A \subseteq B$ then $1_{A,B}$ is a homomorphism if and only if A is a subalgebra of B.*

Proof. Let $1 \le i \le o(\tau)$ and

$$(c_1, \cdots, c_{n_i}) \in (f[A])^{n_i}.$$

For each c_j there exists a_j such that $f(a_j) = c_j$ so

$$f_i(c_1, \cdots, c_{n_i}) = f_i(f(a_1), \cdots, f(a_{n_i})) = f(f_i(a_1, \cdots, a_{n_i})) \in f[A].$$

The algebra $f[A]$ is called a *homomorphic image* of A.

Definition 6. Let $S \subseteq A$. If there exists a smallest subalgebra of A that contains S, then it is called the *subalgebra of A generated by S* and is denoted by $[S]$.

Since the intersection of a set of subalgebras is again a subalgebra, except when the intersection is empty, $[S]$ exists whenever $S \neq \varnothing$. If S is $\varnothing$ then $[S]$ exists if the intersection of all of the subalgebras of A is non-empty. Thus, one method of showing that a subset $X \subseteq A$ is equal to $[S]$ is to prove that $S \subseteq X$, X is closed under the operations, and any subalgebra which contains S also contains X.

An algebra A is said to be *finitely generated* if there exists a finite set $S \subseteq A$ such that $[S] = A$.

Lemma 7. *If $f, g: [S] \to B$ are homomorphisms and $f|S = g|S$ then $f = g$.*

Proof. Let $T = \{x \in [S] : f(x) = g(x)\}$. Now T is a subalgebra of $[S]$ for if $1 \leq i \leq o(\tau)$ and

$$(x_1, \cdots, x_{n_i}) \in T^{n_i},$$

then

$$\begin{aligned} f(f_i(x_1, \cdots, x_{n_i})) &= f_i(f(x_1), \cdots, f(x_{n_i})) \\ &= f_i(g(x_1), \cdots, g(x_{n_i})) \\ &= g(f_i(x_1, \cdots, x_{n_i})) \end{aligned}$$

so

$$f_i(x_1, \cdots, x_{n_i}) \in T.$$

But $S \subseteq T \subseteq [S]$ and $[S]$ contains no proper subalgebra that contains S so $T = [S]$.

5. Direct products

Definition 1. The *direct product of a family* $((A_s, F))_{s \in S}$ *of algebras* of type τ is the algebra $(\times_{s \in S} A_s, F)$ where for each $1 \leq i \leq o(\tau)$ and

$$(a_1, \cdots, a_{n_i}) \in \Big(\bigtimes_{s \in S} A_s\Big)^{n_i},$$

$$f_i(a_1, \cdots, a_{n_i})(s) = f_i(a_1(s), \cdots, a_{n_i}(s))$$

for each $s \in S$.

The projections are onto homomorphisms, for if

$$(a_1, \cdots, a_{n_i}) \in \Big(\bigtimes_{s \in S} A_s\Big)^{n_i}$$

then for each $s \in S$,

$$\begin{aligned} p_s(f_i(a_1, \cdots, a_{n_i})) &= f_i(a_1, \cdots, a_{n_i})(s) \\ &= f_i(a_1(s), \cdots, a_{n_i}(s)) \\ &= f_i(p_s(a_1), \cdots, p_s(a_{n_i})). \end{aligned}$$

In the case $S = \varnothing$, the direct product is a one-element algebra in which each of the operations f_i are constant maps and there are no projections.

Lemma 2. *If $(g_s : A \to B_s)_{s \in S}$ is a family of homomorphisms and $S \neq \varnothing$ then the mapping $g: A \to \times_{s \in S} B_s$, $g(a)(s) = g_s(a)$ is a homomorphism.*

Proof. Let $(a_1, \cdots, a_{n_i}) \in A^{n_i}$. Then for each $s \in S$,

$$
\begin{aligned}
f_i(g(a_1), \cdots, g(a_{n_i}))(s) &= f_i(g(a_1)(s), \cdots, g(a_{n_i})(s)) \\
&= f_i(g_s(a_1), \cdots, g_s(a_{n_i})) \\
&= g_s(f_i(a_1, \cdots, a_{n_i})) \\
&= g(f_i(a_1, \cdots, a_{n_i}))(s).
\end{aligned}
$$

We will call g the *product* of $(g_s\colon A \to B_s)_{s\in S}$ and denote it by $g = \times_{s\in S}\, g_s$.

6. Congruence relations

Definition 1. Let A be an algebra of type τ. A *congruence relation on A* is an equivalence relation θ on A which satisfies the *substitution property*:

For each $i \in \{1, \cdots, o(\tau)\}$, *if*

(i) $(a_j, a_j') \in \theta$ for $j = 1, \cdots, n_i$ *then* $(f_i(a_1, \cdots, a_{n_i}), f_i(a_1', \cdots, a_{n_i}')) \in \theta$.

Note that if f_i is nullary then (1) is vacuously satisfied.

For a homomorphism $h\colon A \to B$, define the *kernel* of h (Ker h) to be the relation on A defined by $(a, a') \in \operatorname{Ker} h$ when $h(a) = h(a')$. Ker h is a congruence relation. For the substitution property, suppose $h(a_j) = h(a_j')$ for $j = 1, \cdots, n_i$. Then

$$
\begin{aligned}
h(f_i(a_1, \cdots, a_{n_i})) &= f_i(h(a_1), \cdots, h(a_{n_i})) \\
&= f_i(h(a_1'), \cdots, h(a_{n_i}')) \\
&= h(f_i(a_1', \cdots, a_{n_i}'))
\end{aligned}
$$

so

$$(f_i(a_1, \cdots, a_{n_i}), f_i(a_1', \cdots, a_{n_i}')) \in \operatorname{Ker} h.$$

The set of congruence relations on A is partially ordered (by inclusion) and has a least element 0 ($= \{(x, x)\colon x \in A\}$) and a greatest element 1 ($= A \times A$); 0 is sometimes called the *trivial congruence relation* on A. This partially ordered set is denoted by $\mathfrak{C}(A)$. Observe also that the intersection of congruence relations is a congruence relation.

Given a congruence relation θ on an algebra A, we construct the *quotient algebra*

$$A/\theta = (\{[x]_\theta\colon x \in A\}, F)$$

such that for $1 \le i \le o(\tau)$,

$$f_i([a_1]_\theta, \cdots, [a_{n_i}]_\theta) = [f_i(a_1, \cdots, a_{n_i})]_\theta$$

where

$$(a_1, \cdots, a_{n_i}) \in A^{n_i}.$$

The substitution property insures that f_i is well defined on A/θ. The *natural homomorphism* $\nu_\theta\colon A \to A/\theta$, $x \mapsto [x]_\theta$ is clearly a homomorphism since

$$\begin{aligned}\nu_\theta(f_i(a_1, \cdots, a_{n_i})) &= [f_i(a_1, \cdots, a_{n_i})]_\theta \\ &= f_i([a_1]_\theta, \cdots, [a_{n_i}]_\theta) \\ &= f_i(\nu_\theta(a_1), \cdots, \nu_\theta(a_{n_i})).\end{aligned}$$

Also $\operatorname{Ker} \nu_\theta = \theta$ and ν_θ is onto.

Theorem 2. (The homomorphism theorem) *If $f\colon A \to B$ is a homomorphism of A onto B and θ is a congruence relation on A such that $\theta \subseteq \operatorname{Ker} f$, then there exists a unique onto homomorphism $g\colon A/\theta \to B$ such that $g \circ \nu_\theta = f$. If, in addition, $\theta = \operatorname{Ker} f$ then g is an isomorphism.*

Proof. If $a, a' \in A$ and $[a]_\theta = [a']_\theta$ then $(a, a') \in \theta \subseteq \operatorname{Ker} f$ so $f(a) = f(a')$. Hence the assignment $g\colon A/\theta \to B$, $g([a]_\theta) = f(a)$ is well defined and obviously onto. It is a homomorphism, for if

$$(a_1, \cdots, a_{n_i}) \in A^{n_i}$$

then

$$\begin{aligned}g(f_i([a_1]_\theta, \cdots, [a_{n_i}]_\theta)) &= g([f_i(a_1, \cdots, a_{n_i})]_\theta) \\ &= f(f_i(a_1, \cdots, a_{n_i})) \\ &= f_i(f(a_1), \cdots, f(a_{n_i})) \\ &= f_i(g([a_1]_\theta), \cdots, g([a_{n_i}]_\theta)).\end{aligned}$$

Also for $a \in A$,

$$(g \circ \nu_\theta)(a) = g([a]_\theta) = f(a)$$

so $g \circ \nu_\theta = f$. Clearly, g is unique. Now suppose $\theta = \operatorname{Ker} f$. Then if $g([a]_\theta) = g([a']_\theta)$ we have $f(a) = f(a')$ so $(a, a') \in \operatorname{Ker} f$ and therefore $[a]_\theta = [a']_\theta$.

7. Classes of algebras

In this book, we will be concerned with subclasses of the class of all algebras of a certain similarity type. Such subclasses will be denoted by boldface Roman capitals. If $\mathbf{K}$ is a class of algebras of type τ then for $A \in \mathbf{K}$, a subalgebra B of A is well defined. In this situation we will also call B a $\mathbf{K}$-subalgebra of A, but note that B need not be a member of $\mathbf{K}$. Similarly, for $A, B \in \mathbf{K}$ a homomorphism $f\colon A \to B$ is also called a $\mathbf{K}$-homomorphism. Finally, for $S \subseteq A$, $[S]$ (if it exists) will also be denoted by $[S]_{\mathbf{K}}$. This convention is especially useful as will be exemplified in the next paragraph.

Let $\mathbf{K}$ be a class of algebras of type $\tau = (n_1, \cdots, n_{o(\tau)})$. For each subset $I = \{i_1, \cdots, i_p\} \subseteq \{1, \cdots, o(\tau)\}$ we can form a new class $\mathbf{K}(I)$ of algebras of similarity type $(n_{i_1}, \cdots, n_{i_p})$ by taking for each algebra of $\mathbf{K}$ the underlying set and by disregarding the operations f_j for $j \notin \{i_1, \cdots, i_p\}$. Precisely, $\mathbf{K}(I)$ is the class of algebras $(A, (f_{i_1}, \cdots, f_{i_p}))$ where $(A, (f_1, \cdots, f_{o(\tau)})) \in \mathbf{K}$.

If $A \in \mathbf{K}$ we will often write $A \in \mathbf{K}(I)$ to mean that $(A, (f_{i_1}, \cdots, f_{i_p})) \in \mathbf{K}(I)$. In this way an algebra of type $(n_1, \cdots, n_{o(\tau)})$ can also be considered as an algebra of type $(n_{i_1}, \cdots, n_{i_p})$.

Throughout this book we will deal with classes of algebras **B** and **A** which are related to each other in the same way that $\mathbf{K}(I)$ is related to **K**, respectively. For the purpose of illustration, assume $\mathbf{A} = \mathbf{K}$ and $\mathbf{B} = \mathbf{K}(I)$. Now let $A \in \mathbf{A}$ and suppose $S \subseteq$ A. Then, in accordance with the convention made above $[S]_{\mathbf{B}}$ denotes the smallest subset of A closed under the operations $f_{i_1}, \cdots, f_{i_p}$ together with the operations $f_{i_1}, \cdots, f_{i_p}$, but $[S]_{\mathbf{A}}$ denotes the smallest subset of A closed under all of the operations $f_1, \cdots, f_{o(\tau)}$, together with $f_1, \cdots, f_{o(\tau)}$. Similarly a **B**-homomorphism preserves only $f_{i_1}, \cdots, f_{i_p}$ but an **A**-homomorphism preserves $f_1, \cdots, f_{o(\tau)}$ (similarly for isomorphisms, congruence relations, etc.).

Since many theorems (e.g. Theorem 12.4) hold for classes which include all subalgebras, homomorphic images, and direct products of members of the class, we adopt the following notation. For a class **K** of similar algebras:

H(K) denotes the class of isomorphic copies of homomorphic images of members of **K**.
S(K) denotes the class of isomorphic copies of subalgebras of members of **K**.
P(K) denotes the class of isomorphic copies of direct products of non-empty families of members of **K**.

Definition 1. A class **K** of similar algebras is *equational* if $\mathbf{H}(\mathbf{K}) \subseteq \mathbf{K}$, $\mathbf{S}(\mathbf{K}) \subseteq \mathbf{K}$ and $\mathbf{P}(\mathbf{K}) \subseteq \mathbf{K}$.

Note that there is an algebra A, consisting of a single element, for each similarity type τ. The operations f_i on A are all constant maps $f_i: A^{n_i} \to A$. This one-element algebra of type τ is a member of every equational class **K** of algebras of type τ and, in fact, it is a homomorphic image of each member of **K**. Observe that the class of algebras of type τ and the class of all one-element algebras of type τ are both equational. The latter class we will call the *trivial class*. Let **K** be a non-empty class of algebras of type τ. Since the intersection of equational classes of type τ which contain **K** is again equational, there is a smallest equational class which contains **K**. This class is called the *equational class generated by* **K**.

Now it can be verified that $\mathbf{H}(\mathbf{H}(\mathbf{S}(\mathbf{P}(\mathbf{K})))) \subseteq \mathbf{H}(\mathbf{S}(\mathbf{P}(\mathbf{K})))$, $\mathbf{S}(\mathbf{H}(\mathbf{S}(\mathbf{P}(\mathbf{K}))))$ $\subseteq \mathbf{H}(\mathbf{S}(\mathbf{P}(\mathbf{K})))$ and $\mathbf{P}(\mathbf{H}(\mathbf{S}(\mathbf{P}(\mathbf{K})))) \subseteq \mathbf{H}(\mathbf{S}(\mathbf{P}(\mathbf{K})))$, so we have that $\mathbf{H}(\mathbf{S}(\mathbf{P}(\mathbf{K})))$ is itself an equational class. Since it must be contained in any equational class that contains **K**, it is, in fact, the equational class generated by **K**.

Definition 2. A class **K** of similar algebras has the *congruence extension property* if for each subalgebra B of $A \in \mathbf{K}$ and each congruence relation θ on B, there exists a congruence relation Φ on A such that $\Phi \cap B^2 = \theta$.

If **K** is an equational class then the *congruence extension property* holds if and only if whenever $f: A \to B$ is a one-one homomorphism and $g: A \to C$ is an onto homomorphism then there exists an onto homomorphism $h: B \to D$ and a one-one homomorphism $k: C \to D$ such that $h \circ f = k \circ g$.

8. The duality principles

The terminology that is necessary for a precise statement of the duality principles can only be formulated within the context of mathematical logic, which is beyond the scope of this book. We will present here less sophisticated versions of these principles, but which will nevertheless enable the reader to deal with all situations which call for an application of one of these principles.

Let **K** be a class of algebras of type $\tau = (n_1, \cdots, n_{o(\tau)})$ and suppose that $\pi = (i_1, \cdots, i_{o(\tau)})$ is a permutation of $(1, \cdots, o(\tau))$. The class **K** is said to be *π-weakly dual* if for any algebra $(A, (f_1, \cdots, f_{o(\tau)})) \in \mathbf{K}$, the algebra $(A, (f_{i_1}, \cdots, f_{i_{o(\tau)}})) \in \mathbf{K}$. In particular, $n_j = n_{i_j}$ for $j = 1, \cdots, o(\tau)$.

Again, let $\tau = (n_1, \cdots, n_{o(\tau)})$ be a similarity type. By a logical formula $P = P(x, y_1, \cdots, y_{o(\tau)})$, in the variables $x, y_1, \cdots, y_{o(\tau)}$, we will mean a statement that involves only logical and set theoretical expressions and the variables $x, y_1, \cdots, y_{o(\tau)}$. We will say that *P is true in an algebra* $(A, (f_1, \cdots, f_{o(\tau)}))$ if $P(A, f_1, \cdots, f_{o(\tau)})$ is, in fact, true. Also define the logical formula $D_\pi(P)$ by

$$D_\pi(P(x, y_1, \cdots, y_{o(\tau)})) = P(x, y_{i_1}, \cdots, y_{o(\tau)}).$$

The class **K** of algebras is said to satisfy the *weak duality principle with respect to π* if for any logical formula P, which is true for all algebras in **K**, $D_\pi(P)$ is also true for all algebras in **K**. We then have:

A π-weakly dual class **K** *satisfies the weak duality principle with respect to π.*

Indeed, suppose **K** is π-weakly dual and that $(A, (f_1, \cdots, f_{o(\tau)}) \in \mathbf{K}$. To prove that $D_\pi(P)$ is true in $(A, (f_1, \cdots, f_{o(\tau)}))$ consider the algebra $(A, (f_{i_1}, \cdots, f_{i_{o(\tau)}}))$. Since the latter algebra is a member of **K**, it satisfies P. But $D_\pi(P(A, f_1, \cdots, f_{o(\tau)}))$ is the same as $P(A, f_{i_1}, \cdots, f_{i_{o(\tau)}})$ which establishes the result.

Note however that in general, if P holds for a specific member of a π-weakly dual class, $D_\pi(P)$ may not hold for this same member. This leads us to the notion of strong duality. The class **K** of algebras satisfies the *strong duality principle with respect to π* if for any logical formula P and any algebra $A \in \mathbf{K}$ we have: P is true in $A \Rightarrow D_\pi(P)$ is true in A.

For the class **K** and permutation π described above, we call **K** *π-strongly dual* if for each $(A, (f_1, \cdots, f_{o(\tau)})) \in \mathbf{K}$, the algebra $(A, (f_1, \cdots, f_{o(\tau)}))$ is isomorphic with $(A, (f_{i_1}, \cdots, f_{i_{o(\tau)}}))$.

A π-strongly dual class **K** *satisfies the strong duality principle with respect to π.*

To prove this let $(A, (f_1, \cdots, f_{o(\tau)})) \in \mathbf{K}$, P a logical formula which is true in $(A, (f_1, \cdots, f_{o(\tau)}))$ and assume that **K** is π-strongly dual with respect to π. But saying that P is true in $(A, (f_1, \cdots, f_{o(\tau)}))$ is exactly the same as saying that $D_\pi(P)$ is true in $(A, (f_{i_1}, \cdots, f_{i_{o(\tau)}}))$. By hypothesis $(A, (f_1, \cdots, f_{o(\tau)}))$ and $(A, (f_{i_1}, \cdots, f_{i_{o(\tau)}}))$ are isomorphic so $D_\pi(P)$ is true in $(A, (f_1, \cdots, f_{o(\tau)}))$.

Applications of the duality principles occur in sections II.3 and II.6.

9. Subdirect products

Definition 1. An algebra A of type τ is said to be a *subdirect product* of a family $(A_s)_{s \in S}$ of type τ if there exists an embedding $f: A \to \times_{s \in S} A_s$ such that for each $s \in S$, $p_s \circ f$ is onto. An algebra A of type τ is *subdirectly irreducible* if (i) $|A| > 1$, and (ii) if A is a subdirect product of $(A_s)_{s \in S}$ with embedding f, then $p_s \circ f$ is an isomorphism for some $s \in S$.

A classic theorem of G. Birkhoff, which will be referred to as the *subdirect product theorem*, states:

If **K** is an equational class then every algebra in **K** is a subdirect product of subdirectly irreducible algebras which are in **K** (see Theorem 4).

Thus, in order to prove that a property holds for every algebra in **K**, it is sufficient to prove that it holds for the subdirectly irreducible algebras in **K**, and that it is preserved under the formation of subalgebras and direct products. In the investigation of a particular equational class, it is therefore often useful to determine the subdirectly irreducible algebras.

An easy but useful result is:

Theorem 2. *An algebra A is a subdirect product of a family $(A_s)_{s \in S}$ of algebras if and only if there is a family $(g_s: A \to A_s)_{s \in S}$ of onto homomorphisms such that for $a, b \in A$, $a \neq b$ there exists an $s \in S$ such that $g_s(a) \neq g_s(b)$.*

The proof is left to the reader.

The following theorem may sometimes be applied in order to characterize subdirectly irreducibles.

Theorem 3. *An algebra A is subdirectly irreducible if and only if it has a least non-trivial congruence relation.*

Proof. ($\Rightarrow$) Since $|A| > 1$, $\mathfrak{C}(A)$ contains at least two members. Let $\theta_0 = \cap(\mathfrak{C}(A) \sim \{0\})$. Clearly $\theta_0 \subseteq \theta$ for every non-trivial congruence θ, so it suffices to show that $\theta_0 \neq 0$. Suppose however that $\theta_0 = 0$. Now $(\nu_\theta: A \to A/\theta)_{\theta \in \mathfrak{C}(A) \sim \{0\}}$ is a family of onto homomorphisms. If $a \neq a'$ but $\nu_\theta(a) = \nu_\theta(a')$ for each $\theta \in \mathfrak{C}(A) \sim \{0\}$ then $(a, a') \in \cap(\mathfrak{C}(A) \sim \{0\})$, a contradiction. Thus, by Theorem 2, A is a subdirect product of $(A/\theta)_{\theta \in \mathfrak{C}(A) \sim \{0\}}$. By hypothesis there exists $\theta_1 \in \mathfrak{C}(A) \sim \{0\}$ such that $p_{\theta_1} \circ f$ is an isomorphism, where f is the embedding. For any $(x, y) \in \theta_1$, $(p_{\theta_1} \circ f)(x) = f(x)(\theta_1) = [x]_{\theta_1} = [y]_{\theta_1} = f(y)(\theta_1) = (p_{\theta_1} \circ f)(y)$ so $x = y$, contradicting $\theta_1 \neq 0$.

($\Leftarrow$) Suppose there exists a smallest non-trivial congruence relation θ_0. Then $|A| > 1$. Suppose also that A is a subdirect product of $(A_s)_{s \in S}$ with embedding f. We must show $p_s \circ f$ is one-one for some $s \in S$. If not, then $\mathrm{Ker}(p_s \circ f) \neq 0$ for each $s \in S$ so $\theta_0 \subseteq \bigcap_{s \in S} \mathrm{Ker}(p_s \circ f)$. Let $(x, y) \in \theta_0$ and $x \neq y$. Then $(p_s \circ f)(x) = (p_s \circ f)(y)$ for all $s \in S$, thus $f(x) = f(y)$, contradicting the assumption that f is an embedding.

We can now prove the subdirect product theorem:

Theorem 4. (G. Birkhoff [5]) *If* **K** *is an equational class then every algebra in* **K** *is a subdirect product of subdirectly irreducible algebras in* **K**.

Proof. Let $A \in \mathbf{K}$. If $|A| = 1$ then A is the subdirect product of an empty family of subdirectly irreducible algebras. So suppose $|A| > 1$. For each pair $(a, a') \in A^2$ such that $a \neq a'$, let $\mathfrak{C}(a, a')$ be the set of all congruence relations θ on A such that $(a, a') \notin \theta$. Now $\mathfrak{C}(a, a')$ is partially ordered by inclusion, and it is easily verified that if $\mathfrak{C}$ is a chain in $\mathfrak{C}(a, a')$ then $\cup\mathfrak{C}$ is a congruence relation and $(a, a') \notin \cup\mathfrak{C}$. Thus, by Zorn's lemma, $\mathfrak{C}(a, a')$ has a maximal element. Let $\theta(a, a')$ be a maximal element in $\mathfrak{C}(a, a')$.

Now in $A/\theta(a, a')$, $[a] \neq [a']$ so there are non-trivial congruence relations on $A/\theta(a, a')$. Let Φ be one of these. Then $(x, y) \in \varphi \Leftrightarrow ([x], [y]) \in \Phi$ defines a congruence relation φ on A. Now $\theta(a, a') \subseteq \varphi$ for

$$(u, v) \in \theta(a, a') \Rightarrow [u] = [v] \Rightarrow ([u], [v]) \in \Phi \Rightarrow (u, v) \in \varphi.$$

But $([a], [a']) \in \Phi$ for

$$([a], [a']) \notin \Phi \Rightarrow (a, a') \notin \varphi \Rightarrow \theta(a, a') = \varphi$$

by the maximality of $\theta(a, a')$ which contradicts the fact that Φ is non-trivial. So

$$([a], [a']) \in \cap(\mathfrak{C}(A/\theta(a, a')) \sim \{0\}) = \Psi,$$

and since $[a] \neq [a']$, Ψ is the smallest non-trivial congruence relation on $A/\theta(a, a')$. By Theorem 3, $A/\theta(a, a')$ is subdirectly irreducible. Since **K** is equational we obtain $A/\theta(a, a') \in \mathbf{K}$.

For each pair a, a' in A, $a \neq a'$, $\nu_{\theta(a,a')}: A \to A/\theta(a, a')$ is an onto homomorphism and

$$\nu_{\theta(a,a')}(a) = [a] \neq [a'] = \nu_{\theta(a,a')}(a')$$

so by Theorem 2, A is a subdirect product of $(A/\theta(a, a'))_{\{(a,a') \in A^2 : a \neq a'\}}$.

10. Free algebras

Definition 1. Let **K** be a class of similar algebras. An algebra $A \in \mathbf{K}$ is said to be *free over* **K** if there exists a set $S \subseteq A$ such that

(i) $[S] = A$.

(ii) If $B \in \mathbf{K}$ and $f: S \to B$ is a function then there exists a homomorphism $g: A \to B$ such that $g|S = f$.

Note that by Lemma 4.7 g is uniquely determined.

In Definition 1, the set S is said to *freely generate* A and is called a *free generating set*.

An algebra A which is free over **K** is determined up to isomorphism, by the cardinality of any free generating set. Indeed, if S_1 and S_2 freely generate $[S_1]$ and $[S_2]$ over **K** and $|S_1| = |S_2|$ then there exists a one-one correspondence f from S_1

onto S_2. By (ii) there exist homomorphisms $f_1: [S_1] \to [S_2]$ such that $f_1|S_1 = f$ and $f_2: [S_2] \to [S_1]$ such that $f_2|S_2 = f^{-1}$. Thus, $(f_2 \circ f_1)|S_1 = 1_{S_1}$ and if follows that $f_2 \circ f_1 = 1_{[S_1]}$. Similarly, $f_1 \circ f_2 = 1_{[S_2]}$ so f_1 is an isomorphism. In particular, if f is a permutation of S then it can be extended to an automorphism of A.

For each cardinal α we pick any one of the isomorphic copies of a free algebra over **K** with α free generators and call it the *free* **K**-*algebra on* α *free generators* and denote it by $\mathscr{F}_{\mathbf{K}}(\alpha)$, or if the free generating set is specified, by $\mathscr{F}_{\mathbf{K}}(S)$. Note also that if $X_1 \subseteq X$, and X is a free generating set and $[X_1] \in \mathbf{K}$ then X_1 is also a free generating set.

An important result concerning free algebras is their existence in an equational class (see Theorem 12.4).

11. Polynomials and identities

Since many classes of algebras are defined by "identities" we will make the concept precise (see also G. Grätzer [2]).

Definition 1. Let A be an algebra of similarity type τ and α an ordinal.

(i) The projections $p_\gamma: A^\alpha \to A$ are called α-*ary polynomials on* A, for each $\gamma < \alpha$.

(ii) If $q_1, \cdots, q_{n_i}$ are α-ary polynomials on A then the functions $f_i(q_1, \cdots, q_{n_i}): A^\alpha \to A$ defined by

$$f_i(q_1, \cdots, q_{n_i})(u) = f_i(q_1(u), \cdots, q_{n_i}(u))$$

$u \in A^\alpha$ is an α-*ary polynomial on* A.

(iii) The α-ary polynomials on A are exactly those functions which can be obtained by a finite number of applications of (i) and (ii).

Definition 2. Let τ be a similarity type and α an ordinal.

(i) The symbols $\{\mathbf{x}_\gamma : \gamma < \alpha\}$ are called α-*ary polynomial symbols* and for any algebra A of type τ they *induce* the α-ary polynomials $\{p_\gamma : \gamma < \alpha\}$ on A.

(ii) If $\mathbf{q}_1, \cdots, \mathbf{q}_{n_i}$ are α-ary polynomial symbols which induce the α-ary polynomials $q_1, \cdots, q_{n_i}$ on A, then $\mathbf{f}_i(\mathbf{q}_1 \cdots, \mathbf{q}_{n_i})$ is an α-*ary polynomial symbol* and *induces* the α-ary polynomial $f_i(q_1, \cdots, q_{n_i})$ on A.

(iii) The α-ary polynomial symbols are exactly those symbols which can be obtained by a finite number of applications of (i) and (ii).

In practice we will often replace $\mathbf{x}_0, \mathbf{x}_1, \mathbf{x}_2$ by $\mathbf{x}, \mathbf{y}, \mathbf{z}$ and also omit "$\cdot$" from polynomials and omit "$\boldsymbol{\cdot}$" from polynomial symbols, whenever possible.

The polynomial q on an algebra A, induced by a polynomial symbol $\mathbf{q}$, is uniquely determined. Indeed, this is immediate from the definition. However, in discussing algebras A_1 and A_2 it is sometimes necessary to say that $\mathbf{q}$ induces q_1 and q_2 on A_1 and A_2 respectively, in order to avoid confusion.

Definition 3. For α-ary polynomial symbols $\mathbf{q}$ and $\mathbf{r}$ of type τ, the symbol $\mathbf{q} \equiv \mathbf{r}$ is called an *identity* and is said to be satisfied in a class $\mathbf{K}$, of algebras of type τ, if for every algebra $A \in \mathbf{K}$ the induced polynomials q and r on A, are identical. In particular if $\mathbf{K} = \{A\}$ we say that A *satisfies* the identity.

Example 4. Consider the class $\mathbf{G}$ of groups which consist of all algebras $(A, (\cdot, e, ^{-1}))$ of type $\tau = (2, 0, 1)$ which satisfy the 3-ary identities:

(i) $\mathbf{x_0(x_1x_2) \equiv (x_0x_1)x_2}$.

(ii) $\mathbf{ex_0 \equiv x_0,\ x_0e \equiv x_0}$.

(iii) $\mathbf{x_0x_0^{-1} \equiv e,\ x_0^{-1}x_0 \equiv e}$.

Suppose e_o is the element of A selected by the nullary operation e. Then the identity $\mathbf{ex_0 \equiv x_0}$ implies that for any $a \in A$, $e_oa = a$. Indeed, $\mathbf{e \cdot x_0}$ induces the polynomial $e \cdot p_0$ and $\mathbf{x_0}$ induces p_0 so

$$(e \cdot p_0)(a_0, a_1, a_2) = e_op_0(a_0, a_1, a_2) = e_oa_0 = a_0 = p_0(a_0, a_1, a_2).$$

For a more complicated identity consider the 4-ary polynomial symbols $\mathbf{x_1^{-1}x_3}$ and $\mathbf{(x_1^{-1}x_0)(x_0^{-1}x_3)}$. Now $\mathbf{G}$ satisfies the identity $\mathbf{(x_1^{-1}x_0)(x_0^{-1}x_3) \equiv x_1^{-1}x_3}$ for if G is a group then these two polynomial symbols induce the polynomial $(p_1^{-1}p_0)(p_0^{-1}p_3)$ and $p_1^{-1}p_3$ on G. But these functions are identical on G for if $a_0, a_1, a_2, a_3 \in G$ then

$$\begin{aligned}
&((p_1^{-1}p_0)(p_0^{-1}p_3))(a_0, a_1, a_2, a_3) \\
&\quad = [(p_1(a_0, a_1, a_2, a_3))^{-1}p_0(a_0, a_1, a_2, a_3)][(p_0(a_0, a_1, a_2, a_3))^{-1}p_3(a_0, a_1, a_2, a_3)] \\
&\quad = (a_1^{-1}a_0)(a_0^{-1}a_3) = a_1^{-1}a_3 = (p_1(a_0, a_1, a_2, a_3))^{-1}p_3(a_0, a_1, a_2, a_3)) \\
&\quad = (p_1^{-1}p_3)(a_0, a_1, a_2, a_3).
\end{aligned}$$

Another application which the reader might consider is the "generalized associative law". Given an algebra A with an associative binary operation f on A; let us, as usual, write ab for $f(a, b)$. Roughly, any arrangement of the parentheses between members of $a_1, \cdots, a_n$, in that order, which yields an element of A, will also determine the same element of A by any other such rearrangement. Thus, for an associative binary operation the parentheses of any product, of the type described, will be omitted. In connection with this, it is easy to prove that for an associative commutative binary operation "$\cdot$" on A, $a_1 \cdots a_n = a_{i_1} \cdots a_{i_n}$ for $\{a_1, \cdots, a_n\} \subseteq A$ and any permutation $(i_1, \cdots, i_n)$ on $\{1, 2, \cdots, n\}$.
The next result shows that homomorphisms preserve identities.

Theorem 5. *Let A and B be algebras of type τ, $\mathbf{q}$ an α-ary polynomial symbol of type τ and $f\colon A \to B$ a homomorphism. Then for each $u \in A^\alpha$ we have $f \circ u \in B^\alpha$ and $f(q(u)) = q(f \circ u)$.*

Proof. First, let $\mathbf{q}$ be $\mathbf{x_\gamma}$ where $\gamma < \alpha$. Then $\mathbf{x_\gamma}$ induces the γ-th projection on A and on B so we have

$$f(p_\gamma(u)) = f(u(\gamma)) = (f \circ u)(\gamma) = p_\gamma(f \circ u).$$

Suppose $1 \leq i \leq o(\tau)$ and $\mathbf{q}_1, \cdots, \mathbf{q}_{n_i}$ are polynomial symbols for which the statement is true. Then for $u \in A^\alpha$:

$$\begin{aligned} f(f_i(q_1, \cdots, q_{n_i})(u)) &= f(f_i(q_1(u), \cdots, q_{n_i}(u))) \\ &= f_i(f(q_1(u)), \cdots, f(q_{n_i}(u))) \\ &= f_i(q_1(f \circ u), \cdots, q_{n_i}(f \circ u)) \\ &= f_i(q_1, \cdots, q_{n_i})(f \circ u). \end{aligned}$$

Lemma 6. *If* $\mathbf{q}$ *is a polynomial symbol that induces* q_s *on* A_s *for each member of a family* $(A_s)_{s \in S}$ *of algebras, then the polynomial induced on* $\times_{s \in S} A_s$ *by* $\mathbf{q}$ *is the function* $q : (\times_{s \in S} A_s)^\alpha \to \times_{s \in S} A_s$ *such that for* $a \in (\times_{s \in S} A_s)^\alpha$ *and* $a_s \in (A_s)^\alpha$ *defined by* $a_s(\gamma) = (a(\gamma))(s)$, $(q(a))(s) = q_s(a_s)$.

Proof. $\mathbf{x}_\gamma$ induces $(p_\gamma)_{A_s}$ on A_s so

$$(q(a))(s) = (p_\gamma)_{A_s}(a_s) = a_s(\gamma) = (a(\gamma))(s).$$

Thus $q(a) = a(\gamma)$. It follows that $q = p_\gamma$.

Suppose that for $j = 1, \cdots, n_i$, $\mathbf{q}^j$ induces q_s^j on A_s for each $s \in S$ and induces q^j on $\times_{s \in S} A_s$ defined by $q^j : (\times_{s \in S} A_s)^\alpha \to \times_{s \in S} A_s$ where for each $a \in (\times_{s \in S} A_s)^\alpha$ and $a_s \in (A_s)^\alpha$ defined by $a_s(\gamma) = (\alpha(\gamma))(s)$, $(q^j(a))(s) = q_s^j(a_s)$. Then the polynomial on $\times_{s \in S} A_s$, induced by $\mathbf{f}_i(\mathbf{q}^1, \cdots, \mathbf{q}^{n_i})$ is defined by

$$\begin{aligned} ((f_i(q^1, \cdots, q^{n_i}))(a))(s) &= (f_i(q^1(a), \cdots, q^{n_i}(a)))(s) \\ &= f_i((q^1(a))(s), \cdots, (q^{n_i}(a))(s)) \\ &= f_i(q_s^1(a_s), \cdots, q_s^{n_i}(a_s)) \\ &= (f_i(q_s^1, \cdots, q_s^{n_i}))(a_s) \end{aligned}$$

where $a_s \in (A_s)^\alpha$, $a_s(\gamma) = (a(\gamma)(s))$. This completes the proof.

Corollary 7. *If every algebra in a class* $\mathbf{K}$ *satisfies an identity then so does every algebra in the equational class generated by* $\mathbf{K}$.

Proof. Clearly the direct product of a family of algebras which satisfy an identity $\mathbf{q}_1 = \mathbf{q}_2$ also satisfies $\mathbf{q}_1 = \mathbf{q}_2$, so all the members of $\mathbf{P}(\mathbf{K})$ satisfy the identity. Similarly, the members of $\mathbf{H}(\mathbf{S}(\mathbf{P}(\mathbf{K})))$ satisfy $\mathbf{q}_1 = \mathbf{q}_2$.

Corollary 8. *If the subdirectly irreducible algebras of an equational class* $\mathbf{K}$ *satisfy an identity then so do all the members of* $\mathbf{K}$.

Proof. By the subdirect product theorem, $\mathbf{K}$ is generated by its subdirectly irreducible members so Corollary 7 applies.

12. Birkhoff's characterization of equational classes

In this section the existence of free algebras is proven for equational classes and this enables us to give a characterization of equational classes.

Let τ be a similarity type and $\alpha > 0$ an ordinal. We form the *algebra of polynomial symbols* $P^\alpha(\tau)$ of type τ by defining operations on the set of all polynomial symbols of type τ:

$$f_i(\mathbf{q_1}, \cdots, \mathbf{q_{n_i}}) = \mathbf{f_i}(\mathbf{q_1}, \cdots, \mathbf{q_{n_i}}), \qquad 1 \leq i \leq o(\tau).$$

Lemma 1. *Let A be an algebra of type τ and $s \in A^\alpha$. Then* $[\{s(\gamma) : \gamma < \alpha\}] = \{q(s) : \mathbf{q}$ *is an α-ary polynomial symbol of type* $\tau\}$.

Proof. Let T denote the right side of the above stated equality. Then for each $\gamma < \alpha$, $s(\gamma) = p_\gamma(s)$ so $\{s(\gamma) : \gamma < \alpha\} \subseteq T$. Also, if $1 \leq i \leq o(\tau)$ and $\mathbf{q_1}, \cdots, \mathbf{q_{n_i}}$ are α-ary polynomial symbols then

$$f_i(q_1(s), \cdots, q_{n_i}(s)) = f_i(q_1, \cdots, q_{n_i})(s) \in T$$

so T is a subalgebra of A that contains $\{s(\gamma) : \gamma < \alpha\}$ and since it is clear that any subalgebra of A that contains $\{s(\gamma) : \gamma < \alpha\}$ also contains T, we have $[\{s(\gamma) : \gamma < \alpha\}] = T$.

Lemma 2. *If A is an algebra of type τ and $A = [\{a(\gamma) : \gamma < \alpha\}]$ where $a \in A^\alpha$ then there exists a congruence relation θ_a on $P^\alpha(\tau)$ such that $P^\alpha(\tau)/\theta_a \cong A$.*

Proof. Define f: $P^\alpha(\tau) \to A$ by $f(\mathbf{q}) = q(a)$. Then f is a homomorphism for

$$\begin{aligned} f(f_i(\mathbf{q_1}, \cdots, \mathbf{q_{n_i}})) &= f(\mathbf{f_i}(\mathbf{q_1}, \cdots, \mathbf{q_{n_i}})) \\ &= (f_i(q_1, \cdots, q_{n_i}))(a) \\ &= f_i(q_1(a), \cdots, q_{n_i}(a)) \\ &= f_i(f(\mathbf{q_1}), \cdots, f(\mathbf{q_{n_i}})). \end{aligned}$$

Also f is onto for if $u \in A$ then by Lemma 1, there exists $\mathbf{q}$ such that $q(a) = u$ so $f(\mathbf{q}) = q(a) = u$.

By the homomorphism theorem $P^\alpha(\tau)/\mathrm{Ker} f \cong A$.

Lemma 3. Let A and B be algebras of type τ and $a \in A^\alpha$, $b \in B^\alpha$. *There exists a homomorphism* f: $[\{a(\gamma) : \gamma < \alpha\}] \to B$ *such that $f \circ a = b$ if and only if for any two α-ary polynomial symbols* $\mathbf{q_1}, \mathbf{q_2}$: $q_1(a) = q_2(a) \Rightarrow q_1(b) = q_2(b)$.

Proof. ($\Rightarrow$) Since $a \in [\{a(\gamma) : \gamma < \alpha\}]^\alpha$, Theorem 11.5 implies $q_1(b) = q_1(f \circ a) = f(q_1(a)) = f(q_2(a)) = q_2(f \circ a) = q_2(b)$.

($\Leftarrow$) Define f: $[\{a(\gamma) : \gamma < \alpha\}] \to B$ by $f(q(a)) = q(b)$ for every α-ary polynomial $\mathbf{q}$. f is well defined by Lemma 1 and by the hypothesis; also $(f \circ a)(\gamma) = f(a(\gamma)) = f(p_\gamma(a)) = p_\gamma(b) = b(\gamma)$ so $f \circ a = b$. Now for $1 \leq i \leq o(\tau)$,

$$\begin{aligned} f(f_i(q_1(a), \cdots, q_{n_i}(a))) &= f(f_i(q_1, \cdots, q_{n_i})(a)) \\ &= (f_i(q_1, \cdots, q_{n_i}))(b) \\ &= f_i(q_1(b), \cdots, q_{n_i}(b)) \\ &= f_i(f(q_1(a)), \cdots, f(q_{n_i}(a))) \end{aligned}$$

for α-ary polynomial symbols $\mathbf{q_1}, \cdots, \mathbf{q_{n_i}}$. So f is the required homomorphism.

Theorem 4. *If* **K** *is an equational class which is not trivial then* $\mathscr{F}_{\mathbf{K}}(\alpha)$ *exists for each cardinal* $\alpha > 0$.

Proof. Let $S = \{a \in A^{\alpha} : A \in \mathbf{K}, A = [\{a(\gamma) : \gamma < \alpha\}]\}$.
For each $a \in S$, let θ_a be the congruence relation defined in Lemma 2. Then $B = \mathsf{X}_{a\in S} P^{\alpha}(\tau)/\theta_a \in \mathbf{K}$ since **K** is equational. For each $\gamma < \alpha$, let $x_{\gamma} \in B$ be defined by $x_{\gamma}(a) = [\mathbf{x}_{\gamma}]_{\theta_a}$. Now the subalgebra C of B, generated by $\{x_{\gamma} : \gamma < \alpha\}$ is a member of **K** and we will show that the x_{γ} are distinct. Indeed, suppose $x_{\gamma_1} = x_{\gamma_2}, \gamma_1 \neq \gamma_2$. By hypothesis, there exists a member of **K** with distinct elements u_1, u_2. So there exists $a_o \in \{u_1, u_2\}^{\alpha}$ such that $a_o(\gamma_1) = u_1, a_o(\gamma_2) = u_2$. But $[\{u_1, u_2\}] \in \mathbf{K}$, and $[\{u_1, u_2\}] = [\{a_o(\gamma) : \gamma < \alpha\}]$ so $a_o \in S$. By Lemma 2,

$$x_{\gamma_1} = x_{\gamma_2} \Rightarrow x_{\gamma_1}(a_o) = x_{\gamma_2}(a_o) \Rightarrow [\mathbf{x}_{\gamma_1}]_{\theta_{a_o}} = [\mathbf{x}_{\gamma_2}]_{\theta_{a_o}} \Rightarrow (\mathbf{x}_{\gamma_1}, \mathbf{x}_{\gamma_2}) \in \theta_{a_o}$$

$$\Rightarrow p_{\gamma_1}(a_o) = p_{\gamma_2}(a_o) \Rightarrow u_1 = a_o(\gamma_1) = p_{\gamma_1}(a_o) = p_{\gamma_2}(a_o) = a_o(\gamma_2) = u_2,$$

a contradiction. Hence $|\{x_{\gamma} : \gamma < \alpha\}| = \alpha$.

Next, we prove that $\{x_{\gamma} : \gamma < \alpha\}$ freely generates C. Let $A \in \mathbf{K}$ and $a \in A^{\alpha}$. It suffices to show that there exists a homomorphism $g: C \to A$ such that $g(x_{\gamma}) = a(\gamma)$ for $\gamma < \alpha$. Let $\mathbf{x} \in \{\mathbf{x}_{\gamma} : \gamma < \alpha\}^{\alpha}$ be defined by $\mathbf{x}(\gamma) = \mathbf{x}_{\gamma}$ for $\gamma < \alpha$; $x \in \{x_{\gamma} : \gamma < \alpha\}^{\alpha}$ by $x(\gamma) = x_{\gamma}$ for $\gamma < \alpha$, and $y_a \in \{x_{\gamma}(a) : \gamma < \alpha\}^{\alpha}$ by $y_a(\gamma) = x_{\gamma}(a)$ for $\gamma < \alpha$.

Claim. If $\mathbf{q}_1$ and $\mathbf{q}_2$ are α-ary polynomial symbols and $q_1(x) = q_2(x)$ then $q_1(a) = q_2(a)$: For each $\gamma < \alpha$,

$$y_a(\gamma) = x_{\gamma}(a) = [\mathbf{x}_{\gamma}]_{\theta_a} = \nu_{\theta_a}(\mathbf{x}_{\gamma}) = (\nu_{\theta_a} \circ \mathbf{x})(\gamma)$$

so $y_a = \nu_{\theta_a} \circ \mathbf{x}$. By Lemma 11.6 $(q_i(x))(a) = q_i(y_a)$, for $i = 1, 2$, so

$$q_1(x) = q_2(x) \Rightarrow q_1(y_a) = q_2(y_a) \Rightarrow q_1(\nu_{\theta_a} \circ \mathbf{x}) = q_2(\nu_{\theta_a} \circ \mathbf{x}).$$

Thus by Theorem 11.5 we have

$$\nu_{\theta_a}(q_1(\mathbf{x})) = \nu_{\theta_a}(q_2(\mathbf{x})).$$

So by the isomorphism of Lemma 2, $q_1(a) = q_2(a)$.

By Lemma 3, there exists a homomorphism f: $[\{x_{\gamma} : \gamma < \alpha\}] = C \to A$ such that $f \circ x = a$, that is $f(x_{\gamma}) = a(\gamma)$ for each $\gamma < \alpha$.

Theorem 5. (G. Birkhoff [3]) *A class* **K** *of similar algebras is equational if and only if there exists a set* Ω *of identities such that* **K** *is exactly the class of algebras that satisfies all the identities in* Ω.

Proof. $(\Leftarrow)$ Since identities are preserved under the formation of homomorphic images, subalgebras, and direct products, algebras formed in these ways from members of **K** will again satisfy all of the identities in Ω and so, by hypothesis, will be members of **K**.

$(\Rightarrow)$ Let Ω be the set of all identities which are satisfied by every member of **K**. Since every member of **K** satisfies all of the identities of Ω, it remains to show that if A satisfies all of the identities in Ω then $A \in \mathbf{K}$. For this purpose suppose $|A| = \alpha$ and consider the free algebra $\mathscr{F}_{\mathbf{K}}(\alpha)$. (If $\alpha = 0$ then clearly $A \in \mathbf{K}$.) We verify the hypothesis of Lemma 3 for a set $\{x_\gamma : \gamma < \alpha\}$ of free generators of $\mathscr{F}_{\mathbf{K}}(\alpha)$ and $\{a_\gamma : \gamma < \alpha\} = A$. Now if $\mathbf{q}_1$ and $\mathbf{q}_2$ are α-ary polynomial symbols and $q_1(x) = q_2(x)$ for $x \in (\mathscr{F}_{\mathbf{K}}(\alpha))^\alpha$, $x(\gamma) = x_\gamma$ then, since $\{x_\gamma : \gamma < \alpha\}$ freely generates $\mathscr{F}_{\mathbf{K}}(\alpha)$, the polynomials induced by $\mathbf{q}_1$ and $\mathbf{q}_2$ are identical for every algebra in $\mathbf{K}$, so $\mathbf{q}_1 \equiv \mathbf{q}_2 \in \Omega$. So A satisfies $\mathbf{q}_1 \equiv \mathbf{q}_2$ and hence the hypothesis of Lemma 3 is satisfied and there exists a homomorphism of $\mathscr{F}_{\mathbf{K}}(\alpha)$ onto A. But $\mathbf{H}(\mathbf{K}) \subseteq \mathbf{K}$ and $\mathscr{F}_{\mathbf{K}}(\alpha) \in \mathbf{K}$ so $A \in \mathbf{K}$.

Corollary 6. *Let* $\mathbf{K}$ *be a class of similar algebras and* Ω *the set of identities which are satisfied by every member of* $\mathbf{K}$. *Then an algebra A is a member of the equational class generated by* $\mathbf{K}$ *if and only if A satisfies every identity in* Ω.

Proof. Let $\mathbf{K}_1$ be the equational class generated by $\mathbf{K}$ and Ω_1 the set of identities which are satisfied by each member of $\mathbf{K}_1$. Since $\mathbf{K} \subseteq \mathbf{K}_1$, $\Omega_1 \subseteq \Omega$.

($\Rightarrow$) Let $A \in K_1$. By Corollary 11.7, A satisfies every identity in Ω.

($\Leftarrow$) Suppose $A \notin \mathbf{K}_1$. By Theorem 5 there exists an identity $\mathbf{p} \equiv \mathbf{q}$ in Ω_1 which is not satisfied by A. But $\Omega_1 \subseteq \Omega$ so $\mathbf{p} \equiv \mathbf{q} \in \Omega$ is not satisfied by A, which violates the hypothesis.

Part 3. Categories

13. Definition of a category

Many of the results in the theory of distributive lattices can be expressed most succinctly in the language of category theory. For this reason the remainder of this chapter consists primarily of definitions, terminology, and elementary results from category theory. In particular the results concerning algebraic and equational categories illustrate the connection between those parts of universal algebra and category theory which will be used throughout the book.

Definition 1. A *category* $\mathscr{A}$ is a class $\mathrm{Ob}\,\mathscr{A}$ whose members are called *objects* of $\mathscr{A}$ together with:

(i) A class of disjoint sets $[A, B]_{\mathscr{A}}$ where $A, B \in \mathrm{Ob}\,\mathscr{A}$ whose members are called the *morphisms* of $\mathscr{A}$.

(ii) A function which assigns to each triple $(A, B, C) \in (\mathrm{Ob}\,\mathscr{A})^3$ a function from $[B, C]_{\mathscr{A}} \times [A, B]_{\mathscr{A}} \to [A, C]_{\mathscr{A}}$.

(iii) If $f \in [A, B]_{\mathscr{A}}$ and $g \in [B, C]_{\mathscr{A}}$ then the value of the function in (ii) at (g, f) will be denoted by $g \circ f$ and is called the *composition* of f and g.

(iv) If $A, B, C, D \in \mathrm{Ob}\,\mathscr{A}, f \in [A, B]_{\mathscr{A}}, g \in [B, C]_{\mathscr{A}}$ and $h \in [C, D]_{\mathscr{A}}$ then $h \circ (g \circ f) = (h \circ g) \circ f$.

(v) For each $A \in \mathrm{Ob}\,\mathscr{A}$ there is a morphism $1_A \in [A, A]_{\mathscr{A}}$ called the *identity morphism* of A which satisfies, for each $f \in [B, A]_{\mathscr{A}}$ and $g \in [A, C]_{\mathscr{A}}$, the conditions:

$1_A \circ f = f$ and $g \circ 1_A = g$. If no confusion is likely to arise then we will also write $[A, B]$ instead of $[A, B]_{\mathscr{A}}$.

Categories will always be denoted by $\mathscr{A}, \mathscr{B}, \mathscr{C}, \cdots$, objects by $A, B, C, \cdots$ and morphisms by $f, g, h, \cdots$. We will also write $f\colon A \to B$ for a morphism $f \in [A, B]_{\mathscr{A}}$ and call A the *domain* of f, written $A = \operatorname{Dom} f$ and call B the *codomain* of f, written $B = \operatorname{Codom} f$. It is easy to see that the identities are unique. Indeed, if 1_A and $1'_A$ are identities then $1_A \circ 1'_A = 1'_A$ and $1_A \circ 1'_A = 1_A$.

For examples, consider: (1) The category $\mathscr{S}$ whose objects are sets and whose morphisms are mappings between sets. (2) The category of topological spaces and continuous maps. (3) The category of (compact) Hausdorff spaces and continuous maps. (4) The category of groups and (group) homomorphisms. (5) Every poset $(P, \leq)$ can be considered as a category whose objects are the members of P and where $[p,q] = \varnothing$ if $p \nleq q$ and $[p, q]$ has exactly one element if $p \leq q$.

***Definition* 2.** Let $\mathscr{A}$ be a category. The *dual category* of $\mathscr{A}$, denoted by $\mathscr{A}^*$, is the category whose class of objects is the same as the class of objects of $\mathscr{A}$ and such that for $A, B \in \operatorname{Ob} \mathscr{A}$, $[A, B]_{\mathscr{A}^*} = [B, A]_{\mathscr{A}}$. If $f\colon A \to B$ and $g\colon B \to C$ are morphisms in $\mathscr{A}^*$ then the composition $g \circ f$ in $\mathscr{A}^*$ is defined to be the composition $f \circ g$ in $\mathscr{A}$.

***Definition* 3.** A category $\mathscr{B}$ is a *subcategory* of a category $\mathscr{A}$ if

(i) $\operatorname{Ob} \mathscr{B} \subseteq \operatorname{Ob} \mathscr{A}$.

(ii) $[A, B]_{\mathscr{B}} \subseteq [A, B]_{\mathscr{A}}$ for all $A, B \in \operatorname{Ob} \mathscr{B}$.

(iii) The identity morphisms in $\mathscr{B}$ are the same as the identity morphisms in $\mathscr{A}$.

(iv) Composition of morphisms in $\mathscr{B}$ is the same as in $\mathscr{A}$.

If in addition, $\mathscr{B}$ satisfies the condition:

(v) $[A, B]_{\mathscr{B}} = [A, B]_{\mathscr{A}}$ for all $A, B \in \operatorname{Ob} \mathscr{B}$ then $\mathscr{B}$ is called a *full subcategory* of $\mathscr{A}$.

14. Special morphisms

A morphism f in a category $\mathscr{A}$ is a *monomorphism* if $f \circ g = f \circ h$ implies $g = h$ for any morphisms g and h in $\mathscr{A}$; f is an *epimorphism* if $g \circ f = h \circ f$ implies $g = h$ for any morphisms g and h in $\mathscr{A}$. These morphisms are also referred to as being *monic* and *epic*, respectively. For diagrams, we employ the symbols:

$A \rightarrowtail B$	$A \twoheadrightarrow B$
monomorphism	epimorphism

A morphism $f\colon A \to B$ is an *isomorphism* if there exists a morphism $f'\colon B \to A$ such that $f' \circ f = 1_A$ and $f \circ f' = 1_B$. In this case we write $f' = f^{-1}$ and say that A and B are *isomorphic*. Clearly f' is also an isomorphism. Note that the composition of two isomorphisms is again an isomorphism. Obviously an isomorphism is both a monomorphism and an epimorphism. The converse is not, in general, the case. For example, in the category of Hausdorff spaces and continuous maps, the isomorphisms are the homeomorphisms, but a one-one continuous map $f: X \to Y$ between Hausdorff spaces such that $f[X]$ is dense in Y is both monic and epic but not necessarily a homeomorphism.

Theorem 1. *If $\mathscr{A}$ is a category in which the morphisms are functions then for any* $f \in [A, B]$:

(i) *If f is one-one then f is monic.*

(ii) *If f is onto then f is epic.*

Proof. (i) Suppose $f \circ g = f \circ h$, where $g, h \in [C, A]$. For any $x \in C, f(g(x)) = f(h(x))$ so $g(x) = h(x)$ and hence $g = h$.

(ii) Suppose $g \circ f = h \circ f$, where $g, h \in [B, C]$. For any $b \in B$, there exists $a \in A$ such that $f(a) = b$ so $g(b) = g(f(a)) = h(f(a)) = h(b)$ and thus $g = h$.

Definition 2. An object B is a *retract* of an object A if there exist morphisms $f: A \to B$ and $g: B \to A$ such that $f \circ g = 1_B$.

Notice that f is necessarily epic and g monic. Also, if the morphisms are functions, then g is one-one and f is onto, as can easily be shown.

Definition 3. A *subobject* of an object A is a monomorphism $f: B \to A$. Two subobjects $f_1: B_1 \to A$ and $f_2: B_2 \to A$ are *isomorphic* if there exists an isomorphism $g: B_1 \to B_2$ such that $f_2 \circ g = f_1$.

The relation of "isomorphic" in the class of subobjects of an object can be thought of as an equivalence relation on this class.

Definition 4. An *extension* of an object A is a monomorphism $f: A \to B$. Two extensions $f_1: A \to B_1$ and $f_2: A \to B_2$ are *isomorphic* if there exists an isomorphism $g: B_1 \to B_2$ such that $g \circ f_1 = f_2$. An extension $f: A \to B$ is *proper* if f is not an isomorphism. An *essential extension* of an object A is an extension $f: A \to B$ such that if $g \circ f$ is monic then g is monic.

Note that if $f: A \to B$ is an isomorphism, then f is an essential extension of A. Also if $f_1: A \to B_1$ and $f_2: A \to B_2$ are isomorphic extensions, then f_1 is essential if and only if f_2 is essential. Finally, if $f: A \to B$ is an essential extension of A and $g: B \to C$ is an essential extension of B then $g \circ f$ is an essential extension of A.

Let $\mathscr{A}$ and $\mathscr{B}$ be categories. The *product category* $\mathscr{A} \times \mathscr{B}$ is defined by $\text{Ob}\, \mathscr{A} \times \mathscr{B} = \text{Ob}\, \mathscr{A} \times \text{Ob}\, \mathscr{B}$ and

$$[(A, B), (A_1, B_1)]_{\mathscr{A} \times \mathscr{B}} = [A, A_1]_{\mathscr{A}} \times [B, B_1]_{\mathscr{B}}.$$

If (f, g) is a morphism in $\mathscr{A} \times \mathscr{B}$, then it is easy to see that (f, g) is an isomorphism (monomorphism, epimorphism) if and only if f and g are isomorphisms (monomorphisms, epimorphisms).

15. Products and coproducts. Limits and colimits

Definition 1. Let $(A_s)_{s \in S}$ be a family of objects of a category $\mathscr{A}$. A *product* is a family $(p_s: A \to A_s)_{s \in S}$ of morphisms in $\mathscr{A}$ such that if $(f_s: B \to A_s)_{s \in S}$ are morphisms in $\mathscr{A}$, then there exists a unique morphism $f: B \to A$ such that $p_s \circ f = f_s$ for each $s \in S$. The morphisms $p_s: A \to A_s$ are called *projections.*

We will demonstrate that for two products $(p_s: A \to A_s)_{s \in S}$ and $(q_s: B \to A_s)_{s \in S}$, there exists an isomorphism $f: A \to B$ such that $q_s \circ f = p_s$ for each $s \in S$. Indeed, there exist morphisms $f: A \to B$ and $g: B \to A$ such that $q_s \circ f = p_s$ and $p_s \circ g = q_s$ for each $s \in S$. Thus, $q_s \circ f \circ g = q_s$ and $p_s \circ g \circ f = p_s$ for each $s \in S$. But $q_s \circ 1_B = q_s$ and $p_s \circ 1_A = p_s$ so by uniqueness, $f \circ g = 1_B$ and $g \circ f = 1_A$. Thus, f is an isomorphism. In view of this isomorphism, we will often talk of *the* product of a family $(A_s)_{s \in S}$ (if such a product exists) by just picking one of the products.

The dual notion of product is that of *coproduct* (also called *sum*).

Definition 2. Let $(A_s)_{s \in S}$ be a family of objects of a category $\mathscr{A}$. A *coproduct* is a family $(j_s: A_s \to A)_{s \in S}$ of morphisms in $\mathscr{A}$ such that if $(f_s: A_s \to B)_{s \in S}$ are morphisms in $\mathscr{A}$, then there exists a unique morphism $f: A \to B$ such that $f \circ j_s = f_s$ for each $s \in S$. The morphisms $j_s: A_s \to A$ are called *injections.*

As in the case of products, coproducts are again isomorphic in an obvious sense and we will therefore often talk about *the* coproduct (if it exists) of a family of objects.

If the product (coproduct) of $(A_s)_{s \in S}$ exists for $S = \varnothing$ then there are no projections (injections). In this case there exists an object A such that for each object B, there exists exactly one morphism $f: B \to A$ $(f: A \to B)$.

If $(p_s: A \to A_s)_{s \in S}$ is a product of $(A_s)_{s \in S}$ we will sometimes use the term "product" to refer to the object A itself. This will be done only if no confusion is likely to arise. A similar convention holds for coproducts.

A generalization of the notion of product which we will be interested in, is that of limit. We first need the definition of a partially ordered system.

Definition 3. A *partially ordered system* in a category $\mathscr{A}$ is a family $(A_s)_{s \in S}$ of objects indexed by a poset S, together with a family of morphisms $f_{st}: A_s \to A_t$, for each $s \leq t$, satisfying

(i) $f_{tu} \circ f_{st} = f_{su}$ for all $s \leq t \leq u$.

(ii) $f_{ss} = 1_{A_s}$ for each $s \in S$.

For brevity we say that $((A_s)_{s \in S}, (f_{st})_{s \leq t})$ is a partially ordered system in $\mathscr{A}$.

Definition 4. A *limit* of a partially ordered system $((A_s)_{s\in S}, (f_{st})_{s\le t})$ in $\mathscr{A}$ is a family $(g_s: A \to A_s)_{s\in S}$ of morphisms such that

(i) $f_{st} \circ g_s = g_t$ for $s \le t$.

(ii) If $(h_s: B \to A_s)_{s\in S}$ is a family of morphisms such that $f_{st} \circ h_s = h_t$ for $s \le t$, then there exists a unique $h: B \to A$ with $g_s \circ h = h_s$ for all $s \in S$.

Definition 5. A *colimit* of a partially ordered system $((A_s)_{s\in S}, (f_{st})_{s\le t})$ is a family $(g_s: A_s \to A)_{s\in S}$ of morphisms such that

(i) $g_t \circ f_{st} = g_s$ for $s \le t$.

(ii) If $(h_s: A_s \to B)_{s\in S}$ is a family of morphisms such that $h_t \circ f_{st} = h_s$ for $s \le t$ then there exists a unique $h: A \to B$ such that $h \circ g_s = h_s$ for all $s \in S$.

Remark. It is immediate from the definitions that if S is a totally unordered poset, then the limit of a partially ordered system $(A_s)_{s\in S}$ (with $f_{ss} = 1_{A_s}$ for each $s \in S$) reduces to the product of the family $(A_s)_{s\in S}$ and dually, the colimit reduces to the coproduct. Limits and colimits are also called *inverse limits* and *direct limits*, respectively.

16. Injectives and projectives

Definition 1. An object A of a category $\mathscr{A}$ is *injective* if for every monomorphism $f: B \to C$ and every morphism $g: B \to A$, there exists a morphism $h: C \to A$ with $h \circ f = g$. An object A is *projective* if for every epimorphism $f: C \to B$ and every morphism $g: A \to B$, there exists a morphism $h: A \to C$ such that $f \circ h = g$.

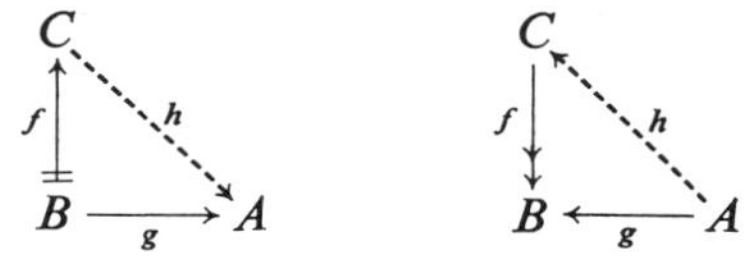
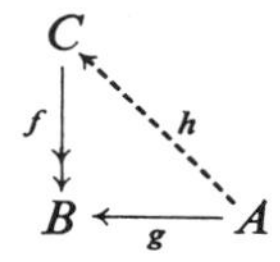

By letting $A = B$ and $g = 1_A$ in these diagrams, we note that if A is injective (projective) and $f: A \to C$ is monic ($f: C \to A$ is epic), then A is a retract of C.

Theorem 2. (i) *A retract of an injective object is injective.*

(ii) *A retract of a projective object is projective.*

Proof. (i) Suppose D is injective and A is a retract of D with morphisms $f_1: D \to A, f_2: A \to D$ and $f_1 \circ f_2 = 1_A$.

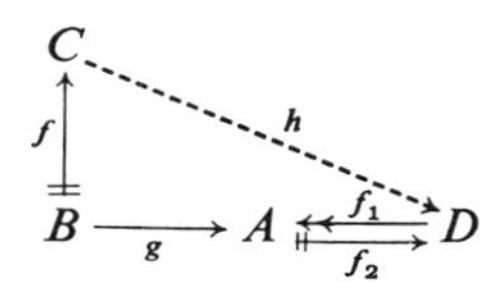

To prove A is also injective, let $f: B \to C$ be monic and $g \in [B, A]$. There exists a morphism $h: C \to D$ such that $h \circ f = f_2 \circ g$. Then the required morphism is $h' = f_1 \circ h$ since $h' \circ f = f_1 \circ h \circ f = f_1 \circ f_2 \circ g = 1_A \circ g = g$. The proof of (ii) is similar.

Theorem 3. *Products of injectives are injective and coproducts of projectives are projective.*

Proof. Let $(p_s: A \to A_s)_{s \in S}$ be a product of a family $(A_s)_{s \in S}$ of injectives. Suppose $f: B \to C$ is monic and $g: B \to A$ is a morphism. Then for each $s \in S$, $p_s \circ g \in [B, A_s]$ so the injectivity of A_s implies the existence of a morphism $h_s: C \to A_s$ such that $h_s \circ f = p_s \circ g$. Hence there exists a unique morphism $h: C \to A$ such that $p_s \circ h = h_s$ for each $s \in S$. But $p_s \circ h \circ f = h_s \circ f = p_s \circ g$ for each $s \in S$ so by uniqueness $h \circ f = g$. The proof of the second part is obtained by dualizing the above argument.

Definition 4. A category $\mathscr{A}$ is said to have *enough injectives* provided that for each object A there exists an extension $f: A \to B$ for which B is injective.

Definition 5. An *injective hull* of an object is an essential extension $f: A \to B$ for which B is injective.

Theorem 6. *Any two injective hulls of an object A are retracts of each other.*

Proof. Let $f_1 \in [A, B_1]$, $f_2 \in [A, B_2]$ be injective hulls of A.

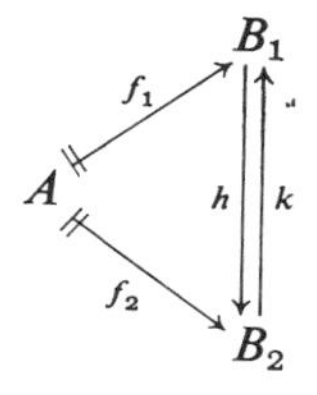

Since B_2 is injective there exists $h \in [B_1, B_2]$ such that $h \circ f_1 = f_2$. But $h \circ f_1$ is monic and f_1 is essential so h is monic. Also B_1 is injective so there exists $k \in [B_2, B_1]$ such that $k \circ h = 1_{B_1}$. Similarly, B_2 is a retract of B_1.

17. Functors

In order to compare various categories the notion of functor is defined:

Definition 1. Let $\mathscr{A}$ and $\mathscr{B}$ be categories. A *covariant functor* $\mathscr{F}: \mathscr{A} \to \mathscr{B}$ is a function which assigns to each $A \in \operatorname{Ob} \mathscr{A}$ an object $\mathscr{F}(A)$ of $\mathscr{B}$ and to each morphism $f: A \to B$ of $\mathscr{A}$ a morphism $\mathscr{F}(f): \mathscr{F}(A) \to \mathscr{F}(B)$ of $\mathscr{B}$ such that

(i) For every $A \in \operatorname{Ob} \mathscr{A}$, $\mathscr{F}(1_A) = 1_{\mathscr{F}(A)}$.

(ii) If $f \circ g$ is defined in $\mathscr{A}$ then $\mathscr{F}(f \circ g) = \mathscr{F}(f) \circ \mathscr{F}(g)$.

Definition 2. A *contravariant functor* $\mathscr{F}:\mathscr{A}\to\mathscr{B}$ is a function which assigns to each $A\in\operatorname{Ob}\mathscr{A}$ an object $\mathscr{F}(A)$ of $\mathscr{B}$ and to each morphism $f\colon A\to B$ of $\mathscr{A}$, a morphism $\mathscr{F}(f)\colon\mathscr{F}(B)\to\mathscr{F}(A)$ of $\mathscr{B}$ such that

(i) For every $A\in\operatorname{Ob}\mathscr{A}$, $\mathscr{F}(1_A)=1_{\mathscr{F}(A)}$.

(ii) If $f\circ g$ is defined in $\mathscr{A}$, then $\mathscr{F}(f\circ g)=\mathscr{F}(g)\circ\mathscr{F}(f)$.

The term functor will always mean covariant functor. Notice that if $\mathscr{F}:\mathscr{A}\to\mathscr{B}$ is a covariant functor, then $\mathscr{F}:\mathscr{A}\to\mathscr{B}^*$ is a contravariant functor.

For a simple example, consider the category $\mathscr{C}$ whose objects are algebras (A, F) of a certain similarity type and whose morphisms are homomorphisms between the algebras. Then the functor $\mathscr{G}:\mathscr{C}\to\mathscr{S}$ defined by $\mathscr{G}((A,F))=A$ and $\mathscr{G}(f)=f$ where $\mathscr{S}$ is the category of sets and maps, is called the *forgetful functor*.

If $\mathscr{F}:\mathscr{A}\to\mathscr{B}$ is a functor, then $\mathscr{A}$ is called the *domain* of $\mathscr{F}$, $\mathscr{A}=\operatorname{Dom}\mathscr{F}$ and $\mathscr{B}$ the *codomain* of $\mathscr{F}$, $\mathscr{B}=\operatorname{Codom}\mathscr{F}$. The *image* of $\mathscr{F}$, $\operatorname{Im}\mathscr{F}$, is the class $\{\mathscr{F}(A):A\in\operatorname{Ob}\mathscr{A}\}$ together with the class of morphisms

$\{\mathscr{F}(f):f$ is a morphism of $\mathscr{A}\}$.

Observe that $\operatorname{Im}\mathscr{F}$ need not be a subcategory of $\mathscr{B}$ for if $f\in[A,B]$, $g\in[C,D]$ and $\mathscr{F}(B)=\mathscr{F}(C)$ then $\mathscr{F}(g)\circ\mathscr{F}(f)$ need not belong to $\operatorname{Im}\mathscr{F}$.

A functor $\mathscr{F}:\mathscr{A}\to\mathscr{B}$ is *constant* if there exists $B\in\operatorname{Ob}\mathscr{B}$, such that $\mathscr{F}(A)=B$ for all $A\in\operatorname{Ob}\mathscr{A}$ and $\mathscr{F}(f)=1_B$ for all morphisms f in $\mathscr{A}$. If $\mathscr{B}$ is a subcategory of $\mathscr{A}$ then $1_{\mathscr{B},\mathscr{A}}\colon\mathscr{B}\to\mathscr{A}$ defined in the obvious way, is called the *inclusion functor* from $\mathscr{B}$ into $\mathscr{A}$. However, the *identity functor* on $\mathscr{A}$ is denoted by $1_{\mathscr{A}}$. Of special importance is the *embedding functor*; that is a functor $\mathscr{F}:\mathscr{A}\to\mathscr{B}$ which satisfies: The function from $[A,B]_{\mathscr{A}}$ to $[\mathscr{F}(A),\mathscr{F}(B)]_{\mathscr{B}}$ induced by $\mathscr{F}$ is one-one for every pair A, B in $\operatorname{Ob}\mathscr{A}$.

If $\mathscr{F}:\mathscr{A}\to\mathscr{B}$ and $\mathscr{G}:\mathscr{B}\to\mathscr{C}$ are functors, then the *composition* $\mathscr{G}\circ\mathscr{F}:\mathscr{A}\to\mathscr{C}$ can be defined in a natural way by

$(\mathscr{G}\circ\mathscr{F})(A)=\mathscr{G}(\mathscr{F}(A))$ for $A\in\operatorname{Ob}\mathscr{A}$ and

$(\mathscr{G}\circ\mathscr{F})(f)=\mathscr{G}(\mathscr{F}(f))$ for a morphism of $\mathscr{A}$.

Obviously, $\mathscr{G}\circ\mathscr{F}$ is again a functor.

Covariant and contravariant functors can also be composed in a similar way. The composition of a covariant (contravariant) functor and a contravariant functor is a contravariant (covariant) functor.

If $\mathscr{F}:\mathscr{A}\to\mathscr{B}$ is a functor, then we say that $\mathscr{F}$ *preserves monomorphisms* (*epimorphisms*) if f monic (epic) implies $\mathscr{F}(f)$ monic (epic). Note that if f is an isomorphism then $\mathscr{F}(f)$ is always an isomorphism. The notion of "$\mathscr{F}$ *preserves limits* (*colimits*)" is now also obvious. In addition, if the objects of $\mathscr{A}$ and $\mathscr{B}$ are sets and the morphisms are maps then we say that $\mathscr{F}$ *preserves onto* (*one-one*) *morphisms* when f onto (one-one) implies $\mathscr{F}(f)$ onto (one-one). Finally, we say that $\mathscr{F}:\mathscr{A}\to\mathscr{B}$ *reflects monomorphisms* (*epimorphisms*) if $\mathscr{F}(f)$ monic (epic) implies f monic (epic). The meaning of "$\mathscr{F}$ *reflects one-one* (*onto*)" maps is again obvious.

We will first use the notion of functor to formulate a concept of "isomorphism" between categories. If $\mathscr{F} : \mathscr{A} \to \mathscr{B}$ is a functor which is one-one and onto between Ob $\mathscr{A}$ and Ob $\mathscr{B}$ and such that the induced maps $[A, B]_{\mathscr{A}} \to [\mathscr{F}(A), \mathscr{F}(B)]_{\mathscr{B}}$ are also one-one and onto, then $\mathscr{A}$ and $\mathscr{B}$ are *isomorphic*. But the notion of isomorphism is often too restrictive for categories and has to be replaced by another notion of "equivalence" which indicates that the categories are essentially the same.

Definition 3. A functor $\mathscr{F} : \mathscr{A} \to \mathscr{B}$ is called an *equivalence* if

(i) For each object B of $\mathscr{B}$ there exists an object A of $\mathscr{A}$ such that $\mathscr{F}(A)$ and B are isomorphic.

(ii) For $A, B \in \text{Ob}\, \mathscr{A}$, the function from $[A, B]_{\mathscr{A}}$ to $[\mathscr{F}(A), \mathscr{F}(B)]_{\mathscr{B}}$ induced by $\mathscr{F}$, is one-one and onto.

$\mathscr{A}$ and $\mathscr{B}$ are called *equivalent* if there exists an equivalence $\mathscr{F} : \mathscr{A} \to \mathscr{B}$. A contravariant functor $\mathscr{F} : \mathscr{A} \to \mathscr{B}$ is called a *coequivalence*, and $\mathscr{A}$ and $\mathscr{B}$ are said to be *coequivalent* if $\mathscr{F} : \mathscr{A} \to \mathscr{B}^*$ is an equivalence.

It is not difficult to prove that the relation of "equivalence" is symmetric, reflexive, and transitive and that the relation of "coequivalence" is symmetric and that the composition of two coequivalences is an equivalence. Also, an equivalence $\mathscr{F} : \mathscr{A} \to \mathscr{B}$ preserves and reflects monomorphisms, epimorphisms, limits, colimits and, of course, isomorphisms.

As an example we will show that an equivalence $\mathscr{F} : \mathscr{A} \to \mathscr{B}$ preserves and reflects monomorphisms. Suppose $f \in [A, B]_{\mathscr{A}}$ is monic and $\mathscr{F}(f) \circ g = \mathscr{F}(f) \circ h$ where $g, h \in [C, \mathscr{F}(A)]_{\mathscr{B}}$. Then by (i) there exists $C' \in \text{Ob}\, \mathscr{A}$ and an isomorphism $i \in [\mathscr{F}(C'), C]_{\mathscr{B}}$. Then $g \circ i, h \circ i \in [\mathscr{F}(C'), \mathscr{F}(A)]_{\mathscr{B}}$ so by (ii) there exists $g', h' \in [C', A]_{\mathscr{A}}$ such that $\mathscr{F}(g') = g \circ i$ and $\mathscr{F}(h') = h \circ i$. We have $\mathscr{F}(f \circ g') = \mathscr{F}(f) \circ \mathscr{F}(g') = \mathscr{F}(f) \circ g \circ i = \mathscr{F}(f) \circ h \circ i = \mathscr{F}(f) \circ \mathscr{F}(h') = \mathscr{F}(f \circ h')$ so again by (ii), since $f \circ g', f \circ h' \in [C', B]_{\mathscr{A}}, f \circ g' = f \circ h'$. Thus $g' = h'$ which implies $g \circ i = \mathscr{F}(g') = \mathscr{F}(h') = h \circ i$ and since i is an isomorphism, $g = h$. On the other hand if $f \in [A, B]_{\mathscr{A}}$ and $\mathscr{F}(f)$ is monic then for $g, h \in [C, A]_{\mathscr{A}}$ and $f \circ g = f \circ h$, we have $\mathscr{F}(f) \circ \mathscr{F}(g) = \mathscr{F}(f \circ g) = \mathscr{F}(f \circ h) = \mathscr{F}(f) \circ \mathscr{F}(h)$ so $\mathscr{F}(g) = \mathscr{F}(h)$. This implies $g = h$.

Remark. If $\mathscr{A}$ and $\mathscr{B}$ are coequivalent then epimorphisms in $\mathscr{A}$ correspond to monomorphisms in $\mathscr{B}$, injectives in $\mathscr{A}$ correspond to projectives in $\mathscr{B}$, etc.

18. Reflective subcategories

Many of the situations that we will encounter can be described by what is called a reflective subcategory of a category.

Definition 1. A subcategory $\mathscr{B}$ of a category $\mathscr{A}$ is *reflective* if there is a functor $\mathscr{R} : \mathscr{A} \to \mathscr{B}$, called a *reflector*, such that for each $A \in \text{Ob}\, \mathscr{A}$ there exists a morphism $\Phi_{\mathscr{R}}(A) : A \to \mathscr{R}(A)$ of $\mathscr{A}$ with the following properties:

(i) If $f \in [A, A']_{\mathscr{A}}$ then $\Phi_{\mathscr{R}}(A') \circ f = \mathscr{R}(f) \circ \Phi_{\mathscr{R}}(A)$.

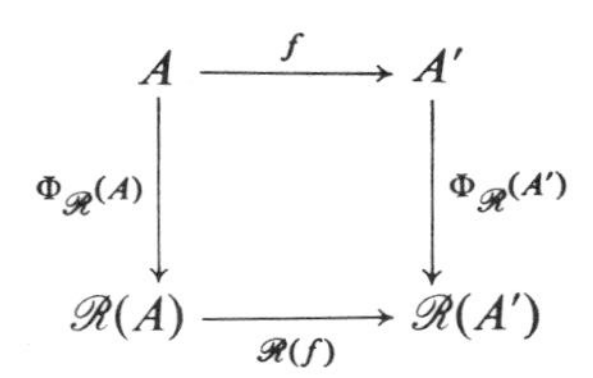

(ii) If $B \in \mathrm{Ob}\,\mathscr{B}$ and $f \in [A, B]_{\mathscr{A}}$ then there exists a unique morphism $f' \in [\mathscr{R}(A), B]_{\mathscr{B}}$ such that $f' \circ \Phi_{\mathscr{R}}(A) = f$.

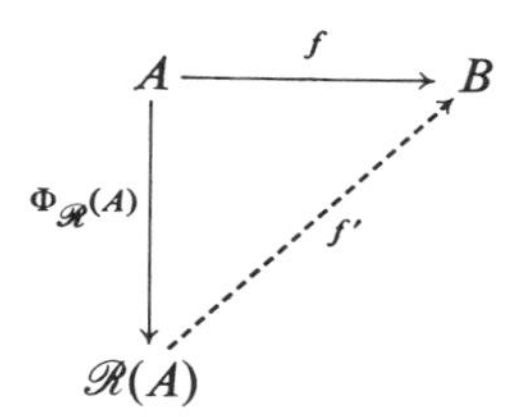

Note that (ii) implies the following: if $g, g' \in [\mathscr{R}(A), B]_{\mathscr{B}}$ and $g \circ \Phi_{\mathscr{R}}(A) = g' \circ \Phi_{\mathscr{R}}(A)$ then $g = g'$. Indeed, $B \in \mathrm{Ob}\,\mathscr{B}$, $g' \circ \Phi_{\mathscr{R}}(A) \in [A, B]_{\mathscr{A}}$ so there exists a unique morphism $x \in [\mathscr{R}(A), B]_{\mathscr{B}}$ such that $x \circ \Phi_{\mathscr{R}}(A) = g' \circ \Phi_{\mathscr{R}}(A)$; it follows that $g = g'$.

If $\mathscr{B}$ is a subcategory of $\mathscr{A}$ and $\mathscr{R}: \mathscr{A} \to \mathscr{B}$ and $\mathscr{S}: \mathscr{A} \to \mathscr{B}$ are functors which satisfy the conditions (i) and (ii) of Definition 1, then it is easy to see that $\mathscr{R}$ and $\mathscr{S}$ are "naturally equivalent". This means that for each $A \in \mathrm{Ob}\,\mathscr{A}$ there exists an isomorphism $\Psi(A): \mathscr{R}(A) \to \mathscr{S}(A)$ such that for each $f \in [A, B]_{\mathscr{A}}$ the following diagram commutes.

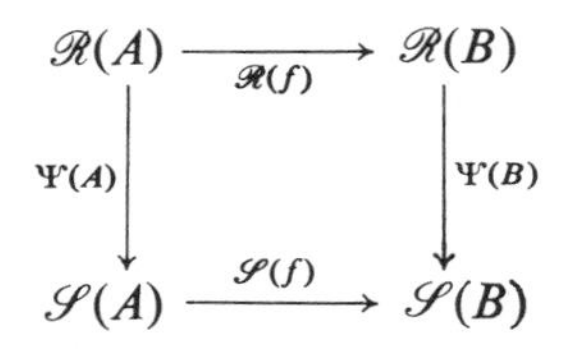

Therefore if $\mathscr{B}$ is reflective subcategory of $\mathscr{A}$ then we will often talk about *the* reflector $\mathscr{R}: \mathscr{A} \to \mathscr{B}$.

The following theorem will be an important tool for proving that a subcategory is reflective.

Theorem 2. *Let $\mathscr{B}$ be a subcategory of $\mathscr{A}$. $\mathscr{B}$ is a reflective subcategory of $\mathscr{A}$ if and only if there exists a function which assigns to every object A of $\mathscr{A}$ an object $\mathscr{R}(A)$ of $\mathscr{B}$ and a function which assigns to every object A of $\mathscr{A}$ a morphism $\Phi_{\mathscr{R}}(A): A \to \mathscr{R}(A)$ of $\mathscr{A}$ such that for every $B \in \mathrm{Ob}\,\mathscr{B}$ and $f \in [A, B]_{\mathscr{A}}$ there exists a unique morphism $f' \in [\mathscr{R}(A), B]_{\mathscr{B}}$ such that $f' \circ \Phi_{\mathscr{R}}(A) = f$.*

Proof. ($\Rightarrow$) Immediate.

($\Leftarrow$) We extend the assignment $\mathscr{R}$ from $\mathrm{Ob}\,\mathscr{A}$ to $\mathrm{Ob}\,\mathscr{B}$ to a functor $\mathscr{R}: \mathscr{A} \to \mathscr{B}$ as follows. For $f \in [A, A_1]_{\mathscr{A}}$ define $\mathscr{R}(f) \in [\mathscr{R}(A), \mathscr{R}(A_1)]_{\mathscr{B}}$ to be the unique morphism in $\mathscr{B}$ for which $\mathscr{R}(f) \circ \Phi_{\mathscr{R}}(A) = \Phi_{\mathscr{R}}(A_1) \circ f$. Thus (i) and (ii) of Definition 1 are satisfied and it remains to show that $\mathscr{R}$ is a functor. Let $f \in [A, A_1]_{\mathscr{A}}$, $g \in [A_1, A_2]_{\mathscr{A}}$. Then $\mathscr{R}(f) \circ \Phi_{\mathscr{R}}(A) = \Phi_{\mathscr{R}}(A_1) \circ f$ and $\mathscr{R}(g) \circ \Phi_{\mathscr{R}}(A_1) = \Phi_{\mathscr{R}}(A_2) \circ g$, so $\mathscr{R}(g) \circ \mathscr{R}(f) \circ \Phi_{\mathscr{R}}(A) = \mathscr{R}(g) \circ \Phi_{\mathscr{R}}(A_1) \circ f = \Phi_{\mathscr{R}}(A_2) \circ g \circ f$ and by uniqueness, $\mathscr{R}(g) \circ \mathscr{R}(f) = \mathscr{R}(g \circ f)$. Finally $1_A \in [A, A]_{\mathscr{A}}$ so $\mathscr{R}(1_A) \circ \Phi_{\mathscr{R}}(A) = \Phi_{\mathscr{R}}(A) \circ 1_A = \Phi_{\mathscr{R}}(A)$, so by uniqueness $1_{\mathscr{R}(A)} = \mathscr{R}(1_A)$.

Theorem 3. *Let $\mathscr{R}: \mathscr{A} \to \mathscr{B}$, $\mathscr{S}: \mathscr{B} \to \mathscr{C}$ be reflectors. Then $\mathscr{S} \circ \mathscr{R}: \mathscr{A} \to \mathscr{C}$ is a reflector.*

Proof. We check that the conditions of Theorem 2 are satisfied. For $A \in \operatorname{Ob} \mathscr{A}$, let $\Phi_{\mathscr{S}\circ\mathscr{R}}(A) = \Phi_{\mathscr{S}}(\mathscr{R}(A)) \circ \Phi_{\mathscr{R}}(A)$. If $C \in \operatorname{Ob} \mathscr{C}$ and $f \in [A, C]_{\mathscr{A}}$, then there exists a unique $f' \in [\mathscr{R}(A), C]_{\mathscr{B}}$ such that $f' \circ \Phi_{\mathscr{R}}(A) = f$ and there exists a unique $f'' \in [(\mathscr{S} \circ \mathscr{R})(A), C]_{\mathscr{C}}$ such that $f'' \circ \Phi_{\mathscr{S}}(\mathscr{R}(A)) = f'$. It follows easily that $f'' \circ \Phi_{\mathscr{S}\circ\mathscr{R}}(A) = f$. For uniqueness: if $g \circ \Phi_{\mathscr{S}\circ\mathscr{R}}(A) = f$ for $g \in [(\mathscr{S} \circ \mathscr{R})(A), C]_{\mathscr{C}}$ then $g \circ \Phi_{\mathscr{S}}(\mathscr{R}(A)) \circ \Phi_{\mathscr{R}}(A) = f$ so $f' = g \circ \Phi_{\mathscr{S}}(\mathscr{R}(A))$ and then $g = f''$.

Theorem 4. *If $\mathscr{B}$ is a reflective, full subcategory of $\mathscr{A}$, then there exists a reflector $\mathscr{S}: \mathscr{A} \to \mathscr{B}$ which is the identity functor on $\mathscr{B}$.*

Proof. We apply Theorem 2. Let $\mathscr{R}: \mathscr{A} \to \mathscr{B}$ be a reflector and define a function which assigns to each object A of $\mathscr{A}$ the object $\mathscr{R}(A)$ if $A \notin \operatorname{Ob} \mathscr{B}$ and A if $A \in \operatorname{Ob} \mathscr{B}$. Also define a function which assigns to an object A of $\mathscr{A}$ the morphism $\Phi_{\mathscr{R}}(A): A \to \mathscr{R}(A)$ if $A \notin \operatorname{Ob} \mathscr{B}$ and $1_A: A \to A$ if $A \in \operatorname{Ob} \mathscr{B}$. Suppose $f \in [A, B]_{\mathscr{A}}$ where $B \in \operatorname{Ob} \mathscr{B}$. If $A \in \operatorname{Ob} \mathscr{B}$ then f itself is the required morphisms for clearly, f is the only member $x \in [A, B]_{\mathscr{A}}$ which satisfies $x \circ 1_A = f$. Since $\mathscr{B}$ is full, $f \in [A, B]_{\mathscr{B}}$. Finally, for $A \notin \operatorname{Ob} \mathscr{B}$, $\mathscr{R}(f)$ is the unique $x \in [\mathscr{R}(A), B]_{\mathscr{B}}$ for which $x \circ \Phi_{\mathscr{R}}(A) = f$.

Theorem 5. *Suppose $\mathscr{R}: \mathscr{A} \to \mathscr{B}$ is a reflector. Then* (i) *$\mathscr{R}$ preserves epimorphisms;* (ii) *$\mathscr{R}$ preserves colimits of partially ordered systems and hence $\mathscr{R}$ preserves coproducts.*

Proof. (i) Suppose $f \in [A, A']_{\mathscr{A}}$ is an epimorphism and $g \circ \mathscr{R}(f) = h \circ \mathscr{R}(f)$ where $g, h \in [\mathscr{R}(A'), B]_{\mathscr{B}}$. Then $g \circ \Phi_{\mathscr{R}}(A') \circ f = g \circ \mathscr{R}(f) \circ \Phi_{\mathscr{R}}(A) = h \circ \mathscr{R}(f) \circ \Phi_{\mathscr{R}}(A) = h \circ \Phi_{\mathscr{R}}(A') \circ f$ so $g \circ \Phi_{\mathscr{R}}(A') = h \circ \Phi_{\mathscr{R}}(A')$. By uniqueness $g = h$. (ii) Let $((A_s)_{s\in S}, (f_{st})_{s\leq t})$ be a partially ordered system in $\mathscr{A}$ and let $(g_s: A_s \to A)_{s\in S}$ be its colimit. To show that $(\mathscr{R}(g_s): \mathscr{R}(A_s) \to \mathscr{R}(A))_{s\in S}$ is a colimit in $\mathscr{B}$ of $((\mathscr{R}(A_s))_{s\in S}, (\mathscr{R}(f_{st})_{s\leq t})$ we observe first that for $s \leq t$, $\mathscr{R}(g_t) \circ \mathscr{R}(f_{st}) = \mathscr{R}(g_t \circ f_{st}) = \mathscr{R}(g_s)$. Next, suppose $(h_s: \mathscr{R}(A_s) \to B)_{s\in S}$ is a family of morphisms in $\mathscr{B}$ such that $h_t \circ \mathscr{R}(f_{st}) = h_s$ for $s \leq t$. Thus $h_t \circ \Phi_{\mathscr{R}}(A_t) \circ f_{st} = h_t \circ \mathscr{R}(f_{st}) \circ \Phi_{\mathscr{R}}(A_s) = h_s \circ \Phi_{\mathscr{R}}(A_s)$ for $s \leq t$. So there exists a unique morphism $h_1 \in [A, B]_{\mathscr{A}}$ such that $h_1 \circ g_s = h_s \circ \Phi_{\mathscr{R}}(A_s)$ for all $s \in S$. Also there exists a unique morphism $h \in [\mathscr{R}(A), B]_{\mathscr{B}}$ such that $h \circ \Phi_{\mathscr{R}}(A) = h_1$. Now $h \circ \mathscr{R}(g_s) \circ \Phi_{\mathscr{R}}(A_s) = h \circ \Phi_{\mathscr{R}}(A) \circ g_s = h_1 \circ g_s = h_s \circ \Phi_{\mathscr{R}}(A_s)$ so $h \circ \mathscr{R}(g_s) = h_s$. For uniqueness, suppose $h' \circ \mathscr{R}(g_s) = h_s$ for all $s \in S$,

$h' \in [\mathscr{R}(A), B]_{\mathscr{B}}$. Then $h' \circ \mathscr{R}(g_s) \circ \Phi_{\mathscr{R}}(A_s) = h_s \circ \Phi_{\mathscr{R}}(A_s)$ so $h' \circ \Phi_{\mathscr{R}}(A) \circ g_s = h_s \circ \Phi_{\mathscr{R}}(A_s)$ and the uniqueness of h_1 implies $h' \circ \Phi_{\mathscr{R}}(A) = h_1$. Then the uniqueness of h implies $h = h'$.

Although reflectors do not, in general, preserve monomorphisms, we have the following theorem.

Theorem 6. *Suppose $\mathscr{R}: \mathscr{A} \to \mathscr{B}$ is a reflector which preserves monomorphisms. If B is an injective object in $\mathscr{B}$, then it is also injective in $\mathscr{A}$.*

Proof. Suppose $f: A \to C$ is a monomorphism in $\mathscr{A}$ and $g \in [A, B]_{\mathscr{A}}$. There exists $h \in [\mathscr{R}(A), B]_{\mathscr{B}}$ such that $h \circ \Phi_{\mathscr{R}}(A) = g$. Now $\mathscr{R}(f)$ is monic in $\mathscr{B}$ so there exists $k \in [\mathscr{R}(C), B]_{\mathscr{B}}$ such that $k \circ \mathscr{R}(f) = h$. The required morphism in $\mathscr{A}$ is $k \circ \Phi_{\mathscr{R}}(C): C \to B$ since $k \circ \Phi_{\mathscr{R}}(C) \circ f = k \circ \mathscr{R}(f) \circ \Phi_{\mathscr{R}}(A) = h \circ \Phi_{\mathscr{R}}(A) = g$.

19. Categories of algebras

In this book we will be concerned mainly with categories whose objects are algebras of a certain similarity type and whose morphisms are the homomorphisms between these algebras. We therefore introduce the following terminology.

Definition 1. Suppose **A** is a class of similar algebras. Then the category $\mathscr{A}$ is defined by Ob $\mathscr{A}$ = **A** and the morphisms of $\mathscr{A}$ are *all* of the homomorphisms $f: A \to B$, where $A, B \in \text{Ob}\,\mathscr{A}$. In this case $\mathscr{A}$ is called the *algebraic category associated with* **A**.

We note that, in view of the terminology established in Section 7, we can say that the morphisms of $\mathscr{A}$ are exactly the **A**-homomorphisms between members of **A**.

To denote the algebraic category associated with a class of algebras, we will use the following procedure. If the class of algebras is represented by a given bold face capital letter, then we will use the corresponding bold face script capital letter to denote the associated category. Conversely, if $\mathscr{A}, \mathscr{B}, \cdots$ denote algebraic categories, then the classes of algebras with which they are associated will be denoted by **A**, **B**, $\cdots$.

There are some notions that at this time need our special attention. First, the term "isomorphism" has been defined between algebras of a certain similarity type but also as certain morphisms in a category. Thus if $\mathscr{A}$ is an algebraic category, it seems that we must distinguish between "algebraic isomorphism" and "categorical isomorphism". Fortunately this distinction need not be made, as the following theorem shows.

Theorem 2. *Let $\mathscr{A}$ be an algebraic category and let $f \in [A, B]$. Then f is an isomorphism in $\mathscr{A}$ if and only if f is an* **A**-*isomorphism.*

Proof. ($\Rightarrow$) There exists $g \in [B, A]$ such that $f \circ g = 1_A$ and $g \circ f = 1_B$. But then f is one-one and onto so it is an **A**-isomorphism.

($\Leftarrow$) f is one-one and onto, so f^{-1} exists which is also an **A**-isomorphism. Hence $f^{-1} \in [B, A]$ and $f \circ f^{-1} = 1_B$ and $f^{-1} \circ f = 1_A$.

There are two other related notions which, however, have distinct meanings, namely that of *subalgebra* and that of *subobject*.

Suppose $\mathscr{A}$ is an algebraic category, $A \in \text{Ob}\,\mathscr{A}$ and B is a subalgebra of A. Then B need not be an object in $\mathscr{A}$ (unless, for example, **A** is equational) but if $B \in \text{Ob}\,\mathscr{A}$ then the inclusion map $1_{B,A}$ *is* a subobject of A. Conversely, if $A, B \in \text{Ob}\,\mathscr{A}$ and $j: B \to A$ is a monomorphism in $\mathscr{A}$, then Im j is a subalgebra of A but need not be an object of $\mathscr{A}$; and even if Im $j \in \text{Ob}\,\mathscr{A}$, it may be the case that B and Im j are not isomorphic (because j need not be one-one). However, in case **A** is equational, then the situation is much simpler (Section 20).

Suppose $\mathscr{A}$ and $\mathscr{B}$ are algebraic categories such that **B** is of similarity type $\tau = (n_1, \cdots, n_{o(\tau)})$ and **A** has type $(n_{i_1}, \cdots, n_{i_p})$ for some $I = \{i_1, \cdots, i_p\} \subseteq \{1, \cdots, o(\tau)\}$. As has been pointed out in Section 7, we will obtain a class $\mathbf{B}(I)$ of algebras of type $(n_{i_1}, \cdots, n_{i_p})$ by taking for each algebra in **B** its underlying set and by disregarding the operations f_i for $i \notin I$. Now, suppose $\mathbf{B}(I) \subseteq \mathbf{A}$. The assignment $\mathscr{G}: \mathscr{B} \to \mathscr{A}$ defined by $\mathscr{G}(B, (f_1, \cdots, f_{o(\tau)})) = (B, (f_{i_1}, \cdots, f_{i_p}))$ and $\mathscr{G}(h) = h$ for a homomorphism h between algebras in **B**, is obviously a functor. In the cases that we will encounter, the following condition will always be satisfied: if $(B, (f_1, \cdots, f_{o(\tau)}))$ and $(B, (f'_1, \cdots, f'_{o(\tau)}))$ are algebras which belong to **B** such that $f_{i_j} = f'_{i_j}$ for $j \leq p$, then $f_i = f'_i$ for all $i \leq o(\tau)$. Clearly, if this condition holds, then $\mathscr{G}$ is an embedding which is one-one on the objects of $\mathscr{B}$. Note that in this case Im $\mathscr{G}$ is a subcategory of $\mathscr{A}$ and that the induced functor $\mathscr{B} \to$ Im $\mathscr{G}$ is an equivalence and in fact an isomorphism. Therefore, whenever this situation arises and when we are dealing with categorical properties, we will identify the category $\mathscr{B}$ with the subcategory Im $\mathscr{G}$ of $\mathscr{A}$. This will enable us to avoid a cumbersome formulation of results which are essentially categorical in nature.

Since our general approach is both categorical and algebraic, the terminology that we have defined so far, will play an important role in distinguishing between categorical and algebraic notions and also in indicating our approach to a given problem.

20. Equational categories

Among the algebraic categories that we will consider will be those whose class of objects is an equational class of algebras. Because of their importance, such categories will be referred to as *equational categories*. An equational category $\mathscr{A}$ is *trivial* if $|A| = 1$, for each $A \in \text{Ob}\,\mathscr{A}$.

Of course, all of the remarks of the preceding section apply to equational categories; certain situations, however, become simpler.

Theorem 1. *Let $\mathscr{A}$ be an equational category. Then the monomorphisms of $\mathscr{A}$ are exactly the one-one* **A***-homomorphisms.*

Proof. We need only prove (cf. Theorem 14.1) that **A**-monomorphisms are one-one. Thus, suppose $f \in [A, B]$ is a monomorphism and $f(x) = f(y)$. Since we can assume that $\mathscr{A}$ is non-trivial, $\mathscr{F}_{\mathbf{A}}(\{s\})$ exists and is an object of $\mathscr{A}$ so there exist $f_0, f_1 \in [\mathscr{F}_{\mathbf{A}}(\{s\}), A]$ such that $f_0(s) = x$ and $f_1(s) = y$. It follows from $f(f_0(s)) = f(x) = f(y) = f(f_1(s))$ that $x = f_0(s) = f_1(s) = y$.

Corollary 2. *Let A be an algebra in an equational category $\mathscr{A}$. Then A is injective in $\mathscr{A}$ if and only if every morphism $g \in [B, A]$ from a subalgebra B of an algebra C can be extended to a morphism $h: C \to A$.*

Proof. Immediate.

The distinction between subalgebras and subobjects in an equational category is, contrary to the general situation, practically non-existent. Indeed, suppose $\mathscr{A}$ is an equational category and $A \in \text{Ob}\,\mathscr{A}$. Suppose B is a subalgebra of A. Then $B \in \text{Ob}\,\mathscr{A}$ so $1_{B,A}$ is a subobject of A. Conversely, if $j: B \to A$ is a subobject of A, then $\text{Im}\, j$ is a subalgebra of A (and isomorphic with B). So from each class of isomorphic subobjects of A, a representative can be chosen. If we take for the representative that subobject which is the inclusion map, then we can identify subalgebras and subobjects in $\mathscr{A}$.

Suppose $\mathscr{A}$ is an equational category. If B is a subalgebra of A, $A \in \text{Ob}\,\mathscr{A}$ and $h: A \to B$ is an epimorphism such that $h|B = 1_B$ then since $1_{B,A}$ is a morphism in $\mathscr{A}$, we infer that B is a retract of A. Conversely, suppose $A, B \in \text{Ob}\,\mathscr{A}, f \in [A, B]$, $g \in [B, A]$ and $f \circ g = 1_B$. Then $g': B \to g[B]$, defined by $g'(x) = g(x)$ for $x \in B$, is an isomorphism and $g' \circ f: A \to g[B]$ is an epimorphism in $\mathscr{A}$ with the property that $g' \circ f|g[B] = 1_{g[B]}$. Thus, if we say $h: A \to B$ is a retract then it is to be assumed that B is a subalgebra of A and h is a morphism such that $h|B = 1_B$.

We now give some results which relate algebraic and categorical concepts in equational categories.

Theorem 3. *If $\mathscr{A}$ is an equational category with enough injectives then $\mathscr{A}$ has the congruence extension property.*

Proof. Let $f \in [A, B]$ be monic, $g \in [A, C]$ onto.

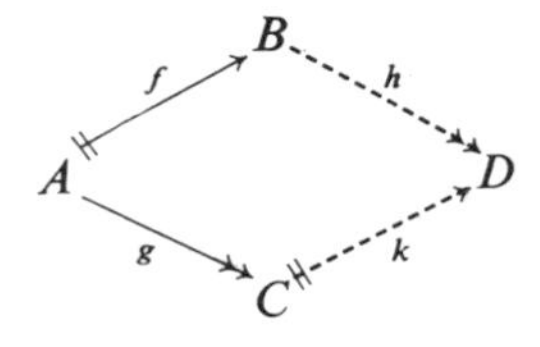

By hypothesis there is an extension $k \in [C, D]$ of C with D injective. But the injectivity of D implies there exists $h \in [B, D]$ such that $h \circ f = k \circ g$. Then $h': B \to h[B]$, given by $h'(x) = h(x)$ for all x, is in $\mathscr{A}$. Also $k[C] \subseteq h[B]$ so by redefining the codomain of k to be $h[B]$, we have the desired result: $h' \circ f = k \circ g$.

In Section 14, we have defined the concepts of extensions, essential extensions, etc. For an equational category $\mathscr{A}$ it will sometimes be convenient to identify isomorphic extensions of an object. Therefore, without loss of generality, when we talk about an extension $f: A \to B$ we will assume that A is subalgebra of $B, f = 1_{A,B}$ and we will say that B is an extension of A.

The meaning is also clear when we say that B is an essential extension of A: every morphism $g \in [B, C]_{\mathscr{A}}$ such that $g|A$ is monic is itself monic. Finally under this convention, B is a proper extension of A if A is a proper subalgebra of B.

It is easy to see that an extension B of A is essential if and only if any congruence relation θ on B for which $\theta \cap A^2$ is trivial is itself trivial.

Theorem 4. *Consider the following statements about an algebra A of an equational category $\mathscr{A}$.*

(i) *A is injective.*

(ii) *A is a retract of each of its extensions.*

(iii) *A has no proper essential extensions.*
Then (i) $\Rightarrow$ (ii) $\Rightarrow$ (iii). *Moreover if $\mathscr{A}$ has enough injectives then* (i), (ii), (iii) *are equivalent.*

Proof. (i) $\Rightarrow$ (ii) Trivial.

(ii) $\Rightarrow$ (iii) Suppose B is an essential extension of A. By hypotheses there exists a morphism $f: B \to A$ such that $f|A = 1_A$. But since B is an essential extension of A, if follows that f is one-one so $A = B$.

Now suppose that there are enough injectives. We prove (iii) $\Rightarrow$ (i). Since $\mathscr{A}$ has enough injectives we can assume that A is a subalgebra of an injective B. By Zorn's Lemma there is a congruence relation θ on B which is maximal subject to the condition that $\theta \cap A^2$ is the trivial congruence relation on A. We will show that $\nu_\theta \circ 1_{A,B}: A \to B/\theta$ is an essential extension of A.

Suppose first that $\nu_\theta \circ 1_{A,B}(a_1) = \nu_\theta \circ 1_{A,B}(a_2)$. Then $(a_1, a_2) \in \theta$ so $a_1 = a_2$, and $\nu_\theta \circ 1_{A,B}$ is an extension of A.

Suppose $f \in [B/\theta, C]_{\mathscr{A}}$ and $f \circ (\nu_\theta \circ 1_{A,B})$ is monic. Now it is easy to see that $\theta \subseteq \text{Ker}(f \circ \nu_\theta)$ and $\text{Ker}(f \circ \nu_\theta) \cap A^2$ is trivial; so by the maximality of θ, $\theta = \text{Ker}(f \circ \nu_\theta)$. It follows that f is monic. So $\nu_\theta \circ 1_{A,B}$ is essential.

By hypothesis $\nu_\theta \circ 1_{A,B}$ is an isomorphism. The formula $((\nu_\theta \circ 1_{A,B})^{-1} \circ \nu_\theta) \circ 1_{A,B} = 1_A$ shows that A is a retract of B. But B is injective and therefore so is A.

Theorem 5. *Let $\mathscr{A}$ be an equational category with enough injectives. Then each algebra $A \in \text{Ob}\,\mathscr{A}$ has an injective hull.*

Proof. Since $\mathscr{A}$ has enough injectives we can assume that A is a subalgebra of an injective algebra C. Let $\mathfrak{S}$ be the set of subalgebras of C which contain A and are essential extensions of A. Note that $\mathfrak{S} \neq \varnothing$. By Zorn's Lemma $\mathfrak{S}$, partially ordered under inclusion, contains a maximal member B.

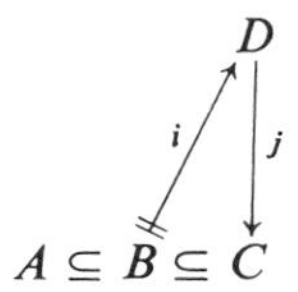

Now suppose B is not injective. Then by Theorem 4(iii) there is a proper essential extension $i: B \to D$ of B. Since C is injective there is a homomorphism $j: D \to C$ such that $j \circ i = 1_{B,C}$. But $1_{B,C}$ is monic and i is essential so j is monic. Define $j': D \to j[D]$ by $j'(d) = j(d)$ for each $d \in D$. Then j' is an isomorphism and hence an essential extension of D. But then $1_{A,B}$, i, and j' are each essential extensions and hence $j' \circ i \circ 1_{A,B}$ is an essential extension of A. For $a \in A$, $j' \circ i \circ 1_{A,B}(a) = j \circ i(a) = 1_{B,C}(a) = a$, so $1_{A,j[D]} = j' \circ i \circ 1_{A,B}$. Thux, $j[D]$ is an essential extension of A so $j[D] \in \mathfrak{S}$. Moreover, since i is a proper extension of B, $B = 1_{B,C}[B] = j \circ i[B] \subset j[D]$, contradicting the maximality of B.

Any two injective hulls of an object in an equational category are isomorphic. Indeed, let $f_1 \in [A, B_1]$, $f_2 \in [A, B_2]$ be injective hulls of A. We observed in the proof of Theorem 16.6 that there exist morphisms $h \in [B_1, B_2]$, $k \in [B_2, B_1]$ such that $k \circ h = 1_{B_1}$ and $h \circ f_1 = f_2$. Clearly k is onto. But $k \circ f_2 = k \circ h \circ f_1 = f_1$ is monic and f_2 is essential so k is monic and therefore one-one.

In equational categories, we can also describe products and coproducts more explicitly.

Theorem 6. *Let $\mathscr{A}$ be an equational category and* $(A_s)_{s\in S}$ be a family of objects in $\mathscr{A}$. Then $(p_t: \times_{s\in S} A_s \to A_t)_{t\in S}$ is a product of $(A_s)_{s\in S}$ in $\mathscr{A}$.

Proof. Precisely, we must show that the projections $p_t: \times_{s\in S} A_s \to A_t$ satisfy the conditions of Definition 15.1. Notice that since $\mathscr{A}$ is equational, p_t is a morphism for all $t \in S$. Suppose $(f_t: B \to A_t)_{t\in S}$ is a family of morphisms in $\mathscr{A}$. Define $f: B \to \times_{s\in S} A_s$ by $f = \times_{t\in S} f_t$. Then, by Lemma 5.2, f is a morphism in $\mathscr{A}$ and $p_t \circ f = f_t$ for each $t \in S$. Suppose $f' \in [B, \times_{s\in S} A_s]$ and $p_t \circ f' = f_t$ for all $t \in S$. Then for $b \in B$ and $t \in S$, $f(b)(t) = p_t(f(b)) = f_t(b) = p_t(f'(b)) = f'(b)(t)$ so $f = f'$.

Theorem 7. *Let $\mathscr{A}$ be an equational category and $(j_s: A_s \to A)_{s\in S}$ a family of morphisms such that if $(f_s: A_s \to B)_{s\in S}$ is also a family of morphisms, then there exists a morphism $h: A \to B$* such that *$h \circ j_s = f_s$ for all $s \in S$. Then $(j_s: A_s \to A)_{s\in S}$ is a coproduct of $(A_s)_{s\in S}$ if and only if $\bigcup_{s\in S} j_s[A_s]$ generates A.*

Proof. ($\Leftarrow$) The uniqueness of h follows from Lemma 4.7.

($\Rightarrow$) Let $A' = [\bigcup_{s\in S} j_s[A_s]]$. For each $s \in S$ define $j_s': A_s \to A'$ by $j_s'(x) = j_s(x)$ for each $x \in A_s$. Notice that all of these maps are morphisms in $\mathscr{A}$ since $\mathscr{A}$ is equational. By hypothesis there exists a morphism $h: A \to A'$ such that $h \circ j_s = j_s'$ for all $s \in S$. Hence $1_{A',A} \circ h \circ j_s = 1_{A',A} \circ j_s' = j_s$ for all $s \in S$. But also $1_A \circ j_s = j_s$ for all $s \in S$. Hence, by uniqueness, $1_{A',A} \circ h = 1_A$. This shows that $1_{A',A}$ is onto and therefore $A' = A$.

In most cases, we will give a direct construction to establish the existence of coproducts for the categories with which we will be concerned. Therefore we will state, without proof, the following general existence theorem for coproducts (The proof can be found, for example, in R. S. Pierce [3], p. 107, Corollary 2.8.).

Theorem 8. *Let $\mathscr{A}$ be an equational category and let $(A_s)_{s \in S}$ be a family of objects in $\mathscr{A}$. Suppose that for each $s \in S$, A_s is a subalgebra of B_s for some $B_s \in \mathrm{Ob}\,\mathscr{A}$ and there exist homomorphisms $h_{s,t}\colon A_s \to B_t$ for each pair s, t. Then the coproduct of $(A_s)_{s \in S}$ exists.*

We observe that if, for example, A_s has a one-element subalgebra for each $s \in S$, then we can let $B_s = A_s$ and take $h_{s,t}\colon A_s \to A_t$ to be the homomorphism that maps A_s onto the one-element subalgebra of A_t. So coproducts exist in all such equational categories.

In Definition 10.1 we introduced free algebras. We will now give a definition of free algebras in more categorical terms.

Definition 9. An algebra A in an algebraic category $\mathscr{A}$ is *free over* $\mathscr{A}$ and has *free generating set* S if there exists a function $f\colon S \to A$ such that for each function $g\colon S \to B$, $B \in \mathrm{Ob}\,\mathscr{A}$ there exists a unique $h \in [A, B]$ such that $h \circ f = g$.

Under rather mild restrictions, we have that in an equational category the algebraic and categorical definitions of freeness are equivalent.

Theorem 10. *Let $\mathscr{A}$ be a non-trivial equational category and S a non-empty subset of $A \in \mathrm{Ob}\,\mathscr{A}$. Then A is free over* **A** *with free generating set S if and only if A is free over $\mathscr{A}$ with free generating set S.*

Proof. ($\Rightarrow$) Let $f = 1_{S,A}$. Clearly, the extension property holds and uniqueness follows from Lemma 4.7.

($\Leftarrow$) It suffices to show that f is one-one and that $f[S]$ generates A. Suppose $f(x) = f(y)$ for $x \neq y$. Now by Theorem 12.4 $\mathscr{F}_{\mathbf{A}}(S)$ exists (where $\mathscr{F}_{\mathbf{A}}(S)$ as usual, denotes the free algebra on S generators in the algebraic sense). Again by hypothesis there exists a morphism $h\colon A \to \mathscr{F}_{\mathbf{A}}(S)$ such that $h \circ f = 1_{S,\mathscr{F}_{\mathbf{A}}(S)}$. But $f(x) = f(y)$ and it follows that $x = y$. Contradiction. This proves that f is one-one. Since $S \neq \varnothing$, we can let $A' = [f[S]]$ and define $f'\colon S \to A'$ by $f'(s) = f(s)$ for all $s \in S$. Since $A' \in \mathrm{Ob}\,\mathscr{A}$ there exists a morphism $h\colon A \to A'$ such that $h \circ f = f'$. Now $1_{A',A} \circ h \circ f = 1_{A',A} \circ f' = f$. By uniqueness $1_{A',A} \circ h = 1_A$ and thus $1_{A',A}$ is onto, proving that $A' = A$.

Theorem 11. *Let $\mathscr{A}$ be a non-trivial equational category and S a set, $|S| > 0$. Suppose $(j_s\colon \mathscr{F}_{\mathbf{A}}(\{s\}) \to A)_{s \in S}$ is a coproduct of $(\mathscr{F}_{\mathbf{A}}(\{s\}))_{s \in S}$ then $A \cong \mathscr{F}_{\mathbf{A}}(S)$.*

Proof. Let $f\colon S \to A$ be defined by $f(s) = j_s(s)$. Suppose $B \in \mathrm{Ob}\,\mathscr{A}$ and $g\colon S \to B$ is a function. There exists, for each $s \in S$, a unique homomorphism $g_s\colon \mathscr{F}_{\mathbf{A}}(\{s\}) \to B$ with $g_s(s) = g(s)$ for each $s \in S$. Again, there exists a unique homomorphism

$h: A \to B$ with $h \circ j_s = g_s$ for each $s \in S$ and it follows that $h \circ f = g$. It is obvious that h is unique and it follows from Theorem 7 that $A \cong \mathscr{F}_{\mathbf{A}}(S)$.

Even in an equational category epimorphisms need not be onto. The following definition will be useful in characterizing epimorphisms in equational categories.

Definition 12. Let $\mathscr{A}$ be an equational category and $A \in \text{Ob}\,\mathscr{A}$. A subalgebra B of A is called *epic in A* if the inclusion map $1_{B,A}$ is an epimorphism.

Notice that B is epic in A if and only if $f|B = g|B$ implies $f = g$ for each pair $f, g \in [A, C]$. Also, every epimorphism is onto if and only if there are no proper epic subalgebras.

Since the notion of a projective object involves the term epimorphism, it is natural to consider the following related objects in an equational category.

Definition 13. Let $\mathscr{A}$ be an equational category. Then $A \in \text{Ob}\,\mathscr{A}$ is *weakly projective* if for each morphism $f \in [C, B]$ such that f is onto and every morphism $g \in [A, B]$ there exists a morphism $h \in [A, C]$ such that $f \circ h = g$.

It is clear that a projective object is weakly projective. The proofs of the following statements are analogous to those in Section 16.

(1) *A retract of a weakly projective object is weakly projective.*

(2) *If B is weakly projective and $h \in [A, B]$ is onto then B is a retract of A.*

In equational categories, there is also a relation between free and weakly projective objects. First note that if $\mathscr{A}$ is equational and has an epimorphism $g: A \to B$ which is not onto, then no free algebra with one or more free generators is projective. Indeed, suppose such a free algebra F would exist. Let $y \in B$. Then there would exist an $x \in F$ and $f \in [F, B]$ such that $f(x) = y$. By the projectivity of F there would exist $h \in [F, A]$ with $g \circ h = f$ and hence $g(h(x)) = f(x) = y$ which yields the contradiction that g is onto. However, every free algebra in an equational category is weakly projective.

Theorem 14. *Let $\mathscr{A}$ be a non-trivial equational category and $A \in \text{Ob}\,\mathscr{A}$. Then A is weakly projective in $\mathscr{A}$ if and only if A is a retract of an $\mathscr{A}$-free algebra.*

Proof. ($\Rightarrow$) $A \in \text{Ob}\,\mathscr{A}$. Since $\mathscr{A}$ is equational, there exists a free algebra F in $\mathscr{A}$ and a homomorphism of F onto A. So if A is projective, it is then a retract of F.

($\Leftarrow$) We need only prove that free algebras are weakly projective. Thus, suppose F is freely generated by S, $h \in [A, B]$ is onto and $f \in [F, B]$. Define a function $g_1: S \to A$ so that $h(g_1(s)) = f(s)$ for each $s \in S$. Now extend g_1 to a homomorphism $g: F \to A$ and we have $h \circ g = f$.

We next state some sufficient conditions under which limits and colimits exist.

Theorem 15. *Suppose* $((A_s)_{s\in S}, (f_{st})_{s\leq t})$ *is a partially ordered system in an equational category* $\mathscr{A}$. *If*

$$A = \left\{x \in \bigtimes_{s\in S} A_s : p_t(x) = f_{st}(p_s(x)) \text{ for } s \leq t\right\} \neq \varnothing$$

then $(p_s|A: A \to A_s)_{s\in S}$ *is the limit of* $((A_s)_{s\in S}, (f_{st})_{s\leq t})$ *in* $\mathscr{A}$.

Proof. It is easily verified that A is a subalgebra of $\bigtimes_{s\in S} A_s$ so $A \in \text{Ob}\,\mathscr{A}$. Now for $s \leq t$, $f_{st} \circ (p_s|A)(x) = f_{st}(p_s(x)) = (p_t|A)(x)$ for each $x \in A$. Next, suppose $(h_s: B \to A_s)_{s\in S}$ is a family of morphisms such that $f_{st} \circ h_s = h_t$ for $s \leq t$. Define $h: B \to A$ by $h(b)(s) = h_s(b)$. Note that $h(b)$ is, in fact, in A since $f_{st}(p_s(h(b))) = f_{st}(h_s(b)) = h_t(b) = p_t(h(b))$. Again it is clear that $h \in [B, A]$ and $p_s \circ h = h_s$ for all $s \in S$. For uniqueness, suppose $h' \in [B, A]$ and $p_s \circ h' = h_s$ for all $s \in S$. Then for $b \in B$ and $s \in S$, $h'(b)(s) = p_s(h'(b)) = h_s(b) = h(b)(s)$ so $h = h'$.

Corollary 16. *If* $\mathscr{A}$ *is an equational category with a nullary operation, then limits exist.*

Proof. Under the hypothesis, it is clear that $A \neq \varnothing$.

Theorem 17. *Suppose* $((A_s)_{s\in S}, (f_{st})_{s\leq t})$ *is a partially ordered system in an equational category* $\mathscr{A}$. *If coproducts exist in* $\mathscr{A}$ *then the colimit of* $((A_s)_{s\in S}, (f_{st})_{s\leq t})$ *exists in* $\mathscr{A}$.

Proof. Let $(j_s: A_s \to C)_{s\in S}$ be a coproduct of the family $(A_s)_{s\in S}$ and θ be the smallest congruence relation on C which identifies the elements $j_s(x)$ and $j_t(f_{st}(x))$ for $x \in A_s$ and $s \leq t$. Then, since $\mathscr{A}$ is equational, $A = C/\theta \in \text{Ob}\,\mathscr{A}$ and $(\nu_\theta \circ j_s: A_s \to A)_{s\in S}$ is a family of morphisms in $\mathscr{A}$. We will prove that this family is the colimit of $((A_s)_{s\in S}, (f_{st})_{s\leq t})$. First, for $x \in A_s$, $s \leq t$, we have by the definition of θ that $(j_s(x), j_t(f_{st}(x))) \in \theta$ so $(\nu_\theta \circ j_t) \circ f_{st} = \nu_\theta \circ j_s$ for $s \leq t$. Next, let $(h_s: A_s \to B)_{s\in S}$ be a family of morphisms such that $h_t \circ f_{st} = h_s$ for $s \leq t$. Then, since $(j_s: A_s \to C)_{s\in S}$ is a coproduct of $(A_s)_{s\in S}$ there exists a unique homomorphism $h_1: C \to B$ such that $h_1 \circ j_s = h_s$ for all $s \in S$. Now for $x \in A_s$ and $s \leq t$, $h_1(j_t(f_{st}(x))) = h_s(x) = h_1(j_s(x))$ so $(j_t(f_{st}(x)), j_s(x)) \in \text{Ker}\, h_1$. Thus $\text{Ker}\, h_1$ identifies $j_s(x)$ and $j_t(f_{st}(x))$ for $x \in A_s$ and $s \leq t$ and so $\theta \subseteq \text{Ker}\, h_1$. By the homomorphism theorem there exists a (unique) homomorphism $h: A \to B$ such that $h \circ \nu_\theta = h_1$. Thus, for $s \in S$: $h \circ (\nu_\theta \circ j_s) = h_1 \circ j_s = h_s$. For uniqueness, suppose $h': A \to B$ is a homomorphism such that $h' \circ (\nu_\theta \circ j_s) = h_s$ for all $s \in S$. By the uniqueness of h_1, we have $h_1 = h' \circ \nu_\theta$. Thus for $x \in C$, $h([x]_\theta) = h(\nu_\theta(x)) = h_1(x) = h'(\nu_\theta(x)) = h'([x]_\theta)$ so $h = h'$.

We also have the following theorem for equational categories.

Theorem 18. *Let* $\mathscr{A}$ *be an equational category and* $((A_s)_{s\in S}, (f_{st})_{s\leq t})$ *a partially ordered system in* $\mathscr{A}$ *and let* $(j_s: A_s \to A)_{s\in S}$ *be a colimit of this system. Suppose* $(h_s: A_s \to B)_{s\in S}$ *is a family of monomorphisms in* $\mathscr{A}$ *such that* $h_t \circ f_{st} = h_s$ *for* $s \leq t$. *Then* j_s *is a monomorphism for each* $s \in S$. *A dual theorem holds for limits.*

Proof. Immediate.

Definition 19. An equational category $\mathscr{A}$ has the *amalgamation property* provided that if $(f_s: A \to B_s)_{s \in S}$ is a non-empty family of monomorphisms in $\mathscr{A}$ then there exists a family of monomorphisms $(g_s: B_s \to C)_{s \in S}$ and a monomorphism $g: A \to C$ such that $g_s \circ f_s = g$ for all $s \in S$.

Remark. Since $(f_s: A \to B_s)_{s \in S}$ can be considered as a partially ordered system in $\mathscr{A}$, it follows immediately from Theorem 18 that we may assume that the family $(g_s: B_s \to C)_{s \in S}$ together with $g: A \to C$ is the colimit of $(f_s: A \to B_s)_{s \in S}$.

Theorem 20. (R. S. Pierce [3]) *If $\mathscr{A}$ is an equational category with enough injectives then $\mathscr{A}$ has the amalgamation property.*

Proof. Let $(f_s: A \to B_s)_{s \in S}$ be a family of monomorphisms. For each $s \in S$, let $j_s: B_s \to B_s'$ be a monomorphism such that B_s' is injective. Let $(p_s: C \to B_s')_{s \in S}$ be the product of $(B_s')_{s \in S}$ in $\mathscr{A}$.

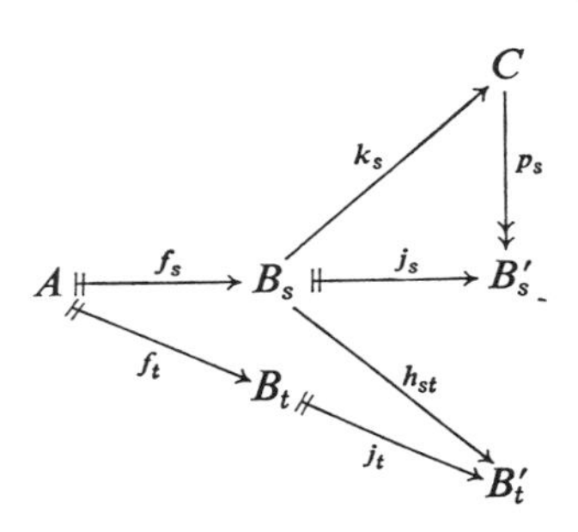

By Theorem 16.3, C is injective. For $s, t \in S$, $s \neq t$ (we may, of course, assume that $|S| > 1$) there exists $h_{st} \in [B_s, B_t']$ such that $h_{st} \circ f_s = j_t \circ f_t$. For each $s \in S$, we define $k_s \in [B_s, C]$ by the conditions $p_t \circ k_s = h_{st}$ for $t \neq s$ and $p_s \circ k_s = j_s$. Now for $s \neq t$, $p_s \circ k_s \circ f_s = j_s \circ f_s = h_{ts} \circ f_t = p_s \circ k_t \circ f_t$. But then $k_s \circ f_s = k_t \circ f_t$. Moreover, since for each $s \in S$, j_s is monic and $p_s \circ k_s = j_s$, k_s is monic and thus $k_s \circ f_s$ is monic for each $s \in S$. This proves the theorem.

II

Lattices

1. Partially ordered sets (continued)

In Section I.2, we defined a partially ordered set and introduced some elementary notions. We resume our discussion of this topic.

Let P be a poset. It is sometimes convenient to write $y \geq x$ instead of $x \leq y$. An element y is said to *cover* x provided that $x < y$ but there is no z for which $x < z$ and $z < y$.

Exercise 1. Let $(P, \leq)$ be a poset. Show that the relation "$<$" defined above satisfies:

(i) $x \nless x$ for each $x \in P$.

(ii) $x < y, y < z \Rightarrow x < z$.

(iii) $x < y \Rightarrow y \nless x$.

Conversely, prove that if $<$ is a relation on a set $S \neq \varnothing$ which satisfies (i), (ii), and (iii), then $(S, \leq)$ is a poset, where $\leq$ is defined by $x \leq y \Leftrightarrow x < y$ or $x = y$.

It is evident that any non-empty subset of a poset can be made into a poset by restricting the partial ordering to the elements of the subset. Two other ways of forming new posets from old ones will now be defined.

If $(P_s)_{s \in S}$ is a family of mutually disjoint posets and S is a chain, then the *ordinal sum* $\bigoplus_{s \in S} P_s$ is defined by $\bigoplus_{s \in S} P_s = (\bigcup_{s \in S} P_s, \leq)$ where $x \leq y$ in $\bigoplus_{s \in S} P_s$ when either

(1) $x \in P_s$, $y \in P_t$ and $s < t$, or

(2) $\{x, y\} \subseteq P_s$ and $x \leq y$ in P_s.

If S is a finite chain, say $S = \{1, \cdots, n\}$ then $\bigoplus_{s \in S} P_s$ is replaced by $P_1 \oplus \cdots \oplus P_n$. Also, by replacing each member x of P_s, $s \in S$ by (x, s), if necessary, we can always assume that the posets are disjoint so that the ordinal sum can always be formed.

The cartesian product $\times_{s \in S} P_s$, of a family $(P_s)_{s \in S}$ of posets, becomes a poset if for $f, g \in \times_{s \in S} P_s$ we define $f \leq g$ when $f(s) \leq g(s)$ for all $s \in S$. This poset is called the *direct product* of $(P_s)_{s \in S}$ and each P_s is a *direct factor*. As usual, if $P_s = P$ for all $s \in S$, we write $\times_{s \in S} P_s = P^S$. Note that for $S = \varnothing$, the direct product reduces to a one element poset.

Let P_1 and P_2 be posets. A function $f: P_1 \to P_2$ is said to *preserve order* if $x \leq y$ implies $f(x) \leq f(y)$. A function $g: P_1 \to P_2$ is an *order embedding* provided

$x \leq y \Leftrightarrow g(x) \leq g(y)$. Note that an embedding is one-one. An embedding g is an *order isomorphism* if g is onto. (If there is no danger of confusion we will use the term isomorphism.) Since the definition of isomorphism implies that g is one-one, an isomorphism is a one-one correspondence for which both g and g^{-1} are order preserving. Although every isomorphism is order preserving, one-one and onto, the converse is false (consider a one-one function from a totally unordered poset onto a chain). Clearly, if $g: P_1 \to P_2$ is an isomorphism, then so is $g^{-1}: P_2 \to P_1$. In this case we say that P_1 and P_2 are *isomorphic* and write $P_1 \cong P_2$ (or equivalently $P_2 \cong P_1$). Thus, for a set S, there is an isomorphism between the power set of S and $\mathbf{2}^S$ which is defined by $T \mapsto f_T$ where

$$f_T(s) = \begin{cases} \mathbf{1} \text{ if } s \in T \\ \mathbf{0} \text{ if } s \notin T. \end{cases}$$

An *anti-isomorphism* $f: P_1 \to P_2$ is an onto function such that $x \leq y \Leftrightarrow f(y) \leq f(x)$. Note that the concept of order isomorphism coincides with the concept of isomorphism in the category of posets and order preserving maps (cf. Sec. I.14).

Example 2. Let $f: X \to Y$, $X \neq \varnothing$ be a function. Then f induces an order preserving map f_1 from the power set of X to the power set of Y defined by $f_1(A) = f_1[A]$, and an order preserving map f_2 from the power set of Y to the power set of X defined by $f_2(B) = f^{-1}[B]$. Show that if f_1 is an isomorphism then f is one-one and onto.

It is often necessary to consider relations which are only reflexive and transitive. Such a relation, $\prec$, on a non-empty set S is called a *quasi-order* and defines, in a natural way, a poset $S/\equiv$ as follows: Define $a \equiv b$, when $a \prec b$ and $b \prec a$. Then $\equiv$ is an equivalence relation, and a well-defined partial ordering is provided for the set $\{[s] : s \in S\}$ of equivalence classes by defining $[s] \leq [t] \Leftrightarrow s \prec t$. Then the natural map $\nu: S \to S/\equiv$ satisfies $x \prec y \Leftrightarrow \nu(x) \leq \nu(y)$.

Exercise 3. Let S be a non-empty set with a quasi-order $\prec$. Let $\nu: S \to S/\equiv$ be the natural map. Suppose P is a poset and $h: S \to P$ is an "order" preserving map (i.e. $x \prec y \Rightarrow h(x) \leq h(y)$). Then there exists a uniquely determined order preserving map $g: S/\equiv \to P$ such that $g \circ \nu = h$.

Exercise 4. Verify that ordinal sums and direct products of posets are, in fact, posets as stated in this section.

Exercise 5. Verify the above statements concerning quasi-order.

Exercise 6. Show that the divisibility relation $|$ on $Z \sim \{0\}$ is a quasi-order and that $(Z \sim \{0\}, |)/\equiv \cong (N, |)$ where $a \equiv b$ means: $a|b$ and $b|a$.

It is often possible to describe posets graphically by the following device. Represent the members of the poset by small circles, $\circ$, in such a way that if $a < b$, then the circle representing a is lower in the diagram than that representing b. Now connect these circles with straight lines when b covers a.

This procedure can always be carried out when the poset is finite, and even in the infinite case the structure of the poset can sometimes be indicated. Below are diagrams of posets which will be of subsequent interest. Figure 6 requires some explanation. Let U_2 be the two element totally unordered poset and Z^- the negative integers with the usual order. Then Figure 6 represents $U_2 \oplus Z^-$.

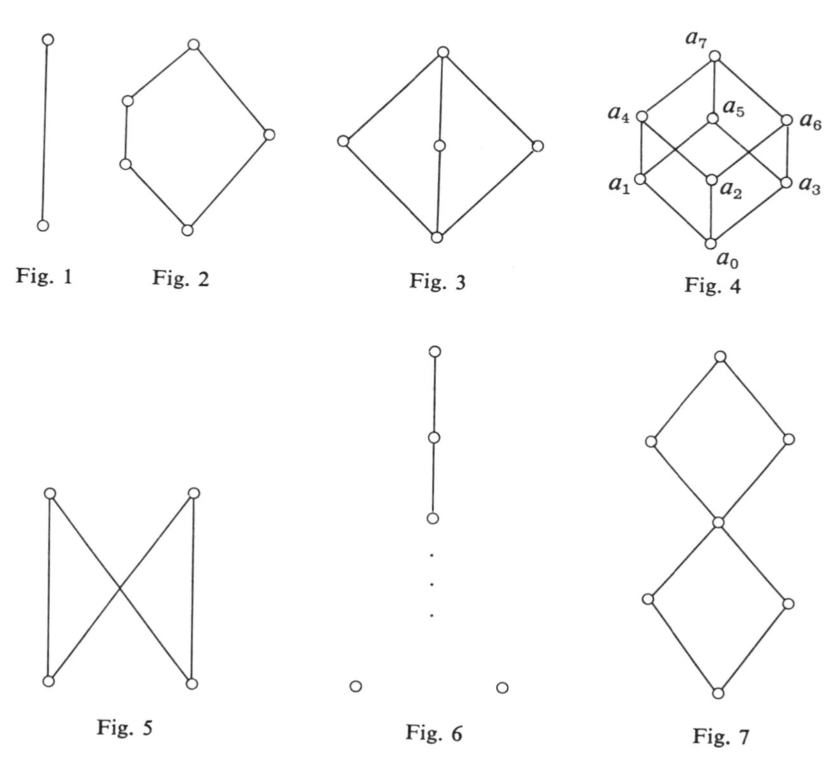

Fig. 1 Fig. 2 Fig. 3 Fig. 4

Fig. 5 Fig. 6 Fig. 7

Let P be a poset. In Section I.2 we defined the terms maximal element, minimal element, 0, 1, upper bound and lower bound. In addition, two elements x, y of P are called *disjoint* if P has a 0 and $z \leq x, z \leq y$ imply $z = 0$. A *node* in P is an element which is comparable with each element of P. Figure 7 has three nodes.

Now suppose $\varnothing \neq Q \subseteq P$; $a, b \in Q$ and $a \leq b$. The *interval* $[a, b]_Q$ is defined to be $\{x \in Q : a \leq x \leq b\}$. Similar definitions are made for $[a, b)_Q$, $(a, b]_Q$, and $(a, b)_Q$. Define $(a]_Q = \{x \in Q : x \leq a\}$ and $[a)_Q = \{x \in Q : x \geq a\}$. In case $Q = P$, the subscripts are generally omitted. In Figure 4, $[a_0, a_5] = \{a_0, a_1, a_3, a_5\} = (a_5]$. Finally, a subset S of P is *convex* if $a, b \in S$ implies $[a, b] \subseteq S$.

Exercise 7. Construct the diagram for the poset $(\{1, 2, \cdots, 10\}, |)$ and locate its maximal elements.

Exercise 8. Prove that the poset depicted in Figure 4 is isomorphic with the power set of a set of three elements.

One of the basic goals of lattice theory is to represent the elements of a lattice as sets in some natural way. The following elementary result is the first theorem of this type.

Theorem 9. (Set representation of posets.) *Every poset is isomorphic with a set of subsets of some set.*

Proof. For a poset P, $\{(x] : x \in P\}$ is a set of subsets of P and since $x \leq y \Leftrightarrow (x] \subseteq (y]$, the isomorphism $x \mapsto (x]$ yields the result.

2. Definition of a lattice

Suppose S is a subset of a poset P. As previously defined, an element u_o of P is an *upper bound* for S if $s \leq u_o$ for each $s \in S$. A *least upper bound* for S is an upper bound u_o with the property that $u_o \leq u$ for all upper bounds u of S. It is evident that there is at most one least upper bound for S and, when it exists, it will be denoted by $\sum^P S$ or simply $\sum S$. If $S = (x_t)_{t \in T}$ then $\sum_{t \in T} x_t$ is a convenient notation. Note that P has a 0 if and only if $\sum \varnothing$ exists and in this case $0 = \sum \varnothing$. The terms *lower bound* and *greatest lower bound* are defined by replacing $\leq$ with $\geq$ in the above definition. The greatest lower bound of S is denoted by $\prod^P S$ or simply $\prod S$ and if $S = (x_t)_{t \in T}$ by $\prod_{t \in T} x_t$. P has a 1 if and only if $\prod \varnothing$ exists and in this case $1 = \prod \varnothing$.

In a poset P, if $S = \{x_1, \cdots, x_n\}$ then we write $\sum_{i=1}^n x_i$ or $x_1 + \cdots + x_n$ for $\sum S$ and $\prod_{i=1}^n x_i$ or $x_1 \cdots x_n$ for $\prod S$. In some cases it will be desirable to write $x_1 \cdot x_2$ instead of $x_1 x_2$. Figures 5, 6 in Section 1 are examples of posets in which there are elements x, y which have upper bounds but no least upper bounds.

Definition 1. A *lattice* is a poset L in which $x + y$ and xy exist for any $x, y \in L$.

The Examples 2, 3, 4, and 5 in Section I.2 are lattices. In Examples I.2.2 and I.2.4, $m + n = \max\{m, n\}$, $m \cdot n = \min\{m, n\}$; in Example I.2.3, $m + n = \text{l.c.m.}\ \{m, n\}$ and $m \cdot n = \text{g.c.d.}\ \{m, n\}$. For elements S and T of the power set of a set X we have $S + T = S \cup T$ and $S \cdot T = S \cap T$. This last example has an important generalization.

Definition 2. A *ring of sets* is a non-empty set R of subsets of a set X such that if $A, B \in R$ then $A \cup B$ and $A \cap B \in R$. Clearly, a ring of sets R is a lattice and for elements $A, B \in R$, $A + B = A \cup B$ and $A \cdot B = A \cap B$.

Example 3. Let L be the convex subsets of the plane. For $A, B \in L$, $A \cdot B = A \cap B$ and $A + B$ is the convex hull of $A \cup B$; that is, the intersection of two convex sets is convex and is therefore the largest convex set contained in both, but the convex hull of the union of two convex sets is the smallest convex set containing both of these sets.

Example 4. Let L be the poset of all subgroups of a group. For $G_1, G_2 \in L$, $G_1 \cdot G_2 = G_1 \cap G_2$ and $G_1 + G_2 = [G_1 \cup G_2]$. In the context of universal algebras, this example can be generalized to the poset of subalgebras of an algebra.

Further examples are given by the open (closed) subsets of a topological space, the ideals of a ring, the normal subgroups of a group, and many others.

Here are some of the elementary properties of a lattice: For elements $x, y, z, x_1, \cdots, x_n, y_1, \cdots, y_n, n \geq 1$ in a lattice L,

(i) $x + y = y \Leftrightarrow x \leq y \Leftrightarrow xy = x$.

(ii) $x(x + y) = x$.

(iii) If $x_i \leq y_i$ for $1 \leq i \leq n$, then $\sum_{i=1}^n x_i \leq \sum_{i=1}^n y_i, \prod_{i=1}^n x_i \leq \prod_{i=1}^n y_i$.

To prove (i), first suppose $x + y = y$. The definition of $x + y$ states, in part, that $x \leq x + y$, so $x \leq x + y = y$. Next suppose $x \leq y$. Then since $x \leq x$, x is a lower bound for x and y. If z were a lower bound for x and y then in particular, $z \leq x$ so $xy = x$. In order to prove $xy = x \Rightarrow x + y = y$, observe that the hypothesis states, in part, that $x \leq y$. But $y \leq y$ so y is an upper bound for x and y. Since any upper bound z for x and y satisfies $y \leq z$ we have $x + y = y$. To prove (ii) we show that x is the greatest lower bound for x and $x + y$. Now $x \leq x$ and $x \leq x + y$ so x is certainly a lower bound. But if z is also a lower bound for x and $x + y$ then in particular $z \leq x$ so $x = x(x + y)$. Finally, to prove (iii) we have $x_i \leq y_i$ for each i, implies $x_i \leq y_i \leq \sum_{j=1}^n y_j$ so $\sum_{j=1}^n y_j$ is an upper bound of $\{x_1, \cdots, x_n\}$. Hence $\sum_{i=1}^n x_i \leq \sum_{j=1}^n y_j$. The second part of (iii) is left to the reader.

Exercise 5. Prove that the following properties hold for elements x, y, z, x_i, y_i in a lattice L.

(i) $x + (y + z) = (x + y) + z$.

(ii) $x + xy = x$.

(iii) If L has a 0 then x, y are disjoint $\Leftrightarrow xy = 0$.

(iv) $\sum_{i=1}^n x_i \leq \prod_{j=1}^n y_j \Leftrightarrow x_i \leq y_j$ for all i, j.

Exercise 6. Suppose $S_1, \cdots, S_n$ are finite non-empty subsets of a lattice L. Prove: $\sum_{i=1}^n (\sum S_i) = \sum(\bigcup_{i=1}^n S_i) = \sum\{u_1 + \cdots + u_n : (u_1, \cdots, u_n) \in \times_{i=1}^n S_i\}$ and similarly for products.

Exercise 7. Let L be a lattice and $T \subseteq L$. Show that if S is the set of maximal elements of T, then $\sum S = \sum T$.

3. Lattices as algebras and categories

So far, we have considered lattices as posets with certain properties. We now wish to view them as algebras of an equational class.

Theorem 1. *For any lattice* $(L, \leq)$, $(x, y) \mapsto x + y$ *and* $(x, y) \mapsto xy$ *are binary operations which define an algebra* $(L, (+, \cdot))$ *that satisfies the identities*

(i) $\mathbf{(x + y) + z = x + (y + z)}$ (i′) $\mathbf{(xy)z = x(yz)}$

(ii) $\mathbf{x + y = y + x}$ (ii′) $\mathbf{xy = yx}$

(iii) $\mathbf{x + x = x}$ (iii′) $\mathbf{xx = x}$

(iv) $\mathbf{x + xy = x}$ (iv′) $\mathbf{x(x + y) = x}$.

Moreover, if $(A, (+, \cdot))$ *is an algebra with two binary operations satsifying* (i)–(iv) *and* (i′)–(iv′), *then*

(v) $u + v = v \Leftrightarrow uv = u$.

The relation $\leq$ *on* A, *defined by* $u \leq v \Leftrightarrow uv = u$, *makes* $(A, \leq)$ *into a lattice in which the least upper bound of* u, v *in* A *is* $u + v$ *and the greatest lower bound of* u, v *in* A *is* uv.

Proof. (i)–(iv) and (i′)–(iv′) are all elementary consequences of the definitions of least upper bound and greatest lower bound. To prove (v) suppose $u + v = v$. Then $uv = u(u + v) = u$ by (iv′). Similarly $uv = u$ implies $u + v = v$. The relation $\leq$ is a partial ordering, for (iii′) implies it is reflexive, (i′) that it is transitive, and (ii′) yields the antisymmetric law. By (iv′) $u, v \leq u + v$; if w is an upper bound for u, v then by (v) $u + w = w$ and $v + w = w$, so $(u + v) + w = (u + w) + (v + w) = w + w = w$ which means $u + v \leq w$ and $u + v$ is the least upper bound for u, v. The proof that uv is the greatest lower bound for u, v is analogous.

Consider the class of all algebras with two binary operations $+, \cdot$ which satisfy (i)–(iv) and (i′)–(iv′). We will denote this equational class by **L** and its associated equational category therefore by $\mathscr{L}$ (cf. Section I.19). Theorem 1 shows that every lattice determines a unique algebra in **L** and each algebra in **L** is determined by a unique lattice in this way, up to isomorphisms. The members of **L** will therefore also be referred to as lattices. Thus, the definitions of **L**-homomorphisms, **L**-subalgebras, etc. apply to lattices. For example, if $L, L_1 \in \text{Ob}\,\mathscr{L} = \mathbf{L}$ then a function $f: L \to L_1$ is an **L**-homomorphism (or equivalently, $f \in [L, L_1]_{\mathscr{L}}$) if and only if $f(x + y) = f(x) + f(y)$ and $f(xy) = f(x)f(y)$. The term *sublattice* is used in this book only to mean an **L**-subalgebra; so S is a sublattice of L if and only if $S \neq \varnothing$ and $x, y \in S$ implies xy and $x + y \in S$.

Since **L** is equational, a number of important results—such as the existence of free lattices—follow from the general theory of universal algebras.

The reader should also note the validity, in **L**, of the generalized associative and commutative laws.

Let $L, L_1 \in \mathbf{L}$ and $f: L \to L_1$ be a function. Then f is an **L**-isomorphism if and only if it is an order isomorphism (caution: see Exercise 3). Indeed, if f is an **L**-isomorphism then $x \leq y \Rightarrow x = xy \Rightarrow f(x) = f(xy) = f(x)f(y) \Rightarrow f(x) \leq f(y)$ and $f(x) \leq f(y) \Rightarrow f(x) = f(x)f(y) = f(xy) \Rightarrow x = xy \Rightarrow x \leq y$. Conversely, $x, y \leq x + y \Rightarrow f(x), f(y) \leq f(x + y)$ and if $f(x) \leq f(u)$ and $f(y) \leq f(u)$

then $x \leq u$ and $y \leq u$; hence $x + y \leq u$ so $f(x + y) \leq f(u)$ and thus $f(x + y) = f(x) + f(y)$. Similarly for products. Thus, in **L**, **L**-isomorphisms and order isomorphisms are simply referred to as isomorphisms.

For a lattice $(L, (\leq)$ consider the relation defined on L by $u \leq' v \Leftrightarrow v \leq u$. It is evident that $\leq'$ partially orders L and in fact $(L, \leq')$ is a lattice in which $u +' v = uv$ and $u \cdot' v = u + v$. This shows that when $(L, (+, \cdot))$ is an algebra in **L**, then so is $(L, (+', \cdot')) = (L, (\cdot, +))$. Observe that this is also evident from the fact that the defining identities (i)–(iv) and (i′)–(iv′) are merely interchanged if the symbols $+$ and $\cdot$ are interchanged. We now apply the weak duality principle with respect to interchanging $+$ and $\cdot$. From now on for $L = (L, (+, \cdot)) \in \mathbf{L}$ we will denote $(L, (\cdot, +))$ by $\breve{L}$. If π is the permutation of operations $+, \cdot$ then for a logical formula P, $\breve{P}$ stands for $D_\pi(P)$. Thus *any lattice theoretic proposition P that is true for all lattices implies the validity of $\breve{P}$ for all lattices.* We call $\breve{P}$ the *dual* of P and $\breve{L}$ the *dual* of L.

Example 2. Consider the following statements about a lattice L:

$P_1 : x(y + z) = xy + xz$ for all $x, y, z \in L$.

$P_2 :$ If $uw = vw$ and $u + w = v + w$, where $u, v, w \in L$ then $u = v$.

Now $P_1 \Rightarrow P_2$ is true in any lattice L, for if P_1 holds as well as the hypothesis of P_2, then $u \leq u(v + w) = uv + uw = uv + vw \leq v$ and similarly $v \leq u$ so $u = v$.

Now the weak duality principle states that $\overbrace{P_1 \Rightarrow P_2}$ is true. But $\overbrace{P_1 \Rightarrow P_2}$ is the same as $\breve{P_1} \Rightarrow \breve{P_2}$ and clearly $\breve{P_2} = P_2$. We conclude that in any lattice, if $x + yz = (x + y)(x + z)$ for all $x, y, z \in L$ then P_2 holds.

The strong duality principle is not applicable to L since $(L, (+, \cdot))$ is not necessarily isomorphic to $(L, (\cdot, +))$. See Figure 1.

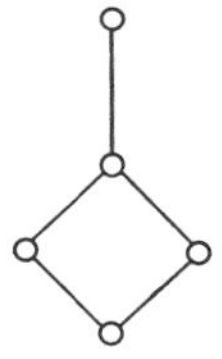

Fig. 1

The notion of weak duality for lattices appears in E. Schröder [1] but that of weak and strong duality in universal algebras is clearly stated in its present form in J. C. C. McKinsey and A. Tarski [2, page 59].

Exercise 3. Show that if L and L_1 are lattices and $f: L \to L_1$ is an order embedding then $f[L]$ need *not* be a sublattice of L_1.

4. Complete lattices and closure operators

A poset P is *complete* if $\sum S$ and $\prod S$ exist for every subset $S \subseteq P$. Actually, for a poset P to be complete it is sufficient that $\prod S$ exists for all $S \subseteq P$. Indeed, if

$S \subseteq P$, then $\sum S = \prod T$, where T is the set of all upper bounds of S. Note that if $S = \varnothing$, then $\sum S = 0$ and $\prod S = 1$. This observation is very useful, for it often occurs that in dealing with a set X of sets, it is given that $\bigcap S \in X$ whenever $S \subseteq X$. We can then conclude that X is a complete poset with $\prod S = \bigcap S$. Note that $\sum S$, although it exists, may not equal $\bigcup S$. Since a complete poset is obviously a lattice, the term *complete lattice* is also used.

The weaker notion of *conditional completeness* refers to a poset in which $\sum S$ exists if $S \neq \varnothing$ and S has an upper bound, and dually.

Example 1. The set of congruence relations on an algebra A is a complete lattice. Indeed, it is easy to see that the intersection of a set of congruence relations on A is a congruence relation (here $\bigcap \varnothing = A \times A$). So $\prod S = \bigcap S$ for any $S \subseteq \mathfrak{C}(A)$. (See Exercise 4.)

Example 2. The regular open sets of a topological space form a complete lattice. Here $\prod^L S = \text{Int}\,(\bigcap S)$ and $\sum^L S = \text{Int}\,(\text{Cl}\,(\bigcup S))$.

Exercise 3. Let L be all finite subsets of an infinite set X together with X itself. Show that L is a complete lattice. Describe explicitly $\sum S$ for $S \subseteq L$.

Exercise 4. Show that in Example 1, $\sum S$ can be explicitly defined by $(x, y) \in \sum S \Leftrightarrow$ there exists a sequence $x = a_1, \cdots, a_n = y$ such that for $1 \leq i \leq n - 1$, $(a_i, a_{i+1}) \in \theta_i$ for some $\theta_i \in S$.

Exercise 5. (A. Tarski [5]) Every order preserving map of a complete lattice into itself has a fixed point. (Hint: For such a map f consider the set $\{x : x \leq f(x)\}$.)

A. Tarski's paper [5] has many interesting corollaries (e.g. Exercise 6, below). A. Davis [1] has proved the converse of Exercise 5; namely, if every order preserving function on a lattice L has a fixed point, then L is complete. See also E. S. Wolk [1] for a proof that avoids the use of ordinals. R. Davies, A. Hayes and G. Rousseau [1] and S. Abian and A. Brown [1] have also obtained results related to fixed point theorems for posets.

Exercise 6. Use Exercise 5 to prove the Schröder-Bernstein Theorem: If X and Y are sets and $f\colon X \to Y$, $g\colon Y \to X$ are one-one functions, then there is a one-one correspondence between X and Y. (Hint: Define a function k from the power set of X to the power set of Y by $k(S) = X \sim g[Y \sim f[S]]$ and apply Exercise 5.)

Exercise 7. If L is a lattice and every directed subset S of L (i.e. $x, y \in S \Rightarrow$ there exists $z \in S$ such that $x \leq z, y \leq z$) has a least upper bound, then L is complete.

Exercise 8. Let P be a poset, $\varnothing \neq S \subseteq Q \subseteq P$. If $\sum^P S$ and $\sum^Q S$ exist, then $\sum^P S \leq \sum^Q S$ and dually, if $\prod^P S$ and $\prod^Q S$ exist, then $\prod^Q S \leq \prod^P S$.

Exercise 9. If P is a poset, $\varnothing \neq S \subseteq Q \subseteq P$ and $\sum^P S$ exists and is in Q, then $\sum^Q S$ exists and $\sum^P S = \sum^Q S$.

In this book it will be convenient to make use of the notion of a closure operator.

Definition 10. Let L be a lattice. A unary operation ^{c} on L is a *closure operator* if ^{c} satisfies the following conditions:

(i) $x \leq y$ implies $x^c \leq y^c$ for $x, y \in L$.

(ii) $x \leq x^c$ for $x \in L$.

(iii) $x^c = x^{cc}$ for $x \in L$.

An element $x \in L$ is *closed* provided $x^c = x$. The operator ^{c} is *additive* if

(iv) $(x + y)^c = x^c + y^c$ for $x, y \in L$.

If ^{c} is a closure operator on a lattice L then for $S \subseteq L$, we denote the set $\{x^c : x \in S\}$ by S^c. It follows that L^c is the set of all closed elements. In fact

(1) If $x, y \in L^c$ then $xy = (xy)^c$.

Indeed $(xy)^c \leq x^c = x$ by (i) and similarly $(xy)^c \leq y$ so $(xy)^c \leq xy \leq (xy)^c$ by (ii).

It follows from (1) that if ^{c} is an additive closure operator then L^c is a sublattice of L.

The following theorem characterizes the closure operators on a lattice L in terms of certain subsets of L.

Theorem 11. *Let L be a lattice and let A be a non-empty subset of L. A is the set of closed elements of a (uniquely determined) closure operator ^{c} on L if and only if:*

(i) $a, b \in A \Rightarrow ab \in A$.

(ii) *For each $a \in L$, the set $[a) \cap A$ has a smallest element.*

If A is a non-empty subset of L, satisfying (i) *and* (ii) *then the corresponding closure operator ^{c} on L is defined by letting a^c be the smallest element of $[a)_L \cap A$ for $a \in L$. Moreover, if A is a sublattice of L satisfying* (ii) *then ^{c} is an additive closure operator.*

Proof. First suppose ^{c} is a closure operator on L. (i) follows from (1). For (ii): if $a \in L$, then $a^c \geq a$ and if $b \in L^c$ and $b \geq a$ then $b = b^c \geq a^c$ so a^c is the smallest element of $[a)_L \cap L^c$. Next, suppose A satisfies (i) and (ii). Define $^c : L \to L$ by defining a^c to be the smallest element of $[a) \cap A$ for $a \in L$. For $a, b \in L$, $a \leq b$ implies $a^c \leq b^c$ and for $a \in L$, $a \leq a^c$. Also, since $a^c \in A$, we have $a^{cc} \leq a^c$ so $a^{cc} = a^c$. That $A = L^c$ is also immediate.

Finally, if A is a sublattice, satisfying (ii) then for $a, b \in L$, $a^c + b^c \in A$. But if $d \in A$, $d \geq a$ and $d \geq b$ then $d \geq a^c$ and $d \geq b^c$ so $d \geq a^c + b^c$. Hence $(a + b)^c = a^c + b^c$.

We will encounter situations where ^{c} is a closure operator on a complete lattice L. In this case L^c is also complete. We have:

Theorem 12. *If L is a complete lattice and $S \subseteq L^c$ then*

(i) $\prod^L S = (\prod^L S)^c = \prod^{L^c} S$.

(ii) $\sum^{L^c} S = (\sum^L S)^c$.

Also, if $S \subseteq L$ then

(iii) $(\sum^L S)^c = (\sum^L S^c)^c$.

Proof. (i) The verification that $\prod^L S = (\prod^L S)^c$ is essentially the same as the proof of (1). But this also shows $\prod^L S \in L^c$ so $\prod^{L^c} S = \prod^L S$. (ii) For each $s \in S$, $s \leq \sum^L S \leq (\sum^L S)^c$. Now suppose $u \in L^c$ is an upper bound for S in L^c. Then for each $s \in S$, $s \leq u$ so $\sum^L S \leq u$ thus $(\sum^L S)^c \leq u^c = u$. (iii) Since $s \leq s^c$ for $s \in S$ we have $\sum^L S \leq \sum^L S^c$ so $(\sum^L S)^c \leq (\sum^L S^c)^c$. Next for $s \in S$, $s \leq \sum^L S$ implies $s^c \leq (\sum^L S)^c$ so $\sum^L S^c \leq (\sum^L S)^c$. Thus $(\sum^L S^c)^c \leq (\sum^L S)^{cc} = (\sum^L S)^c$.

5. The distributive law

In a lattice L the following two statements are equivalent:

(1) $x(y + z) = xy + xz$ for all x, y, z in L.

(2) $u + vw = (u + v)(u + w)$ for all u, v, w in L.

By the weak duality principle, we need only prove (1) $\Rightarrow$ (2). Thus, $(u + v)(u + w) = (u + v)u + (u + v)w = u + (u + v)w = u + uw + vw = u + vw$.

Definition 1. A lattice L is *distributive* if it satisfies one (and hence both) of (1) and (2).

Remark. Clearly, L is distributive if $x(y + z) \leq xy + xz$ for all x, y, z in L or (equivalently) $(u + v)(u + w) \leq u + vw$ for all u, v, w in L.

Example 2. Every chain is a distributive lattice. To verify (1) we need only consider the three cases $x \leq y \leq z$, $y \leq x \leq z$ and $y \leq z \leq x$. For example in the second case, $x(y + z) = xz = x = y + x = xy + xz$.

Example 3. Any ring of sets is a distributive lattice, since for sets A, B, C: $A \cap (B \cup C) = (A \cap B) \cup (A \cap C)$.

Exercise 4. Prove that $(N, |)$ is a distributive lattice (cf. Example I.2.3).

The following formulas are valid when S is a finite non-empty subset of a distributive lattice. The proof is by induction on $|S|$.

$a \sum S = \sum \{as : s \in S\}$

$a + \prod S = \prod \{a + s : s \in S\}$.

By adding the identity:

$\mathbf{x(y + z) = xy + xz}$

to the identities (i)–(iv) and (i′)–(iv′) of Theorem 3.1, we see that distributive lattices form an equational class **D** with associated equational category $\mathscr{D}$. $\mathscr{D}$ is a full subcategory of $\mathscr{L}$.

Another closely related equational class is $\mathbf{D_{01}}$: the distributive lattices with 0, 1. More precisely, $\mathbf{D_{01}}$ consists of algebras $(L, (+, \cdot, 0, 1))$ of similarity type (2, 2, 0, 0) that satisfy a set of identities which define distributive lattices, together with the identities

$$\mathbf{x} + \mathbf{0} = \mathbf{x}, \qquad \mathbf{x} \cdot \mathbf{1} = \mathbf{x}.$$

In view of our discussion in Section I.7 (pages 10 and 11), we can consider $\mathbf{D_{01}}$ as a subclass of **D**. Furthermore, the functor defined in Section I.19 (page 31) from $\mathscr{D}_{01}$ (the equational category associated with $\mathbf{D_{01}}$) to $\mathscr{D}$ is an embedding which is one-one on the objects as can be easily seen. Therefore, in accordance with the convention adopted in Section I.19 (page 70), we will often consider $\mathscr{D}_{01}$ as a subcategory of $\mathscr{D}$. However, $\mathscr{D}_{01}$ is not a full subcategory of $\mathscr{D}$. Also note that $\mathbf{D_{01}}$-isomorphisms are the same as order isomorphisms.

Exercise 5. Show that if $L \in \mathbf{D_{01}}$ and L_1 is a sublattice of L, then

$$[L_1]_{\mathbf{D01}} \cong \begin{cases} L_1 \text{ if } 0_L, 1_L \in L_1 \\ \mathbf{1} \oplus L_1 \text{ if } 0_L \notin L_1,\ 1_L \in \mathrm{L}_1 \\ L_1 \oplus \mathbf{1} \text{ if } 0_L \in L_1,\ 1_L \notin L_1 \\ \mathbf{1} \oplus L_1 \oplus \mathbf{1} \text{ if } 0_L \notin L_1,\ 1_L \notin L_1. \end{cases}$$

(Note that by the definition of a sublattice, L_1 is merely closed under + and $\cdot$.)

We now present a series of unrelated results which will be useful in the sequel.

Lemma 6. *In any distributive lattice L each interval $(a]$, $a \in L$, is a sublattice of L and the mapping $x \mapsto xa$ is a* **D***-homomorphism of L onto $(a]$.*

Proof. Immediate, from the formulas: $(x + y)a = xa + ya$ and $(xy)a = (xa)(ya)$.

In more complicated situations, where it is to be shown that a function $f: L \to L'$ is a **D**-homomorphism, it is usually more convenient to prove:

(1) f preserves order.

(2) $f(x + y) \leq f(x) + f(y)$.

(3) $f(x)f(y) \leq f(xy)$.

These conditions are, of course, sufficient for f to be a **D**-homomorphism. Note that the notions of **D**-isomorphism and order isomorphism coincide.

We also introduce the following notions. Let $L \in \mathrm{Ob}\,\mathscr{D}$; define L_{01} to be the lattice $\mathbf{1} \oplus L \oplus \mathbf{1}$. Also, for $f \in [L, L']_{\mathscr{D}}$ define $f_{01} \in [L_{01}, L'_{01}]_{\mathscr{D}01}$ by $f_{01}|L = f$ and $f(0_{L_{01}}) = 0_{L'_{01}}$ and $f(1_{L_{01}}) = 1_{L'_{01}}$.

Theorem 7. *$\mathscr{D}_{01}$ is a reflective subcategory of $\mathscr{D}$. The reflector $\mathscr{U}:\mathscr{D}\to\mathscr{D}_{01}$ preserves monomorphisms, onto morphisms, and reflects monomorphisms, onto morphisms and epimorphisms.*

Proof. For each $L\in \mathrm{Ob}\,\mathscr{D}$, let $\mathscr{U}(L) = L_{01}$ and let $\Phi_{\mathscr{U}}(L) = 1_{L,\mathscr{U}(L)}$. Now if $L'\in \mathrm{Ob}\,\mathscr{D}_{01}$ and $f\in[L,L']_{\mathscr{D}}$, define $f':\mathscr{U}(L)\to L'$ by $f'|L = f$ and $f'(0)=0$, $f'(1)=1$. Then $f'\in[\mathscr{U}(L),L']_{\mathscr{D}_{01}}$ and $f'\circ\Phi_{\mathscr{U}}(L) = $ f. For uniqueness, note that f' is completely determined on L by $f'\circ\Phi_{\mathscr{U}}(L) = f$. Since it must also preserve 0, 1, it is unique. It follows from Theorem I.18.2 that $\mathscr{U}$ can be extended to a reflector. Suppose $f\in[L,L']_{\mathscr{D}}$ is a monomorphism. Since **D** is equational, f is one-one. Now suppose $u, v\in\mathscr{U}(L)$ and $\mathscr{U}(f)(u) = \mathscr{U}(f)(v)$.

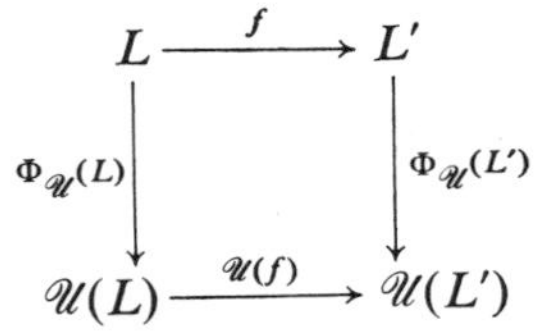

$$\begin{array}{ccc} L & \xrightarrow{f} & L' \\ \Big\downarrow{\scriptstyle \Phi_{\mathscr{U}}(L)} & & \Big\downarrow{\scriptstyle \Phi_{\mathscr{U}}(L')} \\ \mathscr{U}(L) & \xrightarrow{\mathscr{U}(f)} & \mathscr{U}(L') \end{array}$$

If $u\in L$, then $\mathscr{U}(f)(v) = \mathscr{U}(f)(u) = \mathscr{U}(f)(\Phi_{\mathscr{U}}(L)(u)) = \Phi_{\mathscr{U}}(L')(f(u)) = f(u)\in L'$. Thus $v\neq 0, 1$ and so $v\in L$. Hence, $f(v) = \Phi_{\mathscr{U}}(L')(f(v)) = \mathscr{U}(f)(\Phi_{\mathscr{U}}(L)(v))$ $= \mathscr{U}(f)(v) = f(u)$ and hence $u = v$. If $u = 0$, then $\mathscr{U}(f)(v) = \mathscr{U}(f)(u) = \mathscr{U}(f)(0)$ $= 0$ so $v = 0$. Similarly, $v = 1$ if $u = 1$, so $u = v$. It is obvious that if f is onto, then $\mathscr{U}(f)$ is onto and that f reflects monomorphisms and onto morphisms. Finally, suppose $f\in[L,L']_{\mathscr{D}}$ and that $\mathscr{U}(f)$ is epic. Now let $g\circ f = h\circ f$ for $g, h\in[L',L'']_{\mathscr{D}}$. Then $\mathscr{U}(g)\circ\mathscr{U}(f) = \mathscr{U}(h)\circ\mathscr{U}(f)$ so $\mathscr{U}(g) = \mathscr{U}(h)$. But $g = \mathscr{U}(g)|L' = \mathscr{U}(h)|L' = h$.

Lemma 8. *If $(S_i)_{i=1,\cdots,n}$, $n\geq 1$, is a family of non-empty subsets of a distributive lattice then*

$$\prod S_1 + \cdots + \prod S_n = \prod\left\{\sum_{i=1}^{n} f(i) : f\in \mathop{\times}_{i=1}^{n} S_i\right\}$$

Proof. The formula is trivial for $n = 1$. For $n = 2$, we have by distributivity: $\prod S_1 + \prod S_2 = \prod\{(\prod S_1) + y : y\in S_2\} = \prod\{\prod\{x+y : x\in S_1\} : y\in S_2\}$ $= \prod\{x+y : x\in S_1, y\in S_2\} = \prod\{f(1)+f(2) : f\in S_1\times S_2\}$. Now assume the validity of the formula for $1\leq k< n$ and suppose $n>2$. Set $T_1 = S_1, T_2 = S_2,\cdots,$ $T_{n-2} = S_{n-2}$ and for each $x\in S_n$ let $T^x_{n-1} = \{u+x : u\in S_{n-1}\}$. Then by induction,

$$\begin{aligned}\prod\left\{\sum_{i=1}^{n} f(i) : f\in \mathop{\times}_{i=1}^{n} S_i\right\} &= \prod_{x\in S_n}(\prod\{\sum g(i) : g\in T_1\times\cdots\times T_{n-2}\times T^x_{n-1}\})\\ &= \prod_{x\in S_n}(\prod T_1 + \cdots + \prod T_{n-2} + \prod T^x_{n-1})\\ &= (\prod T_1 + \cdots + \prod T_{n-2}) + \prod_{x\in S_n}(\prod T^x_{n-1})\\ &= (\prod T_1 + \cdots + \prod T_{n-2}) + \prod_{x\in S_n}(\prod S_{n-1} + x)\\ &= \prod S_1 + \cdots + \prod S_{n-2} + \prod S_{n-1} + \prod S_n.\end{aligned}$$

Theorem 9. (G. Birkhoff [2]) *A lattice L is distributive if and only if neither of the following are sublattices of L.*

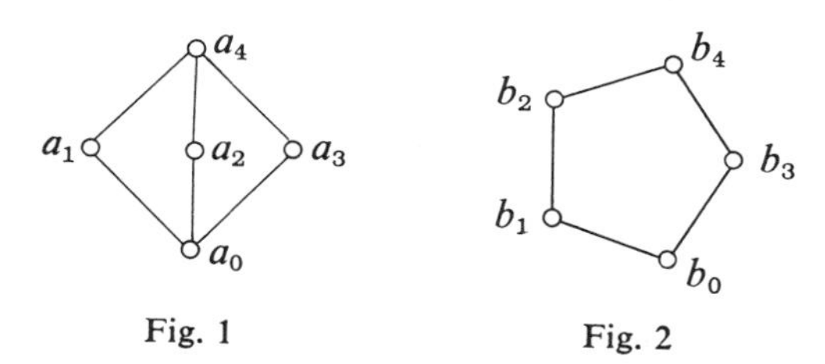

Fig. 1 **Fig. 2**

Proof. ($\Rightarrow$) In Figure 1, $a_1(a_2 + a_3) = a_1a_4 = a_1 \neq a_0 = a_1a_2 + a_1a_3$; and in Figure 2, $b_2(b_1 + b_3) = b_2b_4 = b_2 \neq b_1 = b_1 + b_0 = b_2b_1 + b_2b_3$.

($\Leftarrow$) Suppose first that there exists $\{a, b, c\} \subseteq L$ with the property that $a(b + c) \neq ab + ac$ and $a > c$. Set $b_0 = ab$, $b_1 = c + ab$, $b_2 = a(b + c)$, $b_3 = b$ and $b_4 = b + c$. Then b_0, b_1, b_2, b_3, b_4 are distinct and closed under $+$ and $\cdot$, so they form a sublattice of L which is isomorphic with Figure 2. Thus we may assume that

(4) For all x, y, z in L, if $x > z$ then $x(y + z) = xy + xz$.

But since L is not distributive, there exist p, q, r in L such that $p(q + r) \neq pq + pr$. Let $a_1 = qr + p(q + r)$, $a_2 = rp + q(r + p)$, $a_3 = pq + r(p + q)$, $a_0 = pq + qr + rp$ and $a_4 = (p + q)(q + r)(r + p)$. It is left to the reader to show, by using (4), that a_0, a_1, a_2, a_3, a_4 are distinct and therefore determine a sublattice of L, isomorphic with Figure 1.

Remark. From Theorem 9 and Example 3.2 we see that a lattice L is distributive $\Leftrightarrow uw = vw, u + w = v + w$ implies $u = v$.

Exercise 10. Verify that all of the elements a_i, b_i, defined in the proof of Theorem 9, are distinct.

Exercise 11. Suppose $(L_s)_{s\in S}$ is a family of distributive lattices. Prove that $\times_{s\in S} L_s$ and $\bigoplus_{s\in S} L_s$ are also distributive lattices.

Some group theoretic results can be obtained by considering the lattice of subgroups of a group. For example, O. Ore [1] has proved that the lattice of subgroups of a group G is distributive if and only if G is a generalized cyclic group (i.e. for any two elements $a, b \in G$ there exists $c \in G$ such that a and b are powers of c) cf. M. Suzuki [1].

6. Complements, Boolean algebras

Definition 1. Let L be a lattice with 0, 1. An element $x \in L$ is *complemented* if there exists $y \in L$ such that $xy = 0$ and $x + y = 1$. The set of complemented elements is denoted by $\mathscr{C}(L)$ and is called the *center* of L. L is *complemented* if $\mathscr{C}(L) = L$.

In a distributive lattice, complements are unique by Example 3.2. In fact, if z is a complement of x in L and $xu = 0$ then $u = u1 = u(x + z) = ux + uz = 0 + uz = uz$. So the complement of x, denoted by $\bar{x}$, is the greatest element u in L such that $ux = 0$. It was long conjectured that every uniquely complemented lattice was distributive. This was disproved by R. P. Dilworth [1].

Definition 2. A *Boolean algebra* is a complemented distributive lattice.

Example 3. The chain **2** is a Boolean algebra with $\bar{\mathbf{0}} = \mathbf{1}, \bar{\mathbf{1}} = \mathbf{0}$. **1** and **2** are the only chains which are Boolean algebras.

Example 4. The power set of a set X is a Boolean algebra with $\bar{S} = X \sim S$ for $S \subseteq X$.

Example 5. Consider the set S of formulas in the classical propositional calculus. Define an equivalence relation $\equiv$ on S by $x \equiv y$ when $x \to y$ and $y \to x$ are both provable. Then it is easy to see that $[x] \leq [y] \Leftrightarrow$ "$x \to y$ is provable" defines a partial ordering on $S/\equiv$ which is a Boolean algebra in which $[x] + [y] = [x \vee y], [x][y] = [x \wedge y], 1 = [x \to x]$ and $\bar{x} = [\neg x]$. (The reader interested in pursuing logic—developed using lattice theoretic techniques—should refer to H. Rasiowa and R. Sikorski [1]; also see J. B. Rosser [1] for a simplified lattice theoretic development of P. Cohen's independence proofs.

Example 6. The regular open sets $\mathfrak{R}$ of a topological space X is a Boolean algebra in which $S + T = \text{Int}(\text{Cl}(S \cup T)), S \cdot T = S \cap T$, and $\bar{S} = \widetilde{\text{Cl}(S)}$ for $S \in \mathfrak{R}$.

Example 7. Let X be a set. A subset $S \subseteq X$ is *cofinite* if $X \sim S$ is finite. The poset of all subsets of X which are either finite or cofinite form a Boolean algebra.

Because of the importance of the next example we state it in the form of a definition.

Definition 8. A *field of sets* is a ring of subsets F of a set X such that $X \sim A \in F$ for each $A \in F$.

Obviously, a field of sets is a Boolean algebra with $\bar{A} = X \sim A$.
Next, we derive some of the elementary properties of Boolean algebras.

Theorem 9. *In a Boolean algebra L,*

(i) $xy \leq u + v \Leftrightarrow x\bar{v} \leq \bar{y} + u$; *in particular* $xy = 0 \Leftrightarrow x \leq \bar{y}$.

(ii) $\overline{x + y} = \bar{x}\bar{y}, \overline{xy} = \bar{x} + \bar{y}$ (*de Morgan's laws*).

(iii) *Each interval* $[a, b]$ *is a Boolean algebra under the induced partial ordering of* L.

(iv) *If* $S \subseteq L$ *and* $\sum S$ *exists then for each* $x \in L$, $\sum\{xs : s \in S\}$ *exists and equals* $x \sum S$.

Proof. (i) If $xy \leq u + v$ then $x\bar{v} = x\bar{v}(y + \bar{y}) = x\bar{v}y + x\bar{v}\bar{y} \leq (u + v)\bar{v} + x\bar{v}\bar{y} \leq u + \bar{y}$; conversely, $xy = xy(v + \bar{v}) = xyv + xy\bar{v} \leq xyv + y(\bar{y} + u) \leq v + u$. (ii) Since $(x + y)\bar{x}\bar{y} = x\bar{x}\bar{y} + y\bar{x}\bar{y} = 0$ and $(x + y) + \bar{x}\bar{y} = x + (y + \bar{x})(y + \bar{y}) = x + y + \bar{x} = 1$, the uniqueness of complements implies $\overline{x + y} = \bar{x}\bar{y}$. Similarly $\bar{x} + \bar{y} = \overline{xy}$. (iii) Since $[a, b]$ is a sublattice of L, it is also distributive. Now, if $x \in [a, b]$ then $b\bar{x} + a \in [a, b]$ and $x(b\bar{x} + a) = xa = a$, $x + (b\bar{x} + a) = (x + b)(x + \bar{x}) + a = x + b + a = b$ so $b\bar{x} + a$ is the complement of x in the lattice $[a, b]$. (iv) Clearly, $xs \leq x \sum S$ for each $s \in S$. Now, if $xs \leq y$ for each $s \in S$ then $s \leq \bar{x} + y$ so $\sum S \leq \bar{x} + y$ and consequently $x \sum S \leq y$.

Exercise 10. Prove the dual of Theorem 9 (iv): If $S \subseteq L$ and $\prod S$ exists then for each $x \in L$, $\prod \{x + s : s \in S\}$ exists and equals $x + \prod S$.

Exercise 11. Let L be a Boolean algebra. Prove

(i) If $\sum S$ exists then $\prod \{\bar{s} : s \in S\}$ exists and equals $\overline{\sum S}$.

(ii) If $\prod S$ exists then $\sum \{\bar{s} : s \in S\}$ exists and equals $\overline{\prod S}$.

Theorem 12. *Suppose* L *is a distributive lattice with* 0, 1 *and contains a complemented element* a. *Then* $L \cong (a] \times (\bar{a}] \cong (a] \times [a)$.

Proof. Define $f: L \to (a] \times (\bar{a}]$ by $f(x) = (xa, x\bar{a})$. Then f preserves order. Suppose $f(x) \leq f(y)$; then $xa \leq ya$ and $x\bar{a} \leq y\bar{a}$. Addition yields $x \leq y$. f is onto for if $(u, v) \in (a] \times (\bar{a}]$ then $f(u + v) = ((u + v)a, (u + v)\bar{a}) = (ua + va, u\bar{a} + v\bar{a}) = (u, v)$. Next, define $g: L \to (a] \times [a)$ by $g(x) = (ax, a + x)$. Again g preserves order. If $ax \leq ay$ and $a + x \leq a + y$ then $x \leq a + y$ so $x = x(a + y) = ax + xy \leq ay + xy \leq y$. Finally, suppose $(u, v) \in (a] \times [a)$, then $g(\bar{a}v + u) = (a(\bar{a}v + u), a + (\bar{a}v + u)) = (au, (a + \bar{a})(a + v) + u) = (au, a + v + u) = (u, v)$.

Exercise 13. Prove that if L is a distributive lattice with 0, 1 then $\mathscr{C}(L)$ is a complemented sublattice of L.

Just as in the cases of lattices and distributive lattices, Boolean algebras form an equational class, which we denote by **B**. Here there are two binary operations $+$ and $\cdot$, one unary operation $^-$ and two nullary operations 0, 1. Thus Boolean algebras are exactly the algebras of the form $(L, (+, \cdot, ^-, 0, 1))$ that satisfy a set of identities that define distributive lattices and satisfy

$$\mathbf{1 \cdot x = x} \qquad \mathbf{0 + x = x}.$$

$$\mathbf{x \cdot \bar{x} = 0} \qquad \mathbf{x + \bar{x} = 1}.$$

We can consider **B** as a subclass of $\mathbf{D}_{01}$ (cf. Section I.7). Furthermore, the functor defined in Section I.19 from $\mathscr{B}$ (the equational category associated with **B**) to $\mathscr{D}_{01}$ is an embedding which is one-one on the objects since complementation is unique. Therefore, in accordance with the convention adopted in Section I.19, we will often consider $\mathscr{B}$ as a (full) subcategory of $\mathscr{D}_{01}$.

In particular, f is an order isomorphism if and only if f is a **B**-isomorphism.

Since in a Boolean algebra $(L, \leq)$, $x \leq y \Leftrightarrow \bar{y} \leq \bar{x}$, $(L, \leq)$ is isomorphic with $(L, \leq')$, where $x \leq' y$ means $y \leq x$. Stated in another way, $(L,(+, \cdot))$ is **L**-isomorphic with $(L,(\cdot, +))$. Thus the strong principle of duality holds: *If a lattice-theoretic statement P is true in some lattice* $L \in B$ *then* $\check{P}$ *is also true in L.*

Example 14. Let L be a Boolean algebra. Suppose P is the statement that every element of L is a finite sum of elements x satisfying

(P_1) $x \leq y + z \Rightarrow x \leq y$ or $x \leq z$.

Now, if L satisfies P_1 then every element of L is also a finite product of elements u satisfying

$(\breve{P_1})$ $vw \leq u \Rightarrow v \leq u$ or $w \leq u$.

Note that if L is a distributive lattice satisfying P_1 then it need not satisfy $\breve{P_1}$. Indeed, if L is all finite subsets of an infinite set, then the singletons and $\varnothing$ are the elements which satisfy P_1, but there are no elements at all satisfying $\breve{P_1}$.

Exercise 15. If $f \in [L, L_1]_{\mathscr{D}}$, $L, L_1 \in \text{Ob}\,\mathscr{B}$ and $f(0) = 0, f(1) = 1$, then $f \in [L, L_1]_{\mathscr{B}}$.

Exercise 16. If f is a **B**-homomorphism and $f(x) = 0 \Rightarrow x = 0$, then f is one-one.

Exercise 17. Let α, β be cardinals. Prove that $\mathbf{2}^\alpha$ is a **B**-homomorphic image of $\mathbf{2}^\beta$ if and only if $\beta \geq \alpha$.

Exercise 18. Let $h \in [L, L_1]_{\mathscr{B}}$. Suppose $h(x) \leq \sum_{i=1}^n h(y_i)$, $x, y_i \in L$. Prove there exists $y_i' \in L$, $i = 1, \cdots, n$ such that $h(y_i) = h(y_i')$ and $x \leq \sum_{i=1}^n y_i'$ (Hint: set $y_i' = y_i + x\bar{y}_1 \cdots \bar{y}_n$).

A classic work on Boolean algebras is R. Sikorski [8]. For more concise accounts, see e.g. Ph. Dwinger [2] and P. R. Halmos [3]. For detailed lists of references on this topic the reader is referred to these works.

7. Relatively complemented distributive lattices

Theorem 6.9(iii) suggests a way to generalize the notion of a Boolean algebra.

Definition 1. Let L be a lattice. An element $x \in L$ is *relatively complemented* if x is complemented in every interval $[a, b]$ which contains x.

Exercise 2. Suppose that L is a distributive lattice and there is an element $a \in L$ which is relatively complemented; then $L = (a] \times [a)$ (cf. Theorem 6.12).

Definition 3. A *relatively complemented distributive lattice* is a distributive lattice such that every element is relatively complemented. Such a lattice with 0 is called a *generalized Boolean algebra.*

Before considering relatively complemented distributive lattices in general, we first make some remarks on generalized Boolean algebras.

In a generalized Boolean algebra L, define the *difference*, $x - y$, $x, y \in L$ to be the relative complement of xy in the interval $[0, x]$. It is easy to see that in a Boolean algebra $x - y = x\bar{y}$ (indeed, $xy(x\bar{y}) = 0$ and $xy + x\bar{y} = x(y + \bar{y}) = x$). As an example of a generalized Boolean algebra without 1, take the ring of finite subsets of an infinite set. Here $x - y$ is $x \sim y$. Finally, define *symmetric difference* in a generalized Boolean algebra by $x \oplus y = (x - y) + (y - x)$. Note that in a Boolean algebra $x \oplus y = x\bar{y} + \bar{x}y$. We single out the following formulas for later use:

(1) $x - y \leq x$.

(2) $y + (x - y) = x + y$.

(3) $y(x - y) = 0$.

Exercise 4. A distributive lattice L with 0 is a generalized Boolean algebra if and only if for each $a, b \in L$ there exists an element $c \in L$ such that $ac = 0$ and $a + c = a + b$.

Exercise 5. (M. H. Stone [1]) A *Boolean ring* is a ring (in the classical sense) whose elements are idempotent (i.e. $x^2 = x$ for all elements x).

(i) Show that, if R is a Boolean ring, then $x + x = 0$ for all $x \in R$ and R is commutative.

(ii) If $(L, (+, \cdot, 0))$ is a generalized Boolean algebra, then $(L\,(\oplus, \cdot, 0))$ is a Boolean ring with the property that $x + y = x \oplus y \oplus xy$.

(iii) If $(R, (\oplus, \cdot, 0))$ is a Boolean ring, then $(R, (+, \cdot, 0))$ is a generalized Boolean algebra where $x + y = x \oplus y \oplus xy$. Moreover, $x \oplus y = (x - y) + (y - x)$.

We now return to relatively complemented distributive lattices in general. First, we show that this class of lattices can be made into an equational class of algebras.

Consider the class of algebras of the form $(L, (+, \cdot, \overline{}))$, where $+$ and $\cdot$ are binary operations, $\overline{}$ is a ternary operation and which satisfy a set of identities that define distributive lattices together with the identities:

$$(\mathbf{x} + \mathbf{y}) \cdot \overline{(\mathbf{x}, \mathbf{y}, \mathbf{z})} = \mathbf{x}$$

$$(\mathbf{x} + \mathbf{y}) + \overline{(\mathbf{x}, \mathbf{y}, \mathbf{z})} = \mathbf{x} + \mathbf{y} + \mathbf{z}.$$

It is obvious that L is relatively complemented, since if $a \le b \le c$ then $b\overline{(a, b, c)} = (a + b)\overline{(a, b, c)} = a$ and $b + \overline{(a, b, c)} = (a + b) + \overline{(a, b, c)} = a + b + c = c$. Conversely, if L is a relatively complemented distributive lattice, take $\overline{(a, b, c)}$ to be the complement of $a + b$ in $[a, a + b + c]$.

We can now consider the equational class **R** of relatively complemented distributive lattices as a subclass of **D**. Furthermore, the functor defined in Section I.19 from $\mathscr{R}$ (the equational category associated with **R**) to $\mathscr{D}$ is an embedding which is one-one on the objects, since $\overline{(x, y, z)}$ is the complement of $x + y$ in $[x, x + y + z]$ and is therefore uniquely determined. Therefore, in accordance with the convention adopted in Section I.19 we will often consider $\mathscr{R}$ as a subcategory of $\mathscr{D}$.

Also note that $\mathscr{R}$ is a full subcategory of $\mathscr{D}$. Indeed, let $L, L_1 \in \text{Ob}\,\mathscr{R}$ and $h \in [L, L_1]_{\mathscr{D}}$. Then for $x, y, z \in L$, $(h(x) + h(y))(h(\overline{(x, y, z)})) = h((x + y)\overline{(x, y, z)}) = h(x)$ and $(h(x) + h(y)) + h(\overline{(x, y, z)}) = h(x + y + \overline{(x, y, z)}) = h(x + y + z) = h(x) + h(y) + h(z)$. But $\overline{(h(x), h(y), h(z))}$ is uniquely determined in L_1 so $h(\overline{(x, y, z)}) = \overline{(h(x), h(y), h(z))}$ and hence $h \in [L, L_1]_{\mathscr{R}}$. It follows therefore that h is an order isomorphism between members of **R** $\Leftrightarrow$ h is an **R**-isomorphism.

Exercise 6. Prove that if $h \in [L, L_1]_{\mathscr{D}}$, $L \in \text{Ob}\,\mathscr{R}$ then $h[L] \in \text{Ob}\,\mathscr{R}$.

8. The various categories

In the previous sections we have introduced the categories $\mathscr{D}$, $\mathscr{D}_{01}$, $\mathscr{B}$ and $\mathscr{R}$. At this point we will define some other categories which will be useful.

The category $\mathscr{D}'$ is to be the subcategory of $\mathscr{D}$ such that $\text{Ob}\,\mathscr{D}' = \text{Ob}\,\mathscr{D}$ and a morphism $h \in [L, L_1]_{\mathscr{D}}$ is a morphism of $\mathscr{D}'$ if the following condition is satisfied.

If L has a 0(1) then L_1 has a 0(1) and $h(0) = 0$ $(h(1) = 1)$.

Similarly, we introduce a category $\mathscr{R}'$ whose objects are those of $\mathscr{R}$ and whose morphisms are the morphisms of $\mathscr{R}$ which are also in $\mathscr{D}'$. Note that neither $\mathscr{D}'$ nor $\mathscr{R}'$ are algebraic categories. Also, $\mathscr{D}'$ is a subcategory of $\mathscr{D}$ containing $\mathscr{D}_{01}$ as a full subcategory; and $\mathscr{R}'$ is a subcategory of $\mathscr{R}$ containing $\mathscr{B}$ as a full subcategory.

Exercise 1. Let $f \in [L, L_1]_{\mathscr{D}}$, $L \in \text{Ob}\,\mathscr{B}$. Then $f[L] \in \text{Ob}\,\mathscr{B}$, and the map $f' : L \to f[L]$ defined by $f'(a) = f(a)$ for all $a \in L$ is in $[L, f[L]]_{\mathscr{B}}$.

Exercise 2. Prove that $\mathscr{D}_{01}$ is a reflective subcategory of $\mathscr{D}'$ (The reflector assigns to $L \in \text{Ob}\,\mathscr{D}'$, L itself if L has a 0 and 1, and assigns to L, $L \oplus \mathbf{1}$ if L has a 0 but no 1, etc.).

Since chains will also play an important role in this book, it will often be necessary to consider them as algebras and as categories and also to consider in particular chains with 0 and 1. The following terminology is therefore adopted.

C: The class of algebras L in **D** for which $x + y \in \{x, y\}$ for any x, y in L.

$\mathscr{C}$: The algebraic category associated with **C**. Note that $\mathscr{C}$ is a (full) subcategory of $\mathscr{D}$.

$\mathscr{C}'$: The full subcategory of $\mathscr{D}'$ whose objects are those in $\mathscr{C}$.

$\mathbf{C}_{01}$: The class of algebras L in $\mathbf{D}_{01}$ for which $x + y \in \{x, y\}$ for any x, y in L.

$\mathscr{C}_{01}$: The algebraic category associated with $\mathbf{C}_{01}$. Note that $\mathscr{C}_{01}$ is a (full) subcategory of $\mathscr{D}_{01}$.

A poset P in which $x \cdot y$ (but not necessarily $x + y$) exists for each $x, y \in P$ is called a *semilattice*. Although we will not be dealing with semilattices in the book, it is easy to see that they form an equational class. Moreover, many of the categorical properties with which we will be concerned in **D**, **B**, etc., have similar characterizations in the category of semilattices. See, in particular, G. Bruns and H. Lakser [1], and A. Horn and N. Kimura [1].

9. Ideals and congruence relations

It is evident from universal algebra that congruence relations are of fundamental importance. We begin here to develop some of their properties in the context of distributive lattices. The notion of congruence relation is closely related to that of ideal.

Definition 1. A non-empty subset I of a lattice L is an *ideal* if

(i) $x \leq y, y \in I \Rightarrow x \in I$.

(ii) $x, y \in I \Rightarrow x + y \in I$.

For example, the interval $(a]$ is an ideal for each $a \in L$. Ideals of this form are called *principal ideals*. It is easily verified that for a non-empty set S of a lattice L, the intersection of all the ideals which contain S is again an ideal (note that L itself is an ideal) and is therefore the smallest ideal containing S. This ideal is denoted by $(S]$. In particular, $(\{x\}] = (x]$. A more intrinsic characterization of $(S]$ is given by:

Theorem 2. *If S is a non-empty subset of a lattice L, then* $(S] = \{y \in L : y \leq \sum T \text{ for some } T \subseteq S\}$.

Proof. Let $I = \{y \in L : y \leq \sum T \text{ for some } T \subseteq S\}$ and let $\mathfrak{S}$ be the set of all ideals that contain S. Now I is an ideal for if $x \leq y$ and $y \in I$ then obviously $x \in I$ and if $y_1 \leq \sum T_1, y_2, \leq \sum T_2$ for $T_1 \subseteq S, T_2 \subseteq S$ then $y_1 + y_2 \leq \sum (T_1 \cup T_2)$ and $T_1 \cup T_2 \subseteq S$. Since $S \subseteq I, I \in \mathfrak{S}$ so $(S] = \cap \mathfrak{S} \subseteq I$. On the other hand, if $J \in \mathfrak{S}$ then J must contain all of the elements which are in I so $I \subseteq J$ for all $J \in \mathfrak{S}$. Hence $I \subseteq \cap \mathfrak{S} = (S]$.

Let L be a distributive lattice. Denote by $\mathfrak{I}(L)$, the partially ordered set of all ideals of L.

Theorem 3. *If $L \in \mathbf{D}$, then $\mathfrak{I}(L)$ is a conditionally complete distributive lattice with* 1 *and the map $x \mapsto (x]$ is a* **D***-embedding.*

Proof. Suppose $I \in \mathfrak{I}(L)$ and $\varnothing \neq \mathfrak{X} \subseteq \mathfrak{I}(L)$. If $I \subseteq J$ for each $J \in \mathfrak{X}$, then $\varnothing \neq I \subseteq \cap\mathfrak{X}$ and $\cap\mathfrak{X} \in \mathfrak{I}(L)$ so $\prod^{\mathfrak{I}(L)} \mathfrak{X} = \cap\mathfrak{X}$. It is clear that $(\cup\mathfrak{X}]$ is the smallest ideal that contains J for each $J \in \mathfrak{X}$ so $\sum^{\mathfrak{I}(L)}\mathfrak{X} = (\cup\mathfrak{X}]$. Also if $I, J \in \mathfrak{I}(L)$ then $I \cap J \neq \varnothing$, hence $I \cap J \in \mathfrak{I}(L)$. Thus, $\mathfrak{I}(L)$ is a conditionally complete lattice. Now suppose $I, J, K \in \mathfrak{I}(L)$. If $a \in I(J + K)$ then $a \in I$ and $a \leq \sum S$ for some $S \subseteq J \cup K$. So $a = a\sum S = \sum\{as : s \in S\} \in ((I \cap J) \cup (I \cap K)] = I\cdot J + I\cdot K$ which shows that $\mathfrak{I}(L)$ is distributive. Finally, $(x] \cap (y] = (xy]$ and $(x] + (y] = (x + y]$.

Corollary 4. *If $L \in \mathbf{D}_{01}$ then $\mathfrak{I}(L)$ is a complete lattice and $x \mapsto (x]$* is a $\mathbf{D}_{01}$-*embedding.*

Exercise 5. Show that if $f \in (L, L_1]_{\mathscr{D}}$ is onto and $I \in \mathfrak{I}(L)$ then $f[I] \in \mathfrak{I}(L_1)$.

Exercise 6. For a distributive lattice L and $I, J \in \mathfrak{I}(L)$ show that $I + J = \{x + y : x \in I, y \in J\}$.

Exercise 7. (M. H. Stone [2]) Show that for a lattice L, $\mathfrak{I}(L)$ is distributive if and only if L is distributive.

It is sometimes necessary to consider the concept obtained by dualizing the definition of ideal.

Definition 8. A *filter* is a non-empty subset F of a lattice L satisfying

(i′) $x \leq y, x \in F \Rightarrow y \in F$.

(ii′) $x, y \in F \Rightarrow xy \in F$.

The notion of *principal filter* $[a)$ and the definition of $[S)$ are simply the duals of $(a]$ and $(S]$.

Exercise 9. Prove that a subset of a lattice is convex if and only if it is the intersection of an ideal and a filter.

Exercise 10. Prove directly, the duals of Theorems 2 and 3.

Exercise 11. Give an example of an ideal whose set theoretic complement is not a filter.

Exercise 12. Let $(L_s)_{s \in S}$ be a family in $\mathbf{D}$. Verify that the mapping $f\colon \times_{s \in S} \mathfrak{I}(L_s) \to \mathfrak{I}(\times_{s \in S} L_s)$ defined by $f(u) = \times_{s \in S} u(s)$ is a $\mathbf{D}$-embedding.

Exercise 13. Let I_1, I_2 be ideals and F_1, F_2 filters in a lattice L such that $\varnothing \neq I_1 \cap F_1 \subseteq I_2 \cap F_2$. Then $I_1 \subseteq I_2$ and $F_1 \subseteq F_2$. Thus $\varnothing \neq I_1 \cap F_1 = I_2 \cap F_2 \Rightarrow I_1 = I_2$ and $F_1 = F_2$.

Recall from Section I.6 that the poset of congruence relations of a lattice L is denoted by $\mathfrak{C}(L)$. It is obvious that the intersection of a family of congruence relations is again a congruence relation and it follows from this (cf. Section 4) that $\mathfrak{C}(L)$ is a complete lattice. Thus, if $(\theta_s)_{s\in S}$ is a family of congruence relation of L, then $\prod_{s\in S}\theta_s = \bigcap_{s\in S}\theta_s$. It is not difficult to see that if $(\theta_s)_{s\in S}$ is a family of congruence relations on L, then its sum $\sum_{s\in S}\theta_s$ is given by: $(x, y)\in\sum_{s\in S}\theta_s \Leftrightarrow$ there exist $z_0, z_1, \cdots, z_{n-1}, z_n \in L$ and $s_1, \cdots, s_n \in S$ such that $z_0 = x$, $z_n = y$ and $(z_{i-1}, z_i)\in\theta_{s_i}$ for $i = 1, \cdots, n$.

Exercise 14. Prove the statement in the last paragraph.

The following theorem will be useful.

Theorem 15. (N. Funayama and T. Nakayama [1]) *Let L be a lattice. Then $\mathfrak{C}(L)$ is distributive.*

Proof. Let $\theta_1, \theta_2, \theta_3 \in \mathfrak{C}(L)$. It suffices to show that $\theta_1(\theta_2 + \theta_3) \leq \theta_1\theta_2 + \theta_1\theta_3$. Suppose $(x, y)\in\theta_1(\theta_2 + \theta_3)$. Thus there exist $z_i \in L$, $i = 0, 1, \cdots, n$ such that $x = z_0$, $y = z_n$ and $(z_0, z_1)\in\theta_2$, $(z_1, z_2)\in\theta_3$, $(z_2, z_3)\in\theta_2, \cdots, (z_{n-1}, z_n)\in\theta_2$. Let $u_i = (xy + z_i)(x + y)$, $i = 0, 1, \cdots, n$. Now if $(z_{i-1}, z_i)\in\theta_j$, where $j = 2$ or $j = 3$, then it is clear that $(u_{i-1}, u_i)\in\theta_j$. But $xy \leq u_i \leq x + y$ for $i = 0, 1, \cdots, n$ and also $(x, y)\in\theta_1$. Hence $(xy, x + y)\in\theta_1$ so $(u_{i-1}, u_i)\in\theta_1$ and therefore $(u_{i-1}, u_i)\in\theta_1\theta_j$ where $j = 1$ or $j = 2$. It follows that $(x, y)\in\theta_1\theta_2 + \theta_1\theta_3$.

If L is a lattice and $\theta\in\mathfrak{C}(L)$ then L/θ is again a lattice, since **L** is equational. In particular L/θ is partially ordered by $[x]_\theta \leq [y]_\theta \Leftrightarrow (xy, x)\in\theta$ (or equivalently, $(x, x + y)\in\theta$). Note that an **L**-congruence relation and a **D**-congruence relation on $L\in\mathbf{D}$ are really the same thing; namely, an equivalence relation which enjoys the substitution property with respect to $+$ and $\cdot$.

Finally, suppose $L\in\mathbf{B}$. Obviously, a **B**-congruence relation on L is also an **L**-congruence relation (under the convention of ignoring the unary and nullary operations on L so that $L\in\mathbf{L}$. In fact, the sets of **L**-congruence relations and **B**-congruence relations on L coincide. Indeed, let θ be an **L**-congruence relation on L, we must show $(x, y)\in\theta$ implies $(\bar{x}, \bar{y})\in\theta$. But $(x, y)\in\theta$ implies $(x\bar{y}, 0) = (x\bar{y}, y\bar{y})\in\theta$ so $(\bar{y}, \bar{x}\bar{y}) = (x\bar{y} + \bar{x}\bar{y}, \bar{x}\bar{y})\in\theta$. Similarly $(\bar{x}, \bar{x}\bar{y})\in\theta$ so $(\bar{x}, \bar{y})\in\theta$.

We now begin to investigate the relationship between ideals and congruence relations.

Definition 16. Let I be an ideal in a lattice L. Define a relation $\boldsymbol{\theta}(I)$ by $\boldsymbol{\theta}(I) = \{(x, y)\in L^2 : x + u = y + u \text{ for some } u\in I\}$.

We note the following trivial properties about a lattice L:

(1) If $I\in\mathfrak{I}(L)$ then $(x, y)\in\boldsymbol{\theta}(I)$ if and only if there exist $u, v\in I$ such that $x + u = y + v$.

(2) $I, J\in\mathfrak{I}(L)$, $I\subseteq J \Rightarrow \boldsymbol{\theta}(I)\subseteq\boldsymbol{\theta}(J)$.

(3) If $I = (a]$ for some $a\in L$ then $\boldsymbol{\theta}(I) = \{(x, y)\in L^2 : x + a = y + a\}$.

(4) Suppose L has a 0. Then $\boldsymbol{\theta}(I) = \{(x, x) : x\in L\}$ if and only if $I = (0]$.

Theorem 17. *The relation $\boldsymbol{\theta}(I)$ in Definition 16 is a* **D***-congruence relation when $L \in \mathbf{D}$ and if $a \in I$ then $[a]_{\boldsymbol{\theta}(I)} = I$.*

Proof. Suppose that I is an ideal in L, then $\boldsymbol{\theta}(I)$ is reflexive and symmetric. Let $(a, b), (b, c) \in \boldsymbol{\theta}(I)$. Then there exist $x, y \in I$ such that $a + x = b + x$ and $b + y = c + y$ so $a + (x + y) = b + x + y = c + (x + y)$ and $x + y \in I$ so $(a, c) \in \boldsymbol{\theta}(I)$. If $(a, a'), (b, b') \in \boldsymbol{\theta}(I)$ then there exist x, y in I such that $a + x = a' + x, b + y = b' + y$ so $(a + b) + (x + y) = (a' + b') + (x + y)$ so $(a + b, a' + b') \in \boldsymbol{\theta}(I)$. Also, $ab + (ay + bx + xy) = (a + x)(b + y) = (a' + x)(b' + y) = a'b' + (a'y + b'x + xy)$ and, since $ay + ax + xy$ and $a'y + b'x + xy \in I$, we obtain $(ab, a'b') \in \boldsymbol{\theta}(I)$. The remainder of the statement is immediate.

Exercise 18. (V. K. Balachandran [1]) For u, v in a distributive lattice L and I an ideal in L define $u \equiv v(\text{mod } I)$ provided $(u, v) \in \boldsymbol{\theta}(I)$. Prove the following:

(i) The system $x \equiv a(\text{mod } I)$, $x \equiv b(\text{mod } J)$ has a solution if and only if $a \equiv b(\text{mod}(I + J))$.

(ii) If x is a solution of the system in (i) then y is a solution exactly when $y \equiv x(\text{mod } I \cdot J)$.

Definition 16 and Theorem 17 can of course be dualized. Since we will need both concepts, we adopt similar terminology.

For a filter F in a lattice L, $\boldsymbol{\theta}(F) = \{(x, y) \in L^2 : xu = yu \text{ for some } u \text{ in } F\}$ is an equivalence relation which is in $\mathfrak{C}(L)$ if L is distributive.

For an ideal I in a distributive lattice L, the quotient lattice $L/\boldsymbol{\theta}(I)$ is often denoted by L/I, its elements by $[x]_I$, and the natural homomorphism $\nu_{\boldsymbol{\theta}(I)}$ by ν_I. Similarly, for a filter F, L/F, $[x]_F$ and ν_F replace $L/\boldsymbol{\theta}(F)$, $[x]_{\boldsymbol{\theta}(F)}$ and $\nu_{\boldsymbol{\theta}(F)}$ respectively.

Theorem 19. Suppose I is an ideal in L where $L \in \mathbf{D}$. The natural homomorphism $\nu_I : L \to L/I$ has the following properties:

(i) $[x]_I \leq [y]_I \Leftrightarrow x \leq y + u$ *for some* $u \in I$.

(ii) *L/I has a* 0.

(iii) *If F is a filter in L and $I \cap F = \varnothing$ then $\nu_I[F]$ is a proper filter in L/I.*

Proof. (i) $[x]_I \leq [y]_I \Leftrightarrow [xy]_I = [x]_I \Leftrightarrow (xy, x) \in \boldsymbol{\theta}(I) \Leftrightarrow xy + u = x + u$ for some $u \in I \Leftrightarrow x \leq y + u$ for some $u \in I$. (ii) Let $x_o \in I$, then for any $x \in L$, $x_o \leq x + x_o$ so $[x_o]_I \leq [x]_I$. (iii) By the dual of Exercise 5, we need only show $\nu_I[F] \subset L/I$. Let $x \in I$. Then $[x] \notin \nu_I[F]$ or else there would exist $y \in F$ such that $[x] = \nu_I(y) = [y]$ so $(x, y) \in \boldsymbol{\theta}(I)$ and hence there would exist $z \in I$ such that $x + z = y + z$. But then $y \leq x + z$ implies $y \in I$, a contradiction.

Definition 20. Let $L \in \mathbf{L}$ and θ an **L**-congruence relation on L. Define $\mathbf{I}(\theta)$, to be the set

$$\mathbf{I}(\theta) = \{x \in L : \nu_\theta(x) \leq \nu_\theta(y) \text{ for all } y \in L\}.$$

We make the following observations:

(5) If $\mathbf{I}(\theta) \neq \varnothing$, then it is an ideal.

(6) If L has a 0, then $\mathbf{I}(\theta) = \{x \in L : (x, 0) \in \theta\}$; so in particular, if L has a 0, then $\mathbf{I}(\theta)$ is an ideal.

(7) $\theta \subseteq \theta_1 \Rightarrow \mathbf{I}(\theta) \subseteq \mathbf{I}(\theta_1)$: Indeed $u \in \mathbf{I}(\theta)$ implies $\nu_\theta(u) \leq \nu_\theta(y)$ for all $y \in L$. So for each $y \in L$, $\nu_\theta(uy) = \nu_\theta(u)$ and $(uy, u) \in \theta \subseteq \theta_1$, hence $\nu_{\theta_1}(uy) = v_{\theta_1}(u)$ which implies $\nu_{\theta_1}(u) \leq \nu_{\theta_1}(y)$. Thus $u \in \mathbf{I}(\theta_1)$.

Since all of the lattices, for which it will be necessary to consider the dual of Definition 20, contain 1, we make the following simplifications.

Definition 21. For an **L**-congruence relation θ on a lattice L with 1, define $\mathbf{F}(\theta) = \{x \in L : (x, 1) \in L\}$.

Theorem 22. *Suppose L is a lattice. Then the following are equivalent.*

(i) *L is a distributive lattice.*

(ii) *For each ideal I, $\boldsymbol{\theta}(I)$ is an* **L**-*congruence relation and* $\mathbf{I}(\boldsymbol{\theta}(I)) = I$.

(iii) *Every ideal has the form* $\mathbf{I}(\theta)$ *for some* **L**-*congruence relation θ on L.*

(iv) *Every ideal is of the form* $f^{-1}[\{0\}]$ *for some* **L**-*homomorphism f.*

Proof. (i) $\Rightarrow$ (ii) By Theorem 17, $\boldsymbol{\theta}(I)$ is an **L**-congruence relation. Let $x_0 \in I$. Then by Definition 20, $u \in \mathbf{I}(\boldsymbol{\theta}(I))$ implies $\nu_I(u) \leq \nu_I(x_0)$ so $u \leq x_0 + y$ for some $y \in I$ so $u \in I$. Conversely, $v \in I$ and $v \leq y + v$ imply $\nu_I(v) \leq \nu_I(y)$ for all $y \in L$. Thus $v \in \mathbf{I}(\boldsymbol{\theta}(I))$.

(ii) $\Rightarrow$ (iii) Trivial.

(iii) $\Rightarrow$ (iv) Let I be an ideal and $I = \mathbf{I}(\theta)$ where θ is an **L**-congruence relation on L. So $\nu_\theta : L \to L/\theta$ is an **L**-homomorphism. Now L/θ has a 0 by Theorem 19(ii). Finally, we must show $I = \nu_\theta^{-1}[\{[x_o]\}]$ where $x_o \in I$. But $u \in \nu_\theta^{-1}[\{[x_o]\}] \Leftrightarrow \nu_\theta(u) = [x_o] = \nu_\theta(x_o) \leq \nu_\theta(y)$ for all $y \in L \Leftrightarrow u \in \mathbf{I}(\theta) = I$.

(iv) $\Rightarrow$ (i) If L is not distributive then it contains as a sublattice one of the sublattices in Theorem 5.9. Suppose it is the one in Figure 2 of that theorem. Then $(b_1]_L = f^{-1}[\{0\}]$ for some **L**-homomorphism f. So $f(b_2) \leq f(b_1 + b_3) = f(b_1) + f(b_3) = f(b_3)$ since $f(b_1) = 0$. Hence $f(b_2) = f(b_2)f(b_3) = f(b_2b_3) = f(b_0) \leq f(b_1)$. Thus $f(b_2) = 0$, a contradiction. The argument for Figure 1 is left to the reader.

Exercise 23. Let $L \in \mathbf{D}$ and let I be an ideal in L. Prove that $\boldsymbol{\theta}(I)$ is the smallest **D**-congruence relation θ on L such that $\mathbf{I}(\theta) = I$.

Exercise 24. Let $L \in \mathbf{D}$, $a, b \in L$, $a \leq b$ and let $\boldsymbol{\theta}(a, b) = \boldsymbol{\theta}((b]) \cap \boldsymbol{\theta}([a))$. Prove that $\boldsymbol{\theta}(a, b)$ is the smallest congruence relation θ for which $(a, b) \in \theta$. *(I)*

Lemma 25. *Let L be a generalized Boolean algebra and I an ideal in L. Then* $x \oplus y \in I \Leftrightarrow (x, y) \in \boldsymbol{\theta}(I)$. (See Exercise 7.5.)

Proof. ($\Rightarrow$) From Exercise 7.5, $x + (x \oplus y) = x \oplus (x \oplus y) \oplus x(x \oplus y)$ $= y \oplus x \cdot x \oplus xy = y \oplus x \oplus xy = x + y$. Similarly $y + (x \oplus y) = x + y$. So $x + (x \oplus y) = y + (x \oplus y)$, $x \oplus y \in I$ implies $(x, y) \in \boldsymbol{\theta}(I)$.

($\Leftarrow$) There exists $u \in I$ such that $x + u = y + u$, so $x \leq y + u$. Hence $x - y$ $= x(x - y) \leq (y + u)(x - y) = y(x - y) + u(x - y) \leq 0 + u = u$ so $x - y \in I$. Similarly $y - x \in I$ so $x \oplus y \in I$.

Theorem 26. *In a lattice L the following are equivalent.*

(i) *L is a generalized Boolean algebra.*

(ii) $\mathbf{I}(\theta) \neq \varnothing$ *and* $\boldsymbol{\theta}(\mathbf{I}(\theta)) = \theta$ *for all* $\theta \in \mathfrak{C}(L)$; $\boldsymbol{\theta}(I) \in \mathfrak{C}(L)$ *and* $\mathbf{I}(\boldsymbol{\theta}(I)) = I$ *for all* $I \in \mathfrak{I}(L)$.

(iii) *The map* $\mathbf{I}: \mathfrak{C}(L) \rightarrow \mathfrak{I}(L)$ *defined by* $\theta \mapsto \mathbf{I}(\theta)$ *is an isomorphism with inverse* $\boldsymbol{\theta}: \mathfrak{I}(L) \rightarrow \mathfrak{C}(L)$ *defined by* $I \mapsto \boldsymbol{\theta}(I)$.

Proof. (i) $\Rightarrow$ (ii) Since L has a 0, $\mathbf{I}(\theta) \neq \varnothing$. Now if $(u, v) \in \boldsymbol{\theta}(\mathbf{I}(\theta))$ then there exists $x \in \mathbf{I}(\theta)$ such that $u + x = v + x$. Since $x \in \mathbf{I}(\theta)$, $(x, 0) \in \theta$ so $(u + x, u)$ and $(v + x, v)$ are in θ. Hence $(u, v) \in \theta$. Next suppose $(y, z) \in \theta$. Then $(y - z, 0)$ $= (y(y - z), z(y - z)) \in \theta$. So $y - z \in \mathbf{I}(\theta)$. Similarly, $z - y \in \mathbf{I}(\theta)$ and therefore $y \oplus z \in \mathbf{I}(\theta)$. By Lemma 25, we have $(y, z) \in \theta$. So $\boldsymbol{\theta}(\mathbf{I}(\theta)) = \theta$. The second part follows from Theorem 22 since L is distributive.

(ii) $\Rightarrow$ (iii) Let $\theta_0 \in \mathfrak{C}(L)$. By (ii) $\mathbf{I}(\theta_0) \in \mathfrak{I}(L)$ and the map $\theta \mapsto \mathbf{I}(\theta)$ preserves order. Suppose $\theta_1, \theta_2 \in \mathfrak{C}(L)$ and $\mathbf{I}(\theta_1) \subseteq \mathbf{I}(\theta_2)$ then since $\boldsymbol{\theta}$ preserves order $\theta_1 = \boldsymbol{\theta}(\mathbf{I}(\theta_1)) \subseteq \boldsymbol{\theta}(\mathbf{I}(\theta_2)) = \theta_2$. Now the map is onto for if $I \in \mathfrak{I}(L)$ the $\boldsymbol{\theta}(I) \in \mathfrak{I}(L)$ and $\mathbf{I}(\boldsymbol{\theta}(I)) = I$.

(iii) $\Rightarrow$ (i) For each $I \in \mathfrak{I}(L)$, $\boldsymbol{\theta}(I)$ is an **L**-congruence relation and $\mathbf{I}(\boldsymbol{\theta}(I)) = I$, by hypothesis. So by Theorem 22, L is distributive. Now the trivial congruence relation 0 on L determines the ideal $\mathbf{I}(0)$ which is, by hypothesis, non-empty. But $u \in \mathbf{I}(0)$ implies that for each $y \in L$, $[u]_0 = \nu_0(u) \leq \nu_0(y) = [y]_0$ so $uy \in [uy]_0$ $= [u]_0[y]_0 = [u]_0 = \{u\}$. Thus, $uy = u$ and hence $u \leq y$ for all $y \in L$. Therefore, L has a 0. To prove the existence of relative complements, let $a, b \in L$. Consider the congruence relation $\boldsymbol{\theta}([ab))$ determined by the filter $[ab)$. Since $ab(a + b) = (ab)a$, we have $(a + b, a) \in \boldsymbol{\theta}([ab)) = \boldsymbol{\theta}(\mathbf{I}(\boldsymbol{\theta}([ab))))$. Hence there exists $x \in \mathbf{I}(\boldsymbol{\theta}([ab)))$ such that $a + b + x = a + x$. It follows that $a + b = (a + b)(a + b + x)$ $= (a + b)(a + x) = a + bx$. Now, since $x \in \mathbf{I}(\boldsymbol{\theta}([ab)))$ and L has a 0, $(x, 0) \in \boldsymbol{\theta}([ab))$ so $abx = ab0 = 0$. Thus, $a + bx = a + b$ and $a(bx) = 0$ so L is a generalized Boolean algebra.

Exercise 27. Let L_1 be a subalgebra of a Boolean algebra L and I an ideal in L.

(i) $L_1' = \{a \oplus x : a \in L_1, x \in I\}$ is a subalgebra of L.

(ii) $L_1/(L_1 \cap I) \cong L_1'/I$.

Theorems 22 and 26 are essentially due to J. Hashimoto [1, Theorems 2.2 and 7.2]. Also see G. Grätzer and E. T. Schmidt [4] and [5]. We mention some additional results along these lines which the reader might wish to verify:

(1) (N. Funayama and T. Nakayama [1]) Theorem 15 can be extended to: If $\{\theta\} \cup S \subseteq \mathfrak{C}(L)$ then $\theta \sum S = \sum \{\theta\varphi : \varphi \in S\}$.

(2) (J. Hashimoto [1]) For a lattice L, $\mathfrak{J}(L)$ is a Boolean algebra if and only if L is a generalized Boolean algebra such that if $\cdots \leq c_3 \leq c_2 \leq c_1$ is a chain in L then there exists k_0 for which $c_k = c_{k_0}$ for $k \geq k_0$.

L. Nachbin [2] gives a nice characterization of $\mathfrak{J}(L)$ when L is a generalized Boolean algebra.

10. Subdirect product representation for distributive lattices

The subdirect product theorem is particularly striking when applied to **D** since there is only one subdirectly irreducible algebra in D, namely **2**.

Theorem 1. (G. Birkhoff [5, Theorem 2]) *Every distributive lattice is a* **D**-*subdirect product of copies of* **2**.

Proof. It is sufficient to prove that **2** is the only subdirectly irreducible member of **D**. For this, we apply Theorem I.9.3. Since **2** has only two congruence relations, it is clearly subdirectly irreducible. Conversely, suppose L is subdirectly irreducible and $L \neq \mathbf{2}$. Then there exist elements $a < x < b$ in L. By distributivity $\boldsymbol{\theta}([x)) \cap \boldsymbol{\theta}((x]) = 0$ so $\boldsymbol{\theta}([x)) = 0$ or $\boldsymbol{\theta}((x]) = 0$, a contradiction.

Corollary 2. *Every Boolean algebra is a* **B**-*subdirect product of copies of* **2**.

Proof. Let $L \in \mathbf{B}$. Then, by Theorem 1, there exists a **D**-embedding $f: L \to \mathbf{2}^S$ for some set S such that $p_s \circ f$ is onto for each $s \in S$. Let $s \in S$. Since $p_s \circ f$ is onto, there exists $x \in L$ such that $p_s(f(x)) = \mathbf{0}$. So $f(0)(s) \leq f(x)(s) = p_s(f(x)) = \mathbf{0}$, thus $f(0) = 0$. Similarly, $f(1) = 1$ so f is a **B**-embedding.

The reader has undoubtedly observed that the elements of a distributive lattice behave as if they were sets with $\cup$ and $\cap$ replaced with $+$ and $\cdot$. That this is the general state of affairs is a consequence of Theorem 1 (See also Section III.5).

Remark. The isomorphism $x \mapsto (x]$ of Theorem 1.9 does not, in general, yield the required result, since $\{(x] : x \in L\}$ is not necessarily a ring of sets (see Exercise 5).

Theorem 3. (G. Birkhoff [1, Theorem 25.2]) *Every distributive lattice is isomorphic with a ring of sets.*

Proof. By Theorem 1 there exists a **D**-embedding $f: L \to \mathbf{2}^I$ for some set I. For each $x \in L$, let $g(x) = \{i \in I : f(x)(i) = \mathbf{1}\}$. We leave it to the reader to verify that $\{g(x) : x \in L\}$ is a ring of sets and that $x \mapsto g(x)$ is an isomorphism.

Theorem 4. (M. H. Stone [2]) *Every Boolean algebra is isomorphic with a field of sets.*

Proof. In the case of a Boolean algebra, the ring of sets $\{g(x) : x \in L\}$ in Theorem 3 is, in fact, a field of sets.

Exercise 5. Show that for a poset P, $\{(x] : x \in P\}$ is a ring of sets if and only if P is a chain.

III

The Prime Ideal Theorem

1. Irreducible elements

In lattice theory it is useful, as in classical ring theory, to consider the factorization of elements. We begin by investigating those elements which are irreducible in a sense that we will now define.

Definition 1. Let L be a lattice. An element $a \in L$ is *join irreducible* if $a = x + y$ implies $a = x$ or $a = y$. Dually, $b \in L$ is *meet irreducible* if $b = xy$ implies $b = x$ or $b = y$. If L has a 0, then an element $a \in L$ is an *atom* if $a \neq 0$ but $x \leq a$ implies $x = 0$ or $x = a$.

Denote by $\mathscr{J}(L)$ and $\mathscr{M}(L)$ the join irreducible and meet irreducible elements, respectively. By definition, 0, if it exists, is join irreducible but is not an atom.

Theorem 2. *In a distributive lattice L*

(i) *a is join irreducible if and only if $a \leq x + y$ implies $a \leq x$ or $a \leq y$.*

(ii) *Every atom is join irreducible.*

(iii) *a is an atom if and only if L has a 0, $a \neq 0$, and for each $x \in L$, either $ax = 0$ or $a \leq x$.*

Proof. (i) If a is join irreducible and $a \leq x + y$ then $a = a(x + y) = ax + ay$, so $a = ax$ or $a = ay$. Hence $a \leq x$ or $a \leq y$. For the converse, let $a = x + y$. Then $x \leq a$ and $y \leq a$; but $a \leq x + y$ implies $a \leq x$ or $a \leq y$ so $a = x$ or $a = y$. (ii) Let a be an atom and suppose $a \leq x + y$. Since $ax \leq a$ and $ay \leq a$ we have that either $ax = ay = 0$ or $ax = a$ or $ay = a$. The first case is impossible since $0 \neq a = ax + ay$ so $a \leq x$ or $a \leq y$. (iii) If a is an atom, then $a \neq 0$ and $ax \leq a$ implies $ax = 0$ or $ax = a$. Conversely, if $x < a$ then by hypothesis $ax = 0$ so $x = ax = 0$.

In a chain, all elements are join irreducible. The atoms in the power set of a set X, $X \neq \varnothing$, are the singletons while the join irreducibles are the singletons, together with the empty set.

Exercise 3. Prove that the non-zero join irreducibles in $(N, |)$ are exactly the prime powers.

Definition 4. A distributive lattice L with 0 is *atomic* if each non-zero element of L is a sum of atoms.

Theorem 5. *In a generalized Boolean algebra L*

(i) *a is an atom if and only if it is a non-zero join irreducible.*

(ii) *L is atomic if and only if for each $x \neq 0$ there exists an atom a such that $a \leq x$.*

Proof. (i) By Theorem 2, it suffices to prove that a non-zero join irreducible element a is an atom. Let $x \leq a$. Then $a = x + (a - x)$ so $a = x$ or $a = a - x$. But $a = a - x$ implies $x = xa = x(a - x) = 0$. (ii) The "only if" part is obvious. Conversely, let $x \neq 0$ and $S = \{a \in \mathscr{J}(L) : 0 < a \leq x\}$. Note that by (i) and by the hypothesis, S is non-void and is exactly the set of atoms less than or equal to x. Now x is an upper bound for S. Suppose x_1 is also an upper bound for S but $x \nleq x_1$. Then $x - x_1 \neq 0$ so there exists an $a \in \mathscr{J}(L) \sim \{0\}$ such that $a \leq x - x_1$. Therefore $a \leq x$ so $a \in S$ and hence $a \leq x_1$. But then $a \leq (x - x_1)x_1 = 0$, a contradiction.

Exercise 6. Prove that a Boolean algebra is atomic if and only if 1 is a sum of atoms.

Exercise 7. Show that if a is an atom in a Boolean algebra and $\sum S$ exists, $S \neq \varnothing$ then $a \leq \sum S$ implies $a \leq s$ for some $s \in S$.

2. The descending chain condition

We now state a condition which guarantees a representation of elements as a sum of join irreducibles.

Definition 1. A poset P satisfies the *descending chain condition* if every non-empty subset of P has a minimal element.

Theorem 2. *If a lattice L satisfies the descending chain condition, then each element of L is a sum of a finite non-empty set of mutually incomparable elements of $\mathscr{J}(L)$. If, in addition, L is distributive then this representation is unique.*

Proof. If the first part of the theorem is false, then the descending chain condition implies that the non-empty set S of elements, which can not be so represented, has a minimal element a. But then $a \notin \mathscr{J}(L)$ so there exist $b, c \in L$ such that $a = b + c, b < a$, and $c < a$. By the minimality of a, $\{b, c\} \cap S = \varnothing$ so $b = \sum S_1$ and $c = \sum S_2$ where $S_1, S_2 \subseteq \mathscr{J}(L)$. Now taking the set of S_3 to be the set of maximal elements of $S_1 \cup S_2$ we have $a = b + c = \sum S_1 + \sum S_2 = \sum (S_1 \cup S_2) = \sum S_3$ (cf. Exercise II.2.7). Since the elements of S_3 are incomparable, we have a contradiction and so the first part of the theorem is established.

Next, suppose $\sum S = \sum T$ where S and T are finite non-empty subsets of incomparable elements of $\mathscr{J}(L)$ and L is distributive. By Theorem 1.2(i), for each $s \in S$ there exists $t_s \in T$ such that $s \leq t_s$. Now for the same reason, $t_s \leq s_1$ for some $s_1 \in S$ so $s \leq s_1$. By assumption then, $s = s_1$ so $s = t_s \in T$. Hence $S \subseteq T$ and similarly $T \subseteq S$ so $S = T$.

Corollary 3. *In a finite lattice, each element is a finite sum of join irreducibles.*

Exercise 4. Show that $(N, |)$ satisfies the descending chain condition. What does Theorem 2 say about this lattice?

In contrast to Theorem II.10.3 the proof of the following theorem does not involve the axiom of choice.

Theorem 5. *If L is a distributive lattice satisfying the descending chain condition then the map* $x \mapsto \varphi(x) = \{u \in \mathscr{J}(L) \sim \{0\} : u \leq x\}$ *is an isomorphism onto the ring of sets* $\{\varphi(x) : x \in L\}$.

Proof. By Theorem 2, $x = \sum \varphi(x)$ for each $x \in L$ so $x \leq y \Leftrightarrow \varphi(x) \subseteq \varphi(y)$. To show that $\{\varphi(x) : x \in L\}$ is a ring of sets we need therefore only show (1) $\varphi(x + y) \subseteq \varphi(x) \cup \varphi(y)$ and (2) $\varphi(x) \cap \varphi(y) \subseteq \varphi(xy)$. Now (1) follows from Theorem 1.2(i) (which is where the distributivity is needed) and (2) is immediate.

Corollary 6. *Every finite Boolean algebra L is isomorphic with the power set of some finite set S.*

Proof. The map $x \mapsto \varphi(x)$ of Theorem 5 is an isomorphism from L to a subset of the power set of $\mathscr{J}(L) \sim \{0\}$. We will show that the map is onto. Indeed, let $A \subseteq \mathscr{J}(L) \sim \{0\}$. It suffices to show that $\varphi(\sum A) = A$. If $a \in A$ then $a \leq \sum A$ so $a \in \varphi(\sum A)$. Conversely, if $u \in \mathscr{J}(L) \sim \{0\}$ and $u \leq \sum A$ then $u \leq a$ for some $a \in A$. But a is an atom so $u = a \in A$.

Exercise 7. Show that in a distributive lattice satisfying the descending chain condition, every filter is principal.

3. Prime ideals and maximal ideals

Definition 1. An ideal I in a distributive lattice L is *prime* if

(i) $I \subset L$.

(ii) $xy \in I \Rightarrow x \in I$ or $y \in I$.

The set of all prime ideals of L is denoted by $\mathfrak{P}(L)$. Note that neither $\varnothing$ nor L are members of $\mathfrak{P}(L)$.

It will become evident in what follows that the notion of prime ideal is the natural generalization of meet irreducible and dually for prime filter (see below) and join irreducible. In particular, a principal ideal $(a]$ in L is prime if and only if $a \in \mathscr{M}(L) \sim \{1\}$.

Definition 2. An ideal I in a distributive lattice is *maximal* if it is a maximal member of $\mathfrak{J}(L) \sim \{L\}$. That is, an ideal I is maximal if

(i) $I \subset L$.

(ii) If $I \subseteq J \subset L, J \in \mathfrak{J}(L)$ then $I = J$.

The lattice theoretic duals of prime ideal and maximal ideal are also useful and we state their definitions: a *prime filter* is a proper filter F such that $x + y \in F$ implies $x \in F$ or $y \in F$. A *maximal filter* is a filter which is maximal in the poset of proper filters.

The proof of the following theorem is immediate.

Theorem 3. *An ideal I in a distributive lattice L is prime if and only if $L \sim I$ is a prime filter.*

We also note that a principal filter $[a)$ in a lattice L is prime if and only if $a \in \mathscr{J}(L) \sim \{0\}$.

The next result shows that there is a one-one correspondence between the prime ideals of a lattice $L \in \mathbf{D}$ and the $\mathbf{D}$-homomorphisms of L onto $\mathbf{2}$.

Theorem 4. *Suppose $L \in \mathbf{D}$ and $f: L \to \mathbf{2}$ is a function. Then f is an onto $\mathbf{D}$-homomorphism if and only if $f^{-1}[\{\mathbf{0}\}]$ is a prime ideal of L.*

Proof. $(\Rightarrow) f^{-1}[\{\mathbf{0}\}]$ is a proper and non-empty subset of L since f is onto. If $x \leq y$ and $y \in f^{-1}[\{\mathbf{0}\}]$ then $f(x) \leq f(y) = \mathbf{0}$ so $x \in f^{-1}[\{\mathbf{0}\}]$. If $x, y \in f^{-1}[\{\mathbf{0}\}]$ then $f(x + y) = f(x) + f(y) = \mathbf{0} + \mathbf{0} = \mathbf{0}$ so $x + y \in f^{-1}[\{\mathbf{0}\}]$. If $xy \in f^{-1}[\{\mathbf{0}\}]$ then $f(x)f(y) = f(xy) = \mathbf{0}$ and since 0 is meet irreducible in $\mathbf{2}$, $f(x) = \mathbf{0}$ or $f(y) = \mathbf{0}$.

$(\Leftarrow) f$ preserves order, for if $x \leq y$ and $f(y) \neq \mathbf{1}$ then $f(y) = \mathbf{0}$ so $y \in f^{-1}[\{\mathbf{0}\}]$. Hence $x \in f^{-1}[\{\mathbf{0}\}]$ so $f(x) = \mathbf{0}$. For $x, y \in L$, if $f(x) = \mathbf{1}$ or $f(y) = \mathbf{1}$ then clearly $f(x + y) \leq f(x) + f(y)$. If not then $x, y \in f^{-1}[\{\mathbf{0}\}]$ so $x + y \in f^{-1}[\{\mathbf{0}\}]$ and hence $f(x + y) = \mathbf{0} \leq f(x) + f(y)$. Next, if $f(xy) = \mathbf{1}$ we have $f(x)f(y) \leq f(xy)$ so suppose $xy \in f^{-1}[\{\mathbf{0}\}]$. But $f^{-1}[\{\mathbf{0}\}]$ is prime, so say $x \in f^{-1}[\{\mathbf{0}\}]$. Therefore $f(x)f(y) = \mathbf{0} \cdot f(y) \leq f(xy)$. Finally, f is onto since $f^{-1}[\{\mathbf{0}\}]$ is proper and non-empty.

The following is an easy application of the previous results.

Corollary 5. Let S be a finite set. Then $|\mathscr{F}_{\mathbf{B}}(S)| = 2^{2^{|S|}}$.

Proof. Let m be the number of atoms in $\mathscr{F}_{\mathbf{B}}(S)$. Then it follows from Corollary 2.6 and its proof and from Theorem 1.5 that $|\mathscr{F}_{\mathbf{B}}(S)| = 2^m$. But by Theorem 3 and the remark following this theorem there is a one-one correspondence between the atoms of $\mathscr{F}_{\mathbf{B}}(S)$ and the prime ideals of $\mathscr{F}_{\mathbf{B}}(S)$. But by Theorem 4 there is a one-one correspondence between the prime ideals of $\mathscr{F}_{\mathbf{B}}(S)$ and the morphisms of $[\mathscr{F}_{\mathbf{B}}(S), \mathbf{2}]_{\mathscr{B}}$. Thus m is equal to the number of homomorphisms $\mathscr{F}_{\mathbf{B}}(S) \to \mathbf{2}$. Hence $m = 2^{|S|}$ which completes the proof.

Exercise 6. Show that any Boolean algebra with 2^{2^n} elements is free.

For any prime ideal I in a distributive lattice L, denote by $f_I: L \to \mathbf{2}$ the $\mathbf{D}$-homomorphism defined by $f_I(u) = \mathbf{1}$ if $u \notin I$ and $f_I(u) = \mathbf{0}$ if $u \in I$. f_I is called the $\mathbf{D}$-*homomorphism induced by* I. Note also that if $L \in \mathrm{Ob}\, \mathscr{D}_{01}$ then $f_I \in [L, \mathbf{2}]_{\mathscr{D}_{01}}$ and if $L \in \mathrm{Ob}\, \mathscr{B}$ then $f_I \in [L, \mathbf{2}]_{\mathscr{B}}$. The reader should also note that for $L \in \mathrm{Ob}\, \mathscr{D}$, and I a prime ideal, the homomorphisms $f_I : L \to \mathbf{2}$ and the homomorphism $\nu_I : L \to L/I$ as defined in II.9 are in general not the same. This *is* however the case if $L \in \text{Ob}\,\mathscr{B}$ (cf. Theorem II.9.26).

It is easily verified that if $f \in [L, L_1]_{\mathscr{D}}$ then

(1) if $I \in \mathfrak{P}(L_1)$ then $f^{-1}[I] \in \mathfrak{P}(L) \cup \{\varnothing, L\}$

In (1) each of the following are sufficient for $f^{-1}[I] \in \mathfrak{P}(L)$

(2) f is onto.

(3) $L, L_1 \in \text{Ob}\,\mathscr{D}_{01}, f \in [L, L_1]_{\mathscr{D}_{01}}$.

Theorem 7. (M. H. Stone [2]) *In a distributive lattice L every maximal ideal (filter) is prime.*

Proof. Let I be maximal ideal in L, and suppose $ab \in I$, but $a \notin I$. Then $I \subset (I \cup \{a\}]$ so $(I \cup \{a\}] = L$. Hence there exists $x \in I$ such that $b \leq a + x$ and so $b = b(a + x) = ba + bx \in I$. The proof that a maximal filter is prime is similar.

Corollary 8. *In a relatively complemented distributive lattice L an ideal I is maximal if and only if it is prime.*

Proof. By Theorem 7 we only need to prove that if I is prime then it is maximal. Let $J \in \mathfrak{I}(L)$ and $I \subset J$. Let $x \in J \sim I$ and $y \in I$. To show $J = L$, let $z \in L$. Now x has a complement x' in $[xy, x + z]$. So $xx' = xy \in I$ and since $x \notin I$, $x' \in I \subset J$; hence $x + z = x + x' \in J$. This implies $z \in J$ so $L = J$.

Remark. The converse of Corollary 8 will be proven in Section 6 (Theorem 3).

Exercise 9. Characterize the prime ideals (prime filters) of $(N, |)$.

Exercise 10. Let R be a ring of subsets of a set X and let $x_o \in X$. Show that $\{A \in R : x_o \notin A\}$ is either empty, all of R, or a prime ideal.

Exercise 11. In a Boolean algebra L, prove that an ideal I is prime if and only if $x \in I$ or $\bar{x} \in I$ for all $x \in L$.

Exercise 12. (W. A. J. Luxemburg [2]) Let L be a distributive lattice and L' an *ideal* in L. Let I' be a prime ideal in L' (L' considered as a distributive lattice). Prove that there exists exactly one prime ideal I in L with $I \cap L' = I'$. (Hint: Let $I = \{x \in L : (x]_L \cap L' \subseteq I'\}$).

Exercise 13. Let $I, I_1, \cdots, I_n$ be prime ideals in a distributive lattice and suppose $\bigcap_{j=1}^{n} I_j \subseteq I$. Show that $I_{j_o} \subseteq I$ for some $j_o \in \{1, \cdots, n\}$. Prove the dual statement.

4. The prime ideal theorem

The prime ideal theorem is one of the most important results in the theory of distributive lattices.

We give two proofs of the theorem. The first one is based on the subdirect product theorem for **D** (Theorem II.10.1) and does not involve the axiom of choice, although of course the subdirect product theorem does. The second proof is more direct and is also due to G. Birkhoff [1] as reformulated by M. H. Stone [3].

Theorem 1. (Prime ideal theorem) *Suppose L is a distributive lattice, I is an ideal in L, F is a filter in L, and $I \cap F = \varnothing$. Then there exists a prime ideal J such that $I \subseteq J$ and $J \cap F = \varnothing$.*

1st Proof. Consider the natural homomorphism $\nu_I : L \to L/I$. By Theorem II.9.19, $F_1 = \nu_I[F]$ is a proper filter in L/I. But L/I has a 0 and $(0] \cap F_1 = \varnothing$. So by the dual of the same theorem $\nu_{F_1} : L/I \to (L/I)/F_1 = L'$ is such that L' has a 0, 1 with $0 < 1$. By the subdirect product theorem for distributive lattices we can assume that L' is a ring of subsets of a set X. Let $x_0 \in 1_{L'} \sim 0_{L'}$. Then $J_1 = \{A \in L' : x_0 \notin A\}$ is a prime ideal in L' (cf. Exercise 3.10). But $\nu_{F_1} \circ \nu_I$ is onto so $J = (\nu_{F_1} \circ \nu_I)^{-1}[J_1]$ is a prime ideal in L. Finally, $(\nu_{F_1} \circ \nu_I)[I] = v_{F_1}[\{0_{L/I}\}] = \{0\} \subseteq J_1$ so $I \subseteq J$ and $(\nu_{F_1} \circ \nu_I)[F] \cap J_1 = \nu_{F_1}[F_1] \cap J_1 = \{1_{L'}\} \cap J_1 = \varnothing$ so $F \cap J = \varnothing$.

2nd Proof. Let $\mathfrak{X}$ be the family of all ideals in L which contain I and are disjoint with F. It is easily verified that if $\mathfrak{C}$ is a chain in $\mathfrak{X}$ then $\cup\mathfrak{C} \in \mathfrak{X}$ so by Zorn's Lemma, $\mathfrak{X}$ has a maximal element J. Then $I \subseteq J$ and $F \cap J = \varnothing$; also, J is proper since $F \neq \varnothing$. Suppose $xy \in J$ but $x \notin J$ and $y \notin J$. Then $J \subset (J \cup \{x\}]$ so by the maximality of J, $(J \cup \{x\}] \notin \mathfrak{X}$. But $I \subseteq J \subseteq (J \cup \{x\}]$ so $(J \cup \{x\}] \cap F \neq \varnothing$. Hence there exists $u_1 \in F$ such that $u_1 \leq v_1 + x$ for some $v_1 \in J$. Similarly, there exists $u_2 \in F$ such that $u_2 \leq v_2 + y$ for some $v_2 \in J$. Thus $u_1 u_2 \leq (v_1 + x)(v_2 + y) = v_1 v_2 + v_1 y + x v_2 + xy$ and since $v_1 v_2 + v_1 y + x v_2 + xy \in J$ we have the contradiction $u_1 u_2 \in J \cap F$.

Corollary 2. *In a distributive lattice L, if $x \nleq y$ then there exists a prime ideal containing y and not x.*

Proof. Let $I = (y]$, $F = [x)$ and apply Theorem 1.

Although the axiom of choice implies the prime ideal theorem for distributive lattices, the two are not equivalent; see J. D. Halpern [1] and J. D. Halpern and A. Levy [1]. The prime ideal theorem is, however, equivalent to each of the following:

(i) Every distributive lattice $L \ncong \mathbf{1}$ contains a prime ideal.

(ii) Suppose L is a Boolean algebra, I is an ideal in L, F is a filter in L, and $I \cap F = \varnothing$. Then there exists a prime ideal J such that $I \subseteq J$ and $J \cap F = \varnothing$.

(iii) Every Boolean algebra $L \ncong \mathbf{1}$ contains a prime ideal.

Indeed, to prove that (i) is equivalent to the prime ideal theorem, first suppose $L \not\cong \mathbf{1}$. Then there exists $x, y \in L$ such that $x \nleq y$. By the prime ideal theorem there is a prime ideal containing y and not x. Conversely, the first proof of the prime ideal theorem only requires the existence of a prime ideal in $L' \not\cong \mathbf{1}$. Similarly, (ii) and (iii) are equivalent. Since (i) obviously implies (iii), it will now suffice to verify that (ii) implies (i). Let $L \not\cong \mathbf{1}$ be a distributive lattice. It is known (W. Peremans [1]) that one can construct—without using the axiom of choice—an **L**-embedding $\varphi: L \to L_1$ where L_1 is a Boolean algebra. Now there exists $x \nleq y$ in L and hence $\varphi(x) \nleq \varphi(y)$. By (iii) there exists a prime ideal I in L_1 such that $\varphi(y) \in I$, $\varphi(x) \notin I$. $\varphi^{-1}[I]$ is then the required prime ideal.

Although the existence of a prime ideal is equivalent to the existence of a maximal ideal in a Boolean algebra, the situation is quite different for distributive lattices (with 1)—as is illustrated by the following diagram.

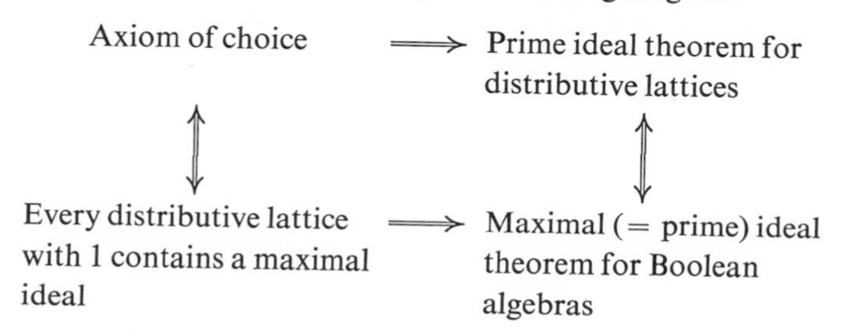

The fact that the existence of a maximal ideal in distributive lattices with 1 implies the axiom of choice was established by G. Klimovsky [1]. See also J. L. Bell and D. H. Fremlin [1].

It is also known (J. Loś and C. Ryll-Nardzewski [1]) that Tychonoff's Theorem for compact Hausdorff spaces is equivalent to the prime ideal theorem. The question of the independence of the prime ideal theorem and other fundamental statements in mathematics such as the Hahn-Banach Theorem, has been studied by many authors (cf. D. Pincus [1] and W. A. J. Luxemburg [1]). See also J. D. Halpern [2], H. Rubin and D. Scott [1], H. Rubin and J. Rubin [1], D. Scott [1] and A. Tarski [2], [3], [4].

Exercise 3. Prove that in a distributive lattice L a prime ideal I is maximal if and only if it is maximal in $\mathfrak{P}(L)$.

Exercise 4. Show that Theorem 1 holds if I is replaced with a non-empty subset of L which is closed under finite sums and is disjoint with F.

5. Representation by sets

Let L be a distributive lattice and for each $x \in L$, let $\hat{x} = \{I \in \mathfrak{P}(L) : x \notin I\}$. Then

(1) $\hat{x} \cup \hat{y} = \widehat{x + y}$ for all $x, y \in L$.

(2) $\hat{x} \cap \hat{y} = \widehat{xy}$ for all $x, y \in L$.

Indeed, suppose $I \in \mathfrak{P}(L)$. Then, using only the definition of prime ideal, we have $I \in \hat{x} \cup \hat{y} \Leftrightarrow I \in \hat{x}$ or $I \in \hat{y} \Leftrightarrow x \notin I$ or $y \notin I \Leftrightarrow x + y \notin I \Leftrightarrow I \in \widehat{x + y}$; $I \in \hat{x} \cap \hat{y} \Leftrightarrow x \notin I, y \notin I \Leftrightarrow xy \notin I \Leftrightarrow I \in xy$. Thus, $\hat{L} = \{\hat{x} : x \in L\}$ is a ring of sets. Since $x \leq y \Rightarrow \hat{x} \subseteq \hat{y}$ and by Corollary 4.2 $\hat{x} \subseteq \hat{y} \Rightarrow x \leq y$, the assignment $x \mapsto \hat{x}$ is an isomorphism between L and $\hat{L}$. So again, we see that every distributive lattice is isomorphic with a ring of sets. If L is a Boolean algebra, then for $x \in L$, $\mathfrak{P}(L) \sim \hat{x} = \hat{\bar{x}}$, so this construction also yields the fact that every Boolean algebra is isomorphic with a field of sets.

Theorem 1. *If L is a distributive lattice and S, T are non-empty subsets of L such that* $\cap \{\hat{x} : x \in S\} \subseteq \cup \{\hat{y} : y \in T\}$ *then there exist* $S' \Subset S$ *and* $T' \Subset T$ *such that* $\cap \{\hat{x} : x \in S'\} \subseteq \cup \{\hat{y} : y \in T'\}$.

Proof. If $[S) \cap (T] = \varnothing$ then there exists, by the prime ideal theorem, a prime ideal I such that $(T] \subseteq I$ and $[S) \cap I = \varnothing$. Since $T \subseteq I$ and $S \cap I = \varnothing$, we obtain the contradiction $I \in \cap \{\hat{x} : x \in S\} \sim \cup \{\hat{y} : y \in T\}$. Hence $[S) \cap (T] \neq \varnothing$ so there exists $u \in [S) \cap (T]$. Therefore there exists $S' \Subset S$, $T' \Subset T$ such that $\prod S' \leq \sum T'$ which, since $x \mapsto \hat{x}$ is an isomorphism, implies $\cap \{\hat{x} : x \in S'\} \subseteq \cup \{\hat{y} : y \in T'\}$.

Exercise 2. A distributive lattice is a generalized Boolean algebra if and only if it is isomorphic with a ring R of sets such that $A, B \in R \Rightarrow A \sim B \in R$.

6. Some corollaries of the prime ideal theorem

Theorem 1. *In a distributive lattice with* 1 *every proper ideal is contained in a maximal ideal.*

Proof. Let $\mathfrak{X}$ be the set of all ideals that contain a given proper ideal I. If $\mathfrak{C}$ is a chain in $\mathfrak{X}$ then $\cup \mathfrak{C}$ is an ideal and it is, furthermore, proper since $1 \notin J_1$ for any $J_1 \in \mathfrak{C}$. By Zorn's lemma there exists an ideal J which is maximal subject to containing I and being proper. Clearly then, I is a maximal ideal.

Theorem 2. *Every proper ideal in a distributive lattice is the intersection of prime ideals.*

Proof. Suppose I is a proper ideal in L. Set $\mathfrak{S} = \{J \in \mathfrak{P}(L) : I \subseteq J\}$. Now $I \subseteq \cap \mathfrak{S}$, but if $x \in \cap \mathfrak{S} \sim I$ then by the prime ideal theorem there exists $J \in \mathfrak{P}(L)$ such that $I \subseteq J$ and $J \cap [x) = \varnothing$. But then $J \in \mathfrak{S}$ so $x \in \cap \mathfrak{S} \subseteq J$, a contradiction.

Remark. Note that the proof implies that $\mathfrak{S} \neq \varnothing$.

In Corollary 3.8 we showed that in a relatively complemented distributive lattice, maximal ideals and prime ideals are the same. That this latter condition implies relative complementation, was first established by L. Nachbin [1] for distributive lattices with 0, 1.

Theorem 3. (Nachbin's theorem) *A distributive lattice is relatively complemented if and only if every prime ideal is maximal.*

Proof. Let L be a distributive lattice in which every prime ideal is maximal. Suppose there exist $a_0, a_1, a_2 \in L$, $a_0 \leq a_1 \leq a_2$, such that a_1 is not complemented in $[a_0, a_2]$. Hence $a_0 < a_1 < a_2$. Let $I_0 = \{x \in L : a_1 x \leq a_0\}$ and $I_1 = \{x \in L : a_2 x \leq a_1 + y \text{ for some } y \in I_0\}$. Note that I_0 and I_1 are ideals and that $a_0 \in I_0$, $a_1 \notin I_0$, $a_1 \in I_1$ and $I_0 \subseteq I_1$. Now $a_2 \notin I_1$. Indeed otherwise, $a_2 \leq a_1 + y$ for some $y \in I_0$. But then if $y' = ya_2 + a_0$, we have $a_1 + y' = a_2$, $a_1 y' = a_0$ which is a contradiction. By the prime ideal theorem there exists a prime ideal J_0 such that $a_2 \notin J_0$, $I_1 \subseteq J_0$. Let $F = [(L \sim J_0) \cup \{a_1\})$. We show that $F \cap I_0 = \varnothing$. Indeed, otherwise there exists $x \in I_0 \cap F_0$; thus $x \geq ya_1$ for some $y \notin J_0$. But $a_1 x \leq a_0$, implies $a_1 y \leq a_1 x \leq a_0$. Therefore $y \in I_0$ and thus $y \in J_0$, a contradiction. Again, by the prime ideal theorem there exists a prime ideal J_1 such that $I_0 \subseteq J_1$, $J_1 \cap F = \varnothing$. But then $J_1 \subset J_0$ and thus J_1 is not maximal, which contradicts the hypothesis of the theorem.

A generalization of Nachbin's theorem to lattices—not necessarily distributive—can be found in D. Adams [1].

In view of Theorems 2 and 3, A. Monteiro [1] posed the question of determining whether a distributive lattice, in which every filter is an intersection of maximal filters and dually, is necessarily relatively complemented. This was answered in the negative, independently by M. E. Adams [1] and R. Balbes [5]. In M. E. Adams [2], it is shown that if L is a distributive lattice without 0 or 1, in which every filter is an intersection of maximal filters and dually, then every sublattice is the intersection of maximal proper sublattices. Moreover, any distributive lattice is the intersection of the maximal proper sublattices of some distributive lattice. See also J. Hashimoto [1] and K. Takeuchi [1].

A more general problem is to determine the extent to which the partial ordering of $\mathfrak{P}(L)$ determines the structure of L. An example of a result of the kind is the following rephrasing of Nachbin's theorem. If L is a distributive lattice such that $\mathfrak{P}(L)$ is totally unordered, then L is relatively complemented. In Section V.3 it will be shown that if $\mathfrak{P}(L) \cong \mathbf{2}^X$ for some $X \neq \varnothing$ then $L \cong \mathscr{F}_{\mathbf{D01}}(X)$. In Chapters VI and X we will return to the question of how $\mathfrak{P}(L)$ determines L, but in a slightly different context.

Another related problem is that of characterizing those posets P for which there exists a distributive lattice L (with 0, 1) such that $\mathfrak{P}(L) \cong P$. The first complete solution was given by T. P. Speed [5] where it was shown that a necessary and sufficient condition is that P be a limit of certain partially ordered systems of finite posets. A more explicit solution has been given recently by B. Davey [2].

Exercise 4. Prove each of the following

(i) If C is a chain then $\mathfrak{P}(C)$ is complete and $(a, b]_{\mathfrak{P}(C)}$ contains an element with an immediate predecessor, when $a < b$.

(ii) If C is a chain and L a distributive lattice such that $\mathfrak{P}(C) = \mathfrak{P}(L)$ then $C \cong L$.

(iii) If C is a complete chain in which each interval $(a, b]$, $a < b$, has an element with an immediate predecessor then $C \cong \mathfrak{P}(C_1)$ for some chain C_1.

Another application of the prime ideal theorem is:

Theorem 5. *Let L' be a sublattice of a distributive lattice L.*

(i) *If I is a prime ideal in L then $I \cap L' \in \mathfrak{P}(L') \cup \{\varnothing, L'\}$.*

(ii) *If I' is a prime ideal in L' then there exists a prime ideal in I in L such that $I \cap L' = I'$.*

Proof. (i) is easily verified. For (ii), first observe that $(I']_L \cap [L' \sim I')_L = \varnothing$, for since I' is closed under sums and $L' \sim I'$ is closed under products, an element x in the intersection would have to satisfy $u \leq x \leq v$ where $v \in I'$ and $u \in L' \sim I'$ which is impossible. Let I be a prime ideal in L such that $(I']_L \subseteq I$ and $I \cap [L' \sim I')_L = \varnothing$. Then $I' \subseteq (I']_L \subseteq I$ and $I' \subseteq L'$ so $I' \subseteq I \cap L'$. Conversely, if $u \in I \cap L'$ then since $I \cap (L' \sim I') = \varnothing$, we have $u \in I'$, so $I' = I \cap L'$.

Corollary 6. *Let $\mathscr{K}$ be one of the categories, $\mathscr{D}$, $\mathscr{D}_{01}$ or $\mathscr{B}$ and suppose $L, L_1 \in \mathrm{Ob}\,\mathscr{K}$ with L_1 a subalgebra of L. Every morphism $f \in [L_1, \mathbf{2}]_{\mathscr{K}}$ can be extended to a morphism $g \in [L, \mathbf{2}]_{\mathscr{K}}$.*

Proof. For $\mathscr{D}_{01}$ and $\mathscr{B}$, $f^{-1}[\{0\}]$ is a prime ideal in L_1 so by Theorem 5 there exists a prime ideal I in L such that $I \cap L_1 = f^{-1}[\{0\}]$. Then $f_I: L \to \mathbf{2}$ is the required morphism. The argument for $\mathscr{D}$ is the same if f is onto, and otherwise g is defined to be the induced constant map.

The question of the uniqueness of these extensions is a more delicate one and is related to the notion of epimorphisms, which will be covered in Section V.8.

Exercise 7. (J. Hashimoto [1]) Prove that for $L \in \mathbf{D}_{01}$, every ideal is a unique intersection of prime ideals if and only if L is a finite Boolean algebra.

IV

Topological Representations

1. Stone spaces

In this chapter we are going to establish a coequivalence between the category $\mathscr{D}_{01}$ and a certain category of topological spaces and maps. These results then yield a coequivalence between the category $\mathscr{B}$ and a suitable category of spaces and maps.

In 1936, M. H. Stone [2] developed a topological representation theory for Boolean algebras and later, in [3], generalized the theory to distributive lattices with 0. Generalizations and extensions of Stone's theory can be found, for example, in G. Grätzer [4], J. Hashimoto [1], A. Nerode [1], L. Rieger [1], and T. P. Speed [1]. Recently H. Priestley [1], [2] has developed a representation theory for distributive lattices with 0, 1 in terms of ordered topological spaces.

The presentation given here drops Stone's requirement that each distributive lattice have a zero but is nevertheless identical with Stone's representation for distributive lattices with 0.

We begin by constructing a topological representation for an arbitrary distributive lattice.

Definition 1. A *Stone* space is a topological space X satisfying:

(i) X is a T_0-space (i.e. for any two distinct points of X, there is an open set containing one and not the other).

(ii) The family of compact open subsets of X is a ring of sets and a basis for X.

(iii) If $(X_s)_{s\in S}$ and $(Y_t)_{t\in T}$ are non-empty families of non-empty compact open sets and $\bigcap_{s\in S} X_s \subseteq \bigcup_{t\in T} Y_t$ then there exist $S' \subseteq S$ and $T' \subseteq T$ such that $\bigcap_{s\in S'} X_s \subseteq \bigcup_{t\in T'} Y_t$.

Remark. The empty set $\varnothing$ is considered here to be a topological space with $\{\varnothing\}$ as topology. This is a trivial example of a Stone space.

We want to associate with each distributive lattice L a Stone space $\mathscr{S}(L)$ called the *representation space* of L. To this end, let $L \in \text{Ob}\,\mathscr{D}$ and define $\mathscr{S}(L)$ to be the topological space whose points are the members of $\mathfrak{P}(L)$ with the topology determined by the basis $\{\varnothing\} \cup \{\hat{x} : x \in L\}$. Recall from Section III.5 that $\hat{x} = \{I \in \mathfrak{P}(L) : x \notin I\}$.

Theorem 2. *If* $L \in \text{Ob}\,\mathscr{D}$ *then* $\mathscr{S}(L)$ *is a Stone space and the compact open sets are exactly the members of* $\{\varnothing\} \cup \{\hat{x} : x \in L\}$.

Proof. Since $\{\varnothing\} \cup \{\hat{x} : x \in L\}$ is a ring of sets, it can indeed serve as a basis for a topology on $\mathfrak{P}(L)$. For (i), let $I, J \in \mathscr{S}(L)$ and $I \neq J$. If $a \in I \sim J$ then $J \in \hat{a}$ and $I \notin \hat{a}$ so $\mathscr{S}(L)$ is a T_0-space. It follows from Theorem III.5.1 that all the sets $\{\varnothing\} \cup \{\hat{x} : x \in L\}$ are compact open sets in $\mathscr{S}(L)$. Furthermore, any non-empty open set has the form $A = \bigcup_{x \in S} \hat{x}$, $\varnothing \neq S \subseteq L$ and so, if A is compact, then there exists $S' \Subset S$ such that $A = \bigcup_{s \in S'} \hat{x}$. Hence $A = \widehat{\sum S'}$. This establishes the validity of (ii). Finally, (iii) follows again from Theorem III.5.1.

Note that it also follows that $\mathscr{S}(L)$ is locally compact.

Exercise 3. For $X \subseteq \mathscr{S}(L)$ show that $\mathrm{Cl}(X) = \{I \in \mathscr{S}(L) : \cap X \subseteq I\}$.

We now proceed to identify the elements of $L \in \mathrm{Ob}\, \mathscr{D}$ with certain compact open subsets of $\mathscr{S}(L)$.

Definition 4. Let X be an arbitrary topological space. A non-empty subset $A \subseteq X$ is called *fundamental* if A is compact open. The empty set, $\varnothing$, is called *fundamental*, provided that for any non-empty family of compact open sets $(X_s)_{s \in S}$, $\bigcap_{s \in S} X_s = \varnothing$ implies that there exists $S' \Subset S$ such that $\bigcap_{s \in S'} X_s = \varnothing$.

The set of fundamental subsets of X will be denoted by $\mathscr{T}(X)$.

Theorem 5. *If X is a Stone space then $\mathscr{T}(X)$ is a ring of sets.*

Proof. Let Y and Z be fundamental sets in X. Clearly, $Y \cup Z$ is fundamental, and if $Y \cap Z \neq \varnothing$ then it is also fundamental by (ii) of Definition 1. Thus, suppose $Y \cap Z = \varnothing$, $Y \neq \varnothing$, $Z \neq \varnothing$. To show that $\varnothing$ is fundamental, suppose $\bigcap_{s \in S} X_s = \varnothing$ where $(X_s)_{s \in S}$ is a non-empty family of compact open sets. If $X_s = \varnothing$ for some $s \in S$, then let $S' = \{s\}$, otherwise apply (iii) of Definition 1 to $\bigcap_{s \in S} X_s \subseteq Y$ and to $\bigcap_{s \in S} X_s \subseteq Z$. So there exist $S_1, S_2 \Subset S$ such that $\bigcap_{s \in S_1} X_s \subseteq Y$ and $\bigcap_{s \in S_2} X_s \subseteq Z$ so $\bigcap_{s \in S_1 \cup S_2} X_s \subseteq Y \cap Z = \varnothing$ which proves that $\varnothing$ is fundamental.

Theorem 6. *For $L \in \mathrm{Ob}\, \mathscr{D}$, the fundamental sets of $\mathscr{S}(L)$ are exactly the sets $\hat{x}$, $x \in L$.*

Proof. Suppose A is fundamental in $\mathscr{S}(L)$. If $A \neq \varnothing$ then Theorem 2 yields the result. Now, suppose $\varnothing$ is fundamental. Since $I \in \bigcap_{x \in L} \hat{x}$ implies $I = \varnothing$ which is not possible for $I \in \mathfrak{P}(L)$, we have $\bigcap_{x \in L} \hat{x} = \varnothing$. But $\varnothing$ is fundamental, so there exists $S \Subset L$ such that $\bigcap_{x \in S} \hat{x} = \varnothing$ and so $\widehat{\prod S} = \varnothing$. Conversely, for each $x \in L$, $\hat{x}$ is fundamental if it is non-empty, so suppose $\hat{a} = \varnothing$ for some $a \in L$. To show that $\varnothing$ is fundamental, suppose $(X_s)_{s \in S}$ is a non-empty family of compact open sets such that $\bigcap_{s \in S} X_s = \varnothing$. Again, if $X_s = \varnothing$ for some s, then there is nothing to prove. So, we may assume that for each $s \in S$, $X_s = \hat{a}_s$ for some $a_s \in L$. Then by Theorem III.5.1, $\bigcap_{s \in S} \hat{a}_s \subseteq \hat{a}$ implies there exists $S' \Subset S$ such that $\bigcap_{s \in S'} X_s = \bigcap_{s \in S'} \hat{a}_s \subseteq \hat{a}$. This shows that $\hat{a} = \varnothing$ is fundamental.

***Corollary* 7.** *$\mathscr{T}(\mathscr{S}(L)) \cong L$.*

Proof. $\mathscr{T}(\mathscr{S}(L)) = \{\hat{x} : x \in L\} \cong L$.

Corollary 8. *A distributive lattice L has a 0 if and only if $\varnothing$ is fundamental in $\mathscr{S}(L)$. Hence in this case, the notions of fundamental and compact open coincide.*

Proof. If L has a 0, then $\varnothing = \hat{0}$ is fundamental. Conversely, if $\varnothing$ is fundamental, then $\varnothing = \hat{a}$ for some $a \in L$. Hence $\hat{a} = \varnothing \subseteq \hat{b}$ for all $b \in L$. So a is the least element of L.

Theorem 9. (M. H. Stone [3]) *For each distributive lattice L, we have $\mathscr{T}(\mathscr{S}(L))) \cong L$ and for each Stone space X, we have $\mathscr{S}(\mathscr{T}(X))$ is homeomorphic with X. Thus there is, up to isomorphisms and homeomorphisms, a one-one correspondence between distributive lattices and Stone spaces.*

Proof. We have already seen that $\mathscr{T}(\mathscr{S}(L)) \cong L$, so it suffices to prove $\mathscr{S}(\mathscr{T}(X))$ is homeomorphic with X.

For each $x \in X$, let $h(x) = \{A \in \mathscr{T}(X) : x \notin A\}$. We will first show that $\varnothing \neq h(x) \subset \mathscr{T}(X)$. Indeed, since $x \in X$ and $\mathscr{T}(X) \cup \{\varnothing\}$ is a basis for X, there exists $A \in \mathscr{T}(X)$ such that $x \in A$ so $h(x) \neq \mathscr{T}(X)$. Now, if $\varnothing$ is fundamental, then $\varnothing \in h(x)$ so suppose that $\varnothing$ is not fundamental. Then there exists a non-empty family $(A_s)_{s \in S}$ of compact open sets such that $\bigcap_{s \in S} A_s = \varnothing$ but $\bigcap_{s \in S'} A_s \neq \varnothing$ for any $S' \subseteq S$. Hence each A_s is fundamental and at least one, say A_{s_0} does not contain x. Thus, $A_{s_0} \in h(x)$. Since $\mathscr{T}(X)$ is a ring of sets, it follows easily that $h(x)$ is a prime ideal in $\mathscr{T}(X)$ so we can define a function $h: X \to \mathscr{S}(\mathscr{T}(X))$ by $x \mapsto h(x)$. Since X is T_0, h is one-one. Our next task is to show that h is onto. Let I be a prime ideal in $\mathscr{T}(X)$.

Claim 1. $\cap \{A \in \mathscr{T}(X) : A \notin I\} \nsubseteq \cup \{B \in \mathscr{T}(X) : B \in I\}$: Suppose not, thus

(1) $\cap \{A \in \mathscr{T}(X) : A \notin I\} \subseteq \cup \{B \in \mathscr{T}(X) : B \in I\}$.

First, if $A = \varnothing$ for some $A \in \mathscr{T}(X)$ then $\varnothing \in \mathscr{T}(X)$ implies $\varnothing \in I$ so we can assume

(2) $A \neq \varnothing$ for all $A \in \mathscr{T}(X) \sim I$.

Now suppose $\{B \in \mathscr{T}(X) : B \in I\} = \{\varnothing\}$, then $\varnothing \in \mathscr{T}(X)$ so by the definition of $\mathscr{T}(X)$, $\varnothing$ is fundamental and we have by (1) that $\cap \{A \in \mathscr{T}(X) : A \notin I\} \subseteq \varnothing$, so there exist $A_1, \cdots, A_n \notin I$ such that $A_1 \cap \cdots \cap A_n = \varnothing$. But then $\varnothing \notin I$, a contradiction. Thus, $\{B \in \mathscr{T}(X) : B \in I\}$ contains some non-empty sets. We can then rewrite (1) as

$$\cap \{A \in \mathscr{T}(X) : A \notin I\} \subseteq \cup \{B \in \mathscr{T}(X) : B \in I, B \neq \varnothing\}$$

where $A \neq \varnothing$ for each $A \in \mathscr{T}(X) \sim I$. Apply (iii) of Definition 1 to obtain $A_1 \cap \cdots \cap A_n \subseteq B_1 \cup \cdots \cup B_m$ where $A_i \notin I$, $B_j \in I$ for $i \leq n, j \leq m$. But then $B_1 \cup \cdots \cup B_m \in I$ so $A_i \in I$ for some i, a contradiction. Hence there exists

(3) $x \in \cap \{A \in \mathscr{T}(X) : A \notin I\} \sim \cup \{B \in \mathscr{T}(X) : B \in I\}$.

Claim 2. $h(x) = I$: Recall $h(x) = \{A \in \mathscr{T}(X) : x \notin A\}$. Suppose $A \in h(x)$ then $x \notin A$. If $A \notin I$ then by (3), $x \in A$ so $A \in I$. Conversely, suppose $B \in I$. Then by (3), $x \notin B$ and so $B \in h(x)$. Hence $h(x) = I$.

Finally, to show that h and h^{-1} are continuous, let A be a non-empty compact open set in X. Then $x \in h^{-1}[\hat{A}] \Leftrightarrow h(x) \in \hat{A} \Leftrightarrow A \notin h(x) \Leftrightarrow x \in A$. Thus $h^{-1}[\hat{A}] = A$ and $\hat{A} = h[A]$.

Many properties of distributive lattices manifest themselves clearly in their representation spaces.

Theorem 10. *Let L be a distributive lattice.*

(i) *L has a* 1 *if and only if $\mathscr{S}(L)$ is compact, and L has a* 0 *if and only if* $\varnothing$ *is fundamental.*

(ii) *L is a complete lattice if and only if the interior of the intersection of any family of fundamental sets in $\mathscr{S}(L)$ is fundamental.*

Proof. (i) If L has a 1, then $\mathscr{S}(L) = \hat{1}$ is compact. On the other hand, if $\mathscr{S}(L)$ is compact and $|L| > 1$, then since it is also open, $\mathscr{S}(L) = \hat{u}$ for some $u \in L$. But then for any $x \in L$, $\hat{x} \subseteq \mathscr{S}(L) = \hat{u}$ so $x \leq u$. (ii) Let $(\hat{a}_s)_{s \in S}$ be a family of fundamental sets in $\mathscr{S}(L)$ and let $a = \prod_{s \in S} a_s$. Then $a \leq a_s$ implies $\hat{a} \subseteq \bigcap_{s \in S} \hat{a}_s$. If $\bigcup_{t \in T} \hat{b}_t$ is any open set contained in $\bigcap_{s \in S} \hat{a}_s$ then $b_t \leq a_s$ for all, t, s. So $b_t \leq a$ for all $t \in T$ which means $\bigcup_{t \in T} \hat{b}_t \subseteq \hat{a}$. Thus, $\text{Int}(\bigcap_{s \in S} \hat{a}_s) = \hat{a}$ is fundamental. Conversely, let $X \subseteq L$. Then by hypothesis $\text{Int}(\cap \{\hat{x} : x \in X\})$ is fundamental. Hence $\text{Int}(\cap \{\hat{x} : x \in X\}) = \hat{a}$ for some $a \in L$. Then $\hat{a} \subseteq \text{Int}(\cap \{\hat{x} : x \in X\}) \subseteq \hat{x}$ for all $x \in X$ so a is a lower bound for X. If a' is also a lower bound for X then $\widehat{a'} \subseteq \cap \{\hat{x} : x \in X\}$ so $\widehat{a'} = \text{Int}(\widehat{a'})$ $\subseteq \text{Int}(\cap \{\hat{x} : x \in X\}) = \hat{a}$ and therefore $a' \leq a$. Thus, $a = \prod X$.

Theorem 11. *The following are equivalent in a distributive lattice L.*

(i) *L is relatively complemented.*

(ii) *$\mathscr{S}(L)$ is a T_1-space (i.e. for any two points $x, y \in \mathscr{S}(L)$, $x \neq y$, there exists an open set A such that $y \in A$, $x \notin A$).*

Proof. (i) $\Rightarrow$ (ii) Suppose L is relatively complemented, $I, J \in \mathscr{S}(L)$ and $I \neq J$. By Theorem III.6.3, $I \nsubseteq J$ so there exists $x \in I \sim J$. Hence $J \in \hat{x}$ and $I \notin \hat{x}$.

(ii) $\Rightarrow$ (i) Suppose $\mathscr{S}(L)$ is T_1 and that I is a prime ideal which is not maximal.

Then there exists an ideal J such that $I \subset J \subset L$. Also there exists a prime ideal J' such that $I \subset J \subseteq J' \subset L$. Since $I \neq J'$ and $\mathscr{S}(L)$ is T_1, there exists a basic open set $\hat{x}$ containing J' but not I. But then $x \in I \sim J'$, a contradiction. This shows that every prime ideal is maximal and hence, by Theorem III.6.3, L is relatively complemented.

Theorem 12. (M. H. Stone [3]) *The following are equivalent in a distributive lattice L.*

(i) *L is generalized Boolean algebra.*

(ii) *$\mathscr{S}(L)$ is a Hausdorff ($= T_2$) space.*

Proof. (i) $\Rightarrow$ (ii) Let I, J be distinct members of $\mathscr{S}(L)$. Then there exist $x \in I \sim J$ and $y \in J \sim I$. Let x' be the complement of x in $[0, x + y]$ then $xx' = 0$ implies $\hat{x} \cap \widehat{x'} = \varnothing, J \in \hat{x}$. Also $I \in \widehat{x'}$ since $y \leq x + x'$. Thus $\mathscr{S}(L)$ is T_2.

(ii) $\Rightarrow$ (i) Since T_2 implies T_1, Theorem 11 implies that L is relatively complemented. If L has no 0 then there are at least two distinct prime ideals I, J and therefore basic open sets $\hat{x}, \hat{y}$ such that $I \in \hat{x}, J \in \hat{y}$ and $\hat{x} \cap \hat{y} = \varnothing$. But this means $xy = 0$.

(Cf. A. Figa-Talamanca and S. Franklin [1]).

Exercise 13. Let $L \in \text{Ob}\, \mathscr{D}$. Prove that if L has no 0 then $\mathfrak{I}(L)$ is isomorphic to the lattice of non-empty open sets of $\mathscr{S}(L)$, and if L has a 0 then $\mathfrak{I}(L)$ is isomorphic to the lattice of open sets of $\mathscr{S}(L)$.

2. Topological representation theory for $\mathscr{D}_{01}$

We will now present a classical result which states that the category $\mathscr{D}_{01}$ and a certain category of topological spaces and maps are coequivalent.

The theory leading up to and including Theorem 4 is that of A. Nerode [1].

Definition 1. A *bounded Stone space* is a compact Stone space in which $\varnothing$ is fundamental.

Definition 2. A function $g: X \to Y$, where X and Y are bounded Stone spaces is *strongly continuous*, provided $g^{-1}[A]$ is compact open for each compact open set A.

Observe that $g: X \to Y$ is strongly continuous if and only if the inverse image of a fundamental set is fundamental.

Exercise 3. Show that if g is strongly continuous then g is continuous.

The category of bounded Stone spaces and strongly continuous maps will be denoted by $\mathscr{S}$. We have already seen that if $L \in \text{Ob}\, \mathscr{D}_{01}$ then $\mathscr{S}(L) \in \text{Ob}\, \mathscr{S}$.

Theorem 4. *The assignment* $L \mapsto \mathscr{S}(L)$ *for* $L \in \text{Ob}\,\mathscr{D}_{01}$ *can be extended to a coequivalence* $\mathscr{S}: \mathscr{D}_{01} \to \mathscr{S}$ *by defining for* $f \in [L, L_1]_{\mathscr{D}01}$, $(\mathscr{S}(f))(I) = f^{-1}[I]$ *for each* $I \in \mathscr{S}(L_1)$. *Moreover,*

(1) $(\mathscr{S}(f))^{-1}[\hat{x}] = \widehat{f(x)}$ *for each* $x \in L$.

Proof. Certainly $\mathscr{S}(f)$ is map from $\mathscr{S}(L_1)$ to $\mathscr{S}(L)$, since the inverse image of a prime ideal in L_1 is a prime ideal in L. To verify (1), we have $I \in (\mathscr{S}(f))^{-1}[\hat{x}]$ $\Leftrightarrow \mathscr{S}(f)(I) \in \hat{x} \Leftrightarrow x \notin \mathscr{S}(f)(I) \Leftrightarrow x \notin f^{-1}[I] \Leftrightarrow f(x) \notin I \Leftrightarrow I \in \widehat{f(x)}$. In particular, it follows that $\mathscr{S}(f)$ is strongly continuous. Next we show that $\mathscr{S}$ is a contravariant functor. For $L \in \text{Ob}\,\mathscr{D}_{01}$ we have for $I \in \mathscr{S}(L)$, $\mathscr{S}(1_L)(I) = (1_L)^{-1}(I) = I = 1_{\mathscr{S}(L)}(I)$. So $\mathscr{S}(1_L) = 1_{\mathscr{S}(L)}$. If $f \in [L, L_1]_{\mathscr{D}01}$ $g \in [L_1, L_2]_{\mathscr{D}01}$ then for $I \in \mathscr{S}(L_2)$, $\mathscr{S}(g \circ f)(I)$ $= (g \circ f)^{-1}[I] = f^{-1}[g^{-1}[I]] = f^{-1}[\mathscr{S}(g)(I)] = \mathscr{S}(f)(\mathscr{S}(g)(I))$. We now show that $\mathscr{S}$ is a coequivalence. Let $X \in \text{Ob}\,\mathscr{S}$. Since X is a bounded Stone space, $\mathscr{T}(X) \in \text{Ob}\,\mathscr{D}_{01}$ and by Theorem 1.9, $\mathscr{S}(\mathscr{T}(X))$ is homeomorphic with X, which in categorical terms means $\mathscr{S}(\mathscr{T}(X))$ is isomorphic with X. Finally, let $L, L_1 \in \text{Ob}\,\mathscr{D}_{01}$. We first show that the function $[L, L_1]_{\mathscr{D}01} \to [\mathscr{S}(L_1), \mathscr{S}(L)]_{\mathscr{S}}$ induced by $\mathscr{S}$ is one-one. Let $f, g \in [L, L_1]_{\mathscr{D}01}, f \neq g$. Then there exists $x \in L$ such that, say $f(x) \nleq g(x)$. Thus there is a prime ideal I of L_1 with $g(x) \in I, f(x) \notin I$. Hence $\mathscr{S}(f)(I) \neq \mathscr{S}(g)(I)$ and so $\mathscr{S}(f) \neq \mathscr{S}(g)$. To prove that the induced map $[L, L_1]_{\mathscr{D}01} \to [\mathscr{S}(L_1), \mathscr{S}(L)]_{\mathscr{S}}$ is onto, let $g \in [\mathscr{S}(L_1), \mathscr{S}(L)]_{\mathscr{S}}$. Now, for each $x \in L$, $\hat{x}$ is fundamental and, since g is strongly continuous, $g^{-1}[\hat{x}]$ is fundamental so $g^{-1}[\hat{x}] = \widehat{y_x}$ for some $y_x \in L_1$. Note that y_x is uniquely determined. Define $f: L \to L_1$ by $f(x) = y_x$. Then $f \in [L, L_1]_{\mathscr{D}01}$. Indeed, $\widehat{f(u+v)} = \widehat{y_{u+v}} = g^{-1}[\widehat{u+v}] = g^{-1}[\hat{u} \cup \hat{v}] = g^{-1}[\hat{u}] \cup g^{-1}[\hat{v}] = \widehat{y_u} \cup \widehat{y_v}$ $= \widehat{f(u)} \cup \widehat{f(v)} = \widehat{f(u) + f(v)}$. Thus, $f(u+v) = f(u) + f(v)$. Similarly, $f(uv) = f(u)f(v)$. Further $\widehat{f(0)} = \widehat{y_0} = g^{-1}[\hat{0}] = g^{-1}[\varnothing] = \varnothing = \hat{0}$, so $f(0) = 0$. Similarly $f(1) = 1$. It remains to show that $\mathscr{S}(f) = g$. Let $I \in \mathscr{S}(L_1)$. Then, to prove $\mathscr{S}(f)(I) = g(I)$ we have $x \in g(I) \Leftrightarrow g(I) \notin \hat{x} \Leftrightarrow I \notin g^{-1}[\hat{x}] \Leftrightarrow I \notin \widehat{f(x)} \Leftrightarrow f(x) \in I \Leftrightarrow x \in f^{-1}[I]$ $\Leftrightarrow x \in \mathscr{S}(f)(I)$. Thus $\mathscr{S}(f)(I) = g(I)$. This completes the proof of the theorem.

Exercise 5. Extend the assignment $\mathscr{T}: \text{Ob}\,\mathscr{S} \to \mathbf{D}_{01}$ to a coequivalence.

Theorem 6. *Let* $f \in [L, L_1]_{\mathscr{D}01}$.

(i) *f is one-one $\Leftrightarrow$ f is a monomorphism $\Leftrightarrow$ $\mathscr{S}(f)$ is an epimorphism $\Leftrightarrow$ $\mathscr{S}(f)$ is onto.*

(ii) *f is an epimorphism $\Leftrightarrow$ $\mathscr{S}(f)$ is one-one $\Leftrightarrow$ $\mathscr{S}(f)$ is a monomorphism.*

(iii) *f is onto $\Leftrightarrow$ $\mathscr{S}(f)$ (with codomain restricted to $\mathscr{S}(f)[\mathscr{S}(L_1)]$) is a homeomorphism.*

Remark. The key question as to which epimorphisms are onto must wait until Section V.8.

Proof. (i) Since $\mathscr{D}_{01}$ is an equational category, monic and one-one mean the same. Assume f is one-one. To show $\mathscr{S}(f)$ is onto, let $I \in \mathscr{S}(L)$. Then $f[I]$ is a prime ideal in $f[L]$ and since $f[L]$ is a sublattice of L_1, by Theorem III.6.5, there exists a prime ideal J in L_1 such that $J \cap f[L] = f[I]$. We show $\mathscr{S}(f)(J) = I$. Indeed, $x \in \mathscr{S}(f)(J) \Leftrightarrow x \in f^{-1}[J] \Leftrightarrow f(x) \in J \cap f(L) \Leftrightarrow f(x) \in f(I) \Leftrightarrow x \in I$. It will now suffice to prove that if $\mathscr{S}(f)$ is epic then f is one-one. But if f were not one-one, then it would not be monic and hence $\mathscr{S}(f)$ could not be epic. (ii) It is enough to prove that if f is epic then $\mathscr{S}(f)$ is one one. Now f epic implies $\mathscr{S}(f)$ is monic. If $\mathscr{S}(f)$ is not one-one then there exist $I, J \in \mathscr{S}(L_1)$, $I \neq J$ such that $\mathscr{S}(f)(I) = \mathscr{S}(f)(J)$. Let $\{x\}$ be a one-element space and $g, h: \{x\} \to \mathscr{S}(L_1)$ be defined by $g(x) = I$, $h(x) = J$. Then $(\mathscr{S}(f)) \circ g = (\mathscr{S}(f)) \circ h$ so $g = h$, a contradiction. (iii) Suppose f is onto. Then f is epic so $\mathscr{S}(f)$ is one-one. Thus it suffices to show that $\mathscr{S}(f)$ with codomain restricted to $\mathscr{S}(f)[\mathscr{S}(L_1)]$, is an open map. But it follows immediately from Theorem 4 that $\mathscr{S}(f)[\widehat{f(x)}] = \mathscr{S}(f)[\mathscr{S}(L_1)] \cap \hat{x}$ for each $x \in L$. Conversely, to prove that f is onto, let $y \in L_1$. Since $\mathscr{S}(f)$ is open, $\mathscr{S}(f)[\hat{y}] = \mathscr{S}(f)[\mathscr{S}(L_1)] \cap A$ where A is open in $\mathscr{S}(L)$. Hence $\hat{y} = \mathscr{S}(f))^{-1}[A]$. Now $A = \bigcup_{s \in S} \widehat{x_s}$ for some $(x_s)_{s \in S} \subseteq L$ and so $\hat{y} = \bigcup_{s \in S} ((\mathscr{S}(f))^{-1})[\widehat{x_s}] = \bigcup_{s \in S} \widehat{f(x_s)}$. Hence there exists $S' \subseteq S$ such that $\hat{y} = \bigcup_{s \in S'} \widehat{f(x_s)}$ and so $y = f(\sum_{s \in S'} x_s)$.

Corollary 7. *If $f \in [L, L_1]_{\mathscr{D}_{01}}$ then $\mathscr{S}(\mathrm{Im} f)$ is homeomorphic with* $\mathrm{Im}(\mathscr{S}(f))$. *In particular, if $f' : X \to Y$ is a strongly continuous map between bounded Stone spaces then* $\mathrm{Im} f'$ *is a bounded Stone space.*

Proof. Let $g: L \to \mathrm{Im} f$ be the onto map induced by f. Then $\mathscr{S}(g) : \mathscr{S}(\mathrm{Im} f) \to \mathscr{S}(L)$ and thus by Theorem 6(iii), $\mathscr{S}(g)$ is a homeomorphism of $\mathscr{S}(\mathrm{Im} f)$ onto $\mathrm{Im}\, \mathscr{S}(g)$. Obviously $\mathrm{Im}\, \mathscr{S}(g) = \mathrm{Im}\, \mathscr{S}(f)$.

Exercise 8. Let $(L_i)_{i=1,\cdots,n}$ be a family of objects in $\mathscr{D}_{01}$ and let $L = \times_{i=1}^{n} L_i$. Prove that $\mathscr{S}(L)$ is the topological union (= disjoint union) of the family $(\mathscr{S}(L_i))_{i=1,\cdots,n}$.

3. Topological representation theory for $\mathscr{B}$

Definition 1. A *Boolean space* is a zero dimensional (i.e. the sets which are both open and closed form a basis for the open sets), compact Hausdorff space.

Boolean spaces arise from Boolean algebras in the same way that Stone spaces arise from distributive lattices. However, Boolean spaces have been studied more extensively and yield a good deal of information about Boolean algebras. In this section we outline some of the basic techniques for applying the theory of Boolean spaces to Boolean algebras.

To begin, let $L \in \mathrm{Ob}\, \mathscr{B}$; then $L \in \mathrm{Ob}\, \mathscr{D}_{01}$ and $\mathscr{S}(L)$ is a Boolean space. Indeed, that $\mathscr{S}(L)$ is compact and Hausdorff follows from Theorems 1.10 and 1.12. Moreover, in a compact Hausdorff space, a set is compact open, if and only if it is open and closed. Hence $\mathscr{S}(L)$ is zero dimensional.

Lemma 2. *If X is a Boolean space then X is a bounded Stone space and $\mathscr{T}(X) \in \mathscr{B}$.*

Proof. To prove that X is a bounded Stone space, we only need to verify Definition 1.1(iii). Let $(X_s)_{s\in S}$, $(Y_t)_{t\in T}$ be non-empty families of compact open sets with $\bigcap_{s\in S} X_s \subseteq \bigcup_{t\in T} Y_t$. Since these sets are open and closed, $\bigcap_{s\in S} X_s$ is closed and therefore compact. Hence, there exists $T' \subseteq T$ such that $\bigcap_{s\in S} X_s \subseteq \bigcup_{t\in T'} Y_t$. Again, $\bigcup_{s\in S} \widetilde{X_s} \supseteq \bigcap_{t\in T'} \widetilde{Y_t}$ and $\bigcap_{t\in T'} \widetilde{Y_t}$ is closed so there exists $S' \subseteq S$ such that $\bigcup_{s\in S'} \widetilde{X_s} \supseteq \bigcap_{t\in T'} \widetilde{Y_t}$ and thus, $\bigcap_{s\in S'} X_s \subseteq \bigcup_{t\in T'} Y_t$. Clearly $\mathscr{T}(X) \in \mathrm{Ob}\,\mathscr{B}$.

Notice that for Boolean spaces the notions of strongly continuous and continuous coincide. Also, a continuous map between Boolean spaces is closed, since a continuous map from a compact space to a Hausdorff space is closed.

Exercise 3. Prove that a closed subspace of a Boolean space is a Boolean space.

We now denote the category of Boolean spaces and continuous maps by $\mathscr{S}_{\mathscr{B}}$. It follows that $\mathscr{S}_{\mathscr{B}}$ is a full subcategory of $\mathscr{S}$. Next, define a contravariant functor $\mathscr{S}_{\mathscr{B}}: \mathscr{B} \to \mathscr{S}_{\mathscr{B}}$ by $\mathscr{S}_{\mathscr{B}}(L) = \mathscr{S}(L)$ for $L \in \mathrm{Ob}\,\mathscr{B}$ and $\mathscr{S}_{\mathscr{B}}(f) = \mathscr{S}(f)$ for $f \in [L, L_1]_{\mathscr{B}}$. The following corollary is an immediate consequence of Theorem 2.4 and Lemma 2.

Corollary 4. *The contravariant functor $\mathscr{S}_{\mathscr{B}}: \mathscr{B} \to \mathscr{S}_{\mathscr{B}}$ is a coequivalence.*

Exercise 5. Prove that for $L \in \mathrm{Ob}\,\mathscr{B}$, $\mathscr{S}_{\mathscr{B}}(L)$ is discrete if and only if L is finite.

The analogue of Theorem 2.6 is:

Theorem 6. *Let $f \in [L, L_1]_{\mathscr{B}}$*

(i) f is one-one $\Leftrightarrow f$ is monic $\Leftrightarrow \mathscr{S}_{\mathscr{B}}(f)$ is epic $\Leftrightarrow \mathscr{S}_{\mathscr{B}}(f)$ is onto.

(ii) f is onto $\Leftrightarrow f$ is epic $\Leftrightarrow \mathscr{S}_{\mathscr{B}}(f)$ is monic $\Leftrightarrow \mathscr{S}_{\mathscr{B}}(f)$ is one-one $\Leftrightarrow \mathscr{S}_{\mathscr{B}}(f)$, with codomain restricted to $\mathscr{S}_{\mathscr{B}}(f)[\mathscr{S}_{\mathscr{B}}(L_1)]$, is a homeomorphism.

Proof. The proof of (i) is immediate from Theorem 2.6. For (ii), first note that if $\mathscr{S}_{\mathscr{B}}(f)$ is one-one, then $\mathscr{S}_{\mathscr{B}}(f): \mathscr{S}_{\mathscr{B}}(L_1) \to \mathscr{S}_{\mathscr{B}}(f)[\mathscr{S}_{\mathscr{B}}(L_1)]$ is a homeomorphism. Indeed, this map is onto, one-one and continuous and such a map from a compact space to a Hausdorff space is a homeomorphism. (ii) now follows from Theorem 2.6(iii), from the coequivalence between $\mathscr{S}_{\mathscr{B}}$ and $\mathscr{B}$ and from the fact that the one-point spaces are in $\mathrm{Ob}\,\mathscr{S}_{\mathscr{B}}$.

Exercise 7. Let L_1, L_2, L_2' be finite Boolean algebras, L_1 a subalgebra of L_2, $f \in [L_2, L_2']_{\mathscr{B}}$, f onto, and let $f' = f|L_1$, $L_1' = \mathrm{Im}\, f'$. Prove that if $g' \in [L_1', L_1]_{\mathscr{B}}$ and $f' \circ g' = 1_{L_1'}$ then there exists $g \in [L_2', L_2]_{\mathscr{B}}$ such that $f \circ g = 1_{L_2'}$ and $g|L_1' = g'$. (Hint: Use the coequivalence between $\mathscr{B}$ and $\mathscr{S}_{\mathscr{B}}$ and Theorem 6 and note that all the representation spaces are discrete. The construction of the desired map is then a trivial matter.)

It follows easily from the results obtained so far, that if $L \in \text{Ob}\,\mathscr{B}$ and L' is a *proper* subalgebra of L, then there exist two distinct prime ideals I_1 and I_2 of L such that $I_1 \cap L' = I_2 \cap L'$. Indeed, by Theorem 6, $\mathscr{S}_{\mathscr{B}}(1_{L',L})$ is onto, but not one-one. Thus, there exist two prime ideals I_1 and I_2 of L, $I_1 \neq I_2$ with $\mathscr{S}_{\mathscr{B}}(1_{L',L})(I_1) = \mathscr{S}_{\mathscr{B}}(1_{L',L})(I_2)$. It follows from the definition of $\mathscr{S}_{\mathscr{B}}(f)$ (cf. Theorem 2.4) that I_1 and I_2 are the desired prime ideals. However, a stronger result can be proved.

Theorem 8. (D. Makinson [1]) *Suppose $L \in \text{Ob}\,\mathscr{B}$ and that L' is a subalgebra of L. If $a \in L \sim L'$ then there exists prime ideals I_1 and I_2 of L with $I_1 \cap L' = I_2 \cap L'$ and $a \in I_2$, $a \notin I_1$.*

Proof. By Theorem 6, $\mathscr{S}_{\mathscr{B}}(1_{L',L}) : \mathscr{S}_{\mathscr{B}}(L) \to \mathscr{S}_{\mathscr{B}}(L')$ is onto. Let $\mathscr{S}_{\mathscr{B}}(1_{L',L})[\hat{a}] = U$, then U is compact, hence closed. Suppose $(\mathscr{S}_{\mathscr{B}}(1_{L',L}))^{-1}[U] = \hat{a}$, then (because $\mathscr{S}_{\mathscr{B}}(1_{L',L})$ is onto) $\mathscr{S}_{\mathscr{B}}(1_{L',L})[\hat{\bar{a}}] = \mathscr{S}_{\mathscr{B}}[L'] \sim U$. But then U is open and closed and thus $U = \hat{b}$ for some $b \in L'$. It then follows from Theorem 2.4 that $1_{L',L}(b) = a$. Contradiction. Thus $(\mathscr{S}_{\mathscr{B}}(1_{L',L}))^{-1}[U] \supset \hat{a}$.

So there exists $I_2 \in (\mathscr{S}_{\mathscr{B}}(1_{L',L}))^{-1}[U]$, $I_2 \notin \hat{a}$. But then $(\mathscr{S}_{\mathscr{B}}(1_{L',L}))(I_2) \in U = \mathscr{S}_{\mathscr{B}}(1_{L',L})[\hat{a}]$ so there exists $I_1 \in \hat{a}$ such that $\mathscr{S}_{\mathscr{B}}(1_{L',L})(I_2) = \mathscr{S}_{\mathscr{B}}(1_{L',L})(I_1)$. I_1 and I_2 are the desired prime ideals.

As an application we prove:

Theorem 9. *Suppose $L \in \text{Ob}\,\mathscr{B}$ and L is infinite. Then $|\mathscr{S}_{\mathscr{B}}(L)| \geq |L|$.*

Proof. Let $S = \{(I_1, I_2) : (I_1, I_2) \in \mathscr{S}_{\mathscr{B}}(L) \times \mathscr{S}_{\mathscr{B}}(L), I_1 \neq I_2\}$. Then there exists a function $f: S \to L$ with $f(I_1, I_2) \in I_1, f(I_1, I_2) \notin I_2$ for $(I_1, I_2) \in S$. Let $L' = [\text{Im}\,f]_{\mathbf{B}}$. Now, suppose $|\mathscr{S}_{\mathscr{B}}(L)| < |L|$, then $|S| < |L|$ (since L is infinite) and thus $|L'| < |L|$ and therefore L' is a proper subalgebra of L. By Theorem 8, there exist prime ideals I_1 and I_2 of L with $I_1 \cap L' = I_2 \cap L'$ and $I_1 \neq I_2$. But $f(I_1, I_2) \in I_1, f(I_1, I_2) \notin I_2$ and $f(I_1, I_2) \in L'$, a contradiction.

Remark. For the proof, we actually did not need Theorem 8, but only the weaker version as formulated prior to Theorem 8.

Exercise 10. (D. Makinson [1]) Give an algebraic proof of Theorem 8.

Exercise 11. (L. Iturrioz and D. Makinson [1]) Let $L \in \text{Ob}\,\mathscr{D}_{01}$ and L' be a $\mathbf{D}_{01}$-subalgebra of L. Suppose $a \in L \sim L'$. Then there exist prime ideals I_1 and I_2 of L with $I_1 \cap L' \subseteq I_2 \cap L'$ and $a \in I_1$, $a \notin I_2$.

Exercise 12. (cf. Theorem V.4.12) Let $L \in \text{Ob}\,\mathscr{D}_{01}$, L infinite. Then $|\mathscr{S}(L)| \geq |L|$.

Exercise 13. Let $L, L_1 \in \text{Ob}\,\mathscr{B}$ and let L_1 be a $\mathbf{B}$-subalgebra of L. Let I be an ideal of L. Let I^* and $(L_1 \cap I)^*$ be the open subsets in $\mathscr{S}_{\mathscr{B}}(L)$ and $\mathscr{S}_{\mathscr{B}}(L_1)$, corresponding with I and $L_1 \cap I$, respectively (cf. Exercise 1.13). Prove: $\widetilde{(L_1 \cap I)^*} = \mathcal{S}_{\mathcal{B}}(1_{L_1,L})[\widetilde{I^*}]$

Exercise 14. (P. D. Bacsich [1]) Let $L, L_1 \in \text{Ob}\,\mathcal{B}$ and let L_1 be a **B**-subalgebra of L. Let I be a prime ideal in L_1, J an ideal in L such that $J \cap L_1 \subseteq I$. Prove that there exists a prime ideal J' in L such that $J \subseteq J'$ and $J' \cap L_1 = I$. (Hint: Use the topological representation and Exercise 13.)

Exercise 15. Suppose L is a Boolean algebra, $A \subseteq L$ and $\sum A$ exists. Prove that $\text{Cl}(\bigcup_{a \in A} \hat{a}) = \widehat{\sum A}$.

V

Extension Properties

1. Subalgebras generated by sets

It will be convenient to introduce the following notation. For a subset S of a Boolean algebra, set $S^- = \{\bar{a} : a \in S\}$. Recall also that if $\mathbf{K}$ is class of algebras of a certain similarity type and $S \subseteq A$, where $A \in \mathbf{K}$, then $[S]_{\mathbf{K}}$ or simply $[S]$ denotes the algebra $\mathbf{K}$-generated by S.

Theorem 1. *Let $L_1 \in \mathbf{D}$, $L_2 \in \mathbf{D}_{01}$ and $L_3 \in \mathbf{B}$. If S_i is a non-empty subset of L_i, $1 \leq i \leq 3$, then*

$$\text{(i) } [S_1]_{\mathbf{D}} = \left\{\sum_{i=1}^{n} (\textstyle\prod T_i) : T_i \Subset S_1, n \geq 1\right\}$$

$$= \left\{\prod_{j=1}^{m} (\textstyle\sum R_j) : R_j \Subset S_1, m \geq 1\right\}.$$

$$\text{(ii) } [S_2]_{\mathbf{D}_{01}} = \left\{\sum_{i=1}^{n} (\textstyle\prod T_i) : T_i \Subset S_2 \cup \{0, 1\}, n \geq 1\right\}$$

$$= \left\{\prod_{j=1}^{m} (\textstyle\sum R_j) : R_j \Subset S_2 \cup \{0, 1\}, m \geq 1\right\}.$$

$$\text{(iii) } [S_3]_{\mathbf{B}} = \left\{\sum_{i=1}^{n} (\textstyle\prod T_i) : T_i \Subset S_3 \cup S_3^-, n \geq 1\right\}$$

$$= \left\{\prod_{j=1}^{m} (\textstyle\sum R_j) : R_j \Subset S_3 \cup S_3^-, m \geq 1\right\}.$$

Proof. We will restrict our attention to the first equality in each case, the others following by duality. Now, since any subalgebra of L_i that contains S_i also contains the set on the right of the first equality (call it X_i), it is sufficient to prove that X_i is closed under the operations. It is clear that each X_i is closed under $+$. That each is closed under $\cdot$ follows from the formula

$$\left(\sum_{i=1}^{n} (\textstyle\prod T_i)\right)\left(\sum_{j=1}^{m} (\textstyle\prod T'_j)\right) = \sum_{\substack{1 \leq i \leq n \\ 1 \leq j \leq m}} \textstyle\prod(T_i \cup T'_j).$$

Thus, the equality for (i) is verified, and since X_2 contains 0, 1, the second is also established. For (iii), first note that $S_3 \neq \varnothing$ implies that there exists $x \in S_3$ and

$\bar{x} \in S_3^-$ so $1 = x + \bar{x} \in X_3$ and $0 = x\bar{x} \in X_3$. Finally, by de Morgan's Law, and the dual of II.5.8

$$\overline{\sum_{i=1}^{n} (\prod T_i)} = \prod_{i=1}^{n} \overline{(\prod T_i)} = \prod_{i=1}^{n} (\sum T_i^-) = \sum \left\{ \prod_{i=1}^{n} f(i) : f \in \bigtimes_{i=1}^{n} T_i^- \right\} \in X_3,$$

so X_3 is also closed under complementation, and hence the final equality holds.

Exercise 2. Suppose L is a Boolean algebra and $L = [S]_{\mathbf{B}}$, $S \neq \varnothing$. Then, considered as a member of $\mathbf{D}$, prove that $L = [S \cup S^-]_{\mathbf{D}}$.

Exercise 3. Obtain a result, analogous to Theorem 1 for generalized Boolean algebras.

2. Extending functions to homomorphisms

With the aid of Theorem 1.1 we obtain simple conditions under which functions can be extended to homomorphisms. The result will be referred to as the *extension theorem.*

Theorem 1. *Let* $\mathbf{K}$ *be one of the classes* $\mathbf{D}$, $\mathbf{D}_{01}$ *or* $\mathbf{B}$. *Suppose* $L, L_1 \in \mathbf{K}$ *and* S *is a non-empty subset of* L. *A function* $f: S \to L_1$ *can be extended to a* $\mathbf{K}$*-homomorphism* $g: [S]_{\mathbf{K}} \to L_1$ *if and only if*:

(i) *In* $\mathbf{D}$: $\prod T_1 \leq \sum T_2 \Rightarrow \prod f[T_1] \leq \sum f[T_2]$ *whenever* $T_1 \subseteq S$ *and* $T_2 \subseteq S$.

(ii) *In* $\mathbf{D}_{01}$: $\prod T_1 \leq \sum T_2 \Rightarrow \prod f[T_1] \leq \sum f[T_2]$ *whenever* $T_1 \cup T_2 \subseteq S$.

(iii) *In* $\mathbf{B}$: $(\prod T_1)(\prod T_2^-) = 0 \Rightarrow) (\prod f[T_1])(\prod (f[T_2])^-) = 0$ *whenever* $T_1 \cup T_2 \subseteq S$.

Remark. Note the cases where $T_1 = \varnothing$ or $T_2 = \varnothing$; for example if $T_2 = \varnothing$ then (ii) reduces to $\prod T_1 = 0 \Rightarrow \prod f[T_1] = 0$ whenever $T_1 \subseteq S$.

Proof. The conditions are clearly necessary in each class. For sufficiency we begin in $\mathbf{D}$. Define $g: [S]_{\mathbf{D}} \to L_1$ as follows. For $S_i \subseteq S$, $1 \leq i \leq n$, $n \geq 1$ define $g(\sum_{i=1}^{n} (\prod S_i)) = \sum_{i=1}^{n} (\prod f[S_i])$. To show that g is well defined, suppose $\sum_{i=1}^{n} (\prod S_i) = \sum_{j=1}^{m} (\prod S_j')$ where $S_i \subseteq S$, $S_j' \subseteq S$, $n, m \geq 1$. Then for each i and each $h \in \bigtimes_{j=1}^{m} S_j'$, $\prod S_i \leq \sum_{j=1}^{m} h(j)$. So by hypothesis, $\prod f[S_i] \leq \sum_{j=1}^{m} f(h(j))$. Thus,

$$\sum_{i=1}^{n} (\prod f[S_i]) \leq \prod \left\{ \sum_{j=1}^{m} f(h(j)) : h \in \bigtimes_{j=1}^{m} S_j' \right\} = \sum_{j=1}^{m} (\prod f[S_j']).$$

Similarly, $\sum_{j=1}^{m} (\prod f[S_j']) \leq \sum_{i=1}^{n} (\prod f[S_i])$, so g is well defined. It is now routine to verify that g is a $\mathbf{D}$-homomorphism. Clearly, $g|S = f$. Next, we consider (ii). If $0 \in S$ then setting $T_1 = \{0\}$ and $T_2 = \varnothing$, the hypothesis of (ii) implies $f(0) = 0$; similarly for 1. In any case we can unambiguously define a function

$f_1 : S \cup \{0,1\} \to L_1$ by $f_1|S = f, f_1(0) = 0, f_1(1) = 1$. But then (ii) can be reformulated as follows

$$\prod T_1 \leq \sum T_2 \Rightarrow \prod f_1[T_1] \leq \sum f_1[T_2] \text{ whenever } T_1 \subseteq S \cup \{0, 1\} \text{ and}$$

$T_2 \subseteq S \cup \{0, 1\}$.

So if we consider L as a distributive lattice, (i) implies that there exists a **D**-homomorphism g: $[S \cup \{0, 1\}]_{\mathbf{D}} \to L_1$ such that $g|S \cup \{0, 1\} = f_1$ and in particular $g|S = f$. Now from (i) and (ii) of Theorem 1.1, it is clear that $[S \cup \{0, 1\}]_{\mathbf{D}} = [S]_{\mathbf{D_{01}}}$ (that is, the underlying sets of these algebras are equal) and since $g(0) = 0$ and $g(1) = 1$, g: $[S]_{\mathbf{D_{01}}} \to L_1$ is the required $\mathbf{D_{01}}$-homomorphism. Let us now prove (iii). Define a function f_1: $S \cup S^- \to L_1$ by $f_1(x) = f(x)$ if $x \in S$ and $f_1(y) = \overline{f(\bar{y})}$ if $y \in S^-$. This function is well defined, for if $u \in S \cap S^-$ then letting $T_1 = \{u, \bar{u}\}$, $T_2 = \varnothing$ in (iii) we have $f(u)f(\bar{u}) = 0$ and then letting $T_1 = \varnothing$, $T_2 = \{u, \bar{u}\}$ we obtain $\overline{f(u)}\,\overline{f(\bar{u})} = 0$ so $\overline{f(\bar{u})} = f(u)$. Thus f_1 is well defined. Also (iii) implies $\prod T_1 \leq \sum T_2$ $\Rightarrow \prod f_1[T_1] \leq \sum f_1[T_2]$ whenever $T_1 \cup T_2 \subseteq S \cup S^-$. So by (ii), if L is considered as a distributive lattice with 0, 1 then there exists a $\mathbf{D_{01}}$-homomorphism g: $[S \cup S^-]_{\mathbf{D_{01}}} \to L_1$ such that $g|S \cup S^- = f_1$. Thus, since $[S \cup S^-]_{\mathbf{D_{01}}} = [S]_{\mathbf{B}}$ and a $\mathbf{D_{01}}$-homomorphism between Boolean algebras preserves complements, we see that g is the desired **B**-homomorphism.

Corollary 2. *Let $L, L_1, L_2 \in \mathbf{B}$ with L a subalgebra of L_2, $a \in L_2 \sim L$, and $b \in L_1$. Suppose f: $L \to L_1$ is a homomorphism. Then a necessary and sufficient condition in order that f can be extended to a homomorphism g: $[L \cup \{a\}]_{\mathbf{B}} \to L_1$ with the property $g(a) = b$, is that*

(i) $ua = 0 \Rightarrow f(u)b = 0$ *for all* $u \in L$.

(ii) $u\bar{a} = 0 \Rightarrow f(u)\bar{b} = 0$ *for all* $u \in L$.

Proof. ($\Rightarrow$) Trivial.

($\Leftarrow$) Extend f to $L \cup \{a\}$ by letting $f(a) = b$. Now apply Theorem 1(iii). Suppose $T_1 \cup T_2 \subseteq L \cup \{a\}$ and $(\prod T_1)(\prod T_2^-) = 0$. If $a \notin T_1 \cup T_2$ then $T_1 \cup T_2 \subseteq L$ and since f is a homomorphism, $\prod f[T_1] \cdot \prod f[T_2]^- = 0$. Suppose $a \in T_1 \sim T_2$. Then $a \cdot \prod(T_1 \sim \{a\}) \cdot \prod T_2^- = 0$ so by (i) of the hypothesis $0 = b \cdot f(\prod(T_1 \sim \{a\}) \cdot \prod T_2^-)$ $= f(a) \cdot \prod f[T_1 \sim \{a\}] \cdot \prod f[T_2]^- = \prod f[T_1] \cdot \prod f[T_2]^-$. Now suppose $a \in T_2 \sim T_1$. Then $(\prod T_1)\bar{a} \prod (T_2^- \sim \{\bar{a}\}) = 0$. Thus $0 = \bar{b} \cdot f((\prod T_1) \cdot \prod (T_2^- \sim \{\bar{a}\}))$ $= \prod f[T_1] \cdot \prod f[T_2]^-$. If $a \in T_1 \cap T_2$ then obviously $\prod f[T_1] \cdot \prod f[T_2]^- = 0$.

Exercise 3. Show that in $\mathbf{D_{01}}$ condition (ii) of Theorem 1 can be replaced by (i) and the conditions

$$\prod T_1 = 0 \Rightarrow f[T_1] = 0 \text{ whenever } T_1 \subseteq S$$

$$\sum T_1 = 1 \Rightarrow \sum f[T_1] = 1 \text{ whenever } T_1 \subseteq S.$$

3. Free algebras

With the extension theorem at our disposal, it is now possible to describe precisely those relations which hold among the generators of the free algebras in **D**, $\mathbf{D}_{01}$ and **B**.

Theorem 1. *Let* **K** *be one of the classes* **D**, $\mathbf{D}_{01}$ *or* **B** *and suppose* $A \in \mathbf{K}$ *and* $A = [S]_{\mathbf{K}}$ *for some non-empty subset* $S \subseteq A$. *Then S freely* **K***-generates A if and only if*:

(i) *In* $\mathbf{D}: \prod T_1 \leq \sum T_2 \Rightarrow T_1 \cap T_2 \neq \varnothing$ *whenever* $T_1 \subseteq S$ *and* $T_2 \subseteq S$.

(ii) *In* $\mathbf{D}_{01}: \prod T_1 \leq \sum T_2 \Rightarrow T_1 \cap T_2 \neq \varnothing$ *whenever* $T_1 \cup T_2 \subseteq S$.

(iii) *In* $\mathbf{B}: \prod T_1 \cdot \prod T_2^- = 0 \Rightarrow T_1 \cap T_2 \neq \varnothing$ *whenever* $T_1 \cup T_2 \subseteq S$.

Proof. The conditions are sufficient by Theorem 2.1. Suppose that the necessity of one of the conditions is false. Then there exist sets T_1 and T_2 which violate (i), (ii) or (iii). Define a function $f: S \rightarrow \mathbf{2}$ by $f(u) = \mathbf{1}$ for $u \in T_1$ and $f(u) = \mathbf{0}$ otherwise. Then there exists a **K**-homomorphism (where **K** is the class for which the condition is assumed to be false) $g: A \rightarrow \mathbf{2}$ such that $g|S = f$. Since g preserves products, sums, and inequalities (and 0, 1 in $\mathbf{D}_{01}$ and **B**), the violated inequality, with g applied to both sides, yields $\mathbf{1} \leq \mathbf{0}$, a contradiction.

Exercise 2. Prove that L is free in $\mathbf{D}_{01}$ if and only if $L \cong \mathbf{2}$ or $L \cong \mathbf{1} \oplus L_1 \oplus \mathbf{1}$ where L_1 is free in **D**.

Exercise 3. Obtain the analogue of Theorem 1 for generalized Boolean algebras.

Exercise 4. Prove that the free algebra with generating set $\varnothing$ does not exist over $\mathscr{D}$ (in the sense of Definition I.20.9), but that $\mathscr{F}_{\mathbf{D}}(\varnothing) \cong \mathbf{1}$ and $\mathscr{F}_{\mathbf{D}_{01}}(\varnothing) \cong \mathscr{F}_{\mathbf{B}}(\varnothing) \cong \mathbf{2}$.

Exercise 5. Let **K** be one of the classes **D**, $\mathbf{D}_{01}$ or **B**. Prove that $\mathscr{F}_{\mathbf{K}}(S)$ has no atoms if S is infinite.

Although the existence of free algebras with $\alpha > 0$ free generators in **D**, $\mathbf{D}_{01}$ and **B** is assured by the fact that these classes are equational (cf. Theorem I.12.4), it is easy to give a direct construction. Let X be a set with cardinality $\alpha > 0$. For each $x \in X$, let $A_x = \{S \subseteq X : x \in S\}$. Then the ring of sets **D**-generated by $T = \{A_x : x \in X\}$ is, in fact, freely generated by T and is therefore $\mathscr{F}_{\mathbf{D}}(\alpha)$. Indeed, if $A_{x_1} \cap \cdots \cap A_{x_n} \subseteq A_{y_1} \cap \cdots \cap A_{y_m}$ $n, m \geq 1$ then $\{x_1, \cdots, x_n\} \in A_{x_1} \cap \cdots \cap A_{x_n} \subseteq A_{y_1} \cap \cdots \cap A_{y_m}$ so, say $\{x_1, \cdots, x_n\} \in A_{y_j}$. Then $y_j \in \{x_1, \cdots, x_n\}$ so say $y_j = x_i$ and hence $A_{x_i} = A_{y_j}$ for some i, j. The $\mathbf{D}_{01}$-free algebra on α free generators is obtained by adjoining a 0 and a 1 to the ring just constructed. For **B**, simply take the field of sets **B**-generated by T and observe that T is again a free set of generators.

Free algebras in the above classes have many interesting properties some of which we will now proceed to explore.

Lemma 6. *In the free distributive lattice* $\mathscr{F}_{\mathbf{D}}(X)$ *suppose* $\{S_i : i = 1, \cdots, n\}$ and $\{T_j : j = 1, \cdots, m\}$, $m, n \geq 1$ *are finite non-empty subsets of X. Then*

$$(1)\ \sum_{i=1}^{n} (\prod S_i) \leq \sum_{j=1}^{m} (\prod T_j)$$

if and only if for each $i \in \{1, \cdots, n\}$ *there exists* $j \in \{1, \cdots, m\}$ *such that* $T_j \subseteq S_i$.

Proof. ($\Leftarrow$) For each $i \in \{1, \cdots, n\}$ there exists $j \in \{1, \cdots, m\}$ such that $\prod S_i \leq \prod T_j \leq \sum_{j=1}^{m} (\prod T_j)$ and so (1) holds.

($\Rightarrow$) Suppose there exists $i_0 \in \{1, \cdots, n\}$ such that for each $j \in \{1, \cdots, m\}$ there exists $x_j \in T_j \sim S_{i_0}$. But then $\prod S_{i_0} \leq x_1 + \cdots + x_n$ implies $x_j \in S_{i_0}$ for some $j \in \{1, \cdots, m\}$, a contradiction.

Theorem 7. *For a free distributive lattice* $\mathscr{F}_{\mathbf{D}}(X)$:

(i) $\mathscr{J}(\mathscr{F}_{\mathbf{D}}(X)) = \{\prod T : T \Subset X\}$.

(ii) $\mathscr{M}(\mathscr{F}_{\mathbf{D}}(X)) = \{\sum T : T \Subset X\}$.

(iii) $\mathscr{J}(\mathscr{F}_{\mathbf{D}}(X)) \cap \mathscr{M}(\mathscr{F}_{\mathbf{D}}(X)) = X$.

(iv) *Every element of* $\mathscr{F}_{\mathbf{D}}(X)$ *is a finite sum of members of* $\mathscr{J}(\mathscr{F}_{\mathbf{D}}(X))$ *and also a finite product of members of* $\mathscr{M}(\mathscr{F}_{\mathbf{D}}(X))$.

(v) $\mathscr{J}(\mathscr{F}_{\mathbf{D}}(X))$ *is closed under finite products and* $\mathscr{M}(\mathscr{F}_{\mathbf{D}}(X))$ *is closed under finite sums.*

Proof. We first establish (i): Let $u \in \mathscr{J}(\mathscr{F}_{\mathbf{D}}(X))$ where $u = \sum_{i=1}^{n}(\prod T_i)$, $n \geq 1$ and $T_i \Subset X$. But then $u = \prod T_i$ for some i. Now let $T \Subset X$; $a, b \in \mathscr{F}_{\mathbf{D}}(X)$ and $\prod T \leq a + b$ where $a = \sum_{i=1}^{n} \prod S_i$ and $b = \sum_{i=n+1}^{m} S_i$, $S_i \Subset X$, $1 \leq i \leq m$. Then by Lemma 6, there exists $j \leq m$ such that $S_j \subseteq T$. This means $\prod T \leq \prod S_j$ so $\prod T \leq a$ or $\prod T \leq b$. By duality (ii) holds. From (i) and (ii) it follows that $X \subseteq \mathscr{J}(\mathscr{F}_{\mathbf{D}}(X)) \cap \mathscr{M}(\mathscr{F}_{\mathbf{D}}(X))$. Finally, if $u \in \mathscr{J}(\mathscr{F}_{\mathbf{D}}(X)) \cap \mathscr{M}(\mathscr{F}_{\mathbf{D}}(X))$, then by (i), $u = \prod T$ for some $T \Subset X$ and so, $u = t_o$ for some $t_o \in T$. (iv) and (v) are now evident.

Part (iii) of Theorem 7 shows that the free generators of a free distributive lattice are uniquely determined.

Exercise 8. Prove that, except for **2**, there is always more than one set of free generators of a free Boolean algebra.

Exercise 9. Prove that $\mathscr{F}_{\mathbf{D}}(X)$ is a Boolean algebra if and only if $1 \leq |X| \leq 2$.

Exercise 10. Verify that $\mathscr{F}_{\mathbf{D}}(\{x, y, z\})$ has the diagram:

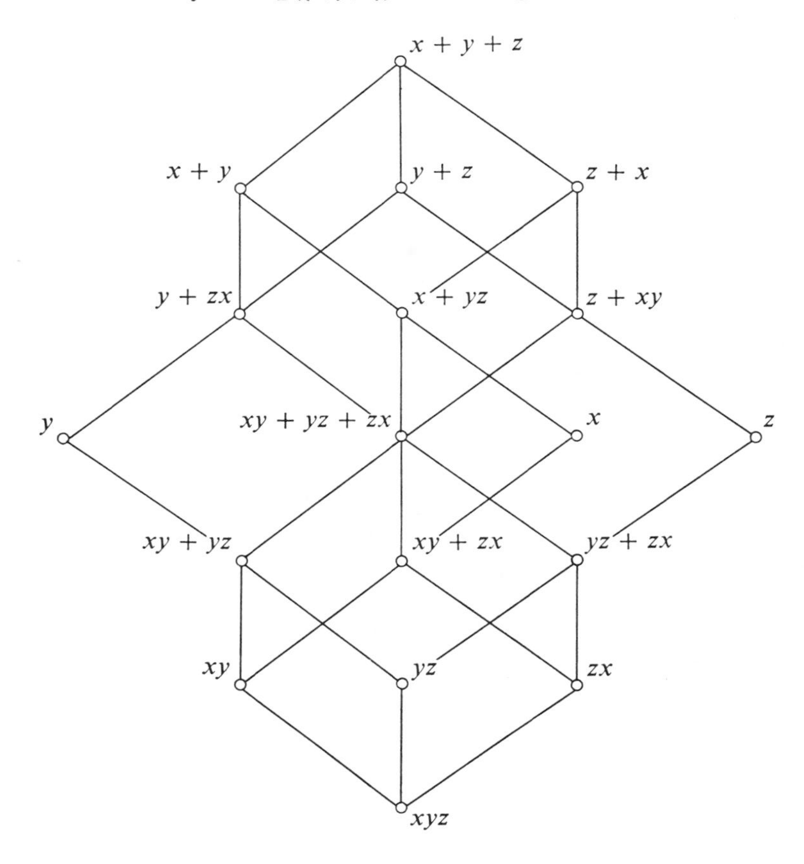

The poset $\mathfrak{P}(L)$ of prime ideals of a distributive lattice L does not in general determine L. The situation for free distributive lattices is now presented.

Lemma 11. *An ideal I in $\mathscr{F}_{\mathbf{D}}(X)$ is prime if and only if $I = (T]$ for some $\varnothing \neq T \subset X$.*

Proof. ($\Rightarrow$) Let $T = I \cap X$. Now $T \neq \varnothing$. Indeed, since $I \neq \varnothing$, there exists $u \in I$ where $u = \prod_{i=1}^{n} \sum T_i$, $T_i \subseteq X$, $n \geq 1$. Since I is prime, $\sum T_i \in I$ for some i and hence $T_i \subseteq I \cap X$. Also $T \neq X$ for otherwise $I = \mathscr{F}_{\mathbf{D}}(X)$. Clearly, $(T] \subseteq I$. Conversely, let $u \in I$; then we have just seen that there exists $T_i \subseteq I \cap X$ with $u \leq \sum T_i$ and thus $u \in (T]$. Therefore $(T] = I$.

($\Leftarrow$) $(T]$ is an ideal. It is proper, for if $x \in X \sim T$ and $x \in (T]$ then $x \leq x_1 + \cdots + x_n$, $x_i \in T$ for $1 \leq i \leq n$, so by Lemma 6, $x = x_i$ for some i and thus $x \in T$, a contradiction. If $ab \in (T]$ then there exists $T_1 \subseteq T$ such that $ab \leq \sum T_1$. By Theorem 7(ii), $\sum T_1 \in \mathscr{M}(\mathscr{F}_{\mathbf{D}}(X))$ so $a \leq \sum T_1$ or $b \leq \sum T_1$. Hence $a \in (T]$ or $b \in (T]$.

Thus we have determined the poset of prime ideals of a free distributive lattice:

Theorem 12. $\mathfrak{P}(\mathscr{F}_{\mathbf{D}}(X)) \cup \{\varnothing, \mathscr{F}_{\mathbf{D}}(X)\}$ *is isomorphic to the power set of X.*

Proof. The assignment $I \mapsto I \cap X$ is a one-one correspondence between the set of prime ideals and the proper non-empty subsets of X.

The fact that $\mathfrak{P}(L)$ determines L when $\mathfrak{P}(L) \cup \{\varnothing, L\}$ is isomorphic to the power set of some set $X \neq \varnothing$ is more difficult and requires some additional lemmas.

Lemma 13. *Let $L \in \mathbf{D}$ and suppose that $P = \mathfrak{P}(L) \cup \{\varnothing, L\}$ is a complete lattice. If $T \subseteq \mathcal{M}(L)$ then $(t] \in \mathfrak{P}(L) \cup \{L\}$ for each $t \in T$, $\sum^P \{(t] : t \in T\} = (\sum T]$ and $\sum T \in \mathcal{M}(L)$.*

Proof. Since $T \subseteq \mathcal{M}(L)$, $(t] \in \mathfrak{P}(L) \cup \{L\}$ for each $t \in T$. If $u \in (\sum T]$ then $u \leq \sum T$ so $t \in (t] \subseteq \sum^P \{(t_1] : t_1 \in T\}$ for each $t \in T$, implying that $\sum T$ and hence u is in $\sum^P \{(t] : t \in T\}$. Conversely, if $u \notin (\sum T]$ then $u \nleq \sum T$ so there exists a prime ideal I such that $\sum T \in I$, $u \notin I$. Thus for each $t \in T$, $t \in I$ so $(t] \subseteq I$ and hence $\sum^P \{(t] : t \in T\} \subseteq I$. Now $u \notin I$ so $u \notin \sum^P \{(t] : t \in T\} \subseteq I$. Since $(\sum T]$ is therefore a prime ideal or all of L, $\sum T$ is meet irreducible.

Lemma 14. *Let $L \in \mathbf{D}$ and suppose $P = \mathfrak{P}(L) \cup \{\varnothing, L\}$ is isomorphic to the power set of a set X, $X \neq \varnothing$. Then I is an atom in P if and only if $I = (a]$ where a is minimal in $\mathcal{M}(L)$.*

Proof. ($\Rightarrow$) Suppose that I has no greatest element. Then for each $u \in I$, there exist $v_u \in I$ and $I(u, v_u) \in \mathfrak{P}(L)$ such that $v_u \nleq u$, $u \in I(u, v_u)$ and $v_u \notin I(u, v_u)$. Let $S = \{I(u, v_u) : u \in I\}$. Since $v_u \in I \sim I(u, v_u)$ we have $I \nsubseteq J$ for all $J \in S$. But I is an atom in P which implies $I \cdot J = 0_P$ for each $J \in S$. Now $I \subseteq \cup S \subseteq \sum^P S$ so by Theorem II.6.9(iv), $I = I \cdot \sum^P S = \sum^P \{I \cdot J : J \in S\} = 0_P$, contradicting the definition of an atom. So I has a greatest element a. It follows that $I = (a]$ and since $I \in P$, $a \in \mathcal{M}(L)$. If $b \leq a$ and $b \in \mathcal{M}(L)$ then $(b] \leq (a]$. But $(a] = I$ is an atom so $(b] = (a]$ and hence $b = a$. Thus, a is minimal in $\mathcal{M}(L)$.

($\Leftarrow$) Suppose a is minimal in $\mathcal{M}(L)$. Then $(a] \in P$. But $\varnothing \neq (a]$ so there is an atom $J \in P$ such that $J \subseteq (a]$. But the converse has shown that there exists $b \in \mathcal{M}(L)$ such that $J = (b]$. Hence $(b] \leq (a]$ and so $b \leq a$. The minimality of a yields $b = a$, so $(a] = (b] = J$ is an atom.

Theorem 15. *If L is a distributive lattice and $\mathfrak{P}(L) \cup \{\varnothing, L\}$ is isomorphic to the power set of some set X, $X \neq \varnothing$, then L is a free distributive lattice on $|X|$ free generators.*

Proof. Let $P = \mathfrak{P}(L) \cup \{\varnothing, L\}$ and let S be the set of minimal elements in $\mathcal{M}(L)$. By Lemma 14, $|S| = |X| > 0$. We first prove that S freely generates $[S]$. Thus, let $T_1 \subseteq S$ and $T_2 \subseteq S$ and $\prod T_1 \leq \sum T_2$. By Lemma 13, $\sum T_2 \in \mathcal{M}(L)$ so there exists $t_1 \in T_1$ such that $t_1 \leq \sum T_2$. Now $(t_1] \subseteq (\sum T_2] = \sum^P \{(t] : t \in T_2\}$ and since $t_1 \in S$, $(t_1]$ is an atom in P so there exists $t_2 \in T_2$ such that $(t_1] \subseteq (t_2]$. The minimality of t_2 and $t_1 \leq t_2$ imply $t_1 = t_2$. By Theorem 1(i), S freely generates $[S]$.

Let $S^\Sigma = \{\sum T : T \subseteq S\}$. Recall from Theorem 7(ii) that $S^\Sigma \subseteq \mathcal{M}(L)$.

Claim. If $I \in \mathfrak{P}(L)$ then $I \cap S^\Sigma \neq \varnothing$ and $\sum^P \{(t] : t \in I \cap S^\Sigma\} = \cup \{(t] : t \in I \cap S^\Sigma\}$: Indeed, since $I \neq \varnothing$, it is a sum of atoms in P. By Lemma 14 there exists $y \in S$ such that $(y] \subseteq I$ so $y \in I \cap S \subseteq I \cap S^{\Sigma}$. For the second part of the claim, it is easily verified that $\cup\{(t] : t \in I \cap S^{\Sigma}\} \in P$.

It remains to prove that $[S] = L$. In fact we will show that each $x \in L$ is a finite product of members of S^{Σ}. The work will be divided into two cases.

First, assume that $S^{\Sigma} \cap [x) \neq \varnothing$. Then

(2) $\hat{x} = \cap\{\hat{y} : y \in S^{\Sigma} \cap [x)\}$.

To see this, let $I \in \hat{x}$. Then $x \notin I$ so $y \in S^{\Sigma} \cap [x)$ implies $y \notin I$ and therefore $I \in \hat{y}$. Conversely, suppose $I \in \cap\{\hat{y} : y \in S^{\Sigma} \cap [x)\} \sim \hat{x}$. Now $I = \sum^{P}\{(t] : t \in S^{\Sigma} \cap I\} = \cup\{(t] : t \in S^{\Sigma} \cap I\}$ and as $x \in I$, $x \in (t]$ for some $t \in S^{\Sigma} \cap I$. So $t \in S^{\Sigma} \cap [x)$ and therefore $I \in \hat{t}$, which is a contradiction. But by (2), $\hat{x} = \hat{y}_1 \cap \cdots \cap \hat{y}_n$, $y_i \in S^{\Sigma}$ and hence $x = y_1 \cdots y_n$, which completes the proof of this case.

Finally, suppose $S^{\Sigma} \cap [x) = \varnothing$. Then $x \nleq t$ for all $t \in S^{\Sigma}$. But

(3) $\hat{x} = \cup\{\hat{s} : s \in S\}$.

Indeed, if $I \notin \hat{x}$ then $x \in I = \cup\{(t] : t \in I \cap S^{\Sigma}\}$ so $x \leq t$ for some $t \in S^{\Sigma}$, a contradiction. If $I \in \hat{x}$ then $x \notin I$ so there is an atom $(s_o]$, $s_o \in S$ such that $(s_o] \nsubseteq I$ and therefore $I \in \cup\{\hat{s} : s \in S\}$. From (3) it follows that $x = s_1 + \cdots + s_n \in S^{\Sigma}$.

Exercise 16. Let L be a distributive lattice. Prove that $L \cong \mathscr{F}_{\mathbf{D}_{01}}(X)$ for some $X \neq \varnothing$ if and only if $\mathfrak{P}(L)$ is isomorphic to a power set of some non-empty set. (Hint: For the "if" part, first show that $L \cong \mathbf{1} \oplus L_1 \oplus \mathbf{1}$ for some distributive lattice L_1; then apply Theorem 15.)

Some of the properties of free Boolean algebras can be easily described by first introducing the notion of "constituent".

Definition 17. Let L be a Boolean algebra and $x \in L$. Define $x^1 = x$, $x^0 = \bar{x}$. For an n-tuple $(x_0, \cdots, x_{n-1}) \in L^n$ and $\sigma \in \mathbf{2}^{\mathbf{n}}$, the element $x_0^{\sigma(0)} \cdots x_{n-1}^{\sigma(n-1)}$ is called a *constituent* of $(x_0, \cdots, x_{n-1})$. (Here $\mathbf{n}$ is considered to be $\{0, 1, \cdots, n-1\}$ and $\mathbf{2} = \{0, 1\}$.)

Lemma 18. *If $n \geq 1$ and $x_0, \cdots, x_{n-1}$ are elements of a Boolean algebra L then* $\sum_{\sigma \in \mathbf{2}^{\mathbf{n}}} x_0^{\sigma(0)} \cdots x_{n-1}^{\sigma(n-1)} = 1$.

Proof. For $n = 1$, the formula reduces to $x_0 + \bar{x}_0 = 1$. Now suppose that the formula is true for $1 \leq k < n$. Then

$$\begin{aligned}
\sum_{\sigma \in \mathbf{2}^{\mathbf{n}}} x_0^{\sigma(0)} \cdots x_{n-1}^{\sigma(n-1)} \\
&= \sum_{\substack{\sigma \in \mathbf{2}^{\mathbf{n}} \\ \sigma(n-1)=0}} x_0^{\sigma(0)} \cdots x_{n-2}^{\sigma(n-2)} \bar{x}_{n-1} + \sum_{\substack{\sigma \in \mathbf{2}^{\mathbf{n}} \\ \sigma(n-1)=1}} x_0^{\sigma(0)} \cdots x_{n-2}^{\sigma(n-2)} x_{n-1} \\
&= \bar{x}_{n-1} \cdot \sum_{\tau \in \mathbf{2}^{\mathbf{n-1}}} x_0^{\tau(0)} \cdots x_{n-2}^{\tau(n-2)} + x_{n-1} \cdot \sum_{\tau \in \mathbf{2}^{\mathbf{n-1}}} x_0^{\tau(0)} \cdots x_{n-2}^{\tau(n-2)} \\
&= \bar{x}_{n-1} + x_{n-1} = 1.
\end{aligned}$$

With the above notation, we see that if S generates a Boolean algebra L, then each member of L is a finite sum of constituents of n-tuples of S.

Corollary 19. *If L is a Boolean algebra and $L = [\{x_0, \cdots, x_{n-1}\}]$, $n \geq 1$, then each member of L is a sum of constituents of $(x_0, \cdots, x_{n-1})$.*

Remark. Thus, in the finite case, we need not deal with constituents of m-tuples of $\{x_0, \cdots, x_{n-1}\}$, $m \leq n$, but only with the constituents of $(x_0, \cdots, x_{n-1})$ itself.

Proof of Corollary 19. Let $a \in L$. Then $a = \sum_{i=1}^{m} (\prod S_i)$ where $S_i \subseteq \{x_0, \cdots, x_{n-1}\} \cup \{\bar{x}_0, \cdots, \bar{x}_{n-1}\}$ so by Lemma 18, $a = \sum_{i=1}^{m} \sum_{\sigma \in \mathbf{2^n}} (\prod S_i) x_0^{\sigma(0)} \cdots x_{n-1}^{\sigma(n-1)}$ and clearly each summand $(\prod S_i) x_0^{\sigma(0)} \cdots x_{n-1}^{\sigma(n-1)}$ is either 0 or $x_0^{\sigma(0)} \cdots x_{n-1}^{\sigma(n-1)}$.

As a first application, we give another proof of Corollary III.3.5. Consider the free algebra $\mathscr{F}_{\mathbf{B}}(S)$ where $S = \{x_0, \cdots, x_{n-1}\}$, $n \geq 1$. We know from Corollary III.2.6 that a finite Boolean algebra has 2^m elements, where m is the number of atoms, so it suffices to prove that $(x_0^{\sigma(0)} \cdots x_{n-1}^{\sigma(n-1)})_{\sigma \in \mathbf{2^n}}$ is a family of distinct atoms of $\mathscr{F}_{\mathbf{B}}(S)$. If $\sigma, \tau \in \mathbf{2^n}$ and $\sigma \neq \tau$ then $x_0^{\sigma(0)} \cdots x_{n-1}^{\sigma(n-1)} \nleq x_0^{\tau(0)} \cdots x_{n-1}^{\tau(n-1)}$. Otherwise, if $\sigma(i) \neq \tau(i)$, then $\tau(i) = 1 - \sigma(i)$ and so $x_0^{\sigma(0)} \cdots x_i^{\sigma(i)} \cdots x_{n-1}^{\sigma(n-1)} \leq x_0^{\tau(0)} \cdots x_n^{\tau(n-1)}$ $\leq x_i^{\tau(i)} = x_1^{1-\sigma(i)}$ and so $x_0^{\sigma(0)} \cdots x_i^{\sigma(i)} \cdots x_{n-1}^{\sigma(n-1)} = 0$, a contradiction. This shows that all 2^n constituents of $(x_0, \cdots, x_{n-1})$ are distinct and that they are atoms. Finally, all atoms are constituents, since atoms are join irreducible.

Definition 20. Let a be an element in a lattice L. A non-empty subset $S \subseteq L$ is *a-disjoint* provided that $xy = a$ whenever x, y are distinct elements of S. S is called *disjoint* if L has a 0 and S is 0-disjoint. A lattice which has no uncountable disjoint subsets is said to satisfy the *countable chain condition.*

The countable chain condition derives its name from the fact that in a complete Boolean algebra L, the countable chain condition holds if and only if every well ordered chain in L is countable (see P. R. Halmos [3]). These conditions are not, in general, equivalent (see Exercises 22 and 23 below). However, in the course of this section we will show that both conditions holds in $\mathscr{F}_{\mathbf{B}}(X)$ and $\mathscr{F}_{\mathbf{D}}(X)$.

Exercise 21. Show that the atoms of a distributive lattice with 0 form a disjoint set.

Exercise 22. Let X be a countable infinite set. Show that $\mathbf{2^X}$ satisfies the countable chain condition but contains a chain isomorphic with the reals.

Exercise 23. Let Y be a set such that $|Y| = 2^{\aleph_0}$ and let L be the Boolean algebra of finite and cofinite subsets of Y (see Example II.6.7). Show that every chain in L is countable but that L does not satisfy the countable chain condition.

Theorem 24. *Every free Boolean algebra satisfies the countable chain condition.*

Proof. Consider the free Boolean algebra $\mathscr{F}_{\mathbf{B}}(X)$. Since the set consisting of one summand from each member of a disjoint set is also disjoint, we need only show that every disjoint set of constituents of n-tuples in X, for arbitrary n, is countable. Thus, let S be a disjoint set of constituents of n-tuples, $n = 1, 2, \cdots$ of X. Let $S_n = S \cap \{x_0^{\sigma(0)} \cdots x_{n-1}^{\sigma(n-1)} : \sigma \in \mathbf{2^n}, x_i \in X\}$. Now S_1 has at most two members for if $\sigma, \tau \in \mathbf{2^1}$ and $\{x_0, y_0\} \subseteq X$ then $x_0^{\sigma(0)} \cdot y_0^{\tau(0)} = 0$ implies $x_0\bar{y}_0 = 0$ or $\bar{x}_0 y_0 = 0$ but in either case, since X freely generates $\mathscr{F}_{\mathbf{B}}(X)$, we have $x_0 = y_0$. Proceeding by induction, suppose $S_1, \cdots, S_n$ are finite. Choose an element $x = x_o^{\sigma_0(0)} \cdots x_n^{\sigma_0(n)}$ in S_{n+1} and for each $i < n + 1$ let

$$S_{n+1}^i = \{x_0^{\tau(0)} \cdots x_{i-1}^{\tau(i-1)} \cdot x_i^{1-\sigma_0(i)} \cdot x_{i+1}^{\tau(i+1)} \cdots x_n^{\tau(n)} : \tau \in \mathbf{2^n}, x_i \in X\} \cap S.$$

Now $S_{n+1}^i \subseteq S$ is a disjoint set and therefore so is the set T_{n+1}^i which is obtained from S_{n+1}^i by deleting the factor $x_i^{1-\sigma_0(i)}$ from each element of S_{n+1}^i.
By induction, we can assume that T_{n+1}^i is finite and therefore S_{n+1}^i is also finite. Now each member $S_{n+1} \sim \{x\}$ is disjoint with x and is therefore in $\bigcup_{i=0}^n S_{n+1}^i$. It follows that $S_{n+1} = (\bigcup_{i=0}^n S_{n+1}^i) \cup \{x\}$ is finite and we conclude that $S = \bigcup_{n<\infty} S_n$ is countable.

Corollary 25. *If L is a free Boolean algebra, then for any $a \in L$, every a-disjoint set is countable.*

Proof. Let $a \in L$ and suppose S is an a-disjoint subset. Then $\{s\bar{a} : s \in S\}$ is disjoint for, if $s_1, s_2 \in S$ and $s_1\bar{a}, s_2\bar{a}$ are distinct, then $s_1 \neq s_2$. Thus $s_1 s_2 = a$ and $(s_1\bar{a})(s_2\bar{a}) = s_1 s_2 \bar{a} = a\bar{a} = 0$. By Theorem 24, $\{s\bar{a} : s \in S\}$ is countable. The proof will be completed by showing that $s_1\bar{a} \neq s_2\bar{a}$ for distinct members $s_1, s_2 \in S$. Now, if s_1, s_2 are distinct in S and $s_1\bar{a} = s_2\bar{a}$ then $s_1\bar{a} \leq s_2$ so $s_1 \leq a + s_2$. Thus, $s_1 = s_1(a + s_2) = s_1 a + s_1 s_2 = s_1 a + a = a$. Similarly, $s_2 = a$ so $s_1 = s_2$, a contradiction.

For a free distributive lattice an even stronger condition holds. We first need a technical result.

Lemma 26. *Let $(Z_i)_{i=1}^m$ and $(T_i)_{i=1}^\infty$ be families of finite non-empty sets such that*

$$Z_1 \nsubseteq T_1, \cdots, Z_m \nsubseteq T_1$$

$$Z_1 \nsubseteq T_2, \cdots, Z_m \nsubseteq T_2$$

$$\vdots \qquad \vdots$$

Then there exist distinct i, j satisfying $Z_1 \nsubseteq T_i \cup T_j, \cdots, Z_m \nsubseteq T_i \cup T_j$.

Proof. The family $(Z_1 \sim T_i)_{i=1}^\infty$ contains only finitely many distinct sets, since Z_1 is finite. So there exists a subfamily $(T_{1,i})_{i=1}^\infty$ of $(T_i)_{i=1}^\infty$ such that $Z_1 \sim T_{1,1}$

$= Z_1 \sim T_{1,2} = \cdots$. Hence $Z_1 \nsubseteq T_{1,1} \cup T_{1,2} \cup \cdots$. Proceeding by induction, suppose $(T_{n,i})_{i=1}^{\infty}$ is a family such that $Z_n \sim T_{n,1} = Z_n \sim T_{n,2} = \cdots$. Now the family $(Z_{n+1} \sim T_{n,i})_{i=1}^{\infty}$ contains only finitely many distinct sets so there is a subfamily $(T_{n+1,i})_{i=1}^{\infty}$ of $(T_{n,i})_{i=1}^{\infty}$ such that $Z_{n+1} \sim T_{n+1,1} = Z_{n+1} \sim T_{n+1,2} = \cdots$. Hence $Z_{n+1} \nsubseteq T_{n+1,1} \cup T_{n+1,2} \cdots$. In particular, $Z_m \nsubseteq T_{m,1} \cup T_{m,2} \cup \cdots$. Now for each $n \in \{1, \cdots, m\}$, $Z_n \nsubseteq T_{n,1} \cup T_{n,2} \cup \cdots$ and since $(T_{m,i})_{i=1}^{\infty}$ is a subfamily of $(T_{n,i})_{i=1}^{\infty}$ we have $Z_n \nsubseteq T_{m,1} \cup T_{m,2}$ for all $n \in \{1, \cdots, m\}$.

Theorem 27. *In a free distributive lattice every a-disjoint set is finite.*

Proof. If there is an infinite a-disjoint set in $\mathscr{F}_{\mathbf{D}}(X)$ then there would be distinct elements $(d_i)_{i=1}^{\infty}$ such that $a < d_i$ for each i, and $d_i d_j = a$ for $i \neq j$. Let $a = \sum_{j=1}^{m} (\prod Z_j)$ and $d_i = \sum_{j=1}^{p(i)} (\prod S_{ij})$. We will show

(5) there exists $n > 0$ such that for each $k \in \{1, \cdots, p(n)\}$, $Z_r \subseteq S_{n,k}$ for some $r \in \{1, \cdots, m\}$.

Assuming the negation of (5) we have that for each i there exists $S_{i,j}$ such that $Z_r \nsubseteq S_{i,j}$ for all $r \in \{1, \cdots, m\}$. By Lemma 26 there exist S_{i,j_i} and S_{k,j_k} $(i \neq k)$ such that $Z_r \nsubseteq S_{i,j_i} \cup S_{k,j_k}$ for all $r \in \{1, \cdots, m\}$. But $\prod (S_{i,j_i} \cup S_{k,j_k}) \leq d_i d_k = a$, so by Lemma 6 there exists r such that $Z_r \subseteq S_{i,j_i} \cup S_{k,j_k}$, a contradiction. Then applying Lemma 6 to (5) we obtain the contradiction $d_n \leq a$, which concludes the proof.

Returning to the question of showing that in $\mathscr{B}$ and $\mathscr{D}$, free algebras contain no uncountable chains, we first mention that F. Galvin and B. Jónsson [1] have proven that the condition holds for free algebras in $\mathscr{L}$. Their proof depends only on the fact that the free generators are uniquely determined and therefore their proof can be applied to establish the validity of the property for free algebras in $\mathscr{D}$. That the property holds in $\mathscr{B}$ was established in 1948 by N. A. Sanin [1], using topological methods. A later proof, in the context of logic, was given by I. Reznikoff [1]. G. W. Day has shown that the property holds in $\mathscr{B}$ by applying measure theoretic methods (oral communication). We present here the proof given by A. Horn [5].

Theorem 28. *Every chain in a free Boolean algebra is countable.*

Proof. Let T be a set of cardinality α and recall that the field generated by $\{A_t : t \in T\}$, defined earlier in this section, is a free Boolean algebra $\mathscr{F}_{\mathbf{B}}(X)$ on α free generators.

Suppose that C is an uncountable chain in $\mathscr{F}_{\mathbf{B}}(X)$ and $0 \notin C$. Then for $x \in C$ there is an n-tuple $(A_{p_1}, \cdots, A_{p_n})$ such that

$$x = \bigcup_{\sigma \in R_1} (A_{p_1}^{\sigma(1)} \cap \cdots \cap A_{p_n}^{\sigma(n)}) \text{ for some } R_1 \subseteq \mathbf{2^n}.$$

So with each $x \in C$ we can associate the pair $(n, |R_1|)$. Now, since C is uncountable, there exist two elements of C which are each associated with the same pair. Let us suppose that x is one of these elements and that the other one is

$$y = \bigcup_{\tau \in R_2} (A_{q_1}^{\tau(1)} \cap \cdots \cap A_{q_n}^{\tau(n)}) \text{ for some } R_2 \subseteq \mathbf{2^n}, |R_1| = |R_2|.$$

We can assume $x \leq y$. Let $P = \{p_1, \cdots, p_n\}$, $Q = \{q_1, \cdots, q_n\}$ and for each $\sigma \in R_1$ and $\tau \in R_2$ set $\sigma^* = \{p_i : \sigma(i) = 1\}$ and $\tau^* = \{q_i : \tau(i) = 1\}$. Note that $|Q \sim P| = |P \sim Q|$.

Now for $S \subseteq T$ we have $S \in x \Leftrightarrow$ there exists $\sigma \in R_1$ such that $S \cap P = \sigma^*$. Since $x \subseteq y$, it follows that for $S \subseteq T$ and $\sigma \in R_1$:

(1) if $S \cap P = \sigma^*$ then there exists $\tau \in R_2$ such that $S \cap Q = \tau^*$.

Let $C_1, \cdots, C_m$ be the distinct members of $\{\sigma^* \cap Q : \sigma \in R_1\}$. For each $K \subseteq Q \sim P$ and $\sigma \in R_1$, set $S = K \cup \sigma^*$ and apply (1) to obtain the fact that for each $1 \leq k \leq m$ there exists $\tau \in R_2$ such that $K \cup C_k = \tau^*$. Hence $|R_1| \geq m \cdot 2^{|q-p|}$. But $|R_1| \leq m \cdot 2^{|q-p|}$ since for each $\sigma \in R_1$, $\sigma^* = C_k \cup K$ for some $k \in \{1, \cdots, m\}$ and $K \subseteq P \sim Q$. Hence,

$$\{\tau^* : \tau \in R_2\} = \{C_k \cup K : 1 \leq k \leq m, K \subseteq Q \sim P\} \text{ and}$$

$$\{\sigma^* : \sigma \in R_1\} = \{C_k \cup K : 1 \leq k \leq m, K \subseteq P \sim Q\}.$$

Thus

$$y = \bigcup_{\tau \in R_2} (A_{q_1}^{\tau(1)} \cap \cdots \cap A_{q_n}^{\tau(n)}) = \bigcup_{\tau \in R_2} \left[\left(\bigcap_{i \in \tau^*} A_i\right) \cap \left(\bigcap_{i \in Q \sim \tau^*} \tilde{A}_i\right)\right]$$

$$= \bigcup_{k=1}^{m} \bigcup_{K \subseteq Q \sim P} \left[\left(\bigcap_{i \in C_k \cup K} A_i\right) \cap \left(\bigcap_{i \in Q \sim (C_k \cup K)} \tilde{A}_i\right)\right]$$

$$= \left\{\bigcup_{k=1}^{m} \left[\left(\bigcap_{i \in C_k} A_i\right) \cap \left(\bigcap_{i \in (P \cap Q) \sim C_k} \tilde{A}_i\right)\right]\right\}$$

$$\cap \left\{\bigcup_{K \subseteq Q \sim P} \left[\left(\bigcap_{i \in K} A_i\right) \cap \left(\bigcap_{i \in (Q \sim P) \sim K} \tilde{A}_i\right)\right]\right\}$$

$$= \bigcup_{k=1}^{m} \left[\left(\bigcap_{i \in C_k} A_i\right) \cap \left(\bigcap_{i \in (P \cap Q) \sim C_k} \tilde{A}_i\right)\right].$$

Similarly,

$$x = \bigcup_{k=1}^{m} \left[\left(\bigcap_{i \in C_k} A_i\right) \cap \left(\bigcap_{i \in (P \cap Q) \sim C_k} \tilde{A}_i\right)\right] \text{ so } x = y.$$

Corollary 29. *Every chain in a free distributive lattice is countable.*

Proof. The field B of sets generated by $\{A_t : t \in T\}$ is a free Boolean algebra on $|T|$ free generator and therefore has no uncountable chains; but it contains the ring of sets D generated by $\{A_t : t \in T\}$ which is the free distributive lattice on $|T|$ free generators.

Whereas we have seen that $|\mathscr{F}_{\mathbf{B}}(\mathbf{n})| = 2^{2^n}$ for all n, the corresponding problem of determining $|\mathscr{F}_{\mathbf{D}}(\mathbf{n})|$ has been outstanding since R. Dedekind proposed it in [1]. He did however show that $|\mathscr{F}_{\mathbf{D}}(\mathbf{4})| = 166$ (clearly $|\mathscr{F}_{\mathbf{D}}(\mathbf{1})| = 1$, $|\mathscr{F}_{\mathbf{D}}(\mathbf{2})| = 4$ and $|\mathscr{F}_{\mathbf{D}}(\mathbf{3})| = 18$). In addition, it is now known that $|\mathscr{F}_{\mathbf{D}}(\mathbf{5})| = 7579$ (R. Church [1]), $|\mathscr{F}_{\mathbf{D}}(\mathbf{6})| = 7{,}828{,}352$ (M. Ward [1]) and $|\mathscr{F}_{\mathbf{D}}(\mathbf{7})| = 2{,}414{,}682{,}040{,}996$. Interesting properties of $|\mathscr{F}_{\mathbf{D}}(\mathbf{n})|$ have been found, in particular, by K. Yamamoto [1], D. Kleitman [1] and N. M. Riviere [1]. For example, Yamamoto showed that for each $\delta > 0$, $2^n \cdot n^{-\frac{1}{2}-\delta} < \log_2 |\mathscr{F}_{\mathbf{D}}(\mathbf{n})| < 2^n \cdot n^{-\frac{1}{2}+\delta}$ when n is sufficiently large.

4. Free Boolean extensions

In order to deal more effectively with the categories $\mathscr{D}$, $\mathscr{D}_{01}$ and $\mathscr{B}$, we have introduced in Section II.8, the category $\mathscr{D}'$. Recall that Ob $\mathscr{D}'$ = Ob $\mathscr{D}$ and that $[L, L_1]_{\mathscr{D}'}$ is the set of those morphisms $f \in [L, L_1]_{\mathscr{D}}$ which satisfy: If L has a 0(1) then L_1 has a 0(1) and $f(0) = (0)$ ($f(1) = 1$). Recall also that $\mathscr{B}$ and $\mathscr{D}_{01}$ are full subcategories of $\mathscr{D}'$.

Morphisms in $\mathscr{D}'$ are one-one for suppose $f \in [L, L_1]_{\mathbf{D}'}$ is monic and $f(x_1) = f(x_2)$. For $i = 1, 2$ let $g_i : \mathscr{F}_{\mathbf{D}}(\aleph_0) \to L$ be **D**-homomorphisms for which $g_i(u) = x_i$ for all free generators u. Since $\mathscr{F}_{\mathbf{D}}(\aleph_0)$ has no 0, or 1, $g_i \in [\mathscr{F}_{\mathbf{D}}(\aleph_0), L]_{\mathscr{D}'}$. But $f(g_1(y)) = f(x_1) = f(x_2) = f(g_2(y))$ for all $y \in \mathscr{F}_{\mathbf{D}}(\aleph_0)$ an hence $g_1(y) = g_2(y)$ for all such y; thus $x_1 = x_2$.

Lemma 1. *Let $L_1 \in$ Ob $\mathscr{B}$, L a sublattice of L_1 and $[L]_{\mathbf{B}} = L_1$. If $L_2 \in$ Ob $\mathscr{B}$ and $g \in [L, L_2]_{\mathscr{D}'}$, then there exists a unique $h \in [L_1, L_2]_{\mathscr{B}}$ such that $h|L = g$.*

Proof. We apply the extension theorem. Thus, suppose $T_1 \cup T_2 \subseteq L$ and $\prod^{L_1} T_1 \cdot \prod^{L_1} T_2^- = 0$. Then $\prod^L T_1 \leq \sum^L T_2$ but $g \in [L, L_2]_{\mathscr{D}'}$ so $\prod g[T_1] \leq \sum g[T_2]$. Note that the cases $T_1 = \varnothing$ and $T_2 = \varnothing$ are also covered. For if, say, $T_1 = \varnothing$, then $\sum^L T_2 = \sum^{L_1} T_2 = 1_{L_1}$ so L has a 1 and $g(1) = 1$. Thus $\prod g[T_1] = 1$ and $\sum g[T_2] = 1$. So there exists $h \in [L_1, L_2]_{\mathscr{B}}$ such that $h|L = g$. The uniqueness follows from the equality of homomorphisms that agree on a set of generators.

Theorem 2. *$\mathscr{B}$ is a reflective subcategory of $\mathscr{D}'$ and the reflector $\mathscr{B} : \mathscr{D}' \to \mathscr{B}$ preserves and reflects monomorphisms. Moreover, $\Phi_{\mathscr{B}}(L)$ is a monomorphism and an epimorphism for each $L \in$ Ob $\mathscr{D}'$.*

Proof. For each $L \in$ Ob $\mathscr{D}'$, let $\mathscr{B}(L) = [\hat{L}]_{\mathbf{B}}$ (i.e. the field of subsets of $\mathfrak{P}(L)$ generated by $\hat{L}$) and let $\Phi_{\mathscr{B}}(L) : L \to \mathscr{B}(L)$ be the map defined by $\Phi_{\mathscr{B}}(L)(x) = \hat{x}$. Obviously $\Phi_{\mathscr{B}}(L) \in [L, \mathscr{B}(L)]_{\mathscr{D}'}$. If $g \in [L, L_1]_{\mathscr{D}'}$, where $L_1 \in$ Ob $\mathscr{B}$, then by Lemma 1 there exists a unique $h \in [\mathscr{B}(L), L_1]_{\mathscr{B}}$ with $h \circ \Phi_{\mathscr{B}}(L) = g$. Since $\Phi_{\mathscr{B}}(L)$ is one-one, it is monic. To prove that $\Phi_{\mathscr{B}}(L)$ is epic, let $g, h \in [\mathscr{B}(L), L_1]_{\mathscr{D}'}$ such that $g \circ \Phi_{\mathscr{B}}(L) = h \circ \Phi_{\mathscr{B}}(L)$. Note that, since $g, h \in \mathscr{D}'$, g and h preserve 0, 1 and also complements. It follows that $L_2 = \{x \in \mathscr{B}(L) : g(x) = h(x)\}$ is a **B**-subalgebra of $\mathscr{B}(L)$ containing $\hat{L}$. Hence $L_2 = \mathscr{B}(L)$ and thus $g = h$.

Now suppose $f \in [L, L_1]_{\mathscr{D}'}$ is monic. To show $\mathscr{B}(f)$ is monic we note that since $\mathscr{B}(L)$ is a Boolean algebra it suffices to prove that $\mathscr{B}(f)(u) = 0 \Rightarrow u = 0$ (cf. Exercise II.6.16). Now we have seen (Theorem 1.1(iii)) that $u \in \mathscr{B}(L)$ can be represented as $u = \sum_{i=1}^{n} (\prod_{x \in S_i} \hat{x})(\prod_{x \in T_i} \bar{\hat{x}})$, $S_i \cup T_i \subseteq L$, $1 \leq i \leq n$.

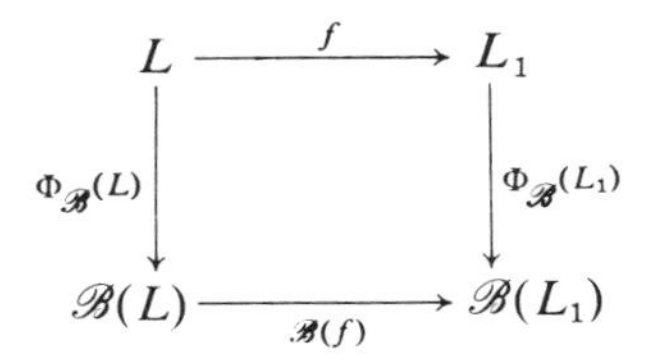

So $0 = \mathscr{B}(f)(u) = \sum_{i=1}^{n}(\prod_{x \in S_i}\mathscr{B}(f)(\hat{x}))(\prod_{x \in T_i}\mathscr{B}(f)(\bar{\hat{x}}))$
$= \sum_{i=1}^{n}(\prod_{x \in S_i}\widehat{f(x)})(\prod_{x \in T_i}\bar{\widehat{f(x)}})$. Consequently, for each i, $\bigcap_{x \in S_i}\widehat{f(x)}$
$\subseteq \bigcup_{x \in T_i}\widehat{f(x)}$ so $\prod f[S_i] \leq \sum f[T_i]$. Note that this is the case even if $S_i = \varnothing$ or T_i
$= \varnothing$. But since f is one-one, we have $\prod S_i \leq \sum T_i$ for each i and consequently $u = 0$.

Finally, suppose $\mathscr{B}(f)$ is monic for $f \in [L, L_1]_{\mathscr{D}'}$. But $\Phi_{\mathscr{B}}(L)$ is one-one and thus $\mathscr{B}(f) \circ \Phi_{\mathscr{B}}(L) = \Phi_{\mathscr{B}}(L_1) \circ f$ is one-one and so f is one-one and hence monic.

Remark. The reflector $\mathscr{B}$ also preserves epimorphisms (Theorem I.18.5) and $\mathscr{B}$ also preserves onto maps since "onto $\Rightarrow$ epic" in $\mathscr{D}$ and "onto $\Leftrightarrow$ epic" in $\mathscr{B}$ (Theorem IV.3.6).

Corollary 3. *$\mathscr{B}$ is a reflective subcategory of $\mathscr{D}_{01}$ and the reflector $\mathscr{B}_{01} : \mathscr{D}_{01} \to \mathscr{B}$ preserves and reflects monomorphisms. Moreover $\Phi_{\mathscr{B}_{01}}(L)$ is monic and epic for each $L \in \text{Ob}\, \mathscr{D}_{01}$.*

Proof. Let $\mathscr{B}_{01}$ be the restriction of $\mathscr{B}$ to $\mathscr{D}_{01}$. Since $\mathscr{D}_{01}$ is a full subcategory of $\mathscr{D}'$, $\mathscr{B}_{01}$ is again a reflector. (Note that for $L \in \text{Ob}\, \mathscr{D}_{01}$, $\Phi_{\mathscr{B}}(L)$ is in $\mathscr{D}_{01}$ and $\Phi_{\mathscr{B}_{01}}(L)$ $= \Phi_{\mathscr{B}}(L)$.) Since, for $L \in \text{Ob}\, \mathscr{D}_{01}$, $\Phi_{\mathscr{B}}(L)$ is monic and epic in $\mathscr{D}'$, $\Phi_{\mathscr{B}_{01}}(L)$ is monic and epic in $\mathscr{D}_{01}$. If $f \in [L, L_1]_{\mathscr{D}_{01}}$ and f is monic then f is one-one and thus monic in $\mathscr{D}'$, hence by Theorem 2, $\mathscr{B}_{01}(f) = \mathscr{B}(f)$ is monic in $\mathscr{B}$. Finally, if $f \in [L, L_1]_{\mathscr{D}_{01}}$ and $\mathscr{B}(f)$ is monic then f is monic in $\mathscr{D}'$, again by Theorem 2, and thus monic in $\mathscr{D}_{01}$.

Remark. $\mathscr{B}_{01}$ also preserves epimorphisms and onto maps.

Corollary 4. *$\mathscr{B}$ is a reflective subcategory of $\mathscr{D}$ and the reflector $\mathscr{B}^* : \mathscr{D} \to \mathscr{B}$ preserves and reflects monomorphisms.*

Proof. Let $\mathscr{B}^* : \mathscr{D} \to \mathscr{B}$ be defined by $\mathscr{B}^* = \mathscr{B} \circ \mathscr{U}$, where $\mathscr{U} : \mathscr{D} \to \mathscr{D}_{01}$ is the reflector defined in Theorem II.5.7. By Theorem I.18.3, $\mathscr{B}^*$ is a reflector. It follows from Theorem II.5.7 and Corollary 3 that $\mathscr{B}^*$ preserves and reflects monomorphisms.

Remark. $\mathscr{B}^*$ also preserves epimorphisms and onto maps.

Definition 5. For each $L \in \text{Ob}\,\mathscr{D}$, a *free Boolean extension* of L is a pair (L_1, f) where L_1 is a Boolean algebra and $f \in [L, L_1]_{\mathscr{D}'}$ is a monomorphism satisfying:

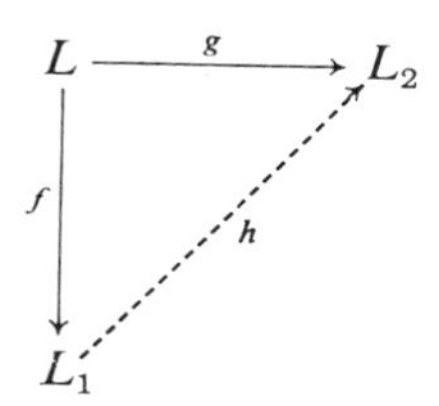

If $g \in [L, L_2]_{\mathscr{D}'}$, $L_2 \in \text{Ob}\,\mathscr{B}$, then there exists a unique $h \in [L_1, L_2]_{\mathscr{B}}$ such that $h \circ f = g$.

Clearly, $(\mathscr{B}(L), \Phi_{\mathscr{B}}(L))$ is a free Boolean extension of L. "Uniqueness" and other categorical properties follow easily from the definition.

In particular,

(1) If (L_1, f) is a free Boolean extension of L, then $[f[L]]_{\mathbf{B}} = L_1$ and f is an epimorphism.

(2) If (L_1, f_1) and (L_2, f_2) are free Boolean extensions of L, then there exists an isomorphism $g: L_1 \to L_2$ such that $g \circ f_1 = f_2$.

(3) If (L_1, f_1) and (L_1', f_1') are free Boolean extensions of L and L' respectively and $h \in [L, L']_{\mathscr{D}'}$ then there exists a unique $h^* \in [L_1, L_1']_{\mathscr{B}}$ such that $f_2 \circ h = h^* \circ f_1$

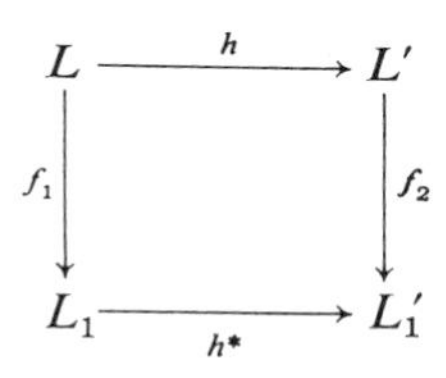

(4) If h is monic (epic) then h^* is monic (epic).

(5) If $h_1^* = h_2^*$ then $h_1 = h_2$.

Exercise 6. Prove statements (1)–(5).

Theorem 7. (A. Nerode [1]) *Let $L \in \text{Ob}\,\mathscr{D}_{01}$ and let (L_1, f) be a free Boolean extension of L. Let X be a closed subspace of $\mathscr{S}_{\mathscr{B}}(L_1)$. Then $\mathscr{S}(f)[X]$ is a bounded Stone space.*

Proof. X is a Boolean space. Let $L_1' = \mathscr{T}(X)$ and let $j: L_1 \to L_1'$ be the onto homomorphism such that $\mathscr{S}_{\mathscr{B}}(j): X \to \mathscr{S}_{\mathscr{B}}(L_1)$ is the inclusion map. Let $L' = \text{Im}(j \circ f)$ and let j' be the homomorphism $j \circ f$ with codomain restricted to L'. Then $1_{L', L_1'} \circ j' = j \circ f$ so by Theorem IV.2.7, $\mathscr{S}(f)[X] = \text{Im}(\mathscr{S}(f) \circ \mathscr{S}_{\mathscr{B}}(j)) = \text{Im}(\mathscr{S}(j') \circ \mathscr{S}(1_{L', L_1'})) = \text{Im}(\mathscr{S}(j'))$ is a bounded Stone space.

If $L \in \mathrm{Ob}\,\mathscr{D}$ and (L_1, f) is a free Boolean extension of L, then we often assume that L is a sublattice of L_1 and that f is the inclusion map. In this case we will simply say that L_1 is the free Boolean extension of L and we will use the symbol B_L instead of L_1. Thus, if $L \in \mathrm{Ob}\,\mathscr{B}$, then its free Boolean extension B_L is L itself.

Theorem 8. *Let (L_1, f) be the free Boolean extension of $L \in \mathrm{Ob}\,\mathscr{D}$. Then L_1 is an essential extension of L in $\mathscr{D}$.*

Proof. Let $g \in [L_1, M]_{\mathscr{D}}$ and $g \circ f$ monic. Define $h \in [L_1, g[L_1]]_{\mathscr{D}}$ by $h(x) = g(x)$ for $x \in L_1$. Then $g \circ f$ monic $\Rightarrow g \circ f$ one-one $\Rightarrow h \circ f$ one-one. Also $g[L_1] \in \mathrm{Ob}\,\mathscr{B}$ and $h \in [L_1, g[L_1]]_{\mathscr{B}}$ so $h^* = h$. Then $h = h \circ 1_{L_1} = h^* \circ f^* = (h \circ f)^*$ is monic since $h \circ f$ is monic.

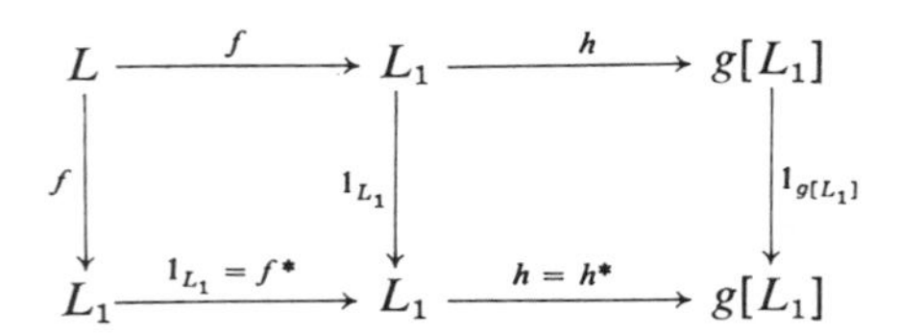

We can also reformulate Lemma 1: If L is a sublattice of L_1, $L_1 \in \mathrm{Ob}\,\mathscr{B}$ and $[L]_{\mathbf{B}} = L_1$ then $1_{L,L_1} \in [L, L_1]_{\mathscr{D}'}$ and L_1 is the free Boolean extension of L; that is, $L_1 = B_L$.

Exercise 9. (B. Banaschewski and G. Bruns [2]) Let $L \in \mathrm{Ob}\,\mathscr{D}$. Prove: B_L is atomic if and only if every proper interval in L contains a *jump*, i.e. if $a, b \in L$, $a < b$ then there exist $c, d \in L$ such that $a \leq c < d \leq b$ and d covers c.

Exercise 10. The free Boolean algebra $\mathscr{F}_{\mathbf{B}}(\alpha)$ is the free Boolean extension of the free distributive lattice $\mathscr{F}_{\mathbf{D01}}(\alpha)$.

Theorem 11. (Also cf. A. Nerode [1]) *Let $L \in \mathrm{Ob}\,\mathscr{D}_{01}$, $L_1 \in \mathrm{Ob}\,\mathscr{B}$, $f \in [L, L_1]_{\mathscr{D}_{01}}$. Then the following are equivalent:*

(i) *(L_1, f) is a free Boolean extension of L.*

(ii) *f is monic and epic.*

(iii) *$\mathscr{S}(f)$ is monic and onto.*

Proof. (i) $\Rightarrow$ (ii) See remarks following Definition 5.

(ii) $\Rightarrow$ (iii) by Theorem IV.2.6.

(iii) $\Rightarrow$ (i) We only need to show that $f[L]$ **B**-generates L_1; but that is true since f is epic. Indeed, suppose that there exists a proper subalgebra L' of L_1 such that $f[L] \subseteq L' \subset L_1$. Hence by Theorem IV.3.8 there exist prime ideals I_1, I_2 of L_1 with $I_1 \neq I_2$, $I_1 \cap L' = I_2 \cap L'$; hence there exist $g, h \in [L_1, \mathbf{2}]_{\mathscr{B}}$, $g \neq h$ with $h \circ f = g \circ f$.

Theorem 11 enables us to give an easy proof of IV.3.12, which we state here as a theorem.

Theorem 12. *If $L \in \mathrm{Ob}\, \mathscr{D}_{01}$ and L is infinite then $|\mathscr{S}(L)| \geq |L|$.*

Proof. Obviously $|B_L| = |L|$. By Theorem IV.3.9, $|\mathscr{S}_{\mathscr{B}}(B_L)| \geq |B_L|$. On the other hand by Theorem 11, $|\mathscr{S}(L)| = |\mathscr{S}_{\mathscr{B}}(B_L)|$. Thus $|\mathscr{S}(L)| \geq |L|$.

Concerning free Boolean extensions, also see W. Peremans [1], G. Grätzer and E. T. Schmidt [3], and C. C. Chen [1]. M. E. Adams [3] has considered the problem of embedding distributive lattices in Boolean algebras such that maximal chains are preserved. See also G. W. Day [1].

5. Embedding of relatively complemented distributive lattices in Boolean algebras

We start by recalling the definition of $\mathscr{R}'$ (Section II.8): $\mathrm{Ob}\, \mathscr{R}' = \mathrm{Ob}\, \mathscr{R}$ and if $L, L_1 \in \mathrm{Ob}\, \mathscr{R}'$ then $h \in [L, L_1]_{\mathscr{R}'} \Leftrightarrow h \in [L, L_1]_{\mathscr{D}'}$.

Monomorphisms of $\mathscr{R}'$ are one-one. Indeed, suppose $f \in [L, L_1]_{\mathscr{R}'}$, f is monic, $f(u) = f(v)$ and $u \neq v$. Consider the prime ideal $(S]$ and the prime filter $[S)$ in a free Boolean algebra $\mathscr{F}_{\mathbf{B}}(S)$ for some infinite set S of free generators. Then $(S] \cap [S) \in \mathrm{Ob}\, \mathscr{R}$ and $(S] \cap [S)$ has neither a 0 nor a 1. Let $h, g \in [\mathscr{F}_{\mathbf{B}}(S), B_L]_{\mathscr{B}}$ be defined by $h(x) = u$ for all $x \in S$ and $g(x) = v$ for all $x \in S$. Now if $y \in (S] \cap [S)$, then $\prod S_1 \leq y \leq \sum S_2$ for some $S_1, S_2 \subseteq S$, hence $h(y) = u$, $g(y) = v$ for all $y \in (S] \cap [S)$. Let $h_1 = h|(S] \cap [S)$ and $g_1 = g|(S] \cap [S)$; then $h_1, g_1 \in [(S] \cap [S), L]_{\mathscr{R}'}$ but $f \circ h_1 = f \circ g_1$ and $h_1 \neq g_1$, a contradiction.

Epimorphisms in $\mathscr{R}'$ are not necessarily onto: Indeed, let L be an object of $\mathscr{R}'$ without 0 and 1, then $1_{L,B_L} \in [L, B_L]_{\mathscr{R}'}$ and is not onto. Now let $f, g \in [B_L, L_1]_{\mathscr{R}'}$ such that $f|L = g|L$. But then L_1 is a Boolean algebra and hence $f = g$. Thus 1_{B_L} is epic but not onto.

Recall that we can consider $\mathscr{R}$ as a subcategory of $\mathscr{D}$ (Section II.7).

Theorem 1. (D. C. Feinstein [1]) *Let $L_1 \in \mathrm{Ob}\, \mathscr{B}$ and suppose L is a proper sublattice of L_1 which is closed under relative complementation. Then L_1 is the free Boolean extension of L if and only if $L = I$ or $L = F$ or $L = I \cap F$, where I is a non-principal prime ideal, F is a non-principal prime filter and $I \cap F \neq \varnothing$.*

We first need a lemma.

Lemma 2. Let L be a relatively complemented distributive lattice. Then L is convex in B_L.

Proof. Suppose $a \leq b \leq c$, $a, c \in L$ and $b \in B_L$. Then we can write $b = u + \sum_{i=1}^{n} u_i \bar{v}_i + \bar{v}$ where $u, v, u_i, v_i \in L$, $1 \leq i \leq n$ but not all of the terms on the right side need actually occur. Now $a \leq b \Rightarrow b = a + b = a + u + \sum_{i=1}^{n} u_i \bar{v}_i + \bar{v}$ and $b \leq c \Rightarrow b = a + cb = (a + cu) + \sum_{i=1}^{n} cu_i \bar{v}_i + c\bar{v}$. Thus we may assume that b can be written as $b = u + \sum_{i=1}^{n} u_i \bar{v}_i$, where on the right side u must occur. If no

$u_i\bar{v}_i$ occurs, then $b \in L$. Now the complement of v_i in $[uv_i, u_i + v_i]$ is in L, for each i. This complement is $(u_i + v_i)\bar{v}_i + uv_i = u_i\bar{v}_i + uv_i$ so $b = u + \sum_{i=1}^{n} u_i\bar{v}_i = u + \sum_{i=1}^{n} (u_i\bar{v}_i + uv_i) \in L$.

Proof of theorem. ($\Rightarrow$) Three cases: (i) $0_{L_1} \in L$. Then $1_{L_1} \notin L$. By Lemma 2, L is an ideal, hence a proper ideal in L_1. Let I be a prime ideal of L_1 such that $I \supseteq L$ and $\nu_L \in [L_1, L_1/L]_{\mathscr{B}}$ the canonical homomorphism (Section II.9); let $\nu' \in [L_1, L_1/L]_{\mathscr{B}}$ be defined by $\nu'(x) = 0$ if $x \in I$ and $\nu'(x) = 1$ otherwise. Then $\nu'|L = \nu_L|L$. Hence, since L_1 is **B**-generated by L, $\nu' = \nu_L$ and thus $L = I$. Finally, the embedding $1_{L,L_1}$ is in $\mathscr{D}'$ so L is not principal. (ii) The proof of the case when $1_{L_1} \in L$, $0_{L_1} \notin L$ is the dual of (i). (iii) Suppose $0_{L_1} \notin L$, $1_{L_1} \notin L$. By Lemma 2, L is convex in L_1. Let $I = (L]_{L_1}$ and $F = [L)_{L_1}$. Now I is a proper ideal in L_1 which is closed under relative complementation. Moreover $I \supseteq L$, thus $[I]_{\mathbf{B}} = L_1$. It follows from (i) that I is a non-principal prime ideal. Similarly, F is a non-principal prime filter. Now $L \subseteq I$, $L \subseteq F$ thus $L \subseteq I \cap F$. Suppose $x \in I \cap F$. Thus $x \in F$ so there exists $a \in L$ such that $a \leq x$. Dually $x \in I$ implies there exists $b \in L$ with $x \leq b$ thus $a \leq x \leq b$. But L is convex, hence $x \in L$. It follows that $L = I \cap F$.

($\Leftarrow$) I, F and $I \cap F$ are obviously sublattices of L_1 which are closed under relative complementation. If $a \in L_1 \sim I$, then $\bar{a} \in I$, hence $[I]_{\mathbf{B}} = L_1$. Similarly, $[F]_{\mathbf{B}} = L_1$. Next, assume $a \notin I \cap F$. If $a \notin I$ and $a \notin F$ then $\bar{a} \in I \cap F$. If $a \notin I$, $a \in F$ then since $I \cap F \neq \varnothing$ there exists $b \in I \cap F$ so $ab \in I \cap F$ and $\bar{a} + b \in I \cap F$. But $a = ab + \overline{(\bar{a} + b)}$ and so $[I \cap F]_{\mathbf{B}} = L_1$. The case for $a \in I$, $a \notin F$ is similar.

Corollary 3. *Let L be a relatively complemented distributive lattice. If L is not a Boolean algebra, then L can be* **D**-*embedded in a Boolean algebra L_1 such that the embedding is in $\mathscr{D}'$ and either $L = I$ or $L = F$ or $L = I \cap F$, where I is a non-principal prime ideal and F is a non-principal prime filter.*

6. Free relatively complemented distributive extensions

Let L be a distributive lattice and, as usual, let B_L denote its free Boolean extension. Denote by R_L, the smallest sublattice of B_L which contains L and which is closed under relative complementation. Note that R_L exists, since the intersection of relatively complemented sublattices is relatively complemented.

Definition 1. If L is a distributive lattice and $S \subseteq L$ then the *convex hull of S in L* is the smallest convex sublattice of L which contains S.

Clearly the convex hull of S exists if and only if $S \neq \varnothing$.

Theorem 2. (D. C. Feinstein [1]) *Let L be a distributive lattice. Then,*

(i) $R_L = \{x \in B_L : x \in L$ *or* $x = a + \sum_{i=1}^{n} a_i\bar{b}_i, n \geq 1, a, a_i, b_i \in L\}$.

(ii) *R_L is the convex hull of L in B_L.*

(iii) *If 0_L exists, then $0_L = 0_{B_L} = 0_{R_L}$; if 0_{R_L} exists then 0_L exists, and dually. Thus $1_{L,R_L}$ and $1_{R_L,B_L}$ are* **D**-*embeddings which are in $\mathscr{D}'$.*

Remark. It follows from (iii) that R_L is a Boolean algebra if and only if L has a 0 and 1; then $R_L = B_L$. We also have that $B_{R_L} = B_L$, and if L has no 0(1) then R_L has no 0(1).

Proof. Let $L_1 = \{x \in B_L : x \in L \text{ or } x = a + \sum_{i=1}^n a_i\bar{b}_i, n \geq 1, a, a_i, b_i \in L\}$. Then L_1 is obviously a sublattice of B_L containing L. Now suppose $x \leq y \leq z$, $x, z \in L_1$, and $y \in B_L$. Then $x = a + \sum_{i=1}^n b_i\bar{c}_i$, $z = e + \sum_{i=1}^m f_i\bar{g}_i$, $m, n \geq 1$, $a, e, b_i, c_i, f_i, g_i \in L$. Since $y \in B_L$, $y = u + \sum_{i=1}^k u_i\bar{v}_i + \bar{v}$; $u, u_i, v_i, v \in L$ where not all terms on the right side need occur. Now $a \leq x \leq y \Rightarrow y = a + y \Rightarrow y = u' + \sum_{i=1}^k u_i\bar{v}_i + \bar{v}$ where $u' \in L$ must occur. Also $y \leq z \Rightarrow y = zy \Rightarrow y = (e + \sum_{i=1}^m f_i\bar{g}_i)(u' + \sum_{i=1}^k u_i\bar{v}_i + \bar{v})$ and thus $y = u'' + \sum_{i=1}^k u_i'\overline{v_i'}$ where $u'' \in L$ must occur, $u_i', v_i' \in L$ and thus $y \in L_1$. Hence L_1 is convex in B_L and thus relatively complemented. Now suppose L' is a sublattice of B_L containing L and which is closed under relative complementation. In order to show that $L' \supseteq L_1$ it suffices to show that for $a, b, c \in L$, $a + b\bar{c} \in L'$. But the complement of $a + c$ in $[a, a + b + c]$ is in L' and this complement is $a + b\bar{c}$. Thus $L_1 = R_L$. (ii) We have already seen from the proof of (i) that R_L is convex in B_L and contains L. Suppose L' is a convex sublattice of B_L containing L; but then L' is relatively complemented and thus $L' \supseteq R_L$. Hence R_L is the convex hull of L in B_L. (iii) If L has a 0_L, then $0_L = 0_{B_L}$ since the embedding of L in B_L is in $\mathscr{D}'$ and thus $0_{B_L} \in R_L$, so $0_{B_L} = 0_{R_L}$. Next, suppose R_L has a 0_{R_L} and suppose that L has no 0_L. Now $0_{R_L} = a$ or $0_{R_L} = a + \sum_{i=1}^n a_i\bar{b}_i$, $n \geq 1$, $a, a_i, b_i \in L$. But there exists $c < a$, $c \in L$ so in both cases $0_{R_L} \cdot c < 0_{R_L}$ and $0_{R_L} \cdot c \in R_L$, a contradiction. Hence L has a 0_L.

Exercise 3. Let $L \in \text{Ob}\,\mathscr{R}$ and suppose L has neither a 0 nor a 1. Suppose L_1 is a sublattice of B_L which is closed under relative complementation and such that L_1 has neither a 0 nor a 1 and $L \subseteq L_1$. Prove $L = L_1$. (Hint: Use Exercise II.9.13 and Theorem 5.1.)

Theorem 4. *Let $L \in \text{Ob}\,\mathscr{D}$, $L_1 \in \text{Ob}\,\mathscr{R}$ and $f \in [L, L_1]_{\mathscr{D}}$. Then there exists a unique $f^* \in [R_L, L_1]_{\mathscr{R}}$ such that $f^*|L = f$. If $f \in [L, L_1]_{\mathscr{D}'}$ then $f^* \in [R_L, L_1]_{\mathscr{R}'}$.*

Proof. Let L' be the convex hull in L_1 of $f[L]$. Now if L has a 0, then $f(0)$ is also the 0 of L'. Similarly, for 1. Thus $f \in [L, L']_{\mathscr{D}'}$. Also note that L' is relatively complemented. There exists a unique $g \in [B_L, B_{L'}]_{\mathscr{B}}$ with $g|L = f$. Let $f^* = g|R_L$ then $f^*|L = f$. We claim that $f^*[R_L] \subseteq L'$. Suppose $x \in R_L$. If $x \in L$, then clearly $f^*(x) \in L'$. If $x \notin L$, then by Theorem 2(i), $x = a + \sum_{i=1}^n a_i\bar{b}_i$; $a, a_i, b_i \in L$. Thus $f^*(x)$ $= f(a) + \sum_{i=1}^n f(a_i)\overline{f(b_i)}$ and again by Theorem 2(i) it follows that $f^*(x)$ belongs to $R_{L'}$. But L' is relatively complemented by Exercise II.7.6, and thus $R_{L'} = L'$. Hence $f^*[R_L] \subseteq L'$. Since $L' \subseteq L_1$ and $\mathscr{R}$ is a full subcategory of $\mathscr{D}$ we have that $f^* \in [R_L, L_1]_{\mathscr{R}}$. Now suppose that $f \in [L, L_1]_{\mathscr{D}'}$. If R_L has a 0 then by Theorem 2(iii) L has a 0_L and $0_L = 0_{R_L} = 0_{B_L}$. Also $f \in [L, L_1]_{\mathscr{D}'}$ so L_1 has a 0_{L_1} and $f(0_L) = f(0_{L_1})$. Thus $f^*(0_{R_L}) = g(0_{R_L}) = g(0_L) = f(0_L) = 0_{L_1}$. Similarly for 1_{R_L}. Hence $f^* \in [R_L, L_1]_{\mathscr{R}'}$. It remains to prove that f^* is unique. Suppose $f_1^* \in [R_L, L_1]_{\mathscr{R}}$ is such that $f_1^*|L = f$. But then $L_2 = \{u \in B_L : f_1^*(u) = f^*(u)\}$ is a sublattice of B_L which contains L and is relatively complemented; thus $R_L = L_2$ so $f_1^* = f^*$.

Theorem 5. *$\mathscr{R}$ is a reflective subcategory of $\mathscr{D}$ and the reflector $\mathscr{R}$ preserves and reflects monomorphisms.*

Proof. It follows immediately from Theorem 4 that $L \mapsto R_L$, $L \in \text{Ob}\,\mathscr{D}$ can be extended to a reflector $\mathscr{R}: \mathscr{D} \to \mathscr{R}$. Now suppose $f \in [L, L_1]_{\mathscr{D}}$ is a monomorphism, $g, h \in [L', R_L]_{\mathscr{R}}$ and $\mathscr{R}(f) \circ g = \mathscr{R}(f) \circ h$. To show $g = h$, let $x \in L'$. By Theorem 2(i), $g(x) = a + \sum_{i=0}^{n} a_i\bar{b}_i$ for some $a, a_i, b_i \in L$, $n \geq 0$ and $h(x) = a' + \sum_{j=0}^{m} a'_j\overline{b'_j}$ for some $a', a'_j, b'_j \in L$, $m \geq 0$. So $f(a) + \sum_{i=0}^{n} f(a_i)\overline{f(b_i)} = \mathscr{R}(f)(g(x)) = (\mathscr{R}(f) \circ g)(x)$ $= (\mathscr{R}(f) \circ h)(x) = \mathscr{R}(f)(h(x)) = f(a') + \sum_{j=0}^{m} f(a'_j)\overline{f(b'_j)}$. Since f is monic it is one-one so $a + \sum_{i=0}^{n} a_i\bar{b}_i = a' + \sum_{j=0}^{m} a'_j\overline{b'_j}$ and thus $g(x) = h(x)$. Therefore $g = h$ and $\mathscr{R}(f)$ is monic. It follows immediately from the commutativity of the diagram below that if $f \in [L, L_1]_{\mathscr{D}}$ and $\mathscr{R}(f)$ is monic then so is f.

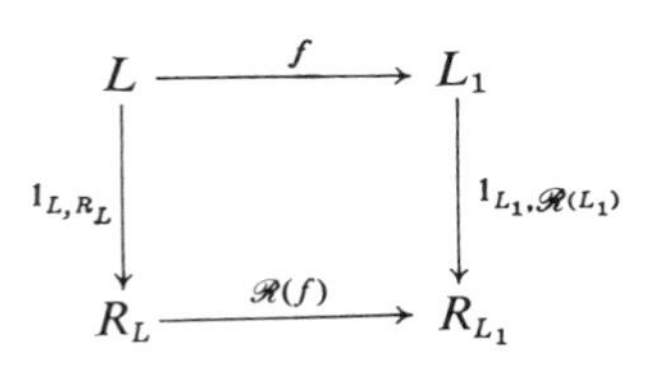

Corollary 6. *$\mathscr{R}'$ is a reflective subcategory of $\mathscr{D}'$ and the corresponding reflector $\mathscr{R}'$ is given by $\mathscr{R}' = \mathscr{R}|\mathscr{D}'$ and $\Phi_{\mathscr{R}} = \Phi_{\mathscr{R}'}$.*

Proof. Immediate from Theorems 2 and 5.

Definition 7. A *free relatively complemented distributive extension* of $L \in \text{Ob}\,\mathscr{D}$ is a pair (L_1, f) where $L_1 \in \text{Ob}\,\mathscr{R}$ and $f \in [L, L_1]_{\mathscr{D}}$ is a monomorphism satisfying: If $L_2 \in \text{Ob}\,\mathscr{R}$ and $g \in [L, L_2]_{\mathscr{D}}$ then there exists a unique $h \in [L_1, L_2]_{\mathscr{R}}$ with $h \circ f = g$ and such that if $g \in [L, L_2]_{\mathscr{D}'}$ then $h \in [L_1, L_2]_{\mathscr{R}'}$.

The existence, uniqueness, and other properties of free relatively complemented distributive extensions follow from Theorem 5 and Corollary 6 and are similar to the properties stated in Section 4 for free Boolean extensions.

We will now make a similar assumption as was made for free Boolean extensions: If $L \in \text{Ob}\,\mathscr{D}$ and (L_1, f) is a free relatively complemented distributive extension of L, then we will often assume that L is a sublattice of L_1 and that f is the inclusion map. In this case we will simply say that L_1 is the free relatively complemented distributive extension of L and in view of the definition of R_L, and the results obtained in this section, we can now say that for $L \in \text{Ob}\,\mathscr{D}$, R_L is the free relatively complemented distributive extension of L.

Theorem 8. *Suppose that $L_1 \in \text{Ob}\,\mathscr{R}$, L is a sublattice of L_1, $1_{L,L_1}$ is in $\mathscr{D}'$ and $[L]_{\mathbf{R}} = L_1$. Then L_1 is the free relatively complemented distributive extension of L.*

Proof. We first show that $B_{L_1} = B_L$. Let L' be a subalgebra of B_{L_1} such that $L' \supseteq L$. But then $L' \cap L_1 \supseteq L$ and $L' \cap L_1$ is closed under the ternary operation defined in II.7. Hence $L' \cap L_1 \supseteq L_1$ and thus $L' \supseteq L_1$. From this we infer that $L' = B_{L_1}$ and it follows that $[L]_{\mathbf{B}} = B_{L_1}$. But the inclusion maps from L into L_1, and from L_1 into B_{L_1} are both in $\mathscr{D}'$. It follows that $B_{L_1} = B_L$. Now R_L is the smallest sublattice of B_{L_1} which contains L and which is closed under relative complementation. But then L_1 is an $\mathbf{R}$-subalgebra of R_L so $L_1 \subseteq R_L$. On the other hand L_1 is a sublattice of $B_{L_1} = B_L$ and L_1 is closed under relative complementation. Hence $L_1 \supseteq R_L$ and thus $L_1 = R_L$.

As an easy application of the preceding results we exhibit the free algebras in **R**.

Theorem 9. $\mathscr{F}_{\mathbf{R}}(S) = \mathscr{R}(\mathscr{F}_{\mathbf{D}}(S))$.

Proof. If $f\colon S \to L$ is a function and $L \in \operatorname{Ob} \mathscr{R}$ then there exists a unique $f_1 \in [\mathscr{F}_{\mathbf{D}}(S), L]_{\mathbf{D}}$ with $f_1|S = f$. Again there exists a unique $f_2 \in [\mathscr{R}(\mathscr{F}_{\mathbf{D}}(S)), L]_{\mathscr{R}}$ with $f_2|\mathscr{F}_{\mathbf{D}}(S) = f_1$, thus $f_2|S = f$. Uniqueness is obvious.

Another useful characterization of $\mathscr{F}_{\mathbf{R}}(S)$ is the following:

Theorem 10. $\mathscr{F}_{\mathbf{R}}(S) = I \cap F$ *where I is the prime ideal and F is the prime filter, in* $\mathscr{F}_{\mathbf{B}}(S)$, *generated by S, respectively.*

Proof. Let $L \in \operatorname{Ob} \mathscr{R}$ and let $f\colon S \to L$ be a function. There exists an $f^* \in [\mathscr{F}_{\mathbf{B}}(S), B_L]_{\mathscr{B}}$ with $f^*|S = f$. Let $g = f^*|I \cap F$, then $g \in [I \cap F, B_L]_{\mathscr{R}}$. It is easy to see that by the convexity of L in B_L that $g[I \cap F] \subseteq L$. So, $g \in [I \cap F, L]_{\mathscr{R}}$ and $g|S = f$. It remains to show that g is unique. Suppose $g' \in [I \cap F, L]_{\mathscr{R}}$. Let $L_1 = 0_{\mathscr{F}_{\mathbf{B}}(S)} \oplus I \cap F \oplus 1_{\mathscr{F}_{\mathbf{B}}(S)}$. Extend g' to $g'' \in [L_1, B_L]_{\mathbf{D}_{01}}$ by $g''(x) = x$ for each $x \in I \cap F$ and $g''(0_{\mathscr{F}_{\mathbf{B}}(S)}) = 0_{B_L}$ and $g''(1_{\mathscr{F}_{\mathbf{B}}(S)}) = 1_{B_L}$. Now $\mathscr{F}_{\mathbf{B}}(S) = B_{L_1}$, so we can extend g'' to $g^* \in [\mathscr{F}_{\mathbf{B}}(S), B_L]_{\mathscr{B}}$. But $g^*|S = f^*|S$, so $g^* = f^*$ and thus $g = f^*|I \cap F = g^*|I \cap F = g'$.

The previous theorem enables us to characterize the finite free algebras in **R**.

Theorem 11. *If S is finite and* $|S| = n$, *then* $\mathscr{F}_{\mathbf{R}}(S) \cong \mathbf{2}^{2^n - 2}$.

Proof. By Theorem 10, $\mathscr{F}_{\mathbf{R}}(S) = \{x \in \mathscr{F}_{\mathbf{B}}(S) : x \in I \text{ and } x \notin L \sim F\}$. Thus $\widehat{\mathscr{F}_{\mathbf{R}}(S)} = \{\hat{x} \in \widehat{(\mathscr{F}_{\mathbf{B}}(S))} : I \notin \hat{x}, L \sim F \in \hat{x}\}$. But by Theorem III.3.5, $\mathscr{F}_{\mathbf{B}}(S) \cong \mathbf{2}^{2^n}$, so $|\mathscr{S}_{\mathbf{B}}(\mathscr{F}_{\mathbf{B}}(S))| = 2^n$ and it follows that $\widehat{\mathscr{F}_{\mathbf{R}}(S)}$ and thus $\mathscr{F}_{\mathbf{R}}(S)$ is isomorphic to $\mathbf{2}^{2^n - 2}$.

7. Boolean algebras generated by a chain. Countable Boolean algebras

In this section we will consider Boolean algebras which are **B**-generated by a chain and apply the results obtained in Section 4 to this situation. Boolean algebras generated by chains were investigated first by A. Mostowski and A. Tarski [1] and later in greater detail by R. D. Mayer and R. S. Pierce [1]. Our treatment of this subject is necessarily restricted to some of the more basic results and to some special features. We will, in particular, be concerned with countable Boolean algebras. The reader who is more interested in this topic is referred to R. D. Mayer and R. S. Pierce [1], and will find some more recent results in B. Rotman [1].

We start our discussions by summarizing Theorem 4.2 and Corollary 4.4, specialized to the case of a Boolean algebra generated by a chain.

Theorem 1. *Let* $L \in \mathrm{Ob}\,\mathscr{B}$ *and let* $L = [C]_{\mathbf{B}}$ *where C is a chain in L. Suppose* $C_1 = C \cup \{0_L, 1_L\}$ *and* $C_2 = C \sim \{0_L, 1_L\}$. *Then*

(i) *If* $1_{C,L} \in \mathscr{D}'$, *then* $L = B_C$.

(ii) $L = B_{C_1}$.

(iii) *If* $C_2 \neq \varnothing$ *then every* **D**-*homomorphism from* C_2 *to a Boolean algebra can be uniquely extended to a* **B**-*homomorphism.*

Remark. The result that the free Boolean extension of a distributive lattice L is uniquely determined (up to isomorphisms) was proved first for the case that L is a chain by A. Mostowski and A. Tarski [1].

One of the attractive aspects of Boolean algebras which are generated by chains is that they can be easily characterized as "interval algebras" in the following way. Suppose $L \in \mathrm{Ob}\,\mathscr{B}$ and $L = [C]_{\mathbf{B}}$ where C is a subchain of L. Without loss of generality we will assume that 0_L and $1_L \in C$. (Thus L is the free Boolean extension B_C of C.) Let $C' = C \sim \{1_L\}$. Then the map $c \mapsto [0_L, c)_C$, $c \in C$ is an isomorphism from C onto a ring C^* of subsets of C'. If L^* is the field of subsets of C' generated by C^*, then L^* is isomorphic to L and of course $L^* = B_{C^*}$. Moreover, it is easy to see that L^* consists of all finite unions of half closed intervals $[c, c')$, $c, c' \in C$, $c \leq c'$, and therefore L is called the *interval algebra* of C. It is also immediate that if $x \in L^*$, $x \neq \varnothing$, then x has a (unique) representation $x = [c_0, x_1) \cup \cdots \cup [c_{n-1}, c_n)$, $n \geq 1$, where $c_0, c_1, \cdots, c_n \in C$ and $c_0 < c_1 < \cdots < c_n$. It therefore follows that every Boolean algebra L which is **B**-generated by a chain C can be represented as the interval algebra of C (assuming that 0_L and 1_L have been adjoined to C). We also note that L has an atom if and only if C has elements c, c' such that c' covers c (cf. Exercise 4.9). Thus L is atomless if and only if C is *dense in itself*, i.e. C has no elements c, c' for which c' covers c.

As an application of the previous discussion we prove:

Theorem 2. (R. D. Mayer and R. S. Pierce [1]) *Suppose* $L = [C]_{\mathbf{B}}$, $L \in \mathrm{Ob}\,\mathscr{B}$ *and C is a chain. If L is infinite, then L is not complete.*

Proof. We will assume that $0_L, 1_L \in C$ and use the terminology and notation introduced above. Since C is infinite, it has either a subchain of type ω or of type ω^*. Since the proofs are similar in either case we will assume the former. Thus C has a subchain $c_0 < c_1 < \cdots$. Let $a_n = [c_0, c_1) \cup [c_2, c_3) \cup \cdots \cup [c_{2n}, c_{2n+1})$ for $n = 0, 1, \cdots$. Suppose $a \in L^*$ and $a \geq a_n$ for each n. If

$$a = [d_0, d_1) \cup [d_2, d_3) \cup \cdots [d_{m-1}, d_m),\ d_0, d_1, \cdots, d_m \in C,\ d_0 < d_1 < \cdots < d_m,$$

then there must exist two consecutive intervals say, $[c_i, c_{i+1})$ and $[c_{i+2}, c_{i+3})$ which are both contained in some interval $[d_{j-1}, d_j)$ for some j, $0 \leq j \leq m-1$. But then if $a' = a \sim [c_{i+1}, c_{i+2})$ we have that $a' < a$ and $a' \geq a_n$ for each n. It follows that the set $\{a_n : n = 0, 1, \cdots\}$ has no sum in L.

Theorem 3. *Suppose $L \in \text{Ob}\, \mathscr{B}$, $L_1 \in \text{Ob}\, \mathscr{D}$, L_1 is countable and $L = [L_1]_{\mathbf{B}}$. Then L_1 has a subchain C such that $[C]_{\mathbf{B}} = L$. In particular, every countable Boolean algebra is* **B***-generated by a chain.*

Remark. The first part of the theorem is due to J. R. Büchi (G. Grätzer [4]).

Proof. Let $L_1 = \{x_1, x_2, \cdots\}$. We construct subchains $C_1, C_2, \cdots$ of L_1 by induction. Let $C_1 = \{y_1\}$ where $y_1 = x_1$. Suppose C_n has been constructed and suppose

(1) $C_n = \{y_1 \leq y_2 \leq \cdots \leq y_m\}$

Let C_{n+1} be $y_1 x_{n+1} \leq y_1 \leq y_1 + y_2 x_{n+1} \leq y_2 \leq y_2 + y_3 x_{n+1} \leq \cdots \leq y_{m-1} \leq y_{m-1} + y_m x_{n+1} \leq y_m \leq y_m + x_{n+1}$. Obviously $C_i \subseteq C_{i+1}$ for all i. Let $C = \bigcup_{i=1}^{\infty} C_i$, then C is a subchain of L_1 and let $[C]_{\mathbf{B}} = L'$. We will show that $L' = L$. It suffices to show that $L_1 \subseteq L'$. Suppose $x \in L_1$. If $x = x_1$ then $x \in C$ and thus $x \in L'$. Suppose $x = x_{n+1}$ for $n \geq 1$ and let C_n be represented by (1). Then

$$y_1 x_{n+1} + \bar{y}_1(y_1 + y_2 x_{n+1}) + \bar{y}_2(y_2 + y_3 x_{n+1}) + \cdots + \bar{y}_{m-1}(y_{m-1} + y_m x_{n+1}) + \bar{y}_m(y_m + x_{n+1}) = x_{n+1}.$$ Hence $x_{n+1} \in L'$.

Corollary 4. *If $L \in \text{Ob}\, \mathscr{D}$ and L is countable, then L has a subchain C such that $B_L \cong B_C$.*

Theorem 5. *Let $L \in \text{Ob}\, \mathscr{B}$ and let C be a chain such that $[C]_{\mathbf{B}} = L$. If $f \in [L, L_1]_{\mathscr{B}}$ is epic, then L_1 is a retract of L. In particular, if L is countable, then every* **B***-homomorphic image of L is a retract of L.*

Proof. We may assume that $0_L, 1_L \in C$ and thus $L = B_C$. Let $C_1 = f[C]$ and let $f_1 = f|C$. Then $f_1 \in [C, C_1]_{\mathscr{D}'}$. Define $g_1 \in [C_1, C]_{\mathscr{D}'}$ such that $f_1 \circ g_1 = 1_{C_1}$. Obviously $g_1 \in [C_1, L]_{\mathscr{D}'}$ and $[C_1]_{\mathbf{B}} = L_1$ and $1_{C_1, L_1} \in \mathscr{D}'$. Hence $L_1 = B_{C_1}$ and thus there exists $g \in [L_1, L]_{\mathscr{B}}$ such that $g|C_1 = g_1$. Now $(f \circ g)|C_1 = f_1 \circ g_1 = 1_{C_1}$. By uniqueness, $f \circ g = 1_{L_1}$. The second part follows from Theorem 3.

Corollary 6. *Every countable Boolean algebra is a retract of $\mathscr{F}_{\mathbf{B}}(\aleph_0)$ and is therefore projective.*

Proof. Immediate from Theorem 5, Theorem I.20.14 and the fact that weakly projective and projective mean the same in $\mathscr{B}$ (cf. Theorem IV.3.6).

The topological version of Corollary 6 has been known for a long time. Indeed we will see that the representation space of $\mathscr{F}_{\mathbf{B}}(\aleph_0)$ is the topological product of $\aleph_0$

copies of the discrete space of 2 points, and this space is (homeomorphic to) the Cantor discontinuum. Thus the topological version of Corollary 6 is: every closed subspace of the Cantor discontinuum is a retract of it (cf. Exercise VII.7.7).

We note that for the proof of Corollary 6 we used the second part of Theorem 5. However, it is also possible to give a more direct proof. We proceed as follows:

Let $L \in \mathrm{Ob}\,\mathscr{B}$ be countable. Then there exists $f \in [\mathscr{F}_{\mathscr{B}}(\aleph_0), L]_{\mathscr{B}}$ with f epic. Now $\mathscr{F}_{\mathbf{B}}(\aleph_0) = \bigcup_{n=0}^{\infty} \mathscr{F}_{\mathbf{B}}(\mathbf{n})$. For each n, let $f_n = f|\mathscr{F}_{\mathbf{B}}(\mathbf{n})$ and let $L_n = \mathrm{Im} f_n$. We can define by induction for each n, $g_n \in [L_n, \mathscr{F}_{\mathbf{B}}(\mathbf{n})]_{\mathscr{B}}$ such that $f_n \circ g_n = 1_{L_n}$ and $g_{n+1}|L_n = g_n$. Indeed, let g_0 be defined by $g_0(0) = 0$ and $g_0(1) = 1$. Suppose $g_0, g_1, \cdots, g_n$, $n > 0$, have been defined such that $f_i \circ g_i = 1_{L_i}$ for $0 \leq i \leq n$ and $g_{i+1}|L_i = g_i$ for $0 \leq i \leq n-1$. Then the existence of g_{n+1} follows from Exercise IV.3.7. Since $L = \bigcup_{n=0}^{\infty} L_n$ there exists $g \in [L, \mathscr{F}_{\mathbf{B}}(\aleph_0)]_{\mathscr{B}}$ with $f \circ g = 1_L$.

In the uncountable case much less is known about projective Boolean algebras. Obviously, for a Boolean algebra L to be projective, it is necessary and sufficient that L be a retract of a free Boolean algebra, but little is known about these retracts. In recent years, subalgebras of free Boolean algebras have been studied from a topological point of view. The dual space of such an algebra is a *dyadic space*, i.e. a space which is a continuous image of a Cantor space. We refer the interested reader to the literature (e.g. R. Engelking and A. Pelczynsky [1]). We also note that it follows from Theorem 3.24 and 3.28 that a projective Boolean algebra B must satisfy the countable chain condition and the condition that every chain in B is countable.

Exercise 7. (D. Scott and H. F. Trotter [1, page 117]) Prove that an infinite complete Boolean algebra is not projective.

Exercise 8. Prove that an infinite complete Boolean algebra cannot be embedded in a projective Boolean algebra.

Exercise 9. The Boolean algebra of all subsets of an infinite set modulo the ideal of finite subsets is not projective (and not complete).

Exercise 10. If C_1 and C_2 are chains with 0 and $C_i' = C_i \oplus \mathbf{1}$ then $B_{C_1 \oplus C_2'} \cong B_{C_1'} \times B_{C_2'}$.

We have seen that countable Boolean algebras have various special properties. We close this section with a characterization of the free Boolean algebra $\mathscr{F}_{\mathbf{B}}(\aleph_0)$. Recall (Exercise 3.5) that no infinite free Boolean algebra has atoms. We will show conversely: If L is an infinite countable Boolean algebra without atoms then L is isomorphic to $\mathscr{F}_{\mathbf{B}}(\aleph_0)$.

We first need a lemma.

Lemma 11. *Let C be a countable chain with* 0 *and* 1 *which is dense in itself. Then C is isomorphic to the unit interval* [0, 1] *of rational numbers.*

Proof. Let $C = \{c_0, c_1, \cdots\}$, $c_0 = 0$, $c_1 = 1$, $[0, 1] = \{r_0, r_1, \cdots\}$, $r_0 = 0$, $r_1 = 1$. Define $f\colon C \to [0, 1]$ as follows: $f(c_0) = r_0$, $f(c_1) = r_1$. Suppose f has been defined on $c_{i_1}, c_{i_2}, \cdots, c_{i_n}$, $i_1 = 0$, $i_2 = 1$, $n \geq 2$. If n is even proceed as follows:

Let $i_{n+1} = \min\{j : c_j \neq c_{i_k}, 1 \leq k \leq n\}$

Let $u = \max\{c_{i_k} : c_{i_k} < c_{i_{n+1}}, 1 \leq k \leq n\}$

$v = \min\{c_{i_k} : c_{i_k} < c_{i_{n+1}}, 1 \leq k \leq n\}$.

Then $u < c_{i_{n+1}} < v$ and there exists an $r \in [0, 1]$ with $f(u) < r < f(v)$. Let $f(c_{i_{n+1}}) = r$.

If n is odd, proceed as follows:

Let $i_{n+1} = \min\{j : r_j \neq f(c_{i_k}), 1 \leq k \leq n\}$

Let $u = \max\{f(c_{i_k}) : f(c_{i_k}) < r_{i_{n+1}}, 1 \leq k \leq n\}$

$v = \min\{f(c_{i_k}) : f(c_{i_k}) > r_{i_{n+1}}, 1 \leq k \leq n\}$.

Then $u < r_{i_{n+1}} < v$. Thus there exists an element $c \in C$ with $f^{-1}(u) < c < f^{-1}(v)$. Let $c_{i_{n+1}} = c$ and define $f(c_{i_{n+1}}) = r_{i_{n+1}}$. It is not difficult to see that f, so defined, is an isomorphism.

Theorem 12. *A countable atomless Boolean algebra is free.*

Proof. Let $L \in \text{Ob}\,\mathscr{B}$ and suppose L is countable and atomless. By Theorem 3, L is **B**-generated by a chain C (with 0, 1). By our remarks earlier in this section (page 106) and by hypothesis, C is dense in itself. Hence by Lemma 11, C is isomorphic to the interval [0, 1] of rationals. Thus all countable atomless Boolean algebras are isomorphic. But $\mathscr{F}_{\mathbf{B}}(\aleph_0)$ is atomless (Exercise 3.5). Hence $L \cong \mathscr{F}_{\mathbf{B}}(\aleph_0)$.

8. Monomorphisms and epimorphisms

In Theorem IV.3.6 we gave a topological proof of the fact that if $f \in [L, L_1]_{\mathscr{B}}$ is epic, then it is onto. We begin this section with an algebraic proof.

Theorem 1. *There are no proper epic subalgebras in* Ob $\mathscr{B}$.

Proof. Suppose L_1 is a proper subalgebra of L, $a \in L \sim L_1$ and $L, L_1 \in \text{Ob}\,\mathscr{B}$. Now $[a) \cap L_1$ is a filter in L_1 which is disjoint with the ideal $(a]_L \cap L_1$, so there exists a prime ideal I in L_1 such that $(a] \cap L_1 \subseteq I$ and $[a)_L \cap L_1 \cap I = \varnothing$. Then $f_I \in [L_1, \mathbf{2}]_{\mathscr{B}}$. By Corollary 2.2, f_I can be extended to a **B**-homomorphism $g_0\colon [L_1 \cup \{a\}] \to \mathbf{2}$ such that $g_0(a) = \mathbf{0}$ since $u \in L_1$, $ua = 0 \Rightarrow f_I(u)\mathbf{0} = \mathbf{0}$ and $u \in L_1$, $u\bar{a} = 0 \Rightarrow u \leq a \Rightarrow u \in I \Rightarrow f_I(u) = \mathbf{0} \Rightarrow f_I(u)\cdot\bar{\mathbf{0}} = f_I(u) = \mathbf{0}$. Similarly, there exists a homomorphism $g_1\colon [L_1 \cup \{a\}] \to \mathbf{2}$ such that $g_1(a) = \mathbf{1}$. It is evident that $g_0|L_1 = g_1|L_1$ but $g_0 \neq g_1$. Finally, by Corollary III.6.6, g_0 and g_1 can be extended to **B**-homomorphisms $g_0', g_1'\colon L \to \mathbf{2}$ such that $g_0'|L_1 = g_1'|L_1$ and $g_0' \neq g_1'$. So L_1 is not an epic subalgebra of L.

Corollary 2. *In $\mathscr{B}$ every epimorphism is onto.*

The notions of epimorphisms in $\mathscr{D}$ and $\mathscr{D}_{01}$ are more interesting because, in these categories, not every epimorphism is onto. Consider the four element Boolean algebra $L = \{0, a, \bar{a}, 1\}$ and the sublattice $L_1 = \{0, a, 1\}$ in the diagram below.

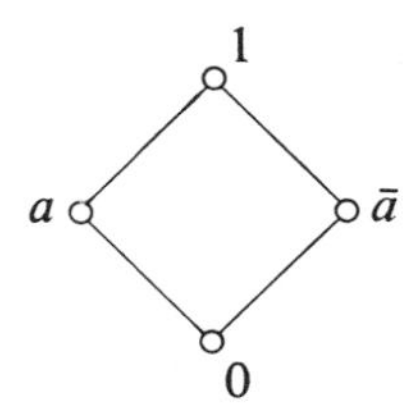

Then L_1 is a **D**-epic subalgebra of L, for if f, g are morphisms in $\mathscr{D}$ with domain L and they agree on L_1, then since $\bar{a}$ is the complement of a in $[0, 1]$, $f(\bar{a})$ and $g(\bar{a})$ will be complements of $f(a)$ in $[f(0), f(1)]$. But complements are unique, so $f(\bar{a}) = g(\bar{a})$ and $f = g$.

For a general characterization, we need the notion of "Boolean combination".

Definition 3. An element x of a distributive lattice L is said to be a *Boolean combination* of the set $\varnothing \neq S \subseteq L$, provided there exists $n > 0$ and $S_i \cup T_i \subseteq S$ for $i = 1, \cdots, n$ satisfying

(i) $x \prod\{f(i) : f(i) \in T_i, i \leq n\} \leq \sum\{f(i) : f(i) \in S_i, i \leq n\}$ for every $f \in \times_{i=1}^{n} (S_i \cup T_i)$, and

(ii) $\prod S_i \leq x + \sum T_i$ for every $i \in \{1, \cdots, n\}$.

Remark. We note the possibility: $S_i = \varnothing$ or $T_i = \varnothing$.

The next lemma justifies the term "Boolean combination".

Lemma 4. *If L is a sublattice of a Boolean algebra and S is a non-empty subset of L, then* (i) *and* (ii) *of Definition 3 are equivalent to*

(iii) $x = \sum_{i=1}^{n} (\prod S_i)(\prod T_i^{-})$

Proof. Clearly, (ii) is equivalent to $(\prod S_i)(\prod T_i^{-}) \leq x$ for each $i \in \{1, \cdots, n\}$ and hence also to $\sum_{i=1}^{n} (\prod S_i)(\prod T_i^{-}) \leq x$. For the equivalence of the reverse inequality and (i), first note that the inequality (i) is equivalent to

$$x \leq (\sum\{\overline{f(i)} : f(i) \in T_i, i \leq n\}) + (\sum\{f(i) : f(i) \in S_i, i \leq n\})$$

and so by Lemma II.5.8, (i) is equivalent to $x \leq \sum_{i=1}^{n} (\prod S_i)(\prod T_i^{-})$.

For the next theorem we rely heavily on the properties of free Boolean extensions. In particular see the remarks following Definition 4.5.

Theorem 5. *Let $L, L_1 \in \text{Ob}\, \mathscr{D}_{01}$ with L a subalgebra of L_1. The following are equivalent.*

(i) *L is an epic subalgebra of L_1.*

(ii) *The morphism $(1_{L,L_1})^*: B_L \to B_{L_1}$ is an isomorphism.*

(iii) *Each element of L_1 is a Boolean combination of L.*

(iv) *For prime ideals I_1, I_2 in L_1, $I_1 \cap L = I_2 \cap L \Rightarrow I_1 = I_2$.*

Proof. (i) $\Rightarrow$ (ii) By hypothesis, the inclusion $1_{L,L_1} \in [L, L_1]_{\mathscr{D}_{01}}$ is epic in $\mathscr{D}_{01}$ and is therefore epic in $\mathscr{D}'$. Consequently, $(1_{L,L_1})^* \in [B_L, B_{L_1}]_{\mathscr{B}}$ is epic in $\mathscr{B}$. But then, $(1_{L,L_1})^*$ is onto and since $1_{L,L_1}$ is monic, $(1_{L,L_1})^*$ is monic and is therefore an isomorphism.

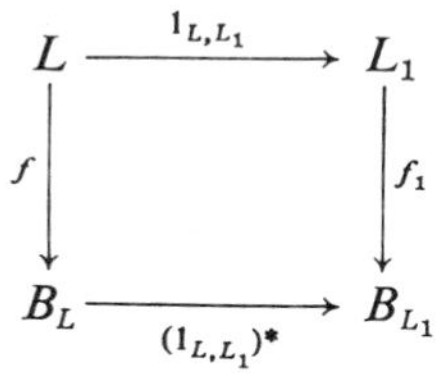

$$\begin{array}{ccc} L & \xrightarrow{1_{L,L_1}} & L_1 \\ {\scriptstyle f}\downarrow & & \downarrow{\scriptstyle f_1} \\ B_L & \xrightarrow[(1_{L,L_1})^*]{} & B_{L_1} \end{array}$$

(ii) $\Rightarrow$ (iii). Let $x \in L_1$ and (B_L, f) and (B_{L_1}, f_1) free Boolean extensions of L and L_1. Since $(1_{L,L_1})^*$ is onto there exists $u \in B_L$ such that $(1_{L,L_1})^*(u) = f_1(x)$. But $[f[L]]_{\mathbf{B}} = B_L$ so there exist $x_i, y_i \in L$ such that $u = \sum_{i=1}^n f(x_i)(\overline{f(y_i)})$. So

$$\begin{aligned} f_1(x) = (1_{L,L_1})^*(u) &= (1_{L,L_1})^*\left(\sum_{i=1}^n f(x_i)(\overline{f(y_i)})\right) \\ &= \sum_{i=1}^n ((1_{L,L_1})^* f(x_i))((1_{L,L_1})^* \overline{f(y_i)}) \\ &= \sum_{i=1}^n ((1_{L,L_1})^* f(x_i))(\overline{(1_{L,L_1})^* f(y_i)}) \\ &= \sum_{i=1}^n f_1(1_{L,L_1}(x_i))(\overline{f_1(1_{L,L_1}(y_i))}) \\ &= \sum_{i=1}^n f_1(x_i)\overline{f_1(y_i)}. \end{aligned}$$

Hence $f_1(x_i) \leq f_1(x) + f_1(y_i)$ for $i = 1, \cdots, n$ and

$$f_1(x) \prod \{f_1(y_j) : j \in J\} \leq \sum \{f_1(x_k) : k \in \{1, \cdots, n\} \sim J\}$$

for each $J \subseteq \{1, \cdots, n\}$. But f_1 is a monomorphism so we conclude that

(1) $x_i \leq x + y_i$ for $i = 1, \cdots, n$ and

(2) $x \prod \{y_j : j \in J\} \leq \sum \{x_k : k \in \{1, \cdots, n\} \sim J\}$ for each $J \subseteq \{1, \cdots, n\}$.

(iii) $\Rightarrow$ (iv) Assume the hypothesis of (iv) and further suppose that $x \in I_2 \sim I_1$ where x satisfies (1) and (2). Let $J_o = \{j < n : y_j \notin I_1\}$. Then by (2), with $J = J_o$, we have $x_{j_o} \notin I_1$ for some $j_o \notin J_o$. Therefore $x_{j_o} \notin I_2$ and hence $y_{j_o} \notin I_2$. Thus $y_{j_o} \notin I_1$ and this contradicts the definition of J_o.

(iv) $\Rightarrow$ (i) Suppose that L is not an epic subalgebra of L_1. Then there exist distinct $h_1, h_2 \in [L_1, L_2]_{\mathscr{D}_{01}}$ such that $h_1|L = h_2|L$. Since $h_1 \neq h_2$, there exist $x \in L_1 \sim L$ such that $h_1(x) \neq h_2(x)$ say $h_1(x) \nleq h_2(x)$. So there exists a prime ideal I in L_2 such that $h_2(x) \in I$ and $h_1(x) \notin I$. Now $h_1^{-1}[I]$ and $h_1^{-1}[I]$ are distinct prime ideals in L_1 but $h_1^{-1}[I] \cap L = h_2^{-1}[I] \cap L$ since $h_1|L = h_2|L$.

Corollary 6. For a morphism $f \in [L, L']_{\mathscr{D}_{01}}$, the following are equivalent: (i) f is an epimorphism. (ii) $(1_{f[L],L'})^*$ is an isomorphism. (iii) Each member of L' is a Boolean combination of $f[L]$.

To obtain the analogous results for $\mathscr{D}$ recall the definition of L_{01} and f_{01} from Section II.5.

Corollary 7. *Let $L, L' \in \mathrm{Ob}\,\mathscr{D}$ and suppose L is a subalgebra of L'. The following are equivalent:*

(i) *L is an epic subalgebra of L.*

(ii) *Each member of L' is a Boolean combination of members of L_{01}.*

(iii) $I_1, I_2 \in \mathfrak{P}(L') \cup \{\phi, L'\}, I_1 \cap L = I_2 \cap L \Rightarrow I_1 = I_2$.

Proof. (i) $\Rightarrow$ (ii) Since L is an epic subalgebra of L' in the category $\mathscr{D}$, L_{01} is an epic subalgebra of L'_{01} in the category $\mathscr{D}_{01}$. (We are identifying the adjoined 0(1) of L_{01} with the corresponding element of L'_{01}, i.e. $0_{L_{01}} = 0_{L'_{01}}$ and $1_{L_{01}} = 1_{L'_{01}}$.) From the proof of Theorem 5 we see that for each $x \in L'$ there exist $x_i, y_i \in L_{01}$ such that $x = \sum_{i=1}^n x_i\bar{y}_i$ where the complements are taken in $B_{L'_{01}}$. Since x_i and y_i may be $0_{L_{01}}$ or $1_{L_{01}}$, x can be written in the form $x = (\sum_{i=1}^n x_i'\overline{y_i'}) + x_0' + \overline{y_0'}$ where x_0', y_0', x_i', $y_i' \in L$, but neither x_0' nor $\overline{y_0'}$ need actually occur. It is now easy to show that x is a Boolean combination of L.

(ii) $\Rightarrow$ (iii) and (iii) $\Rightarrow$ (i) are proved in the same way as Theorem 5.

Exercise 8. Use Corollary 7 to verify the example given before Definition 3.

9. Injectives in $\mathscr{D}$, $\mathscr{D}_{01}$ and $\mathscr{B}$

One of the significant results in the theory of Boolean algebras is Sikorski's theorem [1, Lemma 2] that every complete Boolean algebra is injective. This is, in fact, the key result needed for determining injectives in $\mathscr{D}$, $\mathscr{D}_{01}$ and $\mathscr{B}$.

Lemma 1. *Let $\mathscr{K}$ be one of the categories $\mathscr{D}$, $\mathscr{D}_{01}$ or $\mathscr{B}$. If L_1 is a $\mathscr{K}$-retract of L and L is complete then L_1 is also complete.*

Proof. Let $f: L \to L_1$ and $g: L_1 \to L$ be **K**-homomorphisms such that $f \circ g = 1_{L_1}$. Now let $S \subseteq L_1$. Then, since L is complete, $x = \sum^L g[S]$ exists. We will show that $f(x)$ is the least upper bound of S in L_1. Indeed for $s \in S$, $g(s) \leq x$ so $s = f(g(s)) \leq f(x)$. If $s \leq y$ for all $s \in S$, $y \in L_1$ then $g(s) \leq g(y)$ for all $s \in S$ so $x \leq g(y)$ thus $f(x) \leq f(g(y)) = y$.

Theorem 2. (R. Sikorski [1]) *An algebra is injective in $\mathscr{B}$ if and only if it is complete.*

Proof. If L is injective in $\mathscr{B}$, then the map $g: L \to Y$ where Y is the power set of $\mathfrak{P}(L)$ and such that $g(x) = \hat{x}$ is clearly monic in $\mathscr{B}$ so L is a retract of Y. Hence by Lemma 1, L is complete. Now suppose that L is a complete Boolean algebra, L_0 is a subalgebra of L_1 and $f \in [L_0, L]_{\mathscr{B}}$. Of the subalgebras of L_1 which are domains of extensions of f, there exists, by Zorn's lemma, a maximal such subalgebra L'. Let $f_1 \in [L', L]_{\mathscr{B}}$ be such an extension. If $L' \subset L_1$ then choose an element $a \in L_1 \sim L'$ and set $b = \sum^L f_1[(a]_{L_1} \cap L']$. Then

(1) $ua = 0, u \in L' \Rightarrow f_1(u)b = 0$ and

(2) $u\bar{a} = 0, u \in L' \Rightarrow f_1(u)\bar{b} = 0$.

Indeed, if the hypothesis of (1) holds, then clearly $uv = 0$ for any $v \in (a]_{L_1} \cap L'$ so by Theorem II.6.9(iv), $f_1(u)b = f_1(u) \sum^L f_1[(a]_{L_1} \cap L'] = \sum^L \{f_1(uv) : v \in (a]_{L_1} \cap L'\} = 0$; and if the hypothesis of (2) holds, then $u \leq a$ so $f_1(u) \in f_1[(a]_{L_1} \cap L']$ and hence $f_1(u) \leq b$. But (1) and (2) imply, by Corollary 2.2, that there exists a subalgebra of L_1 which contains L' properly and to which f_1 can be extended. This is, of course, a contradiction and we have therefore that $L' = L_1$.

The injectives in $\mathscr{D}$ and $\mathscr{D}_{01}$ were determined in R. Balbes [1]; cf. B. Banaschewski and G. Bruns [2].

Theorem 3. *An algebra is injective in $\mathscr{D}_{01}$ if and only if it is a complete Boolean algebra.*

Proof. Suppose L is injective in $\mathscr{D}_{01}$. As before, the map $x \mapsto \hat{x}$ from L into the power set Y of $\mathfrak{P}(L)$ is monic in $\mathscr{D}_{01}$ so L is a retract of Y. It follows again from Lemma 1 that L is complete. But L is also a Boolean algebra since it is a homomorphic image of a Boolean algebra. Conversely, by Corollary 4.3 the reflector $\mathscr{B}_{01}: \mathscr{D}_{01} \to \mathscr{B}$ preserves monomorphisms so by Theorem I.18.6 the injective objects in $\mathscr{B}$ are also injectives in $\mathscr{D}_{01}$. But complete Boolean algebras are injective in $\mathscr{B}$ and therefore are also injective in $\mathscr{D}_{01}$.

Corollary 4. *An algebra is injective in $\mathscr{D}$ if and only if it is a complete Boolean algebra.*

Proof. ($\Rightarrow$) Analogous to the proof in Theorem 3.

($\Leftarrow$) By Theorem II.5.7 the reflector $\mathscr{U}: \mathscr{D} \to \mathscr{D}_{01}$ preserves monomorphisms so the complete Boolean algebras, being injectives in $\mathscr{D}_{01}$, are injectives in $\mathscr{D}$, again by Theorem I.18.6.

Corollary 5. *$\mathscr{D}$, $\mathscr{D}_{01}$ and $\mathscr{B}$ have enough injectives.*

Proof. Let $\mathscr{K}$ be one of the categories $\mathscr{D}$, $\mathscr{D}_{01}$, or $\mathscr{B}$. Now we have seen that each algebra $A \in \mathscr{K}$ can be embedded in the power set of prime ideals of A. The power set, being a complete Boolean algebra, is injective in $\mathscr{K}$.

Consequently $\mathscr{D}$, $\mathscr{D}_{01}$ and $\mathscr{B}$ have the congruence extension property, the amalgamation property and injective hulls exist for each of the algebras in these categories. For an alternative proof of the amalgamation property for $\mathscr{D}$, $\mathscr{D}_{01}$ and $\mathscr{B}$ see Theorem VII.8.4. For a characterization of these injective hulls, see Section XII.3.

Exercise 6. Prove that the injectives in $\mathscr{R}$ are the complete Boolean algebras.

Exercise 7. Use Corollary 4 to prove Theorem III.6.5(ii).

For a discussion of the independence of Sikorski's theorem and the axiom of choice, see e.g. W. A. J. Luxemburg [1]. Also in connection with Sikorski's theorem see A. Monteiro [6] and R. Cignoli [2].

Other results concerning injectives in $\mathscr{D}$ and $\mathscr{B}$ can be found in B. Banaschewski and G. Bruns [2] and A. Day [1].

10. Weak projectives in $\mathscr{D}$ and $\mathscr{D}_{01}$

The results in this section are mainly due to R. Balbes and A. Horn [2] (see also G. Grätzer and B. Wolk [1]). The main results are Theorem 7 and Corollary 8.

Since, in $\mathscr{D}$ and $\mathscr{D}_{01}$, not every epimorphism is onto, the concepts of projective and weakly projective differ.

Theorem 1. *There are no projectives in $\mathscr{D}$ and the only projective in $\mathscr{D}_{01}$ is* **2**.

Proof. **2** is projective in $\mathscr{D}_{01}$ since 0, 1 are nullaries. Let $g: \mathbf{3} \to \mathbf{2} \times \mathbf{2}$ be defined by $g(\mathbf{0}) = (\mathbf{0}, \mathbf{0})$, $g(\mathbf{1}) = (\mathbf{1}, \mathbf{0})$, $g(\mathbf{2}) = (\mathbf{1}, \mathbf{1})$. Then g is epic in both $\mathscr{D}$ and $\mathscr{D}_{01}$. For any $L \in \mathrm{Ob}\,\mathscr{D}$ let $f: L \to \mathbf{2} \times \mathbf{2}$ be the **D**-homomorphism that sends all elements into $(\mathbf{0}, \mathbf{1})$. Then, if there exists $h \in [L, \mathbf{3}]_{\mathscr{D}}$ such that $g \circ h = f$, we would have for $u \in L$, $g(h(u)) = f(u) = (\mathbf{0}, \mathbf{1})$ contradicting $g[\mathbf{3}] = \{(\mathbf{0}, \mathbf{0}), (\mathbf{1}, \mathbf{0}), (\mathbf{1}, \mathbf{1})\}$.

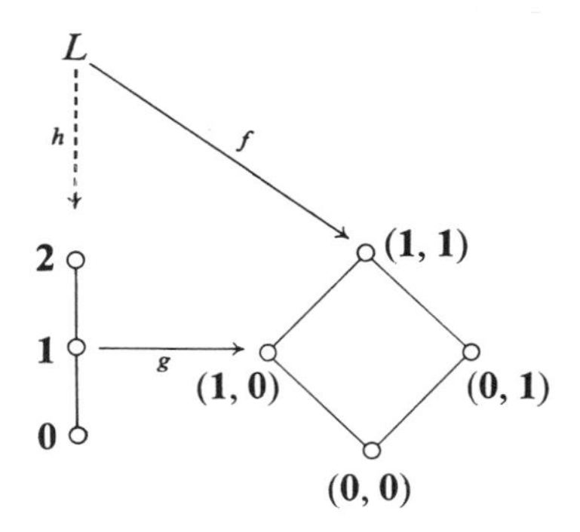

So there are no projectives in $\mathscr{D}$. Next, suppose L is projective in $\mathscr{D}_{01}$, $L \neq \mathbf{2}$. Now $L \neq \mathbf{1}$ since $\mathbf{1}$ is not a $\mathscr{D}_{01}$-retract of $\mathbf{2}$ so L has at least two prime ideals say $I_1 \neq I_2$. Let $x \in I_1 \sim I_2$. Define $f: L \to \mathbf{2} \times \mathbf{2}$ by $f = f_{I_1} \times f_{I_2}$. Then $f \in [L, \mathbf{2} \times \mathbf{2}]_{\mathscr{D}_{01}}$ and if there exists $h \in [L, \mathbf{3}]_{\mathscr{D}_{01}}$ such that $g \circ h = f$ then $g(h(x)) = f(x) = (\mathbf{0}, \mathbf{1})$, again a contradiction.

The problem of determining the weak projectives in $\mathscr{D}_{01}$ can be reduced to the corresponding problem in the category $\mathscr{D}$.

Theorem 2. *An object L of $\mathscr{D}_{01}$ is weakly projective if and only if $L = \mathbf{2}$ or $L \sim \{0, 1\}$ is a sublattice of L which is weakly projective in $\mathscr{D}$.*

Proof. $(\Rightarrow)$ Clearly $L \neq \mathbf{1}$, so we assume $L \neq \mathbf{1}, \mathbf{2}$. Let $f \in [\mathscr{F}_{\mathbf{D}_{01}}(|L|), L]_{\mathscr{D}_{01}}$ be onto. Since L is weakly projective in $\mathscr{D}_{01}$ there exists, by Theorem I.20.14, $g \in [L, \mathscr{F}_{\mathbf{D}_{01}}(|L|)]_{\mathscr{D}_{01}}$ such that $f \circ g = 1_L$. Suppose, $x, y \in L \sim \{0, 1\}$. If $x + y = 1$, then $1 = g(1) = g(x + y) = g(x) + g(y)$. But 1 is join irreducible in $\mathscr{F}_{\mathbf{D}_{01}}(|L|)$ so say $g(x) = 1$. Thus $1 = f(g(x)) = x$. Similarly, $xy \neq 0$. Let

$$h \in [\mathscr{F}_{\mathbf{D}}(|L \sim \{0, 1\}|), L \sim \{0, 1\}]_{\mathscr{D}}$$

be onto. Then $h_{01} \in [\mathscr{F}_{\mathbf{D}_{01}}(|D \sim \{0, 1\}|), L]_{\mathscr{D}}$ (for the definition of h_{01} we refer to Section II.5). But L is weakly projective in $\mathscr{D}_{01}$ so there exists $g \in [L, \mathscr{F}_{\mathbf{D}_{01}}(|L \sim \{0, 1\}|)]_{\mathscr{D}_{01}}$ such that $h_{01} \circ g = 1_L$. But g is one-one and $g(0) = 0$, $g(1) = 1$ so $(h \circ g)|L \sim \{0, 1\} = 1_{L \sim \{0,1\}}$. Hence $L \sim \{0, 1\}$ is projective in $\mathscr{D}$.

$(\Leftarrow)$ $L \sim \{0, 1\}$ is a retract in $\mathscr{D}$ of a free algebra L' thus L is a retract in $\mathscr{D}_{01}$ of L'_{01}, which is free in $\mathbf{D}_{01}$.

We now consider the weak projectives in $\mathscr{D}$. As in $\mathscr{B}$ (see Section 7) there is, at this time, a satisfactory characterization of weak projectivity only for the countable case.

Our first step is to determine those properties of free distributive lattices which are inherited by their retracts. The setting for the remainder of this section will be the equational class $\mathbf{D}$ and the category $\mathscr{D}$. Prefixes and subscripts etc. pertaining to $\mathbf{D}$ and $\mathscr{D}$ will often be omitted.

Consider the following conditions on a distributive lattice L.

(1) Every element of L is a finite sum of members of $\mathscr{J}(L)$.

(2) Every element of L is a finite product of elements of $\mathscr{M}(L)$.

(3) $\mathscr{J}(L)$ is closed under finite products.

(4) $\mathscr{M}(L)$ is closed under finite sums.

Theorem 3. *Suppose L and L_1 are distributive lattices and L_1 is a retract of L, then*

(i) *If L satisfies* (1) *then L_1 satisfies* (1).

(ii) *If L satisfies* (1) *and* (3) *then L_1 satisfies* (4).

Proof. Let $f: L \to L_1$ and $g: L_1 \to L$ be homomorphisms such that $f \circ g = 1_{L_1}$. To show (i) let $x \in L_1$. Then, since L satisfies (1), there exists $S \subseteq \mathscr{J}(L)$ such that $g(x) = \sum S$ and so $x = \sum f[S]$. Let T be the maximal elements of $f[S]$ and note that

$x = \sum T$ so it will suffice to show that $T \subseteq \mathscr{J}(L_1)$. Thus let $t_o \in T$ and $t_o \leq u + v$. We have $t_o = f(s)$ for some $s \in S$ and

$$s \leq g(x) = \sum g[T] \leq \sum g[T \sim \{t_o\}] + (g(u) + g(v)).$$

If $s \leq g(t)$ for some $t \in T \sim \{t_o\}$ then $t_o = f(s) \leq f(g(t)) = t$, contradicting the maximality of t_o. But $s \in \mathscr{J}(L)$ so $s \leq g(u)$ or $s \leq g(v)$ and hence $t_o = f(s) \leq u$ or $t_o \leq v$. To show (ii) suppose $a, b \in \mathscr{M}(L_1)$ but that there exist $c, d \in L_1$ such that $cd \leq a + b, c \nleq a + b, d \nleq a + b$. By hypothesis there exist $S \subseteq \mathscr{J}(L), T \subseteq \mathscr{J}(L)$ such that $g(c) = \sum S, g(d) = \sum T$. Now if $f(s) \leq a + b$ for all $s \in S$ then $c = f(g(c)) = f(\sum S) = \sum f[S] \leq a + b$, a contradiction; so there exists $s_o \in S$ such that $f(s_o) \nleq a + b$. Similarly there exists $t_o \in T$ such that $f(t_o) \nleq a + b$. But $s_o t_o \leq g(c)g(d) \leq g(a) + g(b)$ and by hypothesis $s_o t_o \in \mathscr{J}(L)$ so $s_o t_o \leq g(a)$ or $s_o t_o \leq g(b)$. Say the former holds, then $f(s_o)f(t_o) \leq a$. Since $a \in \mathscr{M}(L_1), f(s_o) \leq a$ or $f(t_o) \leq a$ and hence $f(s_o) \leq a + b$ or $f(t_o) \leq a + b$, a contradiction.

Corollary 4. *In any distributive lattice* (1) *and* (3) *imply* (4).

Proof. Apply Theorem 3(ii) in the case $L = L_1$.

Theorem 5. *If L is weakly projective, then L satisfies* (1), (2), (3), and (4).

Proof. Since L is a retract of a free algebra A, Theorem 3.7 implies that A satisfies (1), (2), and (3). So by Theorem 3, L satisfies (1) and (4), and by duality it also satisfies (2) and (3).

Lemma 6. *Suppose L and L_1 are distributive lattices and L satisfies* (1). *If $h: \mathscr{J}(L) \to L_1$ is an order preserving map then h can be extended to a function $h_1: L \to L_1$ such that $h_1(x + y) = h_1(x) + h_1(y)$ for all $x, y \in L$. If L satisfies* (3) *and $h(xy) = h(x)h(y)$ for all $x, y \in \mathscr{J}(L)$ then h_1 is a homomorphism.*

Proof. For $x = \sum S, S \subseteq \mathscr{J}(L)$, define $h_1(x) = \sum h[S]$. Now h is well defined for if $\sum S = \sum S_1, S_1 \subseteq \mathscr{J}(L)$ then for each $s \in S, s \leq s_1$ for some $s_1 \in S$ so $h(s) \leq \sum h[S_1]$ and hence $\sum h[S] \leq \sum h[S_1]$. Similarly, $\sum h[S_1] \leq \sum h[S]$. It is easy to verify the remaining conditions.

Consider the following condition on a distributive lattice L.

(5) There exists an onto homomorphism $f: \mathscr{F}_{\mathbf{D}}(\alpha) \to L$ for some α and an order preserving map $h: \mathscr{J}(L) \to \mathscr{F}_{\mathbf{D}}(\alpha)$ such that $f \circ h = 1_{\mathscr{J}(L),L}$.

Theorem 7. *A distributive lattice is weakly projective if and only if it satisfies* (1), (2), (3), *and* (5).

Remark. Condition (5) does not appear to be much of an improvement over the characterization of weak projectives as a retract of a free distributive lattice. But it will be shown that a countable distributive lattice satisfying (1) and (3) automatically satisfies (5).

Proof of Theorem 7. ($\Rightarrow$) (1), (2), and (3) follow from Theorem 5, and clearly (5) is also a property of weak projectives, since they are retracts of free algebras.

($\Leftarrow$) There exists an onto homomorphism $f: \mathscr{F}_{\mathbf{D}}(\alpha) \to L$ for some α. Let $h: \mathscr{J}(L) \to \mathscr{F}_{\mathbf{D}}(\alpha)$ be an order preserving map such that $f \circ h = 1$. By Lemma 6, there exists a function $h_1: L \to \mathscr{F}_{\mathbf{D}}(\alpha)$ such that $h(x) + h(y) = h(x + y)$, $x, y \in \mathscr{J}(L)$. Evidently $f \circ h_1 = 1_L$. For each $x \in \mathscr{J}(L)$, let $m(x)$ be a finite subset of $\mathscr{M}(L)$ such that $x = \prod m(x)$. Define $g: \mathscr{J}(L) \to \mathscr{F}_{\mathbf{D}}(\alpha)$ by $g(x) = \prod h_1[m(x)]$. g is well defined and preserves order for if $\prod m(x) \leq \prod m(y)$ then for each $v \in m(y)$ there exists $u \in m(x)$ such that $u \leq v$ so $h_1(u) \leq h_1(v)$ and consequently $\prod h_1[m(x)] \leq \prod h_1[m(y)]$. Now g also preserves products for we already know that $g(xy) \leq g(x)g(y)$. For the reverse inequality, we have $\prod (m(x) \cup m(y)) = \prod m(x) \prod m(y) = xy = \prod m(xy)$, so for each $u \in m(xy)$ there exists $v \in m(x) \cup m(y)$ such that $v \leq u$ thus $h_1(v) \leq h_1(u)$ and hence $\prod h_1[m(x)] \prod h_1[m(y)] \leq h_1(u)$. Since this is true for each $u \in m(xy)$, we have $g(x)g(y) \leq g(xy)$. The hypotheses of both parts of Lemma 6 are satisfied by g, so it can be extended to a homomorphism $g_1 : L \to \mathscr{F}_{\mathbf{D}}(\alpha)$. Finally, for $x \in \mathscr{J}(L)$, $f(g_1(x)) = f(g(x)) = f(\prod h_1[m(x)]) = \prod f(h_1[m(x)]) = \prod m(x) = x$ and hence $f \circ g_1 = 1_L$.

Corollary 8. *A countable distributive lattice L is weakly projective if and only if it satisfies* (1), (2) *and* (3).

Proof. For sufficiency we only need to prove that (5) holds. There exists an onto homomorphism $f: \mathscr{F}_{\mathbf{D}}(\alpha) \to L$ for $\alpha = |L|$. For each $x \in \mathscr{J}(L)$, let $g(x)$ be a member of $\mathscr{F}_{\mathbf{D}}(\alpha)$ such that $f(g(x)) = x$. We can assume $\mathscr{J}(L) = \{x_i : i = 1, 2, 3, \cdots\}$. Define $h: \mathscr{J}(L) \to \mathscr{F}_{\mathbf{D}}(\alpha)$ by $h(x_1) = g(x_1)$ and for $n \geq 1$

$$h(x_n) = g(x_n) \prod \{h(x_i) : x_i > x_n, i < n\} + \sum \{h(x_i) : x_i < x_n, i < n\}.$$

By induction it can be verified that h preserves order and $f(h(x_n)) = x_n$ for all n.

Corollary 9. *A finite distributive lattice is weakly projective if and only if the product of join irreducible elements is join irreducible.*

Proof. (1) and (2) are satisfied in every finite distributive lattice so Corollary 8 yields the result.

Exercise 10. Show that the distributive lattice in the diagram is not weakly projective.

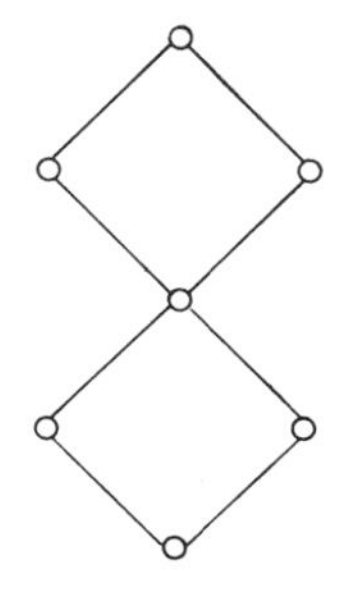

Exercise 11. Prove that the function h defined in Corollary 8 preserves order.

Exercise 12. Show that every a-disjoint set in a weakly projective distributive lattice is finite.

Exercise 13. Let $(L_s)_{s \in S}$ be a family of distributive lattices, $|S| > 1$ and $|L_s| > 1$ for each $s \in S$. Prove that $\times_{s \in S} L_s$ is weakly projective if and only if (i) L_s is weakly projective for each $s \in S$. (ii) S is finite. (iii) Each L_s has a 0 and 1.

VI

Subdirect Products of Chains

1. Introduction

We have seen that every distributive lattice is a subdirect product of copies of **2**. In this chapter we will investigate those distributive lattices which are subdirect products of chains of more than two elements. Although it is not possible to give a full characterization of these lattices, we will see that a good deal can be said about them and that there are important classes of lattices whose members have this property. The basic work in this area is due to F. W. Anderson and R. L. Blair [1].

In order to facilitate our discussions we will introduce some terminology and notation. If K is a chain, $K \in \mathbf{C}$, then the class of all distributive lattices which are subdirect products of copies of K, will be denoted by $\mathbf{C}_K$. The setting for this chapter will be the class $\mathbf{D}$ and the category $\mathscr{D}$. Therefore prefixes and subscripts pertaining to $\mathbf{D}$ and $\mathscr{D}$ will often be omitted.

It should be noted that if $L \in \mathbf{C}_K$ then the embedding and the projections are in $\mathscr{D}'$. Indeed, suppose $f: L \to \bigtimes_{s \in S} K_s$ is the embedding, such that for each $s \in S$, $K_s \cong K$. If $0_L \in L$ then $f(0_L)$ is the zero element of $\bigtimes_{s \in S} K_s$. Otherwise, there would exist an element $a \in \bigtimes_{s \in S} K_s$ such that $a < f(0_L)$. This implies there exists an $s' \in S$ such that $a(s') < f(0_L)(s')$ which again implies the existence of an element $a' \in L$ such that $f(a')(s') = a(s')$, so $a(s') < f(0_L)(s') \leq f(a')(s')$, a contradiction.

For each chain K, $|K| > 2$, $\mathbf{C}_K$ is non-empty ($K \in \mathbf{C}_K$) and it should also be noted that for each such chain K, the lattice $\mathbf{1} \in \mathbf{C}_K$. But on the other hand, if $|K| > 2$, then there are well-known classes of lattices that do not belong to $\mathbf{C}_K$. Indeed, if L is a relatively complemented distributive lattice, $L \neq \mathbf{1}$, in particular if L is in $\mathbf{B}$, then $L \notin \mathbf{C}_K$ for each K, $|K| > 2$. This follows from the fact that if $L \in \mathbf{C}_K$, $|K| > 2$ then K is a homomorphic image of L and therefore L has prime ideals which are not maximal. But by Nachbin's theorem (III.6.3) this implies that L is not relatively complemented. This situation suggests that for a lattice L in $\mathbf{D}$ there is some relationship between the property that L belongs to $\mathbf{C}_K$ ($|K| > 2$) and the property that L has elements which are *not* relatively complemented. In other words, it seems reasonable to conjecture that the "fewer" relatively complemented elements L has, the greater the "probability" is that $L \in \mathbf{C}_K$ for some K, $|K| > 2$. But although the existence of non-relatively complemented elements is a necessary condition for L to belong to some $\mathbf{C}_K$, $|K| > 2$, it is by no means sufficient. For example, we will see that there are distributive lattices without any relatively complemented elements which do not belong to $\mathbf{C}_3$ (and therefore they do not belong to any $\mathbf{C}_K$ for $|K| > 2$). But surprisingly, if L is in $\mathbf{D}_{01}$ and $|L| > 2$ and L has no relatively complemented elements other than 0 and 1, then $L \in \mathbf{C}_3$.

It may already be clear from the previous discussion that the situation is rather complicated, and indeed the problem of characterizing the class $\mathbf{C}_K$ for arbitrary K is by no means completely solved. Even in the case of $\mathbf{C_n}$, $n \geq 3$, where we can say much more, there are still open questions. On the other hand, we will see that the class $\mathbf{C_3}$ can be fully characterized in a quite satisfactory way.

Although a major portion of this chapter will be devoted to the classes $\mathbf{C_n}$, $n \geq 3$, and in particular to $\mathbf{C_3}$, we will also present some results on the infinite case. We will therefore set up the machinery in such a way that it is general enough to be applied to the infinite case. A final remark is in order. We have already mentioned Nachbin's theorem in this context. We will see that this theorem turns out to be a corollary of one of the results in this chapter (in an even somewhat more general form) and we will, in particular, have an opportunity to state some more results on prime ideals (Section 7).

2. Definitions, some basic theorems, and an example

We have already defined, in Section 1, the class $\mathbf{C}_K$ for K in $\mathbf{C}$. However, it will be useful to give a slightly more general definition.

Definition 1. Let $\mathbf{K}$ be a subclass of $\mathbf{C}$. Then $\mathbf{C_K}$ is the class of all algebras in $\mathbf{D}$ which are subdirect products of members of $\mathbf{K}$. In particular, if $\mathbf{K} = \{K\}$ then we will write $\mathbf{C}_K$ instead of $C_{\{K\}}$.

Theorem 2. *Let* $\mathbf{K}$ *be a class of chains. Then* $L \in \mathbf{C_K}$ *if and only if for each* $a < b$ *in* L *there exists* $K \in \mathbf{K}$ *and an onto homomorphism* $h: L \to K$ *such that* $h(a) \neq h(b)$.

Proof. ($\Rightarrow$) Immediate, since the embedding is one-one.

($\Leftarrow$) For every $a < b$ in L there exists, by hypothesis, $K_{a,b} \in \mathbf{K}$ and an onto homomorphism $h_{a,b}: L \to K_{a,b}$ such that $h_{a,b}(a) \neq h_{a,b}(b)$. We apply Theorem I.9.2 to prove that L is a subdirect product of $(K_{a,b})_{a<b}$. Suppose $a \neq b$. Without loss of generality we may assume that $ab < a$. Then $h_{ab,a}(ab) \neq h_{ab,a}(a)$, but this implies that $h_{ab,a}(a) \neq h_{ab,a}(b)$.

As an application of this theorem we now prove:

Theorem 3. $\mathscr{F}_{\mathbf{D}}(S)$, $|S| \geq 2$, *belongs to* $\mathbf{C}_K$ *where* K *is any chain such that* $|K| = |S|$. *In particular,* $\mathscr{F}_{\mathbf{D}}(\mathbf{n}) \in \mathbf{C_n}$ *for* $n \geq 2$.

Proof. Case 1: *S is finite,* $|S| = n \geq 2$. Suppose $a, b \in \mathscr{F}_{\mathbf{D}}(S)$, $a < b$. There exists a prime ideal I of $\mathscr{F}_{\mathbf{D}}(\mathbf{n})$ such that $a \in I$, $b \notin I$. By Theorem V.3.12, I is contained in a chain $I_0 \subset I_1 \subset \cdots \subset I_{n-2}$ of prime ideals. Define $h: \mathscr{F}_{\mathbf{D}}(\mathbf{n}) \to \mathbf{n}$ by $h(x) = \min\{\mathbf{j} : x \in I_j, 0 \leq j \leq n-2\}$ if $x \in \bigcup_{j=0}^{n-2} I_j$ and $h(x) = \mathbf{n} - \mathbf{1}$ if $x \notin \bigcup_{j=0}^{n-2} I_j$. Then $h \in [\mathscr{F}_{\mathbf{D}}(\mathbf{n}), \mathbf{n}]_{\mathscr{D}}$ and $h(a) \neq h(b)$. Apply Theorem 2. Case 2: *S is infinite.* Suppose again, that $a, b \in \mathscr{F}_{\mathbf{D}}(S)$, $a < b$. There exists a subset $S_1 \subseteq S$ such that $a, b \in [S_1]$. Note that S_1 freely generates $[S_1]$. Since $a < b$ there exists a $\mathbf{D}$-homomorphism f of $[S_1]$ onto a two element subchain of K. Define a map $g: S \to K$

which is onto and such that $g|S_1 = f|S_1$. Such a map exists, since S is infinite and $|S| = |K|$. Extend g to a homomorphism $h: \mathscr{F}_{\mathbf{D}}(S) \to K$; then $h|[S_1] = f$ and thus $h(a) \neq h(b)$. Again apply Theorem 2.

3. Basic K-chains

We have already seen that the poset $\mathfrak{P}(L)$ of a distributive lattice L plays a role in determining whether $L \in \mathbf{C}_K$, $|K| \geq 3$. In this section we will state necessary and sufficient conditions in order that this be the case.

Definition 1. Let $L \in \mathbf{D}$, $K \in \mathbf{C}$. A chain $\mathfrak{C}$ of subsets of L is a *basic K-chain* of L if:

(i) $\mathfrak{C}$ is isomorphic to K.

(ii) $A \in \mathfrak{C} \Rightarrow A \in \{L\} \cup \mathfrak{P}(L)$.

(iii) For every $a \in L$ there exists a least $A \in \mathfrak{C}$ such that $a \in A$.

(iv) For every $A \in \mathfrak{C}$ there exists an element $a \in L$ such that A is the least member of $\mathfrak{C}$ containing a.

If $\mathfrak{C}$ is a basic K-chain, then we will often index the members of $\mathfrak{C}$ by the elements of K. Thus $\mathfrak{C} = (A_k)_{k \in K}$ and $k_1 \leq k_2 \Leftrightarrow A_{k_1} \subseteq A_{k_2}$ for $k_1, k_2 \in K$.
The next theorem illustrates the significance of basic K-chains.

Theorem 2. (F. W. Anderson and R. L. Blair [1]) *Let $L \in \mathbf{D}$ and $K \in \mathbf{C}$.*

(i) *If $h: L \to K$ is an onto homomorphism, then $(A_k)_{k \in K}$ is a basic K-chain, where $A_k = \{a \in L : h(a) \leq k\}$ for each $k \in K$.*

(ii) *If $\mathfrak{C}$ is a basic K-chain of L and $h: L \to K$ is defined by letting $h(a)$ be the least k, such that $a \in A_k$, then h is an onto homomorphism.*

Proof. (i) If $k_1 \leq k_2$, $k_1, k_2 \in K$ then obviously $A_{k_1} \subseteq A_{k_2}$. Suppose $k_1 < k_2$. Let $a \in L$ such that $h(a) = k_2$; then $a \in A_{k_2}$ but $a \notin A_{k_1}$ thus $A_{k_1} \subset A_{k_2}$ and so $(A_k)_{k \in K}$ is isomorphic to K. It is immediate that if $A_k \neq L$ then A_k is a prime ideal. It is also easy to see that (iii) and (iv) of Definition 1 are satisfied. (ii) By hypothesis h is well defined and it is obvious that $a_1 \leq a_2 \Rightarrow h(a_1) \leq h(a_2)$. To show that $h(a_1 + a_2) \leq h(a_1) + h(a_2)$ for $a_1, a_2 \in L$ suppose $h(a_1) = k_1$, $h(a_2) = k_2$; then we may assume that $k_1 \leq k_2$. It follows that $a_1, a_2 \in A_{k_2}$ and thus $a_1 + a_2 \in A_{k_2}$ and therefore $h(a_1 + a_2) \leq k_2 = k_1 + k_2$. For products we show $h(a_1 a_2) \geq h(a_1)h(a_2)$. If $h(a_1 a_2) = k$ then $a_1 a_2 \in A_k$. Since either $A_k = L$ or A_k is a prime ideal, we may assume $a_1 \in A_k$, hence $h(a_1) \leq k = h(a_1 a_2)$ thus $h(a_1)h(a_2) \leq h(a_1 a_2)$. Finally, since $\mathfrak{C}$ satisfies (iv) of Definition 1, it follows that h is onto.

The following theorem now easily follows from Theorem 2.2 and Theorem 2.

Theorem 3. *Let* $\mathbf{K}$ *be a class of chains and let* $L \in \mathbf{D}$. *Then* $L \in \mathbf{C_K}$ *if and only if for each* $a, b \in L$, $a < b$ *there exists* $K \in \mathbf{K}$ *and a basic* K*-chain* $\mathfrak{C}$ *of* L *such that* a *and* b *are in distinct members of* $\mathfrak{C}$.

4. The classes $\mathbf{C_n}$ for $\mathbf{n} \geq 3$. Locally separated n-chains

We will now investigate the classes $\mathbf{C_n}$ and assume from now on that n is a (fixed) integer ≥ 3. In accordance with Definition 3.1, a basic $\mathbf{n}$-chain of $L \in \mathbf{D}$ is a chain of subsets $I_0 \subset I_1 \subset \cdots \subset I_{n-1} = L$ where for each i, $0 \leq i \leq n-2$, I_i is a prime ideal of L. Let us also restate Theorems 3.2 and 3.3 for the case that $\mathbf{K} = \{\mathbf{n}\}$.

Theorem 1. *Let* $L \in \mathbf{D}$.

(i) *If* $h: L \to \mathbf{n}$ *is an onto homomorphism and for each* i, $0 \leq i \leq n-1$, $I_i = \{a \in L : h(a) \leq \mathbf{i}\}$ *then the set* $\{I_i : 0 \leq i \leq n-1\}$, *is a basic* $\mathbf{n}$*-chain.*

(ii) *If* $\mathfrak{C} = \{I_i : 0 \leq i \leq n-1\}$ *is a basic* $\mathbf{n}$*-chain of* L *and* $h: L \to \mathbf{n}$ *is defined by letting* $h(a)$ *be the least* $\mathbf{i}$ *for which* $a \in I_i$, *then* h *is an onto homomorphism.*

Theorem 2. *Let* $L \in \mathbf{D}$. *Then* $L \in \mathbf{C_n}$ *if and only if for each* $a, b \in L$, $a < b$ *there exists a basic* $\mathbf{n}$*-chain* $\mathfrak{C}$ *of* L *such that* a *and* b *are in distinct members of* $\mathfrak{C}$.

Exercise 3. Prove that $\mathbf{C_3} \supset \mathbf{C_4} \supset \cdots \supset \mathbf{C_n} \supset \mathbf{C_{n+1}} \cdots$ and if $K \in \mathbf{C}$ and $|K| \geq \aleph_0$ then $\mathbf{C_n} \supset \mathbf{C}_K$ for each n.

Theorem 2 will enable us to give, in algebraic terminology, necessary and sufficient conditions for L to belong to $\mathbf{C_n}$. For this, we define the notion of a *locally separated* $\mathbf{n}$*-chain* which was first introduced by F. W. Anderson and R. L. Blair [1].

Definition 4. Let $L \in \mathbf{D}$. A chain $a_0 \leq a_1 \leq \cdots \leq a_{n-1}$ in L is a *locally separated* $\mathbf{n}$*-chain*, if the system

(1) $a_1 x_0 \leq a_0$

$\cdots$

$a_{i+1} x_i \leq a_i + x_{i-1} \qquad i = 1, 2, \cdots, n-3$

$\cdots$

$a_{n-1} \leq a_{n-2} + x_{n-3}$

has no solution for $x_0, x_1, \cdots, x_{n-3}$ in L.

The significance of locally separated $\mathbf{n}$-chains will become clear from Theorems 5 and 7 below.

Theorem 5. (F. W. Anderson and R. L. Blair [1]) *Let* $L \in \mathbf{D}$. *Then* $a_0 \leq a_1 \leq \cdots \leq a_{n-1}$, $a_i \in L$, $0 \leq i \leq n-1$, *is a locally separated* $\mathbf{n}$*-chain if and only if there exists an onto homomorphism* $h: L \to \mathbf{n}$ *such that* $h(a_i) < h(a_j)$ *for* $i < j$. *In particular, it follows that the elements in a locally separated* $\mathbf{n}$*-chain are distinct.*

Proof. ($\Rightarrow$) We define a sequence $I_0, I_1, \cdots, I_{n-2}$ of ideals of L as follows: $I_0 = \{x \in L : a_1 x \leq a_0\}$. It is easy to see that I_0 is an ideal. Define I_i for $1 \leq i \leq n - 2$ recursively by

$$I_i = \{x \in L : a_{i+1}x \leq a_i + y \text{ for some } y \in I_{i-1}\}.$$

Since I_0 is an ideal, it easily follows that each I_i is an ideal. Furthermore, since $a_{i+1}x \leq a_i + x$ for every $x \in I_{i-1}$, we have $I_{i-1} \subseteq I_i$ for $1 \leq i \leq n - 2$. Also it is immediate that $a_i \in I_i$ for $0 \leq i \leq n - 2$. We will now show that $I_{i-1} \subset I_i$ for $1 \leq i \leq n - 2$ by showing $a_i \notin I_{i-1}$. Indeed, if $a_1 \in I_0$ then $a_1 = a_0$ and the system (1) would have a solution $x_0 = x_1 = \cdots = x_{n-3} = a_{n-1}$. Now, suppose $a_i \in I_{i-1}$ for some i, $2 \leq i \leq n - 2$. Then

$$\begin{aligned} &a_i \leq a_{i-1} + x_{i-2} \text{ for some } x_{i-2} \in I_{i-2} \\ &a_{i-1}x_{i-2} \leq a_{i-2} + x_{i-3} \text{ for some } x_{i-3} \in I_{i-3} \\ &\cdot \quad \cdot \quad \cdot \\ &a_2 x_1 \leq a_1 + x_0 \text{ for some } x_0 \in I_0. \end{aligned}$$

If we now take $x_{i-1} = x_i = \cdots = x_{n-3} = a_{n-1}$ then (1) would have a solution.

We can now construct a basic **n**-chain $J_0 \subset J_1 \subset \cdots \subset J_{n-1} = L$ such that $a_i \in J_i$ for $0 \leq i \leq n - 1$ and $a_i \notin J_{i-1}$ for $1 \leq i \leq n - 1$. Indeed, let J_{n-2} be a prime ideal such that $a_{n-1} \notin J_{n-2}$ and $I_{n-2} \subseteq J_{n-2}$. Such a prime ideal exists, since $a_{n-1} \notin I_{n-2}$. Now suppose $J_{n-1}, J_{n-2}, \cdots, J_i$ have been constructed for some $i > 0$ such that $J_i \subset J_{i+1} \subset \cdots \subset J_{n-1}$, $I_j \subseteq J_j$ and $a_j \in J_j$ for $i \leq j \leq n - 1$ and $a_j \notin J_{j-1}$ for $i + 1 \leq j \leq n - 1$. Let $F_i = [(L \sim J_i) \cup \{a_i\})$. We claim that $F_i \cap I_{i-1} = \varnothing$. If not, then there exists an $x \in I_{i-1}$ and a $y \notin J_i$ such that $x \geq a_i y$. But then $a_i y \in I_{i-1}$. Thus $a_i a_i y \leq a_{i-1} + x_{i-2}$ for some $x_{i-2} \in I_{i-2}$ or $a_i y \leq a_{i-1} + x_{i-2}$ which implies $y \in I_{i-1} \subseteq I_i$. Also by hypothesis $I_i \subseteq J_i$ hence $y \in J_i$, a contradiction. So there exists a prime ideal J_{i-1} such that $I_{i-1} \subseteq J_{i-1}$ and $F_i \cap J_{i-1} = \varnothing$; hence $a_{i-1} \in J_{i-1}$ and $a_i \notin J_{i-1}$. It follows now from Theorem 4.1 that the desired homomorphism exists.

($\Leftarrow$) Suppose $a_0 \leq a_1 \leq \cdots \leq a_{n-1}$ is not locally separated. Then the system (1) would have a solution $x_0, x_1, \cdots, x_{n-3}$. Applying h to both sides of the inequalities of (1) one obtains $h(x_0) = \mathbf{0}$, $h(x_1) \leq \mathbf{1}, \cdots, h(x_{n-3}) \leq \mathbf{n-3}$ and $\mathbf{n-1} \leq (\mathbf{n-2}) + h(x_{n-3}) \leq (\mathbf{n-2}) + (\mathbf{n-3}) = \mathbf{n-2}$, a contradiction.

Exercise 6. Prove that every k-element subchain, $k \geq 3$, of a locally separated **n**-chain is a locally separated **k**-chain.

We are now able to summarize our results in the following theorem.

Theorem 7. *Let $L \in \mathbf{D}$. Then $L \in \mathbf{C_n}$ if and only if for each pair $a, b \in L$, $a < b$ there is a locally separated **n**-chain containing a and b. Moreover, if $L \in \mathbf{C_n}$ and $L \ncong \mathbf{1}$ then $|L| \geq n$.*

Proof. ($\Rightarrow$) By Theorem 2.2 there exists an onto homomorphism $h: L \rightarrow \mathbf{n}$ with $h(a) \neq h(b)$. Hence a and b are contained in a chain $a_0 \leq a_1 \leq \cdots \leq a_{n-1}$ such that $h(a_i) = \mathbf{i}$ for $0 \leq i \leq n - 1$, and by Theorem 5 this chain is a locally separated **n**-chain.

($\Leftarrow$) Again apply Theorem 2.2 and Theorem 5. The last part of the theorem is obvious.

Although $\mathbf{C_n}$ is obviously not an equational class for any n, $\mathbf{C_n}$ is closed under direct products.

Theorem 8. Let $(L_s)_{s \in S}$ be a family in $\mathbf{D}$. Then $L = \times_{s \in S} L_s$ belongs to $\mathbf{C_n}$ if and only if each L_s belongs to $\mathbf{C_n}$.

Proof. ($\Rightarrow$) We only need to show that if $L = L_1 \times L_2, L, L_1, L_2 \in \mathbf{D}$ and $L \in \mathbf{C_n}$ then $L_1 \in \mathbf{C_n}$. Let $a, b \in L_1$, $a < b$. Pick an element $c \in L_2$ then $(a, c) < (b, c)$. By Theorem 7 there exists a locally separated $\mathbf{n}$-chain $(u_0, v_0) < (u_1, v_1) < \cdots < (u_{n-1}, v_{n-1})$ in L which contains (a, c) and (b, c). Now the chain $v_0 \leq v_1 \leq \cdots \leq v_{n-1}$ contains two equal elements, namely c. Hence, by Theorem 5, this is not a locally separated $\mathbf{n}$-chain. Suppose the chain $u_0 \leq u_1 \leq \cdots \leq u_{n-1}$ is not a locally separated $\mathbf{n}$-chain either. Then it is easy to see that $(u_0, v_0) < (u_1, v_1) < \cdots < (u_{n-1}, v_{n-1})$ is not a locally separated $\mathbf{n}$-chain, a contradiction. Hence, $u_0 \leq u_1 \leq \cdots \leq u_{n-1}$ is a locally separated $\mathbf{n}$-chain containing a and b and the result follows from Theorem 7.

($\Leftarrow$) This part is left to the reader.

Corollary 9. *Let $L = L_1 \times L_2, L_1, L_2 \in \mathbf{D}, L_1 \neq \mathbf{1}$. If L_1 has no homomorphisms onto $\mathbf{n}$, then $L \notin \mathbf{C_n}$.*

Proof. Suppose $L \in \mathbf{C_n}$. Then, by Theorem 8, $L_1 \in \mathbf{C_n}$. But $L_1 \neq \mathbf{1}$, thus L_1 has a homomorphism onto $\mathbf{n}$.

Exercise 10. Suppose $L \notin \mathbf{C_n}$. Then there exists an interval $[a, b]$, $a < b$ in L such that there are no homomorphisms from $[a, b]$ onto $\mathbf{n}$.

We have seen that the property $L \in \mathbf{C_n}$ is closely related to the structure of $\mathfrak{P}(L)$. In fact, it follows from Theorem 4.2 that, in order for L to be in $\mathbf{C_n}$, it is sufficient that every prime ideal of L be contained in a chain of $n - 1$ prime ideals. It should however be noted that this condition is not a necessary one. The following example (which is due to D. C. Feinstein [1]) is a member of $\mathbf{C_n}$ which has a prime ideal that is both maximal and minimal in $\mathfrak{P}(L)$.

Let $L = \times_{s \in S} L_s$ where $|S| = \aleph_0$ and for each $s \in S$, $L_s = \mathbf{n}$. For $x \in L$, call x *finite* if $x(s) = \mathbf{0}$ for all but a finite number of $s \in S$ and call x *cofinite* if $x(s) = \mathbf{n-1}$ for all but a finite number of $s \in S$. Let $L_1 = \{x \in L : x \text{ finite or cofinite}\}$. Then L_1 is clearly a $\mathbf{D_{01}}$-sublattice of L and $L_1 \in \mathbf{C_n}$. Let $I = \{x \in L_1 : x \text{ finite}\}$ then I is a prime ideal of L_1. Also I is a maximal prime ideal. Indeed, let $y \in L_1 \sim I$. Then y is cofinite, so if $S_1 = \{s \in S: y(s) \neq \mathbf{n-1}\}$ then S_1 is a finite subset of S. Define $x \in L_1$ by $x(s) = \mathbf{n-1}$ if $s \in S_1$ and $x(s) = \mathbf{0}$ if $x \notin S_1$. Then x is finite, so $x \in I$ and $x + y = 1_L = 1_{L_1}$. Hence I is maximal. Also, I is a minimal prime ideal. Indeed, suppose there exists a prime ideal I_1, $I_1 \subset I$. Let $a \in I \sim I_1$. Now $a \in I$, thus a is finite. Let

$S_1 = \{s \in S : a(s) \neq \mathbf{0}\}$. Define $b \in L_1$ by $b(s) = \mathbf{0}$ if $s \in S_1$ and $b(s) = \mathbf{n-1}$ if $s \notin S_1$. Then b is cofinite so $b \notin I_1$. But $ab = 0_L = 0_{L_1} \in I_L$, a contradiction.

On the other hand, in the finite case, the aforementioned condition is necessary for $L \in \mathbf{C_n}$.

Theorem 11. (D. C. Feinstein [1]) *If $L \in \mathbf{C_n}$ and L is finite, then every prime ideal in L is contained in a chain of $n - 1$ prime ideals.*

We first prove the following lemma.

Lemma 12. *Let $L = \bigtimes_{i=1}^{n} C_i$, where each C_i is a finite chain. Then the prime ideals of L are exactly the sets $A_{i,a}$, $1 \leq i \leq n$, $a \in C_i$, $a \neq 1_{C_i}$, where $A_{i,a} = \{x \in L : x(i) \leq a\}$.*

Proof. Certainly, each $A_{i,a}$ is a prime ideal. Conversely, suppose J is a prime ideal of L. Obviously, the projection of J on each C_i is either a set $p_i[J] = (a]$ for some $a \in C_i$, $a \neq 1_{C_i}$ or it is all of C_i. But the image of one and only one projection is a proper subset. Indeed, suppose the projections of J on C_i and C_j are proper subsets of C_i and C_j, respectively, where $i \neq j$. Let $x, y \in L$ be defined by $x(i) = 1_{C_i}$ and $x(k) = 0_{C_k}$ for $k \neq i$; $y(j) = 1_{C_j}$ and $y(k) = 0_{C_k}$ for $k \neq j$. Then $xy = 0 \in J$ and $x \notin J$ and $y \notin J$.

Proof of Theorem 11. Let L_1 be a subdirect product of $(C_i)_{1 \leq i \leq k}$ such that $C_i = \mathbf{n}$ for each i. We can further assume that the embedding of L_1 in $L = \bigtimes_{i=1}^{k} C_i$ is the inclusion map. Suppose J is a prime ideal of L_1. Let J' be a prime ideal of L with $J' \cap L_1 = J$. By Lemma 12, $J' = A_{i,a}$ for some $1 \leq i \leq k$, $a \in C_i$, $a \neq 1_{C_i}$. Then J' is contained in a chain of $n - 1$ prime ideals of L. Since all of the projections of L on C_i, $1 \leq i \leq k$ are onto, the intersections of these prime ideals with L_1 yield a chain of $n - 1$ prime ideals containing J.

5. The class $\mathbf{C_3}$

We have seen in Corollary 4.9 that, if $L \in \mathbf{D}$ and L has a direct factor L_1 such that $L_1 \neq \mathbf{1}$ and L_1 has no homomorphisms onto $\mathbf{n}$, then $L \notin \mathbf{C_n}$. The converse does not hold. Thus, if $L \notin \mathbf{C_n}$, then L need not have a direct factor L_1, $L_1 \neq \mathbf{1}$ such that L_1 has no homomorphisms onto $\mathbf{n}$. Indeed, take for L the lattice $\mathbf{1} \oplus (\mathbf{2} \times \mathbf{3})$. Then $(\mathbf{0}, \mathbf{0}) < (\mathbf{0}, \mathbf{1})$ is not contained in a locally separated $\mathbf{4}$-chain. Hence $L \notin \mathbf{C_4}$. On the other hand, L has no direct factor except $\mathbf{1}$ and L itself. It is however surprising that for $n = 3$ the converse of Corollary 4.9 does hold. Thus, if $L \notin \mathbf{C_3}$, then L has a direct factor L_1, $L_1 \neq \mathbf{1}$ which has no homomorphisms onto $\mathbf{3}$, and L_1 can even be chosen such that it is a Boolean algebra. This situation will be the subject of this and the next section.

Locally separated $\mathbf{3}$-chains can be characterized easily as the following theorem demonstrates.

Theorem 1. *Let $L \in \mathbf{D}$. A chain $a_0 \leq a_1 \leq a_2$ in L is a locally separated* **3**-*chain if and only if a_1 is not complemented in $[a_0, a_2]$.*

Proof. Suppose the set of inequalities $a_1x \leq a_0, a_2 \leq a_1 + x$ has a solution x_0. Let $y_0 = a_2x_0 + a_0$. Then $a_1y_0 = a_0$ and $a_1 + y_0 = a_2$. Hence a_1 is complemented in $[a_0, a_2]$. Conversely, if a_1 is complemented in $[a_0, a_2]$ then it is obvious that $a_0 \leq a_1 \leq a_2$ is not a locally separated **3**-chain.

Theorem 4.7 now acquires, for $n = 3$, the following simple formulation.

Theorem 2. *Let $L \in \mathbf{D}$. Then $L \in \mathbf{C}_3$ if and only if for each pair, $a, b \in L$, $a < b$ there exists a chain $a_0 \leq a_1 \leq a_2$ in L containing a and b and such that a_1 is not complemented in $[a_0, a_2]$.*

In case L has a 1 (0), we have an even simpler characterization.

Theorem 3. *Suppose $L \in \mathbf{D}$ and suppose L has a* 1. *Then $L \in \mathbf{C}_3$ if and only if for each relatively complemented element $a < 1$ there is an element in $[a, 1]$ which is not complemented in $[a, 1]$.*

We first prove:

Lemma 4. *Suppose that $L \in \mathbf{D}$ and that L has a* 1. *Then $L \in \mathbf{C}_3$ if and only if for every $a < 1$, $\{a, 1\}$ is contained in a locally separated* **3**-*chain.*

Proof. The necessity is obvious. For sufficiency, suppose $a, b \in L$, $a < b$ and suppose $a \leq b \leq 1$ is not contained in a locally separated **3**-chain. Then by Theorem 1, there exists an element b' such that $b + b' = 1$ and $bb' = a$ and, since $a < b$, we have $b' < 1$. By hypothesis, $\{b', 1\}$ is in a locally separated **3**-chain $a_0 \leq a_1 \leq 1$ (i.e. $b' \in \{a_0, a_1\}$). It follows from Theorem 4.5 that there exists a homomorphism h from L onto **3** such that $h(b') < 1$. Suppose $h(a) = h(b)$. But $h(b)h(b') = h(a)$ and thus $h(b) \leq h(b')$. But also $h(b) + h(b') = 1$, implying $h(b') = 1$. Hence $h(a) \neq h(b)$. It follows from Theorem 2.2 that $L \in \mathbf{C}_3$.

Proof of Theorem 3. ($\Rightarrow$) Immediate from Theorems 1 and 2.
($\Leftarrow$) Suppose $L \notin \mathbf{C}_3$. By Lemma 4, there exists an element $a < 1$ such that $\{a, 1\}$ is not contained in a locally separated **3**-chain. It follows from Theorem 1 that every element in $[a, 1]$ has a complement in $[a, 1]$. Hence by hypothesis a is not relatively complemented. On the other hand, since $\{a, 1\}$ is not contained in a locally separated **3**-chain, a is complemented in $[u, 1]$ for every $u \leq a$, so a is relatively complemented, a contradiction.

Theorem 3 has the following corollary, the proof of which we leave to the reader.

Corollary 5. *Suppose $L \in \mathbf{D}$, $|L| > 2$ and L has a* 1 *or a* 0, *but no other relatively complemented elements. Then $L \in \mathbf{C}_3$.*

The converse of Corollary 5 is not true even if $L \in \mathbf{D}_{01}$ as is demonstrated by the example $L = \mathbf{3} \times \mathbf{3}$. On the other hand, Corollary 5 is no longer valid if L has neither a 0 nor a 1. Finally, if $L \in \mathbf{C}_3$, then L may still have an abundance of relatively complemented elements (Exercise 6 and 7).

Exercise 6. Show that $Z \times \mathbf{2}$ has no relatively complemented elements and that $Z \times \mathbf{2} \notin \mathbf{C}_3$.

Exercise 7. Let L be the lattice of all $x \in \mathbf{3}^Z$ such that for some $n_x > 0$, $x(n) = \mathbf{1}$ for $n \geq n_x$, and $x(n) = \mathbf{0}$ for $n \leq -n_x$. Show that $L \in \mathbf{C}_3$ and that L has relatively complemented elements (Note that L has no 0 and no 1.).

6. A structure theorem for $\mathbf{C}_3$

We have seen (cf. Exercise 4.10) that, if $L \in \mathbf{D}$, $L \neq \mathbf{1}$ and $L \notin \mathbf{C}_\mathbf{n}$, then there exists an interval $[a, b]$ in L such that $[a, b]$ has no homomorphisms onto $\mathbf{n}$. But in general $[a, b]$ does not need to be a direct factor of L. In contrast to this situation for $n > 3$, we have the following result.

Theorem 1. (R. Balbes and Ph. Dwinger [1]) *Let $L \in \mathbf{D}$. $L \in \mathbf{C}_3$ if and only if L does not have as a direct factor a Boolean algebra of more than one element.*

Proof. ($\Rightarrow$) Immediate from Theorem 4.8.

For the sufficiency part of the proof we first need a lemma.

Lemma 2. *Let $L \in \mathbf{D}$ and suppose $x_0, y_0 \in L$ are such that $x_0 < y_0$ and for each $x \in L$, y_0 has a complement $f(x)$ in $[x_0, x + y_0]$ and x_0 has a complement $g(x)$ in $[xx_0, y_0]$. Then $f: L \to L$ and $g: L \to L$ are homomorphisms. Moreover*

(i) $f(x)g(y) = x_0 y$.

(ii) $f(x) + g(y) = x + y_0$.

(iii) $f(f(x)) = f(x)$.

(iv) $g(g(x)) = g(x)$.

(v) $f(g(x)) = x_0$.

(vi) $g(f(x)) = y_0$.

Proof. In this proof and in the next, it will be useful to refer to Figure 1.

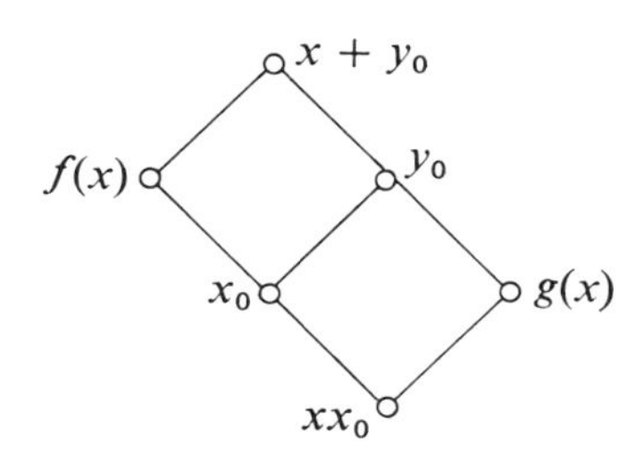

Fig. 1

From the definition of f, we have for $x, y \in L$;

$$f(x)y_0 = x_0, f(x) + y_0 = x + y_0.$$

$$f(y)y_0 = x_0, f(y) + y_0 = y + y_0.$$

Observe first, that f and g are well defined by the uniqueness of relative complementation. Now, by the above formulas, $(f(x) + f(y))y_0 = x_0$ and $(f(x) + f(y)) + y_0 = x + y + y_0$. So, again by the unique complementation $f(x) + f(y) = f(x + y)$. Also $f(x)f(y)y_0 = x_0$ and $f(x)f(y) + y_0 = xy + y_0$ so $f(x)f(y) = f(xy)$. Thus f is a homomorphism. To see that g is a homomorphism, we start with $g(x)x_0 = xx_0$, $g(x) + x_0 = y_0$, $g(y)x_0 = yx_0$, $g(y) + x_0 = y_0$. Then $(g(x) + g(y))x_0 = (x + y)x_0$ and $(g(x) + g(y)) + x_0 = y_0$ so $g(x) + g(y) = g(x + y)$. Also $g(x)g(y)x_0 = xyx_0$ and $g(x)g(y) + x_0 = y_0$. Thus $g(x)g(y) = g(xy)$.

Now we prove (i)–(vi):

(i): $f(x)g(y) = f(x)y_0g(y) = x_0g(y) = yx_0$.

(ii): $f(x) + g(y) = f(x) + x_0 + g(y) = f(x) + y_0 = x + y_0$.

(iii): By definition, $f(f(x))$ is the complement of y_0 in $[x_0, f(x) + y_0] = [x_0, x + y_0]$.

Thus $f(f(x)) = f(x)$.

(iv): $g(g(x))$ is the complement of x_0 in $[g(x)x_0, y_0] = [xx_0, y_0]$.

Hence $g(g(x)) = g(x)$.

(v): $f(g(x))$ is the complement of y_0 in $[x_0, g(x) + y_0] = [x_0, y_0]$.

But so is x_0, so $f(g(x)) = x_0$.

(vi): $g(f(x))$ is the complement of x_0 in $[f(x)x_0, y_0] = [x_0, y_0]$ so $g(f(x)) = y_0$.

We are now ready to prove the sufficiency part of Theorem 1.

Sufficiency: Suppose $L \notin \mathbf{C}_3$, then $L \neq \mathbf{1}$. Now by Theorem 5.2 there exist $x_0, y_0 \in L$, $x_0 < y_0$ such that every element in $[x_0, y_0]$ is complemented in $[x_0, y_0]$. So this interval is a Boolean algebra under the induced partial order and the hypotheses of Lemma 2 are satisfied (again by Theorem 5.2). We will repeatedly use Lemma 2 without explicitly mentioning it each time. Now let $L_1 = \{y + g(y) : y \in L\}$ and let $L_2 = [x_0, y_0]$. The result will be established in a series of steps by showing that L_1 is a sublattice of L and that $L \cong L_1 \times L_2$.

(1) *L_1 is a sublattice of L*: Let $y_1, y_2 \in L$. Then $(y_1 + g(y_1)) + (y_2 + g(y_2)) = (y_1 + y_2) + g(y_1 + y_2) \in L_1$. For products, we have by the definition of $g(y_1)$ and $g(y_2)$ that $g(y_i)x_0 = y_ix_0$ and $g(y_i) + x_0 = y_0$ for $i = 1, 2$. Thus $g(y_2) \leq y_0 \leq g(y_1) + x_0$. Multiply both sides of the last inequality by $y_1g(y_2)$ to obtain $y_1g(y_2) \leq y_1g(y_2)g(y_1) + y_1g(y_2)x_0 \leq g(y_1) + y_1y_2x_0 \leq g(y_1) + y_2$. Thus $y_1g(y_2) \leq (y_1 + g(y_1))(y_2 + g(y_2))(y_1 + g(y_2))(y_2 + g(y_1)) = y_1y_2 + g(y_1)g(y_2)$. Similarly, $y_2g(y_1) \leq y_1y_2 + g(y_1)g(y_2)$. So

$(y_1 + g(y_1))(y_2 + g(y_2)) = (y_1y_2 + g(y_1)g(y_2)) + (y_1g(y_2) + y_2g(y_1))$
$= y_1y_2 + g(y_1)g(y_2) = y_1y_2 + g(y_1y_2) \in L_1$.

(2) *Define* $h: L \to L_1 \times L_2$ *by* $h(u) = (u + g(u), y_0u + x_0))$.

(3) $x \leq y$ *implies* $h(x) \leq h(y)$: g is a **D**-homomorphism so $x + g(x) \leq y + g(y)$ and clearly $y_0x + x_0 \leq y_0y + x_0$ and thus $h(x) \leq h(y)$.

(4) $h(x) \leq h(y)$ *implies* $x \leq y$: By hypothesis $x + g(x) \leq y + g(y)$ and $y_0x + x_0 \leq y_0y + x_0$. Now (see Figure 1) $x \leq x + y_0 = f(x) + y_0$. Thus $x \leq xf(x) + xy_0 \leq xf(x) + (y_0y + x_0) \leq f(x) + y + x_0$. Finally, $x \leq (y + g(y))(f(x) + y + x_0) = y + f(x)g(y) + x_0g(y) = y + x_0y + yx_0 = y$.

(5) h *is onto*: Let $(w, x) \in L_1 \times L_2$. Thus $w = y + g(y)$ for some $y \in L$ and $x_0 \leq x \leq y_0$. We will show that if $p = xy + yf(y) + xg(y)$, then $h(p) = (w, x)$. Indeed, $h(p) = (p + g(p), y_0p + x_0)$. Thus we must prove: (5_1) $p + g(p) = w$ and (5_2) $y_0p + x_0 = x$. For the proof of (5_1) we first calculate $g(p)$. Now $g(p) = g(xy + yf(y) + xg(y)) = g(x)g(y) + g(y)g(f(y)) + g(x)g(g(y))$
$= g(x)g(y) + g(y) = g(y)$. Thus $p + g(p) = xy + yf(y) + xg(y) + g(y) =$
$xy + yf(y) + g(y) = xy + (y + g(y))(f(y) + g(y)) = xy + (y + g(y))(y + y_0)$
$= xy + y + g(y)y_0 = y + g(y)y_0 = y + g(y) = w$. For (5_2), $y_0p + x_0$
$= y_0(xy + yf(y) + xg(y)) + x_0 = y_0xy + y_0yf(y) + y_0xg(y) + x_0$
$= y_0xy + yx_0 + xg(y) + x_0 = y_0xy + xg(y) + x_0 = y_0xy + (x + x_0)(g(y) + x_0)$
$= y_0xy + (x + x_0)y_0 = y_0xy + xy_0 + x_0y_0 = y_0x + x_0 = x$. This completes the proof.

The proof of the previous theorem implies the following corollary.

Corollary 3. *If* $L \in \mathbf{R}$ *then each interval* $[x, y]$ *is a direct factor of* L.

Exercise 4. Give a short proof of the sufficiency part of Theorem 1 in case L has a 0 and a 1 (Hint: Use Theorem 5.3 and Exercise II.7.2.).

7. Maximal ideals and Nachbin's theorem

We first show that Nachbin's theorem (III.6.3) is an immediate consequence of the results obtained in the previous sections.

Theorem 1. *Let L be a distributive lattice. If every prime ideal in L is a maximal ideal, then L is relatively complemented.*

Proof. Suppose $L \notin \mathbf{R}$. Then there exist elements $a \leq b \leq c$ in L such that b is not complemented in $[a, c]$. By Theorem 5.1 the chain $a \leq b \leq c$ is a locally separated **3**-chain. It follows from Theorems 4.1 and 4.5 that there exist prime ideals I_0 and I_1 in L such that $I_0 \subset I_1$. Hence I_0 is not maximal.

The question arises as to whether the existence of a maximal ideal in a distributive lattice L is equivalent to the existence of relatively complemented element.

If L has a relatively complemented element $a \neq 0$, then L has a maximal ideal (Theorem 2, below). On the other hand, L may have maximal ideals, but no relatively complemented elements.

Theorem 2. *Let L be a distributive lattice and suppose L has elements a, b, $b < a$ such that a is complemented in $[b, c]$ for every $c \geq a$. Then L has a maximal ideal. In particular, if L has a non-zero relatively complemented element, then L has a maximal prime ideal.*

Proof. Let I_1 be an ideal of L which is maximal, subject to the condition that $b \in I_1$, $a \notin I_1$. If I_1 were not maximal there would exist a prime ideal I_2 such that $I_1 \subset I_2$. But then $a \in I_2$. Let $c > a$, $c \notin I_2$. But a is relatively complemented in $[b, c]$, a contradiction.

Exercise 3. Let L be a distributive lattice. Then $I = \{(x, \mathbf{0}) : x \in L\}$ is a maximal ideal in $L \times \mathbf{2}$. Note that if $L \times \mathbf{2}$ is relatively complemented, then so is L.

Exercise 4. Let L be a distributive lattice and let I be a prime ideal in L. Prove that I is maximal if and only if for each pair, $a, c \in L$, $a \leq c$, $a \notin I$, there exists an $x \in I$, $x \leq a$, such that a is complemented in $[x, c]$.

8. Subdirect products of infinite chains

The conditions imposed on $\mathfrak{P}(L)$ in order that L belongs to $\mathbf{C}_K$ are essentially different for the case that K is finite and the case that K is infinite. For example, if K is an infinite chain and every prime ideal is contained in a chain of prime ideals which is isomorphic with K, then L need not belong to $\mathbf{C}_K$: take $L = Q =$ rationals. The chain of prime ideals of Q is of the power of the continuum, but Q is not the subdirect product of chains of the power of the continuum. Thus one cannot expect a characterization of $\mathbf{C}_\mathbf{K}$ where $\mathbf{K}$ is a class of infinite chains which is analogous to Theorem 4.7. However, there are important subclasses of $\mathbf{D}$ whose members are subdirect products of infinite chains. One such class consists of the infinite free distributive lattices (cf. Theorem 2.3) and we will encounter some other interesting classes of subdirect products of infinite chains in Sections VII.4 and X.4. In this section we also present some classes of subdirect products of infinite chains. The results here are based on the following theorem which is an immediate consequence of Theorem 3.3.

Theorem 1. *Let $L \in \mathbf{D}$ and let $\mathbf{K}$ be a class of chains. If every prime ideal of L is a member of a basic K-chain for some $K \in \mathbf{K}$ then $L \in \mathbf{C}_\mathbf{K}$.*

Theorem 2. (Ph. Dwinger [8]) *Let $L \in \mathbf{D}$ and suppose $\mathfrak{P}(L)$ has no maximal (minimal) members. Then L is a subdirect product of (dual) limit ordinals.*

Proof. We assume that no prime ideal of L is maximal. By Theorem 1 it suffices to show that if I is a prime ideal of L then I is a member of a basic α-chain where α is a limit ordinal. Let κ be some ordinal with $\kappa > |L|$. Define for each ordinal $\alpha < \kappa$ a prime ideal I_α of L as follows. Let $I_0 = I$. Suppose α is an ordinal such that $\mathbf{0} < \alpha < \kappa$ and such that I_β has been defined for $\beta < \alpha$ and such that $I_{\beta_1} \subseteq I_{\beta_2}$ for $\beta_1 \leq \beta_2 < \alpha$. If $\bigcup_{\beta<\alpha} I_\beta \neq L$, then $\bigcup_{\beta<\alpha} I_\beta$ is a prime ideal and we define I_α to be a prime ideal properly containing $\bigcup_{\beta<\alpha} I_\beta$. If $\bigcup_{\beta<\alpha} I_\beta = L$, let $I_\alpha = L$. It is not difficult to see that there exists an ordinal, and therefore a smallest ordinal γ such that $I_\gamma = L$. Also it easily follows that γ is a limit ordinal and thus $\bigcup_{\alpha<\gamma} I_\alpha = L$ and $I_\alpha \neq L$ for $\alpha < \gamma$. Also if $\alpha_1 < \gamma$ and $\alpha_2 < \gamma$ then $I_{\alpha_1} \subset I_{\alpha_2}$ if and only if $\alpha_1 < \alpha_2$. Now consider the chain $\mathfrak{C} = (I_\alpha)_{\alpha<\gamma}$. We claim that $\mathfrak{C}$ is a basic γ-chain. Certainly $\mathfrak{C}$ and γ are isomorphic. If $a \in L$ then, since $\bigcup_{\alpha<\gamma} I_\alpha = L$, there is a smallest $\alpha = \alpha_0$ such that $a \in I_{\alpha_0}$ and I_{α_0} is the smallest member of $\mathfrak{C}$ which contains a. Finally, let $\alpha < \gamma$. If $\alpha = 0$ and $a \in I_0$ then I_0 is the smallest member of $\mathfrak{C}$ containing a. If $\alpha > 0$, pick an $a \in I_\alpha \sim \bigcup_{\beta<\alpha} I_\beta$, then I_α is the smallest member of $\mathfrak{C}$ containing a. It follows that $\mathfrak{C}$ is a basic γ-chain containing $I = I_0$ as a (first) member. This completes the proof of the theorem.

Exercise 3. Let $L \in \mathbf{D}$ and suppose $\mathfrak{P}(L)$ has no maximal and no minimal members. Prove that L is the subdirect product of chains of the type $\mathfrak{d} \oplus \gamma$ where δ and γ are limit ordinals.

We close this section with the following observation. Let $\mathbf{C^M}$ denote the class of distributive lattices L for which $\mathfrak{P}(L)$ has no maximal and no minimal members and let $\mathbf{C}^\infty$ denote the class of lattices which are subdirect products of chains which have neither a largest nor a smallest element. Then we have that $C^{\mathbf{M}} \subset \mathbf{C}^\infty$. Indeed, it follows from Exercise 3 that $\mathbf{C^M} \subseteq \mathbf{C}^\infty$. On the other hand, if L is an infinite free distributive lattice, then by Theorem 2.3 $L \in \mathbf{C}^\infty$ but by Theorem V.3.12, L has maximal and minimal prime ideals, hence $L \in \mathbf{C}^\infty \sim \mathbf{C^M}$.

Exercise 4. Prove that if $L \in \mathbf{C}^\infty$, then L has no relatively complemented elements.

Exercise 5. (G. Grätzer, oral communication) Prove that $\mathbf{C}^\infty \subset \bigcap_{n=2}^\infty \mathbf{C_n}$.

Exercise 6. Let $\mathbf{C^H}$ be the class of lattices L satisfying the following condition: For each pair, $a, b \in L$ there exists an automorphism f of L such that $f(a) = b$ and such that if $a < b$ then $x < f(x)$, and if $a \nless b$ and $b \nless a$ then $x \nless f(x)$ and $f(x) \nless x$ for all $x \in L$. Prove that L is distributive for $L \in \mathbf{C^H}$, and that $\mathbf{C^H} \subset \mathbf{C^M}$ (oral communication by J. Berman).

VII

Coproducts and Colimits

1. Existence and characterization of coproducts in $\mathscr{D}$, $\mathscr{D}_{01}$ and $\mathscr{B}$

Suppose $(A_s)_{s \in S}$ is a family of objects in an equational category $\mathscr{K}$. In Theorem I.20.7 we observed that a family of morphisms $(j_s: A_s \to A)_{s \in S}$ is a coproduct of $(A_s)_{s \in S}$ if and only if

(1) $\bigcup_{s \in S} j_s[A_s]$ generates A.

(2) If $(f_s: A_s \to B)_{s \in S}$ is a family of morphisms, then there exists an $h \in [A, B]_{\mathscr{K}}$ such that $h \circ j_s = f_s$ for each $s \in S$.

The following theorem gives a workable criterion for dealing with coproducts in $\mathscr{D}_{01}$.

Theorem 1. *Suppose $(L_s)_{s \in S}$ is a family of objects in $\mathscr{D}_{01}$.*

(i) If $|L_{s_0}| = 1$ for some $s_0 \in S$ then a coproduct of $(L_s)_{s \in S}$ is the family $(j_s: L_s \to \mathbf{1})_{s \in S}$ of constant maps.

(ii) If $|L_s| > 1$ for each $s \in S$ then a necessary and sufficient condition for a family $(j_s: L_s \to L)_{s \in S}$ of morphisms to be a coproduct of $(L_s)_{s \in S}$ is

(ii)$_1$ If $S_1 \subseteq S$, $\{a_s, b_s\} \subseteq L_s$ for each $s \in S_1$ and $\prod_{s \in S_1} j_s(a_s) \leq \sum_{s \in S_1} j_s(b_s)$ then $a_s \leq b_s$ for some $s \in S_1$, and

(ii)$_2$ $\bigcup_{s \in S} j_s[L_s]$ $\mathbf{D}_{01}$-generates L.

Remark. Note that (ii)$_1$ implies that j_s is one-one for each $s \in S$.

Proof. (i) Clearly $\bigcup_{s \in S} j_s[L_s]$ generates $\mathbf{1}$. Now suppose that for each $s \in S$, $f_s: L_s \to L_1$ is in $\mathscr{D}_{01}$. Since $L_{s_0} = \mathbf{1}$, $0_{L_1} = f_{s_0}(0_{L_{s_0}}) = f_{s_0}(1_{L_{s_0}}) = 1_{L_1}$, we have $L_1 = \mathbf{1}$. Thus the constant map $h: \mathbf{1} \to \mathbf{1} = L_1$ satisfies $h \circ j_s = f_s$ for each $s \in S$.

(ii) ($\Leftarrow$) It suffices to show that (2) holds. Thus, suppose $(f_s: L_s \to L_1)_{s \in S}$ is a family of morphisms in $\mathscr{D}_{01}$. We first verify: (3) If $S_1 \cup S_2 \subseteq S$, $X_s \subseteq L_s$ for each $s \in S_1$, $Y_s \subseteq L_s$ for each $s \in S_2$, and $\prod_{s \in S_1} (\prod j_s[X_s]) \leq \sum_{s \in S_2} (\sum j_s[Y_s])$ then $\prod_{s \in S_1} (\prod f_s[X_s]) \leq \sum_{s \in S_2} (\sum f_s[Y_s])$. Since

$$(\textstyle\prod_{s \in S_1} (j_s(\prod X_s)))(\prod_{s \in S_2 \sim S_1} j_s(1_{L_s})) \leq (\sum_{s \in S_1 \sim S_2} j_s(0_{L_s})) + (\sum_{s \in S_2} j_s(\sum Y_s))$$

implies, by (ii)$_1$, that $\prod X_s \leq 0_{L_s}$ for some $s \in S_1 \sim S_2$ or $\prod X_s \leq \sum Y_s$ for some $s \in S_1 \cap S_2$ or $1_{L_s} \leq \sum Y_s$ for some $s \in S_2 \sim S_1$. We see that, in each of the three cases, the conclusion of (3) holds. Now define a function $f: \bigcup_{s \in S} j_s[L_s] \to L_1$ by $f(j_s(a)) = f_s(a)$ for $a \in L_s$. Note, by (3), that f is well defined and that by the extension theorem for $\mathbf{D_{01}}$ (Theorem V.2.1), f extends to the required $\mathbf{D_{01}}$-homomorphism.

($\Rightarrow$) Assume (ii)$_1$ is false. Then there exists $S_1 \subseteq S$ and $\{a_s, b_s\} \subseteq L_s$ for each $s \in S_1$ such that for each $s \in S_1$ there is a prime ideal $I(s)$ such that $a_s \notin I(s)$ and $b_s \in I(s)$. So there exist morphisms $f_s: L_s \to \mathbf{2}$ such that $f_s(a_s) = \mathbf{1}$ and $f_s(b_s) = \mathbf{0}$ for each $s \in S_1$. Now $|L_s| > 1$ for each $s \in S$, so there exist morphisms $f_s: L_s \to \mathbf{2}$ for each $s \in S \sim S_1$. Finally, (2) implies the existence of a morphism $h: L \to \mathbf{2}$ such that $h \circ j_s = f_s$ for each $s \in S$. But then

$$\mathbf{1} = \prod_{s \in S_1} f_s(a_s) = \prod_{s \in S_1} h(j_s(a_s)) = h\left(\prod_{s \in S_1} j_s(a_s)\right)$$

$$\leq h\left(\sum_{s \in S_1} j_s(b_s)\right) = \sum_{s \in S_1} h(j_s(b_s)) = \sum_{s \in S_1} f_s(b_s) = \mathbf{0},$$

a contradiction.

The existence of coproducts in $\mathcal{D}_{01}$ is immediate from Theorem I.20.8. Indeed, we may assume that all $|A_s| > 1$ (otherwise the coproduct consists of the constant maps $j_s: A_s \to \mathbf{1}$). For each $s \in S$ let $B_s = A_s$ and $h_{s,t} = 1_{\mathbf{2}, B_t} \circ f_I$ where I is a prime ideal in A_s and $s, t \in S$.

Another proof of the existence of coproducts in $\mathcal{D}_{01}$, will be given in Section 7.

Now that we have established the situation for $\mathcal{D}_{01}$ it is easy to obtain the analogous results for $\mathcal{B}$.

Theorem 2. *Suppose that $(L_s)_{s \in S}$ is a family of objects in $\mathcal{B}$ and $(j_s: L_s \to L)$ is a coproduct in $\mathcal{D}_{01}$. Then this is also a coproduct in $\mathcal{B}$.*

Proof. Since $[\bigcup_{s \in S} j_s[L_s]]_{\mathbf{D_{01}}} = L$ it easily follows that L is a Boolean algebra. So $(j_s: L_s \to L)_{s \in S}$ is a coproduct in $\mathcal{B}$.

Note that if all $|L_s| > 1$, then again all j_s are one-one.

Turning now to $\mathcal{D}$, we state the analogue of Theorem 1, leaving the proof (which is similar to the proof of Theorem 1) to the reader. Note however that in case $|L_s| = 1$ for some $s \in S$, we no longer have a special case.

Theorem 3. *A necessary and sufficient condition that a family of morphisms $(j_s: L_s \to L)_{s \in S}$ be a coproduct in $\mathcal{D}$ of a family $(L_s)_{s \in S}$ of objects in $\mathcal{D}$ is*

(i) If $S_1 \subseteq S$, $S_2 \subseteq S$, $a_s \in L_s$ for each $s \in S_1$ and $b_s \in L_s$ for each $s \in S_2$ and $\prod_{s \in S_1} j_s(a_s) \leq \sum_{s \in S_2} j_s(b_s)$ then there exists $s_0 \in S_1 \cap S_2$ such that $a_{s_0} \leq b_{s_0}$ and

*(ii) $\bigcup_{s \in S} j_s[L_s]$ **D**-generates L.*

Note that in this case the j_s are always one-one, regardless of whether $|A_s| = 1$ for some $s \in S$.

For a direct construction of the coproduct in $\mathscr{D}$ of a family $(L_s)_{s\in S}$ of objects in $\mathscr{D}$, we start by assuming $L_s \cap L_{s'} = \varnothing$ for $s \neq s'$ (this can be done of course, by replacing each member x of L_s by (x, s) for each $s \in S$).

Let $Y = \bigtimes_{s\in S} (\mathfrak{P}(L_s) \cup \{\varnothing, L_s\})$ and for each $x \in \bigcup_{s\in S} L_s$, let $\varphi(x) = \{f \in Y : x \notin \bigcup_{s\in S} f(s), f(s') \neq \varnothing$ for some $s' \in S, f(s'') \neq L_{s''}$ for some $s'' \in S\}$. Denote by R, the ring of subsets of Y which is **D**-generated by $\{\varphi(x) : x \in \bigcup_{s\in S} L_s\}$. Next, for each $s \in S$, let $\varphi_s : L_s \to R$ be defined by $x \mapsto \varphi(x)$. Clearly, $\bigcup_{s\in S} \varphi_s[L_s]$ **D**-generates R. So to prove $(\varphi_s : L_s \to R)_{s\in S}$ is a coproduct of $(L_s)_{s\in S}$ we must verify that each φ_s is a morphism and that (i) of Theorem 3 holds. For $s_o \in S$, $x, y \in L_{s_o}$ and $f \in Y, f \in \varphi_{s_o}(x + y)$ and $f(s') \neq \varnothing, f(s'') \neq L_{s''}$ for some s', s'' in $S \Leftrightarrow f \in \varphi(x + y)$ and $f(s') \neq \varnothing, f(s'') \neq L_{s''}$ for some $s', s'' \Leftrightarrow x + y \notin \bigcup_{s\in S} f(s), f(s') \neq \varnothing$ for some $s' \in S \Leftrightarrow x + y \notin f(s_o), f(s') \neq \varnothing$ for some $s_o, s' \Leftrightarrow x \notin f(s_o)$ or $y \notin f(s_o)$, and $f(s') \neq \varnothing$ for some $s_o, s' \Leftrightarrow x \notin \bigcup_{s\in S} f(s)$ or $y \notin \bigcup_{s\in S} f(s)$ and $f(s') \neq \varnothing$ for some s' $\Leftrightarrow f \in \varphi(x)$ or $f \in \varphi(y) \Leftrightarrow f \in \varphi(x) \cup \varphi(y) = \varphi_{s_o}(x) \cup \varphi_{s_o}(y)$; similarly for products.

Finally, let $S_1 \subseteq S$, $S_2 \subseteq S$, $a_s \in L_s$ for $s \in S_1$, $b_s \in L_s$ for $s \in S_2$ and $\bigcap_{s\in S_1} \varphi_s(a_s) \subseteq \bigcup_{s\in S_2} \varphi_s(b_s)$. Also, assume that if $s \in S_1 \cap S_2$ then $a_s \nleq b_s$. Define a member $f \in \bigtimes_{s\in S} (\mathfrak{P}(L_s) \cup \{\varnothing, L_s\})$ by

$$f(s) = \begin{cases} \varnothing \text{ if } s \in S_1 \sim S_2 \\ I(s) \text{ if } s \in S_1 \cap S_2 \\ L_s \text{ otherwise} \end{cases}$$

where $I(s)$ is a prime ideal containing b_s but not a_s. But this implies the contradiction $f \in \bigcap_{s\in S_1} \varphi_s(a_s) \sim \bigcup_{s\in S_2} \varphi_s(b_s)$.

Exercise 4. Show that the coproduct of a family $(L_s)_{s\in S}$ of objects in $\mathscr{D}$ has a 0(1) if and only if S is finite and L_s has a 0(1) for each $s \in S$.

Exercise 5. Let $(L_s)_{s\in S}$ be a family of objects in $\mathscr{D}'$. Let $(i_s : L_s \to L)_{s\in S}$ be a coproduct of this family in $\mathscr{D}$ and let $(j_s : \mathscr{U}'(L_s) \to L')_{s\in S}$ be a coproduct of $(\mathscr{U}'(L_s))_{s\in S}$ in $\mathscr{D}_{01}$, where $\mathscr{U}' : \mathscr{D}' \to \mathscr{D}_{01}$ is the functor of Exercise II.8.2. Prove the following:

(i) If L_s has no 0 and no 1 for all $s \in S$, then $(i_s : L_s \to L)_{s\in S}$ is a coproduct of $(L_s)_{s\in S}$ in $\mathscr{D}'$.

(ii) If there exists an $s_0 \in S$, and an $s_1 \in S$ such that L_{s_0} has a 0 and L_{s_1} has a 1, then $(j'_s : L_s \to L')_{s\in S}$ is a coproduct of $(L_s)_{s\in S}$ in $\mathscr{D}'$ where $j'_s = j_s | L_s$ for each $s \in S$.

(iii) If there exists an $s_0 \in S$ such that L_{s_0} has a 0 and L_s has no 1 for all $s \in S$, then $(j'_s : L_s \to L'')_{s\in S}$ is the coproduct of $(L_s)_{s\in S}$ in $\mathscr{D}'$, where $L'' = L' \sim \{1_{L'}\}$ and $j'_s = j_s | L_s$ for each $s \in S$, and dually.

Exercise 6. Prove that coproducts exist in $\mathscr{R}'$ (Hint: Apply Corollary V.6.6 and Theorem I.18.5(ii).).

The notion of a coproduct in $\mathscr{D}$ has been generalized by R. Balbes and A. Horn [1] in the following manner: Let $(L_p)_{p\in P}$ be a family of distributive lattices indexed by a poset P. Then there exists $L \in \text{Ob}\,\mathscr{D}$ (called the *order sum* of $(L_p)_{p\in P}$) satisfying

(i) There exist monomorphisms $(f_p: L_p \to L)_{p\in P}$.

(ii) $L = [\bigcup_{p\in P} f_p[L_p]]_{\mathbf{D}}$.

(iii) If $p < q$ and $x \in L_p$, $y \in L_q$ then $f_p(x) < f_q(y)$.

(iv) If $L' \in \text{Ob}\,\mathscr{D}$ and $(f'_p: L_p \to L')_{p\in P}$ is a family in $\mathscr{D}$ such that $f'_p(x) \leq f'_q(y)$ for $x \in L_p$, $y \in L_q$, whenever $p < q$, then there is a morphism $f: L \to L'$ such that $f \circ \varphi_p = f'_p$ for all $p \in P$.

Clearly order sums reduce to coproducts when the poset P is totally unordered. (Order sums have been further generalized by M. Höft in [1] to partially ordered algebras.)

The fact that uncountable chains do not exist in free distributive lattices and free Boolean algebras has been greatly extended in a paper by G. Grätzer and H. Lakser [1]. They define a class $\mathbf{K}$ of lattices to have the property $P(\alpha, \beta)$ provided that if $(L_s)_{s\in S}$ is a family in $\mathbf{K}$ satisfying

(i) $|S| = \beta$ and

(ii) for each $s \in S$, every chain in L_s has cardinality $< \alpha$ then every chain in the coproduct of $(L_s)_{s\in S}$ has cardinality $< \alpha$.

Their main result is: For each of the classes $\mathbf{D}$, $\mathbf{D}_{01}$ and $\mathbf{B}$.

(i) $P(\alpha, \beta)$ holds for $\beta \geq \aleph_0$ if and only if α is a regular cardinal and $\alpha > \aleph_0$.

(ii) $P(\alpha, \beta)$ holds for $1 < \beta < \aleph_0$ if and only if α is either regular or cofinal with $\aleph_0$.

2. The coproduct convention

Let $(A_s)_{s\in S}$ be a family of objects in one of the categories $\mathscr{D}$, $\mathscr{D}_{01}$ or $\mathscr{B}$ (in the latter two cases, assume $|A_s| > 1$ for all $s \in S$). Suppose $(j_s: A_s \to A)_{s\in S}$ is a coproduct of this family. Then there exists a family $(A'_s)_{s\in S}$ of subalgebras of A, such that $A'_s \cong A_s$ for each $s \in S$ and such that $(1_{A'_s,A}: A'_s \to A)_{s\in S}$ is a coproduct of $(A'_s)_{s\in S}$. Indeed, this follows, since each j_s is an embedding, so $j_s[A_s]$ can be taken for A'_s. This justifies the establishment of the *coproduct convention*: If $(1_{A_s,A}: A_s \to A)_{s\in S}$ is a coproduct of a family $(A_s)_{s\in S}$ in $\mathscr{D}$, $\mathscr{D}_{01}$ or $\mathscr{B}$, then we write $A \doteq \sum_{s\in S} A_s$, and if $S = \{1, 2, \cdots, n\}$, then we write $A \doteq A_1 + \cdots + A_n$. Note that in $\mathscr{D}_{01}$ and $\mathscr{B}$ if $A \doteq \sum_{s\in S} A_s$ then either $|A_s| = 1$ for all $s \in S$ or $|A_s| > 1$ for all $s \in S$.

The second part of Theorem 1.1 now reads $L \doteq \sum_{s\in S} L_s$ if and only if

(1) $\bigcup_{s\in S} L_s$ $\mathbf{D}_{01}$-generates L and

(2) If $\{a_{s_i}, b_{s_i}\} \subseteq L_{s_i}$ for $1 \leq i \leq n$, $n \geq 1$ and $\prod_{i=1}^n a_{s_i} \leq \sum_{i=1}^n b_{s_i}$ then $a_{s_i} \leq b_{s_i}$ for some $i \in \{1, \cdots, n\}$.

We have seen in Exercise V.3.16 that $\mathfrak{P}(\mathscr{F}_{\mathbf{D01}}(X))$ is isomorphic to the power set of X. Since $\mathscr{F}_{\mathbf{D01}}(X)$ is the coproduct of copies of $\mathscr{F}_{\mathbf{D01}}(\mathbf{1})$ it seems natural to ask whether a more general result holds for coproducts. We restrict ourselves to the category $\mathscr{D}_{01}$ and we have indeed the following theorem.

Theorem 1. *Let $(L_s)_{s\in S}$ be a family of objects in $\mathscr{D}_{01}$. Then $\mathfrak{P}(\sum_{s\in S} L_s)$ is isomorphic to $\bigtimes_{s\in S} \mathfrak{P}(L_s)$ (where the coproduct is taken in $\mathscr{D}_{01}$). Furthermore, if for each $s\in S$, I_s is a prime ideal in L_s, then the corresponding prime ideal in $\sum_{s\in S} L_s$ is the prime ideal I which is uniquely determined by the property that $I \cap L_s = I_s$ for each $s\in S$.*

Proof. The map $I \mapsto I_s = I \cap L_s$ for each $s\in S$, where $I\in\mathfrak{P}(\sum_{s\in S} L_s)$, obviously defines a map $\mathfrak{P}(\sum_{s\in S} L_s) \to \bigtimes_{s\in S} \mathfrak{P}(L_s)$. For $I, J\in\mathfrak{P}(\sum_{s\in S} L_s)$ and $I \subseteq J$, we obviously have $I_s \subseteq J_s$ for each $s\in S$. If $(I_s)_{s\in S}$ is a family where $I_s \in \mathfrak{P}(L_s)$ for each $s\in S$, then it follows from the definition of coproduct (Definition I.15.2) that there exists a unique $I\in\mathfrak{P}(\sum_{s\in S} L_s)$ such that $I \cap L_s = I_s$ for each $s\in S$. It remains to show that if $I, J\in\mathfrak{P}(\sum_{s\in S} L_s)$ and $I_s \subseteq J_s$ for each $s\in S$ then $I \subseteq J$. But if $x\in I$ then $x = \prod X_1 + \cdots + \prod X_n$ where $X_j \subseteq \bigcup_{s\in S} L_s$ for $1 \le j \le n$. We have $f_I(x) = \mathbf{0}$ $\Rightarrow f_I(\prod X_j) = \mathbf{0}$ for each $j \Rightarrow$ there exists for each j, $s_j\in S$ and $x_{s_j}\in X_j \cap L_{s_j}$ such that $f_I(x_{s_j}) = \mathbf{0}$ for each $j \Rightarrow x_{s_j}\in I_{s_j}$ for each $j \Rightarrow x_{s_j}\in J_{s_j}$ for each $j \Rightarrow f_J(x_{s_j}) = \mathbf{0}$ for each $j \Rightarrow f_J(x_{s_j}) = \mathbf{0}$ for each $j \Rightarrow f_J(\prod X_j) = \mathbf{0}$ for each $j \Rightarrow f_J(x) = \mathbf{0} \Rightarrow x\in J$.

Although Exercise V.3.16 is in the form of an equivalence, we will show in Section X.5 that the converse of Theorem 1 does not hold. Thus, if $L\in\text{Ob}\,\mathscr{D}_{01}$ and $\mathfrak{P}(L)$ is isomorphic to $\bigtimes_{s\in S} P_s$ where P_s is a partially ordered set for each $s\in S$, then L is not necessarily isomorphic to a coproduct of a family $(L_s)_{s\in S}$ where for each $s\in S$, $L_s\in\text{Ob}\,\mathscr{D}_{01}$ and $\mathfrak{P}(L_s)$ is isomorphic to P_s.

3. The center of the coproduct in $\mathscr{D}_{01}$

In this section we will prove that in $\mathscr{D}_{01}$ the center of the coproduct of objects in $\mathscr{D}_{01}$ is the coproduct of their centers. The difficult part of the proof is handled by the following lemma.

Lemma 1. *If $L_1, L_2\in\text{Ob}\,\mathscr{D}_{01}$ and $L \doteq L_1 + L_2$ then $\mathscr{C}(L) \subseteq [\mathscr{C}(L_1) \cup \mathscr{C}(L_2)]_{\mathbf{D01}}$.*

Proof. It suffices to prove that if $x\in\mathscr{C}(L)$ and

(1) $x = \sum_{i=1}^{p} a_i b_i, p \ge 1, a_i b_i \ne 0, a_i\in L_1, b_i\in L_2$

then any a_i that appears in (1) can be replaced by a member of $\mathscr{C}(L_1)$ and still leave (1) valid. Indeed, if this replacement is possible, then each a_i can be successively replaced by members of $\mathscr{C}(L_1)$ and then the process repeated by replacing each b_i by a member of $\mathscr{C}(L_2)$, thus showing that $x\in[\mathscr{C}(L_1) \cup \mathscr{C}(L_2)]_{\mathbf{D01}}$. Now to prove that this replacement is possible, suppose $x \ne 0, 1$ has the representation (1) and that $\bar{x}$ is the complement of x in $\mathscr{C}(L)$. Then $\bar{x} = \sum_{j=1}^{q} c_j d_j$ for some $q \ge 1$, $c_j\in L_1$, $d_j\in L_2$. We will replace a_1 by an element of $\mathscr{C}(L_1)$. If $a_1\in\mathscr{C}(L_1)$, we are done, so suppose $a_1 \notin \mathscr{C}(L_1)$. Then (2) of Section 2 implies, that for each $j\in\{1, \cdots, q\}$, $a_1 c_j = 0$ or $b_1 d_j = 0$. Therefore

(2) $(a_1 + \sum_{i=2}^{p} b_i) + (\sum\{c_j : a_1c_j = 0\}) + (\sum\{d_j : b_1d_j = 0\}) \geq x + \bar{x} = 1.$

Now $a_1\sum\{c_j : a_1c_j = 0\} = 0$ so $a_1 + \sum\{c_j : a_1c_j = 0\} \neq 1$. Applying (2) of Section 2 to (2), we obtain $\sum_{i=2}^{p} b_i + \sum\{d_j : b_1d_j = 0\} = 1$. This implies

(3) $b_1 \leq \sum_{i=2}^{p} b_i$.

Let $\{S_1, \cdots, S_r\}$ be all subsets S of $\{b_2, \cdots, b_p\}$ with the property that $b_1 \leq \sum S$. Note from (3) that $r \geq 1$. Let $T_j = \{a_k : b_k \in S_j\}$ for $j = 1, \cdots, r$. We will show that the required replacement element is: $A = a_1 + \prod T_1 + \cdots + \prod T_r$. Clearly, $x \leq Ab_1 + a_2b_2 + \cdots + a_pb_p$. On the other hand, for each $j \in \{1, \cdots, r\}$, $(\prod T_j)b_1 \leq (\prod T_j)\sum S_j \leq \sum\{a_kb_k : b_k \in S_j\} \leq x$ so $Ab_1 \leq x$ and thus

(4) $x = Ab_1 + a_2b_2 + \cdots + a_pb_p$.

It remains to show that $A \in \mathscr{C}(L_1)$. Let $\mathfrak{F}$ be the collection of all sets T which consist of exactly one member from each of the sets $T_1, \cdots, T_r$. For such a $T \in \mathfrak{F}$, (4) yields

(5) $(A + \sum T + \sum\{b_i : a_i \notin T, i \geq 2\}) + (\sum\{c_j : Ac_j = 0\} + \sum\{d_j : b_1d_j = 0\})$
$\geq x + \bar{x} = 1.$

Now, if $\sum\{b_i : a_i \notin T, i \geq 2\} + \sum\{d_j : b_1d_j = 0\} = 1$, then $b_1 \leq \sum\{b_i : a_i \notin T, i \geq 2\}$ so $\{b_i : a_i \notin T, i \geq 2\} = S_j$ for some $j \in \{1, \cdots, r\}$. But this is impossible, for by the definition of T, there exists $a_{i_o} \in T_j \cap T$ where $i_o \geq 2$, so $b_{i_o} \in S_j \sim \{b_i : a_i \notin T, i \geq 2\} = S_j \sim S_j = \varnothing$. By applying (2) of Section 2 to (5) for each $T \in \mathfrak{F}$: $A + \sum T + \sum\{c_j : Ac_j = 0\} = 1$. Thus,

$$\begin{aligned} 1 &= A + \prod\{\sum T : T \in \mathfrak{F}\} + \sum\{c_j : Ac_j = 0\} \\ &= A + \prod T_1 + \cdots + \prod T_r + \sum\{c_j : Ac_j = 0\} = A + \sum\{c_j : Ac_j = 0\}. \end{aligned}$$

This means that A and $\sum\{c_j : Ac_j = 0\}$ are complements in L_1, which completes the proof.

Theorem 2. (R. Balbes [4]) *If* $L \doteq \sum_{s\in S} L_s$ *then* $\mathscr{C}(L) \doteq \sum_{s\in S} \mathscr{C}(L_s)$.

Proof. It suffices to prove that

(6) $[\bigcup_{s\in S} \mathscr{C}(L)_s]_{\mathbf{D01}} = \mathscr{C}(L)$.

Indeed, (6) implies that condition (1) of Section 2 is satisfied by the family $(\mathscr{C}(L_s))_{s\in S}$ and clearly (2) holds for this family since $\mathscr{C}(L_s) \subseteq L_s$ for each $s \in S$. Now let $x \in [\bigcup_{s\in S} \mathscr{C}(L_s)]_{\mathbf{D01}}$. Then there exist $a_{i_j} \in \mathscr{C}(L_{i_j}), j = 1, \cdots, n_i, i = 1, \cdots, p$ such that $x = \sum_{i=1}^{p} a_{i_1} \cdots a_{i_{n_i}}$. It is easy to verify that x has a complement in L, namely $\prod_{i=1}^{p} (\bar{a}_{i_1} + \cdots + \bar{a}_{i_{n_i}})$. Thus $x \in \mathscr{C}(L)$. For the reverse inclusion, let $x \in \mathscr{C}(L)$. Then $x = \sum_{i=1}^{n} a_{i_1} \cdots a_{i_{n_i}}$ where $a_{i_k} \in L_{i_k}$ for $1 \leq k \leq n_i$ and $\bar{x}$—the complement of x in L—has the form $\bar{x} = \sum_{j=1}^{m} b_{j_1} \cdots b_{j_{n_j}}, b_{j_k} \in L_{j_k}$ for $1 \leq j \leq m, 1 \leq k \leq n_j$. But then $\bar{x}$ and x are complements in $L' = [(\bigcup_{k=1}^{n} L_{i_k}) \cup (\bigcup_{k=1}^{m} L_{j_k})]_{\mathbf{D01}}$. We then have $x \in \mathscr{C}(L')$ and by induction on Lemma 1, $x \in \mathscr{C}(L') \subseteq [(\bigcup_{k=1}^{n} \mathscr{C}(L_{i_k})) \cup (\bigcup_{k=1}^{m} \mathscr{C}(L_{j_k}))]_{\mathbf{D01}} \subseteq [\bigcup_{s\in S} \mathscr{C}(L_s)]$.

W. J. Blok [2] has given a topological proof of Theorem 2.

4. Structure theorems for coproducts of Boolean algebras and chains

We will apply the results of the previous sections to the special case of the coproduct $L \doteq B + C$ in $\mathscr{D}_{01}$ of a Boolean algebra B and a chain C. These coproducts can be characterized by a structure theorem that will be proved in this section. We will also consider the question of the uniqueness of the representation $L \doteq B + C$. The answer here turns out to be surprisingly simple, although by no means trivial. In Chapter X we will apply the results of this section to the special case that $\mathscr{C}$ is a finite chain.

In order to simplify the statements in this section, the symbols $L, L', L_1, L_2, \cdots$, $C, C', C_1, C_2, \cdots$, and $B, B', B_1, B_2, \cdots$ will be used to stand for objects in $\mathscr{D}_{01}, \mathscr{C}_{01}$ and $\mathscr{B}$, respectively. The setting for this chapter is the category $\mathscr{D}_{01}$.

From the results of Section 2, we can say that for $\mathbf{D}_{01}$-subalgebras L_1, L_2 of $L, L \doteq L_1 + L_2$ if and only if

(1) $L_1 \cup L_2$ $\mathbf{D}_{01}$-generates L and

(2) for $a_1, b_1 \in L_1$; $a_2, b_2 \in L_2$, $a_1a_2 \leq b_1 + b_2 \Rightarrow a_1 \leq b_1$ or $a_2 \leq b_2$.

In case $L_1 = B$ and $L_2 = C$, condition (2) can be replaced by

(3) for $a \in B$ and $c, d \in C$, $ac \leq d \Rightarrow a = 0$ or $c \leq d$.

Indeed, $ac \leq d$ can also be written as $ac \leq 0 + d$ and so (3) follows immediately from (2). Conversely, suppose $ac \leq b + d$, $a, b \in B$ and $c, d \in C$ then $a\bar{b}c \leq d$ implies, by (3), that $a\bar{b} = 0$ or $c \leq d$ thus $a \leq b$ or $c \leq d$.

The following theorem is basic for our investigations and is an extension of a result which will be discussed in Chapter X.

Theorem 1. *$L \doteq B + C$ if and only if for each $x \in L$, x has a unique representation $x = \sum_{i=1}^{n} a_ic_i$, $n \geq 1$, where $\{c_1, c_2, \cdots, c_n\} \subseteq C$, $\{a_1, a_2, \cdots, a_n\} \subseteq B$ and $0 = c_1 < c_2 < \cdots < c_n$, $1 = a_1, a_2 > \cdots > a_n > 0$.*

Proof. ($\Leftarrow$) Certainly $[B \cup C]_{\mathbf{D}_{01}} = L$. We show that (3) is satisfied. Suppose $ac \leq d$, $a \in B$, $\{c, d\} \subseteq C$ and suppose $c \nleq d$, $a > 0$. Now $ac \leq d \Rightarrow ad = ac \Rightarrow 1 \cdot 0 + ad = 1 \cdot 0 + ac$. Now $c > 0$ so if d > 0 then the unique representation hypothesis implies $c = d$, a contradiction. Hence $d = 0$ so $ac = 0$. But then $1 \cdot 0 + ac = 0 = 1 \cdot 0 + 0 \cdot c$ so by uniqueness $a = 0$, a contradiction.

($\Rightarrow$) Thus suppose $L \doteq B + C$. If $x \in L$ and $x = 0$ then we have the representation $x = 1 \cdot 0$. Now assume $x \neq 0$. Because $[B \cup C]_{\mathbf{D}_{01}} = L$, x can be written as $x = \sum_{i=1}^{n} a_ic_i$ where $a_i \in B$, $c_i \in C$ for $1 \leq i \leq n$. Since C is a chain, we can assume $c_1 < \cdots < c_n$. Furthermore, if we write, for every i, $1 \leq i \leq n$, $a_i' = \sum_{j=i}^{n} a_j$, then we have $\sum_{i=1}^{n} a_i'c_i = \sum_{i=1}^{n} (\sum_{j=i}^{n} a_j)c_i = \sum_{i=1}^{n} \sum_{j=1}^{i} a_ic_j = \sum_{i=1}^{n} a_ic_i$. Since $a_1' \geq a_2' \cdots \geq a_n'$ we may therefore assume that in the expression $x = \sum_{i=1}^{n} a_ic_i$, $a_1 \geq a_2 \cdots \geq a_n$. Now, by dropping all but the maximal members of $\{a_ic_i : i = 1, \cdots, n\}$ (and since $x \neq 0$) we see that x can be written as $x = \sum_{j=1}^{m} a_{i_j}c_{i_j}$ where $0 < c_{i_1} < c_{i_2} < \cdots < c_{i_m}$, $a_{i_1} > a_{i_2} > \cdots > a_{i_m} > 0$. The required representation is therefore

$x = 1 \cdot 0 + \sum_{j=1}^{m} a_{i_j} c_{i_j}$. Finally, for uniqueness, suppose $x = \sum_{i=1}^{n} a_i c_i = \sum_{j=1}^{m} b_j d_j$, $n \leq m, 0 = c_1 < c_2 \cdots < c_n, 0 = d_1 < d_2 \cdots < d_m, 1 = a_1, a_2 > \cdots > a_n > 0, 1 = b_1, b_2 > \cdots > b_m > 0$. Now $a_1 = b_1 = 1$ and $c_1 = d_1 = 0$. Suppose $a_i = b_i, c_i = d_i$ for $1 \leq i \leq k, k < n$. Then $a_{k+1}c_{k+1} \leq \sum_{j=1}^{m} b_j d_j \leq \sum_{i=1}^{k} d_i + \sum_{i=k+1}^{m} b_i = d_k + b_{k+1} = c_k + b_{k+1}$. Now $c_{k+1} > c_k$ hence by (2) $a_{k+1} \leq b_{k+1}$. Similarly, $b_{k+1} \leq a_{k+1}$, hence $a_{k+1} = b_{k+1}$. Again, $a_{k+1}c_{k+1} \leq \sum_{j=1}^{m} b_j d_j$. If $k + 1 = m$, then we have $\sum_{j=1}^{m} b_j d_j = \sum_{j=1}^{k+1} b_j d_j \leq d_{k+1}$ thus $a_{k+1}c_{k+1} \leq d_{k+1}$, but $a_{k+1} > 0$, so by (3) $c_{k+1} \leq d_{k+1}$. If $k + 1 < m$ then $\sum_{j=1}^{m} b_j d_j \leq \sum_{j=1}^{k+1} d_j + \sum_{j=k+2}^{m} b_j = d_{k+1} + b_{k+2}$. Thus $a_{k+1}c_{k+1} \leq d_{k+1} + b_{k+2}$. But if $a_{k+1} \leq b_{k+2}$ then $b_{k+1} = a_{k+1} \leq b_{k+2}$ contradicting $b_{k+1} > b_{k+2}$. Hence $a_{k+1} \nleq b_{k+2}$. Therefore by (2) we have also, in this case, $c_{k+1} \leq d_{k+1}$. Similarly, $d_{k+1} \leq c_{k+1}$ thus $c_{k+1} = d_{k+1}$. Hence $a_i = b_i, c_i = d_i$ for $i = 1, \cdots, n$. Now suppose $m > n$, then $b_{n+1}d_{n+1} \leq \sum_{i=1}^{n} c_i = c_n$. But $b_{n+1} \neq 0$ thus we have from (3) that $d_{n+1} \leq c_n = d_n$, contradicting $d_{n+1} > d_n$. Hence $m = n$. This completes the proof of the theorem.

We will now present the main structure theorem for coproducts of Boolean algebras and chains in $\mathscr{D}_{01}$. First, an element $x \in C^S$, where C is a chain and S is a set, is called a *diagonal element* if x is a constant map.

Theorem 2. *If $L \doteq B + C$, then L is a* $\mathbf{D}_{01}$*-subdirect product of C^S for some set S. Moreover, L is a* $\mathbf{D}_{01}$*-subalgebra of C^S and is* $\mathbf{D}_{01}$*-generated by the union of a* **B***-subalgebra of C^S and the set D of diagonal elements of C^S. Conversely, any* $\mathbf{D}_{01}$*-subalgebra L of C^S, $S \neq \varnothing$, which is* $\mathbf{D}_{01}$*-generated by the union of a* **B***-subalgebra of C^S and D, is the coproduct of B and D.*

Remark. Part of Theorem 2 states that if $L \doteq B + C$, then $L \in \mathbf{C}_C$ (cf. Sections VI.1 and VI.2). Although we could prove this part of the theorem, using the results of Chapter VI, we prefer to give a direct proof at this point (cf. Exercise 4).

Proof of Theorem 2. Suppose $L \doteq B + C$. Let S be the set of prime ideals of B. We define a homomorphism $h_1 : B \to C^S$ as follows. For each $a \in B$, let $(h_1(a))(I) = 0$ if $a \in I$ and 1 if $a \notin I$ for each $I \in S$. Define $h_2 : C \to C^S$ by $(h_2(c))(I) = c$ for each $I \in S$ and $c \in C$. Note, that h_1 and h_2 are in $\mathscr{D}_{01}$. Let $h : L \to C^S$ be the unique extension of h_1 and h_2 to L. We will show that h is an embedding. Let $x, y \in L$ and suppose $h(x) \leq h(y)$. Then x can be represented by $x = \sum_{i=1}^{n} a_i c_i$ and $y = \prod_{j=1}^{m} (b_j + d_j)$ where $\{a_1, \cdots, a_n, b_1, \cdots, b_m\} \subseteq B$ and $\{c_1, \cdots, c_n, d_1, \cdots, d_m\} \subseteq C$. (Note that we do not use Theorem 1 but only the fact that $B \cup C$ generates L.) Now we have, for each i, j, that $h_1(a_i)h_2(c_i) \leq h(x) \leq h(y) \leq h_1(b_j) + h_2(d_j)$. If $a_i \leq b_j$ then $a_i c_i \leq b_j + d_j$. If $a_i \nleq b_j$ then there exists an $I \in S$ such that $a_i \notin I, b_j \in I$, hence $(h(a_i c_i))(I) = (h(a_i))(I)(h(c_i))(I) = (h_1(a_i))(I)(h_2(c_i))(I) = c_i$. Again, $(h(b_j + d_j))(I) = (h(b_j))(I) + (h(d_j))(I) = (h_1(b_j))(I) + (h_2(d_j))(I) = d_j$. But $h_1(a_i)h_2(c_i) \leq h_1(b_j) + h_2(d_j)$ thus $c_i \leq d_j$ and so, in this case, we also have $a_i c_i \leq b_j + d_j$ for every i, j. Hence $x \leq y$, and it follows that h is an embedding. Since $h[L]$ contains D, we have that L is a subdirect product of C^S. Also, $h[B]$ is a **B**-subalgebra of C^S and is isomorphic to B and $h[C] = D$ which is isomorphic to C and $h[L]$ is the coprodcut of $h[B]$ and D.

Conversely, suppose $S \neq \varnothing$ is a set and L is a $\mathbf{D_{01}}$-subalgebra of C^S, $\mathbf{D_{01}}$-generated by the union of Boolean subalgebra B of C^S and D. We only need to show that (3) holds. Suppose $ax \leq y$, $a \in B$, $x, y \in D$ and suppose $x \nleq y$. Thus since $x, y \in D$, $x(s) \nleq y(s)$ for each $s \in S$. Also $a(s)x(s) \leq y(s)$ for each $s \in S$. But a is a complemented element of C^S, hence $a(s) = 0$ or 1 for each $s \in S$. Therefore $a(s) \neq 1$ for each $s \in S$ and thus $a(s) = 0$ for each $s \in S$ and it follows that $a = 0$.

Theorem 3. *Let $L \doteq B + C$, A a set, $B \cong \mathbf{2}^A$, and C a finite chain. Then L is isomorphic to C^A. Conversely, if C is a finite chain and A is a set, then C^A is the coproduct of its center and C.*

Proof. First, suppose $B \cong \mathbf{2}^A$ and C is a finite chain. We consider B to be the set of all subsets of A and alter the proof of the first part of Theorem 2 as follows. Define $h_1: B \to C^A$ by $(h_1(x))(a) = 0$ if $a \notin x$ and $= 1$ if $a \in x$ for $x \in B$, $a \in A$. Obviously h_1 is a homomorphism. Define $h_2: C \to C^A$ by $(h_2(c))(a) = c$ for $c \in C$ and each $a \in A$. Let $h: B + C \to C^A$ be the unique extension of h_1 and h_2 to L. That h is an embedding can be proved in essentially the same way as in Theorem 2. It remains to show that h is onto. Let $f \in C^A$. For each $c \in C$, let $f^c = \{a \in A : f(a) = c\}$. It is easy to see that $(h_1(f^c))(a) = 0$ if $f(a) \neq c$ and $= 1$ if $f(a) = c$ for each $a \in A$. Also $(h(\sum_{c \in C} f^c))(a) = \sum_{c \in C}((h_1(f^c))(a))((h_2(c))(a)) = f(a)$. Hence $h(\sum_{c \in C} f^c) = f$. For the second part of the theorem, observe that the center of C^A consists of all $f \in C^A$ with $f(a) = 0$ or 1 for each $a \in A$. Furthermore, since C and thus D is finite, $[\mathscr{C}(C^A) \cup D]_{\mathbf{D_{01}}} = C^A$. Using an argument similar to that of the second part of Theorem 2, we see that C^A is the coproduct of its center and D. Finally, note that D is isomorphic to C.

Exercise 4. Prove the first part of Theorem 2, using the results of Chapter VI.

We have seen that coproducts of Boolean algebras and chains are lattices of functions. We will now show that these coproducts are, in fact, the lattices of continuous functions from a Boolean space to discrete chains (i.e. chains endowed with the discrete topology). First note that if X is a Boolean space and C is a discrete chain then the subset of C^X consisting of the continuous functions is obviously a $\mathbf{D_{01}}$-subalgebra of C^X. (We recall our assumption made at the beginning of this section that all of our objects are in $\mathscr{D}_{01}$.) We will denote this sublattice by $\langle X, C\rangle$. The following theorem gives a topological characterization of coproducts of Boolean algebras and chains. We note that the first part of Theorem 2 is actually a corollary of this theorem, and conversely, we could extend the proof of Theorem 2 and obtain this theorem. However, we will give a short independent proof.

Theorem 5. *L is a coproduct of B and C, $B \ncong \mathbf{1}$, $C \ncong \mathbf{1}$ if and only if L is isomorphic to $\langle \mathscr{S}_{\mathscr{B}}(B), C\rangle$.*

Proof. ($\Leftarrow$) Note that C is isomorphic to the chain of constant functions, and we will identify C with this chain in $\langle \mathscr{S}_{\mathscr{B}}(B), C\rangle$. Thus, for $c \in C$ we will—without danger of confusion—also let c denote the constant function $c: \mathscr{S}_{\mathscr{B}}(B) \to C$ for which $c(x) = c$. Also note that

$$\mathscr{C}(\langle \mathscr{S}_{\mathscr{B}}(B), C \rangle) = \{x \in \langle \mathscr{S}_{\mathscr{B}}(B), C \rangle : \text{Im}(x) \subseteq \{0, 1\}\}.$$

If $x \in \langle \mathscr{S}_{\mathscr{B}}(B), C \rangle$, $x \neq 0$ then

$$(4)\ x = a_1 c_1 + \cdots + a_n c_n, \text{ where } a_i \in \mathscr{C}(\langle \mathscr{S}_{\mathscr{B}}(B), C \rangle), c_i \in C, 1 \leq i \leq n,$$

$a_1 > a_2 > \cdots > a_n > 0$, $0 \neq c_1 < c_2 \cdots < c_n$. Indeed, let $\{c_1, c_2, \cdots, c_n\} = \text{Im}(x) \sim \{0\}$ (note that this image is finite, since $\mathscr{S}_{\mathscr{B}}(B)$ is compact) and for $1 \leq i \leq n$, let $a_i \in \mathscr{C}(\langle \mathscr{S}_{\mathscr{B}}(B), C \rangle)$ be defined by $a_i^{-1}[\{1\}] = \bigcup_{j=i}^{n} x^{-1}(c_j)$. Then (4) is easily established. It now follows from (4) that $[\mathscr{C}\langle \mathscr{S}_{\mathscr{B}}(B), C \rangle) \cup C]_{\mathbf{D_{01}}} = \langle \mathscr{S}_{\mathscr{B}}(B), C \rangle$. We only need to check (3) to conclude that $\langle \mathscr{S}_{\mathscr{B}}(B), C \rangle \doteq \mathscr{C}(\langle \mathscr{S}_{\mathscr{B}}(B), C \rangle) + C$; but this is immediate. Finally, note that $B \cong \mathscr{C}(\langle \mathscr{S}_{\mathscr{B}}(B), C \rangle)$.

($\Rightarrow$) If L is a coproduct of B and C, then by the first part, $\langle \mathscr{S}_{\mathscr{B}}(B), C \rangle$, is isomorphic to a coproduct of $\mathscr{C}(\langle \mathscr{S}_{\mathscr{B}}(B), C \rangle)$ and C, and hence to a coproduct of B and C.

Exercise 6. Verify (4).

We turn now to the question of the uniqueness of a representation $L \doteq B + C$. Our first result in answering this question is formulated in the next theorem.

Theorem 7. *Suppose $L \doteq B + C$ and $L \doteq B' + C'$. Then $B = B'$ and C is isomorphic to C'. Moreover an isomorphism $f: C \to C'$ can be chosen such that $f(x) \leq x$ for each $x \in C$.*

Proof. That $B = B'$ follows immediately from the fact that $\mathscr{C}(L) = B = B'$ (Theorem 3.2). Now for each $x \in C$, $x \neq 0$, x can be represented by

$$(5)\ x = \textstyle\sum_{i=1}^{n} a_i c_i',\ n \geq 1 \text{ where } \{c_1', c_2', \cdots, c_n'\} \subseteq C', \{a_1, a_2, \cdots, a_n\} \subseteq B$$

$0 < c_1' < \cdots < c_n'$ and $1 = a_1 > \cdots > a_n > 0$. (Indeed by Theorem 1, for $x \neq 0$, we have $n \geq 2$ and $x \leq c_1' + a_2 + \cdots + a_n = 0 + a_2$ thus by (2), $a_2 = 1$ and $x = \sum_{i=2}^{n} a_i c_i'$, where $n \geq 2$, $1 = a_2 > \cdots > a_n > 0$ and $0 < c_1' < \cdots < c_n'$.) Now define $f: C \to C'$ as follows. Let $f(0) = 0$ and if $x \in C$, $x \neq 0$ with unique representation (5), let $f(x) = c_1' <$. Obviously $f(x) \leq x$ for each $x \in C$. We will show that f is an isomorphism. Let $x, y \in C \sim \{0\}$ where the unique representation for x is given by (5) and the unique representation of y by $y = \sum_{j=1}^{m} a_j' d_j'$, $m \geq 1$ where $\{d_1', \cdots, d_m'\} \subseteq C'$, $\{a_1', \cdots, a_m'\} \subseteq B$, $0 < d_1' < \cdots < d_m'$ and $1 = a_1' > a_2' > \cdots > a_m' > 0$. First, suppose $x \leq y$. If $m = 1$, then $f(x) = c_1' \leq x \leq y \leq d_1' = f(y)$. Suppose $m \geq 2$. Then $c_1' \leq x \leq y \leq d_1' + (a_2' + \cdots + a_m') = d_1' + a_2'$. But $a_2' \neq 1$ so by (2) $c_1' \leq d_1'$. Thus again $f(x) \leq f(y)$. Next, suppose $f(x) \leq f(y)$ and $x \nleq y$. Then $y < x$ and thus $f(y) \leq f(x)$, so $c_1' = f(x) = f(y) = d_1'$. If $n = 1$, then $x = c_1' = d_1' \leq y$, contradicting $y < x$. If $n \geq 2$ then $x \leq c_1' + a_2 = d_1' + a_2 = a_1' d_1' + a_2 \leq y + a_2$. But $y < x$ so by (2), $a_2 = 1$, a contradiction. Thus in either case $x \leq y$. It remains to show that f is onto. Let $y \in C', y \neq 0$. Let y have the unique representation $y = \sum_{i=1}^{n} a_i c_i, n \geq 1, \{c_1, \cdots, c_n\} \subseteq C, \{a_1, \cdots, a_n\} \subseteq B, 0 < c_1 < \cdots < c_n$, $1 = a_1 > a_2 > \cdots > a_n > 0$. Again, let c_n have the unique representation $c_n = \sum_{j=1}^{m} a'_j c'_j, m \geq 1, \{c'_1, \cdots, c'_m\} \subseteq C', \{a'_1, \cdots, a'_m\} \subseteq B, 0 < c'_1 < \cdots < c'_m$, $1 = a'_1 > a'_2 \cdots > a'_n > 0$. We claim that $c'_1 = y$. If $m \geq 2$, then $y \leq c_1 + \cdots + c_n$ $= c_n \leq c'_1 + (a'_2 + \cdots a'_m) = c'_1 + a'_2$. But $a'_2 \neq 1$, thus by (2) $y \leq c'_1$. If $m = 1$, then obviously $y \leq c'_1$. On the other hand $c'_1 = a'_1 c'_1 \leq c_n$ so $a_n c'_1 \leq a_n c_n \leq y$, but $a_n \neq 0$ and by (2) $c'_1 \leq y$ and it follows that $y = c'_1$. Thus, $f(c_n) = c'_1 = y$ completing the proof of the theorem.

Theorem 7 tells us that if $L \doteq B + C$ and $L \doteq B + C'$, then C and C' are isomorphic. However, the reader should observe that this does not necessarily imply that $C = C'$. Therefore, the next natural question to ask is whether there exist chains C, C' and a Boolean algebra B and an $L \in \mathrm{Ob}\, \mathscr{D}_{01}$ such that $L \doteq B + C$ and $L \doteq B + C'$ and such that $C \neq C'$ (although C and C' are necessarily isomorphic). Before this question can be answered we must discuss "rigid" chains. This is the purpose of the following section.

5. Rigid chains

In this section the symbol C will always stand for a chain and the setting will be the category $\mathscr{D}$.

Definition 1. A chain C is *rigid* if it has no proper automorphism.

Definition 2. Let $f: C \to C$ be an automorphism. f is *increasing* (*decreasing*) if $f(x) \geq x (f(x) \leq x)$ for each $x \in C$; f is *strictly increasing* (*strictly decreasing*) if f is increasing (decreasing) and without any fixed points. The group of automorphisms of a chain C will be denoted by $\mathfrak{A}(C)$.

Exercise 3. Let $f \in \mathfrak{A}(C)$. Then there exists $g \in \mathfrak{A}(C)$ which is increasing (decreasing) and such that f and g have the same set of fixed points.

Exercise 4. Let $f, g \in \mathfrak{A}(C)$. Define $f + g: C \to C$ and $fg: C \to C$ by $(f + g)(x)$ $= f(x) + g(x)$ and $(fg)(x) = f(x)g(x)$ for each $x \in C$. Then $f + g, fg \in \mathfrak{A}(C)$.

Exercise 5. Let $(S_i)_{i \in I}$ be a family of mutually disjoint and convex subchains of C and for every $i \in I$ let $f_i: S_i \to S_i$ be an automorphism. Then the family $(f_i)_{i \in I}$ can be extended to an automorphism $f: C \to C$.

In this section we are particularly interested in rigid chains, since they will play an important role in the next section. Our main purpose will be to prove a theorem stating necessary and sufficient conditions in order that a chain be not rigid.

Exercise 6. Every (dual) ordinary is rigid. The ordinal sum of an ordinal and a dual ordinal is rigid.

Theorem 7. *A chain C has the property that each of its subchains is rigid if and only if C is the ordinal sum of an ordinal and a dual ordinal. (Thus* $C = \alpha \oplus \breve{\beta}$, α, β *ordinals.)*

We first prove a lemma.

Lemma 8. *Suppose C has the property that each of its subchains has a smallest or a largest element. Then C is the ordinal sum of an ordinal and a dual ordinal.*

Proof. Let $C_1 = \{x \in C : (x] \text{ an ordinal}\}$. Let $C_2 = \{x \in C : [x) \text{ a dual ordinal}\}$. Then C_1 is an ordinal and C_2 is a dual ordinal. (Note that C_1 or C_2 may be $\mathbf{0}$.) Indeed, let $U \subseteq C_1$, $U \neq \varnothing$. Choose an element $c \in U$. Then $(c] \cap U \subseteq (c]$ and $(c] \cap U \neq \varnothing$. Thus $(c] \cap U$ has a smallest element p, and obviously p is the smallest element of U. Thus C_1 is an ordinal and dually, C_2 is a dual ordinal. Next, we show $C_1 \cup C_2 = C$. Suppose $c \in C$, $c \notin C_1$, $c \notin C_2$. By definition of C_1, there exists $U \subseteq (c]$, $U \neq \varnothing$ such that U has no smallest element. Dually, there exists $V \subseteq [c)$, $V \neq \varnothing$ such that V has no largest element. Note that $u \leq v$ for all $u \in U$, $v \in V$. Hence $U \cup V$ has neither a smallest nor a largest element, a contradiction. Hence $C = C_1 \cup C_2$. In order to show that $C = C_1 \oplus C_2$, we may assume that $C_1 \neq \varnothing$ and $C_2 \neq \varnothing$. If $C_1 \cap C_2 \neq \varnothing$, then there exists an element $c \in C_1 \cap C_2$. Then $(c]$ is an ordinal, $[c)$ is a dual ordinal, and $C = (c] \oplus ([c) \sim \{c\})$. If $C_1 \cap C_2 = \varnothing$, then $C = C_1 \oplus C_2$. Indeed, suppose there exists $a \in C_1$, $b \in C_2$ with $a > b$. But if $U \subseteq (b]$, $U \neq \varnothing$, then $U \subseteq (a]$ thus U has a smallest element, and $(b]$ is therefore an ordinal. Thus $b \in C_1$, and hence $C_1 \cap C_2 \neq \varnothing$, a contradiction.

Proof of Theorem 7. ($\Rightarrow$) Let $S \subseteq C$, $S \neq \varnothing$. If S has no smallest and no largest element, then S contains Z which is not rigid, a contradiction. Now apply Lemma 8.

($\Leftarrow$) Suppose $C = \alpha \oplus \breve{\beta}$. But then every subchain of C is again the ordinal sum of an ordinal and a dual ordinal and each such a subchain is rigid by Exercise 6.

We now state the theorem enunciated above.

Theorem 9. (cf. C. Holland [1]) *The following are equivalent for a chain C.*

(i) *C is not rigid.*

(ii) *C has a convex subchain S which has a strictly increasing (strictly decreasing) automorphism.*

(iii) *C has a convex subchain T such that* $T = \bigoplus_{n \in Z} T_n$ *where* $\{T_n : n \in Z\}$ *is a set of isomorphic subchains of C.*

Proof. (i) $\Rightarrow$ (ii) Let $f\colon C \to C$ be a proper automorphism. Then there exists $c_o \in C$ with $f(c_o) \neq c_o$ and we may assume that $f(c_o) > c_o$. Observe that for every integer n, f^n is an automorphism of C (f^0 = the identity map and $f^1 = f$). Let $S = \{x \in C : f^n(c_o) \leq x < f^{n+1}(c_o) \text{ for some integer } n\}$. Thus $S = \bigcup_{n \in Z} S_n$ where $S_n = [f^n(c_o), f^{n+1}(c_o))$. It is easy to see that S is a convex subchain of C and that $f_1 = f|S$ is a strictly increasing automorphism of S.

(ii) $\Rightarrow$ (iii) Suppose g is a strictly increasing automorphism of S. Pick $c_o \in S$, then $c_o < g(c_o)$. It is again easy to show that S has a convex subchain T such that $T = \bigoplus_{n\in Z} T_n$ where $T_n = \{x \in S : g^n(c_o) \leq x < g^{n+1}(c_o)$ for some integer $n\}$ and the map $g_n: T_n \to T_{n+1}$, defined by $g_n(x) = g(x)$, is obviously an isomorphism for each n.

(iii) $\Rightarrow$ (i) By hypothesis, there exists, for each integer n, an isomorphism $f_n: T_n \to T_{n+1}$. Since $T = \bigoplus_{n\in Z} T_n$, the map $f: T \to T$ defined by $f(x) = f_n(x)$ for $x \in T_n$ is a strictly increasing automorphism. By Exercise 5, this automorphism can be extended to an automorphism of C. Hence C is not rigid.

Theorem 9 can be useful in determining whether a chain is not rigid. Examples of chains which trivially satisfy the conditions of Theorem 9 are: the integers, the rationals, and the reals. For a non-trivial example, we use Theorem 9 to show that the Cantor discontinuum D is not rigid. Let $S = D \sim \{0, 1\}$ and suppose that each member of S is represented by a ternary expansion. For $n = 0, -1, -2, \cdots$ let $S_n = \{x \in S : 2\cdot 3^{n-2} \leq x < 2\cdot 3^{n-1}\}$ and for $n = 1, 2, 3, \cdots$ let

$$S_n = \{x \in S : 2\cdot 3^{-1} + \cdots + 2\cdot 3^{-n} \leq x < 2\cdot 3^{-1} + \cdots + 2\cdot 3^{-(n+1)}\}.$$

Then $S = \bigoplus_{n\in Z} S_n$ and each S_n is isomorphic with $D \sim \{1\}$. Therefore D is not rigid.

Exercise 10. Let f be a proper automorphism of C and let $f(c_o) > c_o$ for some $c_o \in C$. Then c_o is contained in a convex subchain S of C such that the restriction of f to S is a strictly increasing automorphism.

It follows immediately from Theorem 9 that if every convex subchain of C has a least or a largest element then C is rigid (cf. Exercise 6). However, this is not a necessary condition for C to be rigid. The following two exercises provide examples.

Exercise 11. Prove that $\bigoplus_{\alpha<\omega} C_\alpha$ is rigid where $C_\alpha = \breve{\omega}$ for each $\alpha < \omega$.

Exercise 12. Let $(C_n)_{n<\omega}$ be a family of dual ordinals with $|C_n| < |C_{n+1}|$ for $n = 0, 1, 2$ and $|C_0| > \aleph_0$. Prove that $C = \bigoplus_{n\in\omega} C_n$ is rigid. (Hint: use Theorem 9.)

6. Uniqueness of representation of coproducts of Boolean algebras and chains

We now turn our attention to the question raised at the end of Section 4. First recall that, in the setting of the category $\mathscr{D}$, it was shown (Theorem 5.9) that a chain C is not rigid if and only if C has a convex subchain S, which has a strictly increasing automorphism. However the setting for this section, as was the case for Section 4, will be again the category $\mathscr{D}_{01}$. But we note that the result just mentioned also holds for the category $\mathscr{D}_{01}$ in a slightly modified form: C is not rigid if and only if C contains a convex subchain S, such that $S' = S \sim \{0, 1\}$ is not void and such that S has an automorphism which is strictly increasing on S'. In this section, the symbols $L, L', L_1, L_2, \cdots, C, C', C_1, C_2, \cdots$ and $B, B', B_1, B_2, \cdots$ will have the meaning as in Section 4 and $C = C'$ denotes set-theoretic equality.

Theorem 1. *Suppose C is not rigid. If $L \doteq B + C$, where $B \not\doteq \mathbf{2}$, then there exists a chain C' such that $L \doteq B + C'$ and $C' \neq C$.*

Proof. It follows from the hypothesis and from the remark above that C has a convex subchain S such that $S' = S \sim \{0, 1\}$ is non-void and such that there exists an automorphism $f: S \to S$ with $f(x) > x$ for each $x \in S'$. Suppose a is an element of B such that $a \in B \sim \{0, 1\}$. For each $x \in S'$, let $d_x = x + af(x)$ and let $C' = (C \sim S') \cup \{d_x : x \in S'\}$. We will show that C' has the desired properties and we will proceed to do this in steps. (1) *C' is a chain*: Clearly $\{0,1\} \subseteq C'$ and if $\{u, v\} \subseteq C \sim S'$ then they are comparable. If $\{x, y\} \subseteq S'$ and $x \leq y$ then $d_x \leq d_y$. Finally, suppose $u \in C \sim S'$ and $v = d_x$ for some $x \in S'$. If $u < x$, then $u < x + af(x) = d_x$. If $x < u$, then $u \nleq f(x)$. Indeed if $u \leq f(x)$ then $x < u \leq f(x)$ and thus by the convexity of S, $u \in S'$, contradicting $u \in C \sim S'$. Hence $u \nleq f(x)$, so $f(x) \leq u$ and $d_x = x + af(x) \leq x + au \leq x + u \leq u$. (2) *$B \cup C'$ generates L*: It is sufficient to show that S' is a subset of $[B \cup C']_{\mathbf{D_{01}}}$. So let $x \in S'$, then $x = x + f^{-1}(x) = \bar{a}x + f^{-1}(x) + ax = \bar{a}(x + af(x)) + f^{-1}(x) + ax = \bar{a}d_x + d_{f^{-1}(x)}$ which is in $[B \cup C']_{\mathbf{D_{01}}}$. (3) *$L \doteq B + C'$ and $C \neq C'$*: For the first part it is sufficient to show that (3) of Section 4 holds for B and C'. Thus suppose $p, q \in C'$, $b \in B$ and $bp \leq q$. Assume that $b \neq 0$. If $p, q \in C \sim S'$ then it is immediate that $p \leq q$. There are three remaining cases. First, suppose $p \in C \sim S'$, $q = d_x$ for some $x \in S'$: then $bp \leq x + af(x) \leq f(x)$ thus, by (3) of Section 4, since $b \neq 0$, $p \leq f(x)$. But if $x \leq p$, then $x \leq p \leq f(x)$ and thus by the convexity of S, $p \in S'$, a contradiction. Hence $x \nleq p$ and thus $p < x \leq d_x = q$. Next, suppose $p = d_x$ for some $x \in S'$ and $q \in C \sim S'$: then $bx \leq bx + abf(x) = bd_x = bp \leq q$ thus $bx \leq q$ and since $b \neq 0$, we have $x \leq q$. By convexity $q \nleq f(x)$ and therefore $f(x) < q$ and $p = x + af(x) \leq q$. Finally, we have the case that $p = d_x$, $q = d_y$, $x, y \in S'$: suppose $p \nleq q$ then $x \nleq y$. Now $bx + abf(x) = bd_x \leq q = y + af(y)$. Thus also $bx \leq y + a$, but $x \nleq y$ so $b \leq a$ and $bf(x) = abf(x) \leq y + f(y) = f(y)$ which implies $f(x) \leq f(y)$. But f is an isomorphism, so $x \leq y$, a contradiction. The proof will be concluded by showing that $C \neq C'$. Let $x \in S'$ then $d_x \in C'$. Suppose $d_x \in C$, then $c = x + af(x)$ for some $c \in C$. Thus $x \leq c$ and $c \leq x + a$. But $a \neq 1$ so $c \leq x$ and we have $x = c$. But $af(x) \leq c$, $a \neq 0$ thus $f(x) \leq c = x$, a contradiction. Hence $d_x \notin C$, showing $C \neq C'$.

The next natural problem is to determine the class $\mathbf{E}$ of all chains C such that $L \doteq B + C$ and $L \doteq B + C'$ implies $C = C'$ for every Boolean algebra B and for every chain C'. The following theorem characterizes $\mathbf{E}$.

Theorem 2. (R. Balbes and Ph. Dwinger [2]) *C belongs to $\mathbf{E}$ if and only if C is rigid.*

Proof. It follows immediately from Theorem 1 that if $C \in \mathbf{E}$, then C is rigid. For the converse, suppose C is rigid and $C \notin \mathbf{E}$. Then there exist B and C' and L such that $L \doteq B + C$, $L \doteq B + C'$ and $C \neq C'$. By Theorem 4.7, there exist isomorphisms $f: C \to C'$ and $g: C' \to C$ such that $f(x) \leq x$ for each $x \in C$ and $g(x') \leq x'$ for each $x' \in C'$. Since $C \neq C'$ there exists $x_o \in C$ such that $f(x_o) < x_o$.

Let $h = g \circ f \colon C \to C$, then $h(x_o) = g(f(x_o) \leq f(x_o) < x_o$. Hence, h is a proper automorphism of C, a contradiction.

Theorems 1, 2, and Theorem 4.7 yield immediately a corollary which answers completely the question of the uniqueness of the representation $L \doteq B + C$, which was raised at the end of Section 4.

Corollary 3. *Let $L \doteq B + C$ and $L \doteq B' + C'$. Then:*

(i) $B = B'$.

(ii) $C = C'$ *if either* $B \cong \mathbf{2}$ *or* C *is rigid.*

The proof is left to the reader.

Exercise 4. Prove the following. Suppose $L_1 = \langle X_1, C_1 \rangle, L_2 = \langle X_2, C_2 \rangle$, $X_1, X_2 \in \text{Ob}\, \mathscr{S}_{\mathscr{B}}, |C_1| > 2, |C_2| > 2$. Suppose there exists an isomorphism $f \colon L_1 \to L_2$. Then there exist a homeomorphism $g \colon X_1 \to X_2$ and an isomorphism $h \colon C_1 \to C_2$ such that

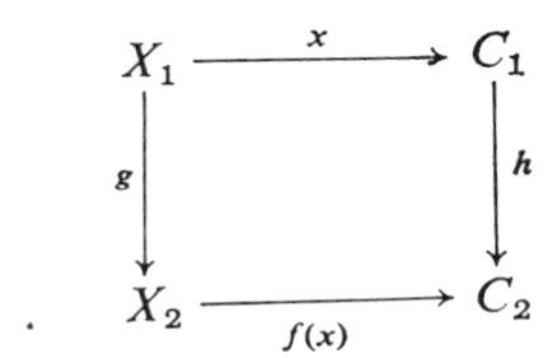

this diagram commutes for each $x \in \mathscr{C}(L_1)$. (Hint: Use Theorems 4.5 and 4.7.)

Exercise 5. Prove the following. C is rigid if and only if the following condition is satisfied: If X_1 and $X_2 \in \text{Ob}\, \mathscr{S}_{\mathscr{B}}$, C' a chain such that there exists an isomorphism $f \colon \langle X_1, C \rangle \to \langle X_2, C' \rangle$ then there exist a homeomorphism $g \colon X_1 \to X_2$ and an isomorphism $h \colon C \to C'$ such that

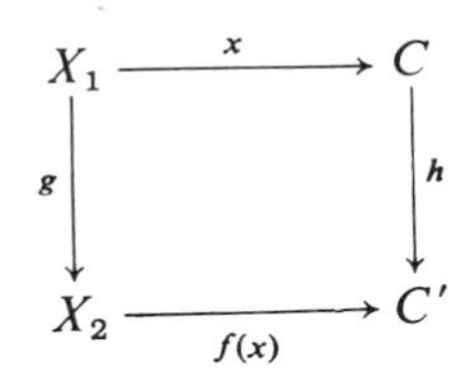

this diagram commutes for each $x \in \langle X_1, C \rangle$. (Hint: Use Theorems 4.5 and 6.2.)

The results that we have obtained can be generalized by replacing the Boolean algebra B in $L \doteq B + C$ by an arbitrary algebra in $\mathbf{D}_{01}$. We will state these generalizations in a series of exercises.

Exercise 6. Check by inspection of the proof of Theorem 4.1 that this theorem can be generalized as follows: $L \doteq L_1 + C$ if and only if for each $x \in L$, x has a unique representation $x = \sum_{i=1}^{n} a_i c_i$, $n \geq 1$, where $\{c_1, \cdots, c_n\} \subseteq C, \{a_1, \cdots, a_n\} \subseteq L_1$ and $0 = c_1 < c_2 \cdots < c_n$, $1 = a_1, a_2 > \cdots a_n > 0$.

Exercise 7. Check by inspection of the proof of Theorem 4.7 that this theorem can be generalized as follows: Suppose $L \doteq L_1 + C$ and $L \doteq L_1 + C'$. Then C is isomorphic to C'. Moreover, the isomorphism between C and C' can be chosen to be decreasing.

Exercise 8. Show by an example that $L \doteq L_1 + C$ and $L \doteq L_1' + C'$ does not necessarily imply $L_1 = L_1'$. Also show that if $L \doteq C_1 + C_2$ and $L \doteq C_1' + C_2'$ then $C_1 = C_i'$ and $C_2 = C_j'$ where $\{i, j\} = \{1, 2\}$. (Hint: Show that $C_1 \cup C_2$—and thus $C_1' \cup C_2'$—is the set of all meet and join irreducible elements of L.)

Exercise 9. Theorem 1 can be generalized as follows: let C be a non-rigid chain. If $L_1 \doteq L + C$ and $\mathscr{C}(L) \not\cong \mathbf{2}$ then there exists a chain C', $C' \neq C$ such that $L_1 \doteq L + C'$.

Exercise 10. Let $\mathbf{E}^*$ be the class of all chains C such that $L_1 \doteq L + C$ and $L_1 \doteq L + C'$ implies $C = C'$ for every L and every C'. Show that $\mathbf{E}^* = \mathbf{E}$.

Exercise 11. Prove that Corollary 3 can be generalized as follows:

(i) Suppose $L \doteq L_1 + C$ and $L \doteq L_1 + C'$. If C is rigid, then $C = C'$.

(ii) Suppose $L \doteq L_1 + C$ and $L \doteq L_1 + C'$ implies $C = C'$ for every C', and suppose $\mathscr{C}(L_1) \not\cong \mathbf{2}$, then C is rigid.

The results of this section have been generalized by S. D. Comer and Ph. Dwinger [1].

7. The representation space of coproducts in $\mathscr{D}_{01}$ and $\mathscr{B}$

Theorem 1. *The topological product of a family of Boolean spaces is a Boolean space.*

Proof. This follows immediately from the compactness of the topological product (Tychonoff's theorem), the definition of product topology and from the definition of Boolean space.

Theorem 2. (A. Nerode [1]) *The topological product of a family of bounded Stone spaces is a bounded Stone space.*

Proof. Let $(X_s)_{s \in S}$ be a family of bounded Stone spaces. For each $s \in S$, let $L_s = \mathscr{T}(X_s)$, where $\mathscr{T}$ is the coequivalence $\mathscr{T}: \mathscr{S} \to \mathscr{D}_{01}$ (cf. Exercise IV.2.5) and let $Y_s = \mathscr{S}_{\mathscr{B}}(B_{L_s})$. Then by Theorem V.4.11 and by Theorem IV.2.6 there exists, for each $s \in S$, a one-one, onto, and continuous map f_s: $Y_s \to X_s$. Hence there exists a one-one, onto, and continuous map f: $Y \to X$ where Y is the topological product of $(Y_s)_{s \in S}$ and where X is the topological product of $(X_s)_{s \in S}$. By Theorem 1, Y is a Boolean space. Certainly X is a compact T_0-space. Let $\mathfrak{B}$ be the set of subsets $p_s^{-1}[U]$, where p_s: $X \to X_s$ are the projections and where U is a basic (= compact open) subset of X_s. Let R be the ring of subsets of X generated by $\mathfrak{B}$, then R is a basis for the open sets of X. If $A \in R$, then $f^{-1}[A]$ is open and closed, hence compact. Therefore A is compact. Again, if A is a compact open subset of X, then A is a finite union of members of $\mathfrak{B}$ and hence $A \in R$ (and $f^{-1}[A]$ is open and closed). Thus, the family of compact open subsets of X is a ring and a basis for X. So conditions (i) and (ii) of Definition IV.1.1 are satisfied. Condition (iii) also holds for X, since the inverse image, under f, of a compact open set is open and closed and (iii) holds for Y, since Y is a Boolean space. Similarly, $\varnothing$ is fundamental, and therefore X is a bounded Stone space.

Theorem 3. *The representation space of a coproduct of a family of objects in $\mathscr{D}_{01}(\mathscr{B})$ is the topological product of the representation spaces of the members of the family.*

Remark. The proof will yield yet another existence proof for coproducts in $\mathscr{D}_{01}$ and $\mathscr{B}$.

Proof of Theorem 3. The statement for $\mathscr{B}$ follows immediately from Theorem 1 and from the coequivalence between $\mathscr{B}$ and $\mathscr{S}_{\mathscr{B}}$ and from the fact that continuous maps between Boolean spaces are strongly continuous. For $\mathscr{D}_{01}$: suppose $(X_s)_{s\in S}$ is a family of bounded Stone spaces whose topological product is X and with projections $p_s\colon X \to X_s$. By Theorem 2, X is a bounded Stone space and it follows from the proof of this theorem that each p_s is strongly continuous. If Y is a bounded Stone space and $f_s \in [Y, X_s]_{\mathscr{S}}$ for each $s \in S$, then there exists a (uniquely determined) continuous map $g\colon Y \to X$ such that $p_s \circ g = f_s$ for each $s \in S$. If U is compact open in some X_s, then it is immediate that $g^{-1}[p_s^{-1}[U]]$ is compact open. But this implies that g is strongly continuous. It follows that X is a product of $(X_s)_{s\in S}$ in $\mathscr{S}$ and by the coequivalence between $\mathscr{D}_{01}$ and $\mathscr{S}$ we have that coproducts exist in $\mathscr{D}_{01}$.

Theorem 3 enables us to exhibit the representation spaces of the free algebras in $\mathbf{D}_{01}$ and $\mathbf{B}$. Indeed $\mathscr{F}_{\mathbf{D}_{01}}(\alpha)$ is the coproduct in $\mathscr{D}_{01}$ of α copies of the free algebra on one generator which is $\mathbf{3}$. The representation space of $\mathbf{3}$ is the 2-points space $\{p_1, p_2\}$ with $\{p_1\}$ as the only proper open subset. Therefore $\mathscr{S}(\mathscr{F}_{\mathbf{D}_{01}}(\alpha))$ is the topological product of α copies of this space. Similarly, $\mathscr{F}_{\mathbf{B}}(\alpha)$ is the coproduct in $\mathscr{B}$ of α copies of the free algebra on one generator, which is $\mathbf{2}^2$. The representation space of $\mathbf{2}^2$ is the 2-points space $\{p_1, p_2\}$ with the discrete topology, and therefore $\mathscr{S}_{\mathscr{B}}(\mathscr{F}_{\mathbf{B}}(\alpha))$ is the topological product of α copies of the 2-points discrete space (these spaces are often called *Cantor* spaces and it is not difficult to show that $\mathscr{S}_{\mathscr{B}}(\mathscr{F}_{\mathbf{B}}(\aleph_0))$ is homeomorphic to the Cantor discontinuum.

Exercise 4. Describe the compact open subsets of the representation space of $\mathscr{F}_{\mathbf{D}_{01}}(\alpha)$.

Exercise 5. Characterize the open and closed subsets of the representation space of $\mathscr{F}_{\mathbf{B}}(\alpha)$.

Exercise 6. By considering their representation spaces, prove that $\mathscr{F}_{\mathbf{D01}}(\alpha)$ and $\mathscr{F}_{\mathbf{B}}(\alpha)$ are atomless if α is infinite.

Exercise 7. (cf. remarks following Corollary V.7.6) Prove that every closed subspace of the Cantor discontinuum is a retract of it (cf. P. R. Halmos [3]).

Exercise 8. (R. Sikorski [8]) Prove that for α infinite, $\mathscr{S}_{\mathscr{B}}(\mathscr{F}_{\mathbf{B}}(\mathbf{2}^{\alpha}))$ contains a dense subset of power α.

Exercise 9. (A. Tarski [1]) Prove that for α infinite, the representation space of $\mathbf{2}^{\alpha}$ has $\mathbf{2}^{2^{\alpha}}$ points. (Hint: $\mathscr{F}_{\mathbf{B}}(\mathbf{2}^{\alpha})$ is a subalgebra of $\mathbf{2}^{\alpha}$. Thus $\mathbf{2}^{2^{\alpha}} = |\mathscr{S}_{\mathscr{B}}(\mathscr{F}_{\mathbf{B}}(\mathbf{2}^{\alpha}))| \leq |\mathscr{S}_{\mathscr{B}}(\mathbf{2}^{\alpha})|$.)

8. Colimits

Let $((L_s)_{s\in S}, (f_{st})_{s\leq t})$ be a partially ordered system in the category $\mathscr{K}$, where $\mathscr{K}$ is one of the categories $\mathscr{D}_{01}$ or $\mathscr{B}$. Then we know that its colimit $(g_s: L_s \to L)_{s\in S}$ exists in $\mathscr{K}$ (Theorem I.20.17). In particular, we will determine the representation spaces of the colimits of these systems. These results will be generalizations of the material in Section 7.

Because of the coequivalences $\mathscr{S}: \mathscr{D}_{01} \to \mathscr{S}$ and $\mathscr{S}_{\mathscr{B}}: \mathscr{B} \to \mathscr{S}_{\mathscr{B}}$ the representation space of the colimit is the limit of the representation spaces. What we will prove is that this limit can be taken in the category of all topological spaces.

We denote the category of all topological spaces and continuous maps by $\mathscr{Top}$.

We first state the following theorem.

Theorem 1. *Let $((X_s)_{s\in S}, (f_{st})_{s\leq t})$ be a partially ordered system of topological spaces. Let X be the topological product of $(X_s)_{s\in S}$ and let $Y = \{y \in X : f_{st}(p_s(y)) = p_t(y), s \leq t, s, t \in S\}$ where $p_s: X \to X_s$ are the projections. Let $p'_s = p_s| Y$ for each $s \in S$. Then $(p'_s: Y \to X_s)_{s\in S}$ is the limit of the system in $\mathscr{Top}$.*

The proof is analogous to that of Theorem I.20.15 and is therefore omitted.

Theorem 2. *Let $((L_s)_{s\in S}, (f_{st})_{s\leq t})$ be a partially ordered system in $\mathscr{B}$. Then the limit of $((\mathscr{S}_{\mathscr{B}}(L_s))_{s\in S}, (\mathscr{S}_{\mathscr{B}}(f_{st}))_{s\leq t})$ in $\mathscr{Top}$ is also the limit in $\mathscr{S}_{\mathscr{B}}$.*

Proof. Let X be the topological product of $(\mathscr{S}_{\mathscr{B}}(L_s))_{s\in S}$, $Y = \{y \in X, f_{st}(p_s(y)) = p_t(y), s \leq t, s, t \in S\}$, and $p'_s = p_s| Y$. Then by Theorem 1 $(p'_s: Y \to (\mathscr{S}_{\mathscr{B}}(L_s)))_{s\in S}$ is the limit of the system $((\mathscr{S}_{\mathscr{B}}(L_s))_{s\in S}, (\mathscr{S}_{\mathscr{B}}(f_{st})_{s\leq t}))$ in $\mathscr{Top}$.

In order to show that it is also the limit in $\mathscr{S}_{\mathscr{B}}$, it suffices to show that Y is a closed subspace of X, since then Y is a Boolean space. Let $x \in X \sim Y$. There exists $s_o, t_o \in S$, $s_o \leq t_o$ such that $f_{s_ot_o}(p_{s_o}(x)) \neq p_{t_o}(x)$. Let U, V be open subsets of $\mathscr{S}_{\mathscr{B}}(L_{t_o})$ such that $f_{s_ot_o}(p_{s_o}(x)) \in U$, $p_{t_o}(x) \in V$, $U \cap V = \varnothing$. Let $W = f_{s_ot_o}^{-1}[U]$ and let $V^* = p_{t_o}^{-1}[V]$ and $W^* = p_{s_o}^{-1}[W]$. Obviously $V^* \cap W^*$ is open and $x \in V^*$. It also easily follows that $x \in W^*$ and thus $x \in V^* \cap W^*$ and $V^* \cap W^* \cap Y = \varnothing$. Hence $X \sim Y$ is open.

Theorem 3. *Let* $((L_s)_{s\in S}, (f_{st})_{s\leq t})$ *be a partially ordered system in* $\mathcal{D}_{01}$. *Then the limit of* $((\mathcal{S}(L_s))_{s\in S}, ((\mathcal{S}(f_{st}))_{s\leq t})$ *in* $\mathcal{T}op$ *is also the limit in* $\mathcal{S}$.

Proof. Let X be the topological product of $(\mathcal{S}(L_s))_{s\in S}$ and let Y be defined as in the proof of Theorem 2. It suffices to show that Y is a bounded Stone space. By Theorem V.4.11 there exist one-one, onto, and continuous maps f_s: $\mathcal{S}_{\mathcal{B}}(B_{L_s}) \to \mathcal{S}(L_s)$ and hence there exist a one-one, onto, and continuous map f: $X' \to X$ where X' is the topological product of $(\mathcal{S}_{\mathcal{B}}(B_{L_s}))_{s\in S}$. Note that by Theorem 7.3, X is the representation space of the coproduct L of $(L_s)_{s\in S}$, and that X' is the representation space of B_L. Let $Y' = f^{-1}[Y]$. Then by the proof of Theorem 7.3, Y' is a Boolean space and hence a closed subspace of X'. But then $\mathcal{S}(1_{L_1,B_L}) = f$ and it follows from Theorem V.4.7 that Y is a bounded Stone space.

As an application of the previous results we give an alternate proof that the amalgamation property holds in $\mathcal{B}$, $\mathcal{D}_{01}$ and $\mathcal{D}$ (see also the remarks following Corollary V.9.5).

Theorem 4. (B. Jónsson [1], A. Daigneault [1], Ph. Dwinger and F. M. Yaqub [1]) *The categories* $\mathcal{B}$, $\mathcal{D}_{01}$ *and* $\mathcal{D}$ *have the amalgamation property.*

Proof. We will first verify that the amalgamation property holds in the category $\mathcal{B}$. By virtue of the coequivalence between $\mathcal{B}$ and $\mathcal{S}_{\mathcal{B}}$, it suffices to prove the dual of the statement for $\mathcal{S}_{\mathcal{B}}$ (cf. Theorem IV.3.4 and the remark following Definition I.20.19). Thus, suppose $(f_s\colon X_s \to Y)_{s\in S}$ is a family of epimorphisms in $\mathcal{S}_{\mathcal{B}}$. Then, by Theorem 8.2, the limit, in $\mathcal{S}_{\mathcal{B}}$, of this sytem is obtained as follows. Let X be the topological product of $(X_s)_{s\in S}$ and Y, and let $(p_s\colon X \to X_s)_{s\in S}$ and $p\colon X \to Y$ be the projections. Let $T = \{x\in X : f_s(p_s(x)) = p(x)\}$. Then $(p_s'\colon T \to X_s, p'\colon T \to Y)_{s\in S}$ is the limit of the system $(f_s\colon X_s \to Y)_{s\in S}$ where $p_s' = p_s|T$ and $p' = p|T$. We only need to prove that the p_s', for each $s\in S$, and p' are epimorphisms, but since $p' = f_s \circ p_s'$ it suffices to show that all p_s' are epic. Let $s_o \in S$ and let $x_{s_0} \in X_{s_0}$. Since all f_s are epic and hence onto, by Theorem IV.3.4, there exists, for each $s\in S$, $s \neq s_o$, a point $x_s \in X_s$ such that $f_s(x_s) = f_s(x_{s_0})$. Let $x \in X$ be determined by $p_s(x) = x_s$ for $s\in S$ and $p(x) = f_{s_0}(x_{s_0})$. Then $x\in T$. Indeed for $s\in S, f_s(p_s(x)) = f_s(x_s) = f_{s_0}(x_{s_0}) = p(x)$ for all $s\in S$. Thus p_{s_0}' is onto and hence epic, which completes the proof of this part of the theorem. The cases $\mathcal{K} = \mathcal{D}$ and $\mathcal{K} = \mathcal{D}_{01}$ follow now immediately since the reflectors $\mathcal{B}^*\colon \mathcal{D} \to \mathcal{B}$, and $\mathcal{B}_{01}\colon \mathcal{D}_{01} \to \mathcal{B}$ preserve colimits and preserve and reflects monomorphisms by Theorem I.18.5 and by Corollaries V.4.3 and V.4.4.

VIII

Pseudocomplemented Distributive Lattices

1. Definitions and examples

The study of pseudocomplemented distributive lattices commenced with a paper by V. Glivenko [1] in 1929. Although there is a reference to what we now call Stone algebras in the 1937 classic paper of M. H. Stone [3], it was G. Grätzer and E. T. Schmidt [1] who first solved Problem 70 of G. Birkhoff [6, 1948 edition] and thereby generated widespread interest in the topic. This, in fact, was the first paper in which the terms Stone lattice and relative Stone lattice were used.

A pseudocomplemented distributive lattice is a distributive lattice L with 0, 1 such that for each $a \in L$ there is a greatest element a^* which is disjoint with a. The problem referred to above is then: What is the most general pseudocomplemented distributive lattice in which $a^* + a^{**} = 1$?

Another problem which attracted interest was posed by O. Frink [1] in 1962: Is every Stone algebra isomorphic to a subalgebra of the algebra of all ideals of a complete atomic Boolean algebra? This problem was solved by G. Grätzer [1] in 1963. More direct proofs were later given by G. Bruns [2] and G. Grätzer [3]. More recently, H. Lakser [2] has shown that every pseudocomplemented distributive lattice can be embedded (so as to preserve $+$, $\cdot$, 0 and *) in the lattice of ideals of an atomic Boolean algebra.

A monograph devoted to Stone algebras by J. Varlet [1] appeared in 1963. In 1969 C. C. Chen and G. Grätzer presented two papers [1], [2] in which Stone algebras were represented as a triple—consisting of two "simple structures" and a "connecting map". The free Stone algebras on finitely many free generators were determined in R. Balbes and A. Horn [3], the injective Stone algebras characterized by R. Balbes and G. Grätzer [1], and the injective hulls of Stone algebras were described using the "triple method" by H. Lakser [1]. In 1970, K. B. Lee [1] gave a characterization of all of the equational subclasses of pseudocomplemented distributive lattices. This was followed by three significant papers: H. Lakser [2] and G. Grätzer and H. Lakser [2], [3] on the structure of pseudocomplemented distributive lattices, dealing in particular with subdirect products, the congruence extension property, amalgamation, and injectivity. A generalization to Stone algebras of order n has been investigated by T. Katriňák and A. Mitschke [1]. Because of their relationship to Post algebras we will cover this topic in Section X.9.

Besides the papers we have mentioned, many others have been published on pseudocomplemented distributive lattices and closely related topics. We mention just a few: J. Berman [1], B. Davey [1], A. Day [2], T. Hecht and T. Katriňák [1], T. Katriňák [4], H. Lakser [3], P. Ribenboim [1], T. P. Speed [2], J. Varlet [2], [4].

In the first part of the chapter we develop the elementary properties of pseudocomplemented distributive lattices, discuss the relation between maximal filters and minimal prime ideals, and characterize the representation spaces of these lattices. The classical theorem of Glivenko is also included in this treatment.

Our approach in describing the equational subclasses is to determine first the subdirectly irreducible in the whole class (a result of H. Lakser [2]) and then proceed to characterize the subclasses: both in terms of the identities that define the subclasses and in terms of conditions imposed on the maximal filters.

The equational subclasses form a chain

$$\mathbf{B}_{-1} \subset \mathbf{B}_0 \subset \mathbf{B}_1 \subset \cdots \subset \mathbf{B}_\omega.$$

In this chain $\mathbf{B}_{-1}$ is the trivial class, $\mathbf{B}_0$ is the class of Boolean algebras, $\mathbf{B}_1$ is the class of Stone algebras, and $\mathbf{B}_\omega$ is the whole class of pseudocomplemented distributive lattices.

Next we deal with the various characterizations of $\mathbf{B}_1$ and present a generalization of Stone algebras; namely, distributive lattices with 1 in which every interval is a Stone algebra.

In the last sections the study of Stone algebras is resumed with a characterization of injectives in $\mathscr{B}_1$ and the free Stone algebras. A summary of recent results concerning various equational subclasses of pseudocomplemented distributive lattices is included in this chapter.

It was observed in Section II.6 that in a Boolean algebra L, the complement $\bar{x}$ of an element x is the largest member of L which is disjoint with x. This is the basis for the next definition.

Definition 1. Let L be a distributive lattice with 0. An element $a \in L$ is *pseudocomplemented* if there is a largest member of L which is disjoint with a. If such an element exists, it is denoted by a^* and is called the *pseudocomplement* of a. A *pseudocomplemented distributive lattice* is a distributive lattice with 0 in which every element has a pseudocomplement.

There are many examples of pseudocomplemented distributive lattices but at this point we only present four.

Example 2. Boolean algebras: $a^* = \bar{a}$.

Example 3. Chains with 0, 1: $a^* = \begin{cases} 0 \text{ if } a \neq 0 \\ 1 \text{ if } a = 0. \end{cases}$

Example 4. $\mathfrak{I}(L)$ is a pseudocomplemented distributive lattice if L is a distributive lattice with 0: For $I \in \mathfrak{I}(L)$, $I^* = \{x \in \mathrm{L} : xi = 0 \text{ for all } i \in I\}$. To see this, note that $0_{\mathfrak{I}(L)} = (0]$ and let $J = \{x \in L : xi = 0 \text{ for all } i \in I\}$. Then clearly, $J \in \mathfrak{I}(L)$. If $I \cap K = (0]$ for some $K \in \mathfrak{I}(L)$, then for any $k \in K$, $ki = 0$ for all $i \in I$, hence $k \in J$, and thus $K \subseteq J$ and $J = I^*$. That $\mathfrak{I}(L)$ is distributive follows from Theorem II.9.3.

Example 5. The lattice of open sets of a topological space X: Here, if U is an open set then $U^* = \operatorname{Int}(\tilde{U}) = \widetilde{\operatorname{Cl} U}$.

If L is a pseudocomplemented distributive lattice, then $x \leq 0^*$ for any $x \in L$, hence L has a largest element $1 = 0^*$. Two important subsets associated with L are $\mathscr{R}g(L) = \{x \in L : x^{**} = x\}$ and $\mathscr{D}s(L) = \{x \in L : x^* = 0\}$. The members of $\mathscr{R}g(L)$ are called *regular* and those of $\mathscr{D}s(L)$ *dense*. Note that $\{0, 1\} \subseteq \mathscr{R}g(L)$ since $0^* = 1$ and $1^* = 0$; and that $1 \in \mathscr{D}s(L)$. L itself is *dense* if $\mathscr{D}s(L) = L \sim \{0\}$.

In Example 5, $\mathscr{R}g(L)$ is the set of regular open sets of X and $\mathscr{D}s(L)$ is the set of dense open sets of X.

Exercise 6. Show that L is dense if and only if 0 is meet irreducible.

Exercise 7. If L is a distributive lattice with 0 and a^* and b^* exist then $(a + b)^*$ exists and is equal to a^*b^*. What about $(ab)^*$?

Exercise 8. Prove that if L is a complete distributive lattice satisfying the infinite distributive law of Theorem II.6.9(iv), then L is pseudocomplemented. Hence, all finite distributive lattices are pseudocomplemented.

Exercise 9. Prove that if L is a pseudocomplemented distributive lattice, then $\mathscr{D}s(L)$ is a filter in L.

2. Basic properties

The following theorem should acquaint the reader with some of the basic properties and techniques used in studying pseudocomplemented distributive lattices.

Theorem 1. *For elements a, b of a pseudocomplemented distributive lattice L*

(i) $aa^* = a^{**}a^* = 0$.

(ii) $ab = 0 \Leftrightarrow a \leq b^*$.

(iii) $a \leq a^{**}$.

(iv) $a^{***} = a^*$.

(v) $ab = 0 \Leftrightarrow a^{**}b = 0$.

(vi) $a \leq b \Rightarrow b^* \leq a^*$.

(vii) $(a + b)^* = a^*b^*$.

(viii) $a^{**}b^{**} = (ab)^{**}$.

(ix) $(a^{**} + b^{**})^{**} = (a + b)^{**}$.

(x) $(a + a^*)^* = 0$.

(xi) $a \in \mathscr{R}g(L) \Leftrightarrow a = c^*$ *for some* $c \in L$.

Proof. (i) By definition of a^* and a^{**}. (ii) The fact that $ab = 0$ implies $a \leq b^*$ is part of the definition of b^* and conversely, if $a \leq b^*$ then $ab \leq b^*b = 0$. (iii) $aa^* = 0$

implies $a \leq (a^*)^*$. (iv) By (iii), $a^* \leq (a^*)^{**}$ and since $aa^{***} \leq a^{**}a^{***} = 0$, we have $a^{***} \leq a^*$. (v) $ab = 0 \Rightarrow b \leq a^* \Rightarrow a^{**}b \leq a^{**}a^* = 0$. Conversely, if $a^{**}b = 0$ then $ab \leq a^{**}b = 0$. (vi) $a \leq b \Rightarrow ab^* \leq bb^* = 0$. Thus $b^* \leq a^*$. (vii) By (vi), $(a + b)^* \leq a^*, b^*$ thus $(a + b)^* \leq a^*b^*$. Conversely, $(a^*b^*)(a + b) = 0$ implies $a^*b^* \leq (a + b)^*$. (viii) Applying (vi) twice, we have $(ab)^{**} \leq a^{**}, b^{**}$ implying $(ab)^{**} \leq a^{**}b^{**}$. For the converse, $(ab)(ab)^* = 0$ implies, by (v), that $ab^{**}(ab)^* = 0$ and, again by (v), $a^{**}b^{**}(ab)^* = 0$. Thus $a^{**}b^{**} \leq (ab)^{**}$. (ix) From (vii) and (iv) we have: $(a^{**} + b^{**})^{**} = (a^{***}b^{***})^* = (a^*b^*)^* = (a + b)^{**}$. (x) Follows from (vii) and (i). (xi) If $a \in \mathscr{R}g(L)$ then $a = a^{**}$ so take $c = a^*$. Conversely, if $a = c^*$ for some $c \in L$ then $a^{**} = c^{***} = c^* = a$.

In this chapter, if we talk about a *minimal prime ideal*, we will always mean a minimal member of $\mathfrak{P}(L)$.

Theorem 2. *Suppose L is a pseudocomplemented distributive lattice and* $\varnothing \neq I \subset L$. *The following are equivalent:*

(i) $L \sim I$ *is a maximal filter.*

(ii) $L \sim I$ *is a prime filter and* $a + a^* \in L \sim I$ *for each* $a \in L$.

(iii) I *is a minimal prime ideal.*

(iv) I *is a prime ideal and* $a \in I \Rightarrow a^{**} \in I$.

(v) I *is a prime ideal and* $I \cap \mathscr{D}_s(L) = \varnothing$.

Proof. (i) $\Rightarrow$ (ii) The first part follows from Theorem III.3.7. For the second part: if $a \notin L \sim I$ then $L \sim I$ is properly contained in $[(L \sim I) \cup \{a\})$ so by hypothesis $[(L \sim I) \cup \{a\}) = L$. It follows that $0 = xa$ for some $x \in L \sim I$ and so $x \leq a^*$. But $x \in L \sim I$ implies $a^* \in L \sim I$ and hence $a + a^* \in L \sim I$.

(ii) $\Rightarrow$ (iii) Since $L \sim I$ is a prime filter, I is a prime ideal. Suppose $J \in \mathfrak{P}(L)$ and $J \subset I$ with $a \in I \sim J$. Then $aa^* = 0 \in J$ and hence $a^* \in J \subset I$, contradicting $a + a^* \notin I$.

(iii) $\Rightarrow$ (iv) Since I is a minimal prime ideal, $L \sim I$ is a maximal (prime) filter. So by (i) $\Rightarrow$ (ii) if $a \in I$, then $a^* \notin I$. But $a^*a^{**} = 0 \in I$ so $a^{**} \in I$.

(iv) $\Rightarrow$ (v) Suppose $a \in I \cap \mathscr{D}_s(L)$ then $a^* = 0 \in I$ thus $1 = a^{**} \in I$, a contradiction.

(v) $\Rightarrow$ (i) Since I is a prime ideal, $L \sim I$ is a proper (prime) filter. Suppose F is a filter and $L \sim I \subset F$ with $a \in F \sim (L \sim I)$. Then by Theorem 1(x), $a + a^* \in \mathscr{D}_s(L)$ and thus $a + a^* \notin I$. But $a \in I$ and therefore $a^* \in L \sim I \subset F$. Also $a \in F$ and thus $0 = aa^* \in F$ which implies that $F = L$.

We will close this section with a characterization of the representation space of a pseudocomplemented distributive lattice.

Theorem 3. *Let L be a distributive lattice. Then L is pseudocomplemented if and only if* $\mathscr{S}(L)$ *is a bounded Stone space and for every compact open subset U of* $\mathscr{S}(L)$, $\widetilde{\mathrm{Cl}\, U}$ *is compact. Moreover,* $\widehat{a^*} = \widetilde{\mathrm{Cl}\, \hat{a}}$ *for each* $a \in L$ *in this case.*

Proof. ($\Rightarrow$) If U is a compact open subset of $\mathscr{S}(L)$, then there is an $a \in L$, such that $U = \hat{a}$. Since $\hat{a} \cap \widehat{a^*} = \varnothing$ and $\widetilde{\widehat{a^*}}$ is closed and contains $\hat{a}$, we have $\mathrm{Cl}\,\hat{a} \subseteq \widetilde{\widehat{a^*}}$. If $p \notin \mathrm{Cl}\,\hat{a}$, then there exists $b \in L$ such that $p \in \hat{b}$ and $\hat{a} \cap \hat{b} = \varnothing$ or $ab = 0$ and thus $b \leq a^*$ so $\hat{b} \leq \widehat{a^*}$. It follows that $p \in \widehat{a^*}$ and $\mathrm{Cl}\,\hat{a} = \widetilde{\widehat{a^*}}$. Thus $\widetilde{\mathrm{Cl}\,\hat{a}} = \widehat{a^*}$ is compact.

($\Leftarrow$) If $a \in L$ then by hypothesis $\widetilde{\mathrm{Cl}\,\hat{a}}$ is compact (and open), hence there exists $b \in L$ such that $\hat{b} = \widetilde{\mathrm{Cl}\,\hat{a}}$. But then $\hat{b} \cap \hat{a} = \varnothing$ so $ba = 0$. If $xa = 0$ for $x \in L$ then $\hat{x} \cap \hat{a} = \varnothing$ and thus $\hat{x} \cap \mathrm{Cl}\,\hat{a} = \varnothing$ or $\hat{x} \subseteq \hat{b}$ and so $x \leq b$. It follows that $b = a^*$.

3. The class $\mathbf{B}_\omega$

Theorem 1. (P. Ribenboim [1]) *The class of pseudocomplemented distributive lattices is the equational class of algebras of type* (2, 2, 1, 0) *with operations* $(+, \cdot, {}^*, 0)$ *and satisfying*

(i) *A set of identities that define distributive lattices with* 0.

(ii) $\mathbf{x \cdot (x \cdot y)^* = x \cdot y^*}$.

(iii) $\mathbf{x \cdot 0^* = x}$.

(iv) $\mathbf{0^{**} = 0}$.

Proof. Clearly, any pseudocomplemented distributive lattice satisfies (i), (iii), and (iv). Also, in any such lattice, $(xy)(xy)^* = 0$ implies $x(xy)^* \leq y^*$ so $x(xy)^* \leq xy^*$. On the other hand, $(xy^*)(xy) = 0$ implies $xy^* \leq (xy)^*$ so $xy^* \leq x(xy)^*$. Conversely, we can assume $(A, (+, \cdot, {}^*, 0))$ is a distributive lattice with 0 under $+, \cdot$ and 0. For $a \in A$, $aa^* = a(a0^*)^* = a(0^*)^* = a0 = 0$, and if $ab = 0$ then $ab^* = a(ab)^* = a0^* = a$ so $a \leq b^*$. Thus A is a pseudocomplemented distributive lattice.

Definition 2. The equational class defined in Theorem 1 is denoted by $\mathbf{B}_\omega$ and is called the class of *pseudocomplemented distributive lattices*. The corresponding category is $\mathscr{B}_\omega$.

Since pseudocomplements can be defined solely in terms of the partial ordering, the concepts of order isomorphism and $\mathbf{B}_\omega$-isomorphism are the same.

Since Boolean algebras are of type (2, 2, 1, 0, 0), by disregarding the nullary operation 1, they form a subclass of $\mathbf{B}_\omega$ in which $(L, (+, \cdot, {}^-, 0, 1)) \in \mathbf{B}$ corresponds to $(L, (+, \cdot, {}^-, 0)) \in \mathbf{B}_\omega$. Thus we can consider $\mathbf{B}$ to be the subclass of all $L \in \mathbf{B}_\omega$ in which $\mathbf{x + x^* = 0^*}$ is an identity.

Now, using a similar argument as in Section II.5 applied to $\mathscr{B}_\omega$ and $\mathscr{B}$, we will often consider $\mathscr{B}$ as a subcategory of $\mathscr{B}_\omega$. By disregarding the nullary operation 0 and the unary operation $*$, we can consider $\mathbf{B}_\omega$ as a subclass of $\mathbf{D}$ and $\mathscr{B}_\omega$ as a subcategory of $\mathscr{D}$. Since every object of $\mathscr{B}_\omega$ has a 0 and 1 and since a $\mathbf{B}_\omega$-homomorphism preserves the zero and one element, $\mathbf{B}_\omega$ can then also be considered as a subclass of $\mathbf{D}_{01}$ and $\mathscr{B}_\omega$ as a subcategory (not full, however) of $\mathscr{D}_{01}$.

Theorem 3. *Suppose* $L \in \mathbf{B}_\omega$ *and* $f: L \to \mathbf{2}$ *is a function.* $f^{-1}[\{0\}]$ *is a minimal prime ideal (equivalently,* $L \sim f^{-1}[\{0\}]$ *is a maximal filter) if and only if* f *is a* $\mathbf{B}_\omega$-*homomorphism.*

Proof. $(\Leftarrow)$ $f(1) = f(0^*) = (f(0))^* = \mathbf{0}^* = \mathbf{1}$ and $f(0) = \mathbf{0}$ so $\varnothing \neq f^{-1}[\{\mathbf{0}\}] \subset L$. Also, f is a $\mathbf{D}$-homomorphism so $f^{-1}[\{\mathbf{0}\}]$ is a prime ideal. If $a \in f^{-1}[\{\mathbf{0}\}]$ then $f(a^{**}) = (f(a))^{**} = \mathbf{0}^{**} = \mathbf{0}$ so $a^{**} \in f^{-1}[\{\mathbf{0}\}]$. By Theorem 2.2(iv), $f^{-1}[\{\mathbf{0}\}]$ is a minimal prime ideal.

$(\Rightarrow)$ Let $f^{-1}[\{\mathbf{0}\}]$ be a minimal prime ideal. Since it is prime, f is a $\mathbf{D}$-homomorphism. Clearly, $f(0) = \mathbf{0}$. Let $a \in L$. If $f(a) = \mathbf{0}$, then by Theorem 2.2(ii) $(f(a))^* = \mathbf{0}^* = \mathbf{1} = f(a + a^*) = f(a) + f(a^*) = f(a^*)$. If $f(a) = \mathbf{1}$, then $(f(a))^* = \mathbf{1}^* = \mathbf{0} = f(aa^*) = f(a)f(a^*) = f(a^*)$.

Theorem 4. *Coproducts exist in* $\mathscr{B}_\omega$.

Proof. Let $(L_s)_{s\in S}$ be a family in $\mathbf{B}_\omega$. If $L_s \cong \mathbf{1}$ for some s, then since $\mathbf{B}_\omega$-homomorphisms preserve 0, 1, the coproduct of $(L_s)_{s\in S}$ in $\mathscr{B}_\omega$ is $\mathbf{1}$. Now suppose $|L_s| > 1$ for each $s \in S$. Apply Theorem I.20.8 with $B_s = A_s$ for each $s \in S$. Indeed, to define $h_{st}: A_s \to A_t$ we select a maximal filter in A_s and by Theorem 3 obtain a homomorphism $h: A_s \to \mathbf{2}$. Let $h_{st} = 1_{\mathbf{2},A_t} \circ h$.

Exercise 5. Let L, L_1 be Boolean algebras $f \in [L, L_1]_{\mathbf{B}}$ and suppose f is one-one. Then the mapping $g: L \oplus \mathbf{1} \to L_1 \oplus \mathbf{1}$, defined by $g|L = f$, $g(1_{L\oplus\mathbf{1}}) = 1_{L_1\oplus\mathbf{1}}$ is in $\mathbf{B}_\omega$.

Exercise 6. If $n \leq m$, then $\mathbf{2}^n \oplus \mathbf{1}$ is a $\mathbf{B}_\omega$-subalgebra of $\mathbf{2}^m \oplus \mathbf{1}$.

Exercise 7. Show that in Exercise II.9.12, if each L_s has a 0, then the map f is a $\mathbf{B}_\omega$-embedding.

Exercise 8. Let $L, L_1 \in \mathbf{B}_\omega$ and let $f: L \to L_1$ be a $\mathbf{D}_{01}$-homomorphism. Prove that f is a $\mathbf{B}_\omega$-homomorphism if and only if $\mathrm{Cl}((\mathscr{S}(f))^{-1}[U]) = (\mathscr{S}(f))^{-1}[\mathrm{Cl}(U)]$ for each compact open subset U if $\mathscr{S}(L)$.

4. Glivenko's theorem and some implications

Theorem 1. *For each element* a *of a pseudocomplemented distributive lattice* L, $(a]$ *is also in* $\mathbf{B}_\omega$, *under the operations induced by the partial ordering of* L, *and the map* $x \mapsto xa$ *is a* $\mathbf{B}_\omega$-*homomorphism onto* $(a]$.

Proof. Clearly, $(a]$ is a sublattice of L with 0, and $x \mapsto xa$ is a $\mathbf{D}$-homomorphism preserving 0. The pseudocomplement of $b \in (a]$ is given by ab^*, for $ab^* \in (a]$, $b(ab^*) = 0$ and if $bx = 0$, where $x \in (a]$ then $x \leq b^*$ so $x \leq ab^*$. Finally, it follows from Theorem 3.1(ii) that $x \mapsto xa$ is a $\mathbf{B}_\omega$-homomorphism.

Although this result will be useful, a deeper theorem, due to V. Glivenko [1], will now be considered. First, from now on we will consider the sets $\mathscr{R}g(L)$ and

$\mathscr{D}s(L)$ of regular and dense elements of a pseudocomplemented distributive lattice L as partially ordered sets, where the partial ordering is that induced by L. Recall (Exercise 1.9) that $\mathscr{D}s(L)$ is a filter in L, and we will see that part of Glivenko's theorem states that $\mathscr{R}g(L)$ is a Boolean algebra. Furthermore, since $\mathscr{D}s(L)$ is a filter, $\boldsymbol{\theta}(\mathscr{D}s(L))$ is a $\mathbf{D}$-congruence relation on L. In fact we have the following lemma.

Lemma 2. *Let $L \in \mathbf{B}_\omega$ and let F be any filter in L. Then $\boldsymbol{\theta}(F)$ is a $\mathbf{B}_\omega$-congruence relation.*

Proof. We only need to show that if $(a, b) \in \boldsymbol{\theta}(F)$, then $(a^*, b^*) \in \boldsymbol{\theta}(F)$. But if $au = bu$ for some $u \in F$, then $b^*au = 0$, thus $b^*u \leq a^*$ and so $b^*u \leq a^*u$. Similarly $a^*u \leq b^*u$, hence $(a^*, b^*) \in \boldsymbol{\theta}(F)$.

Theorem 3. (Glivenko's Theorem) *If $L \in \mathbf{B}_\omega$ then $\mathscr{R}g(L)$ is a Boolean algebra under the operations induced by the partial ordering on L. The operations $\overset{\mathscr{R}g(L)}{+}$, $\overset{\mathscr{R}g(L)}{\cdot}$, $0_{\mathscr{R}g(L)}$, $*^{\mathscr{R}g(L)}$ are given by*

$$u \overset{\mathscr{R}g(L)}{+} v = (u + v)^{**}$$

$$u \overset{\mathscr{R}g(L)}{\cdot} v = uv$$

$$0_{\mathscr{R}g(L)} = 0$$

$$u^{*\mathscr{R}g(L)} = u^*.$$

Furthermore, the map $r_L\colon L \to \mathscr{R}g(L)$ defined by $r_L(x) = x^{**}$ is a $\mathbf{B}_\omega$-homomorphism of L onto $\mathscr{R}g(L)$ and $\mathscr{R}g(L) \cong L/\mathscr{D}s(L)$.

Proof. In this proof we will repeatedly make use of Theorem 2.1, without mentioning this explicitly. If $u, v \in \mathscr{R}g(L)$ then $u, v \leq u + v \leq (u + v)^{**}$ and if $w \geq u, v; w \in \mathscr{R}g(L)$ then $w \geq u + v$ thus $w = w^{**} \geq (u + v)^{**}$. Again if $u, v \in \mathscr{R}g(L)$ then $(uv)^{**} = u^{**}v^{**} = uv$, thus $uv \in \mathscr{R}g(L)$. It follows that $\mathscr{R}g(L)$ is a lattice such that for $u, v \in \mathscr{R}g(L)$, $u \overset{\mathscr{R}g(L)}{+} v = (u + v)^{**}$ and $u \overset{\mathscr{R}g(L)}{\cdot} v = uv$. Also note that $0, 1 \in \mathscr{R}g(L)$. If $x, y \in L$, then $r_L(x + y) = (x + y)^{**} = (x^{**} + y^{**})^{**}$ $= (r_L(x) + r_L(y))^{**} = r_L(x) \overset{\mathscr{R}g(L)}{+} r_L(y)$ and $r_L(xy) = (xy)^{**} = x^{**} \cdot y^{**}$ $= r_L(x) \cdot r_L(y) = r_L(x) \overset{\mathscr{R}g(L)}{\cdot} r_L(y)$. Also, $r_L(0) = 0$ and $r_L(1) = 1$ and r_L is onto $\mathscr{R}g(L)$. It follows that $\mathscr{R}g(L)$ is distributive and r_L is a $\mathbf{D}_{01}$-homomorphism.

In order to show that $\mathscr{R}g(L)$ is a Boolean algebra, let $u \in \mathscr{R}g(L)$ then $u \overset{\mathscr{R}g(L)}{+} u^* = (u + u^*)^{**} = 1$ and $u \overset{\mathscr{R}g(L)}{\cdot} u^* = uu^* = 0$. It follows that $\mathscr{R}g(L)$ is a Boolean algebra and $u^{*\mathscr{R}g(L)} = u^*$. We also have that for $x \in L$, $r_L(x^*) = (r_L(x))^*$ and thus r_L is a $\mathbf{B}_\omega$-homomorphism. It remains to show that $\mathscr{R}g(L) \cong L/\mathscr{D}s(L)$. Thus we must show that $\operatorname{Ker} r_L = \boldsymbol{\theta}(\mathscr{D}s(L))$. First, let $(x, y) \in \boldsymbol{\theta}(\mathscr{D}s(L))$ then there exists $u \in \mathscr{D}s(L)$ such that $xu = yu$. But $u^* = 0$ so $r_L(x) = x^{**} = x^{**}u^{**} = (xu)^{**}$ $= (yu)^{**} = y^{**}u^{**} = y^{**} = r_L(y)$. Next, if $(x, y) \in \operatorname{Ker} r_L$ then $x^{**} = y^{**}$ so

$x^* = y^*$. Then $x(x + x^*)(y + y^*) = x(y + y^*) = xy + xy^* = xy$ $= y(x + x^*)(y + y^*)$. But $x + x^*$ and $y + y^* \in \mathscr{D}s(L)$ and hence $(x + x^*)(y + y^*) \in \mathscr{D}s(L)$. Therefore $(x, y) \in \boldsymbol{\theta}(\mathscr{D}s(L))$.

It easily follows, from the definition of r_L, that it is a closure operator on L and that the set of "closed" elements is $\mathscr{R}g(L)$. The following theorem is an immediate consequence of Theorem 3 and Theorem II.4.12.

Theorem 4. *In Theorem 3, r_L preserves arbitrary sums (i.e. if $\sum^L S$ exists for $S \subseteq L$, then $\sum^{\mathscr{R}g(L)}\{s^{**} : s \in S\}$ exists and equals $(\sum^L S)^{**}$). Moreover, if L is complete then $\mathscr{R}g(L)$ is complete and for $S \subseteq L$,*

(i) $\sum^{\mathscr{R}g(L)}\{s^{**} : s \in S\} = (\sum^L \{s^{**} : s \in s\})^{**} = (\sum^L S)^{**}$.

(ii) $\prod^{\mathscr{R}g(L)}\{s^{**} : s \in S\} = \prod^L\{s^{**} : s \in S\}$.

It follows immediately from Theorem 4 and Example 1.5 that the regular open sets of a topological space form a complete Boolean algebra.

We have seen, in Exercise II.6.13, that if $L \in \mathbf{D}_{01}$, then $\mathscr{C}(L)$ is a $\mathbf{D}_{01}$-subalgebra of L. For $L \in \mathbf{B}_\omega$ we have:

Theorem 5. *For $L \in \mathbf{B}_\omega$,*

(i) $\mathscr{C}(L) = \{x \in L : x + x^* = \mathbf{1}\}$.

(ii) *$\mathscr{C}(L)$ is a $\mathbf{D}_{01}$-subalgebra of $\mathscr{R}g(L)$.*

Proof. (i) If $x + x^* = 1$, then since $xx^* = 0$ we have $x \in \mathscr{C}(L)$. If $x \in \mathscr{C}(L)$, then $x\bar{x} = 0 \Rightarrow \bar{x} \leq x^*$ so $1 = x + \bar{x} \leq x + x^*$. (ii) To prove that $\mathscr{C}(L) \subseteq \mathscr{R}g(L)$ let $x \in \mathscr{C}(L)$, then by (i), $x + x^* = 1$ so $x^{**} = x^{**}(x + x^*) = x^{**}x + x^{**}x^* = x$. Clearly, $0, 1 \in \mathscr{C}(L)$. Now, let $x, y \in \mathscr{C}(L)$. Then $x + y \in \mathscr{C}(L) \subseteq \mathscr{R}g(L)$. Hence, $x \overset{\mathscr{R}g(L)}{+} y = x + y$. Similarly for products.

Exercise 6. Prove that for a Boolean algebra L, $\mathscr{C}(\mathfrak{I}(L))$ is the set of principal ideals of L.

Exercise 7. For $L \in \mathbf{B}_\omega$, define "symmetric difference" by $x \oplus y = x^*y + xy^*$. Prove that the following are equivalent.

(i) $x^{**} = y^{**}$.

(ii) $(x, y) \in \boldsymbol{\theta}(\mathscr{D}s(L))$.

(iii) $(x \oplus y)^* \in \mathscr{D}s(L)$.

Exercise 8. Show that $\mathscr{R}g(L)$ is not, in general, a sublattice of L.

Theorem 9. *The assignment $L \mapsto \mathscr{R}g(L)$ can be extended to a reflector $\mathscr{R}g : \mathscr{B}_\omega \to \mathscr{B}$ and $\mathscr{R}g$ preserves monomorphisms.*

Proof. For $L \in \mathscr{B}_\omega$, let $\Phi_{\mathscr{R}g}(L) = r_L$. We must prove the extension property (cf. Theorem I.18.2). Suppose $f \in [L, L_1]_{\mathscr{B}_\omega}$, where $L_1 \in \text{Ob}\,\mathscr{B}$. Define $f' : \mathscr{R}g(L) \to L_1$ by $f' = f|\mathscr{R}g(L)$. Then $f' \in [\mathscr{R}g(L), L_1]_{\mathscr{B}}$. Indeed, if $u, v \in \mathscr{R}g(L)$, then by Theorem 3, $f'(u \overset{\mathscr{R}g(L)}{+} v) = f'((u + v)^{**}) = f((u + v)^{**}) = (f(u + v))^{**} = \overline{\overline{f(u + v)}} = f(u + v) = f(u) + f(v) = f'(u) + f'(v)$. Again, $f'(u \overset{\mathscr{R}g(L)}{\cdot} v) = f'(uv) = f(uv) = f(u)f(v) = f'(u)f'(v)$. Also $f'(0) = 0$ and $f'(1) = 1$ and for $u \in \mathscr{R}g(L)$, $f'(u^*) = f(u^*) = (f(u))^* = \overline{f(u)} = \overline{f'(u)}$. Next, for $u \in L$, $f'(\Phi_{\mathscr{R}g}(L)(u)) = f'(u^{**}) = f(u^{**}) = (f(u))^{**} = \overline{\overline{f(u)}} = f(u)$. For uniqueness, suppose $f_1, f_2 \in [\mathscr{R}g(L), L_1]_{\mathscr{B}}$ such that $f_1 \circ \Phi_{\mathscr{R}g}(L) = f_2 \circ \Phi_{\mathscr{R}g}(L)$, then for $u \in \mathscr{R}g(L)$, $f_1(u) = f_1(u^{**}) = f_1(\Phi_{\mathscr{R}g}(L)(u)) = f_2(\Phi_{\mathscr{R}g}(L)(u)) = f_2(u^{**}) = f_2(u)$ and thus $f_1 = f_2$. Finally, suppose $g \in [L, L_1]_{\mathscr{B}_\omega}$ is monic, and hence one-one. We show that $\mathscr{R}g(g)$ is one-one. Note that $\mathscr{R}g(g)$ is the unique morphism determined by $\mathscr{R}g(g) \circ r_L = r_{L_1} \circ g$. Hence $\mathscr{R}g(g) = g|\mathscr{R}g(L)$ is one-one and since $\mathscr{R}g(g)$ is a morphism in $\mathscr{B}$, it is monic.

In [1], O. Frink showed that much of the theory of pseudocomplemented (distributive) lattices extends to pseudocomplemented semi-lattices. In particular, Glivenko's Theorem generalizes by defining $u \overset{\mathscr{R}g(L)}{+} v = (u^*v^*)^*$. Results of this kind were anticipated by G. Grätzer and E. T. Schmidt and have also been extended as, for example, in P. V. Venkatanarasimhan [1]. See also T. Katrinák [1], [4].

5. Subdirectly irreducible algebras in $\mathscr{B}_\omega$

The main purpose of this section is to determine the subdirectly irreducibles in $\mathbf{B}_\omega$. This was first accomplished by H. Lakser [2], who showed that $L \in \mathbf{B}_\omega$ is subdirectly irreducible if and only if $L = L_1 \oplus \mathbf{1}$ where $L_1 \in \mathbf{B}$. The "only if" part of his proof depends on a result (see Exercise 4 below) which was used to characterize $\mathfrak{C}(L)$ for $L \in \mathbf{B}_\omega$.

We present here a more direct proof of the "only if" part, which is due to J. Berman.

Theorem 1. (H. Lakser [2]) *L is subdirectly irreducible in $\mathbf{B}_\omega$ if and only if $L = L_1 \oplus \mathbf{1}$ where $L_1 \in \mathbf{B}$.*

Proof. We will apply Theorem I.9.3.

($\Leftarrow$) Since $1_{L_1} < 1_L$, $\boldsymbol{\theta}([1_{L_1}))$ is non-trivial (cf. Lemma 4.2). To show that it is the least non-trivial congruence relation, suppose θ is a non-trivial congruence relation. In order to verify $\boldsymbol{\theta}([1_{L_1})) \subseteq \theta$, let $(u, v) \in \boldsymbol{\theta}([1_{L_1}))$, $u \neq v$. Then $u1_{L_1} = v1_{L_1}$ so $u, v \geq 1_{L_1}$. Thus we can assume that $u = 1_{L_1}$ and $v = 1_L$. Since θ is non-trivial, there exist $(x, y) \in \theta$ and $x \neq y$; we can assume $x \nleq y$. If $x = 1_L, y = 1_{L_1}$ then $(u, v) \in \theta$. If $x = 1_L, y \neq 1_{L_1}$ then $(v, y) \in \theta$ thus $(v, u) = (v, 1_{L_1}) = (v, y + 1_{L_1}) = (v + u, y + u) \in \theta$. Finally, if $x \leq 1_{L_1}, y < 1_{L_1}$ then, denoting the complement of y in L_1 by $\bar{y}$ and that of x by $\bar{x}$, we have: $(x, y) \in \theta \Rightarrow (x\bar{y}, y\bar{y}) \in \theta \Rightarrow (x\bar{y}, 0) \in \theta$

$\Rightarrow ((x\bar{y})^*, 0^*) \in \theta$. But $x\bar{y} \neq 0$ and $x\bar{y} \in L_1$ thus $(\bar{x} + y, 1_L) \in \theta$ and so $(u, v) = (1_{L_1}, 1_L) = (1_{L_1}, 1_{L_1}) + (\bar{x} + y, 1_L) \in \theta$.

($\Rightarrow$) Let θ_o be the least non-trivial congruence relation on L. There exist $p \neq q$, $(p, q) \in \theta_o$. We may assume $p < q$. Now $q = 1$. Indeed if $q < 1$ then $\boldsymbol{\theta}([q))$ is non-trivial. But then $(p, q) \in \theta_o \subseteq \boldsymbol{\theta}([q))$ and thus $q \leq p$, a contradiction. Suppose there exists $x \in L$ such that $p < x < 1$. Then $\boldsymbol{\theta}([x))$ is non-trivial and $(p, 1) = (p, q) \in \boldsymbol{\theta}([x))$, so $p \geq x$. It follows that 1 covers p. Next we prove that 1 is join irreducible. Suppose $1 = x + y$, $x \neq 1$, $y \neq 1$. Then $\theta_o \subseteq \boldsymbol{\theta}([x))$ and $\theta_o \subseteq \boldsymbol{\theta}([y))$ so $px = x$, $py = y$ implying $p = 1$, a contradiction. Since 1 is join irreducible we have $L = (p] \oplus \mathbf{1}$. We will show that $(p]$ is a Boolean algebra. It suffices to show that for $0 < x < p$, $x + x^* = p$. Suppose $x + x^* = s < p$. Since L can also be considered as an algebra in $\mathbf{D}$, $\boldsymbol{\theta}((p])$ and hence $\theta_1 = \boldsymbol{\theta}((p]) \cap \boldsymbol{\theta}([s))$ is a congruence relation of L, where L is considered as an algebra in $\mathbf{D}$. We will show that θ_1 is also a $\mathbf{B}_\omega$-congruence relation on L. We only need to show that if $(u, v) \in \theta_1$ then $u^* + p = v^* + p$. Suppose $u^* + p \neq v^* + p$, but then $\{u^* + p, v^* + p\} = \{p, 1\}$ and we may therefore assume $u^* + p = 1$. But 1 is join irreducible and so $u^* = 1$ and $u = 0$. Hence $(0, v) \in \boldsymbol{\theta}([s))$ and we infer that $0s = vs$ or $v \leq s^*$. But also $s^* = (x + x^*)^* = 0$ and so $v = 0$ implying $u^* + p = v^* + p$. It follows that θ_1 is a congruence relation in $\mathbf{B}_\omega$. Now $(p, 1) \notin \theta_1$ since $p + p \neq 1 + p$, so $\theta_o \nsubseteq \theta_1$. But $\theta_1 \neq 0$ since $(s, p) \in \theta_1$ and $s \neq p$, so $\theta_o \subseteq \theta_1$, a contradiction.

Corollary 2. *Every pseudocomplemented distributive lattice is a subdirect product of algebras of the form $L \oplus \mathbf{1}$ where $L \in \mathbf{B}$.*

Proof. Immediate from Theorem 1 and Birkhoff's subdirect product theorem.

Theorem 3. *Suppose $L \in \mathbf{B}_\omega$, $L = L_1 \oplus \mathbf{1}$ and $L_1 \in \mathbf{B}$. If L' is a $\mathbf{B}_\omega$-subalgebra of L, then $L' = L_1' \oplus 1_L$ where $L_1' \in \mathbf{B}$ and either $L_1' = \{0\}$ or L_1' is a $\mathbf{B}$-subalgebra of L_1.*

Proof. Let $L_1' = \{x \in L' : x \neq 1_L\}$. Clearly then, $L' = L_1' \oplus \{1_L\}$. Suppose $L_1' \neq \{0\}$. Then there exists $x \in L'$ such that $x \notin \{1_L, 0\}$. Since $x \in L_1$ we have $1_{L_1} = x + \bar{x}^{L_1} = x + x^* \in L'$. Now if $x, y \in L_1'$, then since 1_L is join irreducible in L and L' is a $\mathbf{B}_\omega$-subalgebra of L, $x + y \in L_1'$; also $xy \in L_1'$. Finally, if $x \in L_1'$ and $x \neq 0$, then $\bar{x}^{L_1'} = x^* \in L_1'$.

Exercise 4. (H. Lakser [2]) Let $L \in \mathbf{B}_\omega$. Consider $\mathscr{D}\!s(L)$ as an algebra in $\mathbf{D}$ and $\mathscr{R}g(L)$ as an algebra in $\mathbf{B}$. Prove:

(i) $[0, \boldsymbol{\theta}(\mathscr{D}\!s(L))]_{\mathfrak{C}(L)} \cong \mathfrak{C}(\mathscr{D}\!s(L))$.

(ii) $[\boldsymbol{\theta}(\mathscr{D}\!s(L)), 1]_{\mathfrak{C}(L)} \cong \mathfrak{C}(\mathscr{R}g(L))$.

(iii) Identify $[0, \boldsymbol{\theta}(\mathscr{D}\!s(L))]_{\mathfrak{C}(L)}$ with $\mathfrak{C}(\mathscr{D}\!s(L))$ and $[\boldsymbol{\theta}(\mathscr{D}\!s(L)), 1]_{\mathfrak{C}(L)}$ with $\mathfrak{C}(\mathscr{R}g(L))$ and let $f\colon \mathfrak{C}(L) \to \mathfrak{C}(\mathscr{R}g(L)) \times \mathfrak{C}(\mathscr{D}\!s(L))$ be defined by $f(\theta) = (\theta + \boldsymbol{\theta}(\mathscr{D}\!s(L)), \theta \cap \boldsymbol{\theta}(\mathscr{D}\!s(L)))$. Then f is a one-one homomorphism and

$$f[\mathfrak{C}(L)] = \{(\theta_1, \theta_2) \in \mathfrak{C}(\mathscr{R}g(L)) \times \mathfrak{C}(\mathscr{D}\!s(L)) : (a, 1) \in \theta_1, a \in \mathscr{R}g(L) \Rightarrow (u, 1) \in \theta_2 \text{ for each } u \in \mathscr{D}\!s(L), u \geq a\}.$$

(iv) Prove the "only if" part of Theorem 5.1 using (i)—(iii). (Hint: Show that if L is subdirectly irreducible, $L \in \mathbf{B}_\omega$, then either $\mathscr{D}_\delta(L) = \mathbf{1}$ or $\mathscr{D}_\delta(L) = \mathbf{2}$.)

Exercise 5. (J. Berman) Prove that for $L \in \mathbf{B}_\omega$, $\mathscr{D}_\delta(L)$ is the intersection of all maximal filters of L.

Exercise 6. (H. Lakser [2]) Prove that every subdirectly irreducible member L of $\mathbf{B}_\omega$ can be embedded in the lattice of ideals of an atomic Boolean algebra. (Hint: Let Y be an infinite set. Show that $\mathbf{2}^X \oplus \mathbf{1}$ can be embedded in the lattice of ideals of the atomic Boolean algebra of finite and cofinite subsets of $X \times Y$.)

Exercise 7. (H. Lakser [2]) Prove that for each $L \in \mathbf{B}_\omega$ there exists an atomic Boolean algebra L_1 and an embedding of L into $\mathfrak{I}(L_1)$. (Hint: See Exercise 3.7.)

Further results can be found in J. Berman [1] and G. Grätzer and H. Lakser [2] and H. Lakser [2] and [3].

6. The equational subclasses of $\mathbf{B}_\omega$

Let $\mathbf{B}_{-1}$ be the members of $\mathbf{B}_\omega$ which contain one element. For a non-negative integer n, let $\mathbf{B}_n$ be the equational subclass of $\mathbf{B}_\omega$ which is generated by $\mathbf{2}^n \oplus \mathbf{1}$. We will show, in Theorem 6, that these are *all* of the equational subclasses of $\mathbf{B}_\omega$. For $n = 0$, $\mathbf{2}^n \oplus \mathbf{1}$ is $\mathbf{2}$ and since every Boolean algebra is a subdirect product of copies of $\mathbf{2}$ and $\mathbf{B}$ itself is an equational subclass of $\mathbf{B}_\omega$, we have, therefore, $\mathbf{B}_0 = \mathbf{B}$. For $n > 0$ we first need a lemma.

Lemma 1. *If $L \in \mathbf{B}_\omega$ and there exists an integer $n \geq 1$ such that every prime filter is contained in at most n maximal filters, then L is a subdirect product of algebras of the form $\mathbf{2}^k \oplus \mathbf{1}$ where $k \leq n$.*

Proof. It suffices to show (cf. Theorem I.9.2) that for $a, b \in L$, $a \neq b$ there exists $f \in [L, \mathbf{2}^k \oplus \mathbf{1}]_{\mathscr{B}_\omega}$, for some $k \leq n$, such that h is onto and $f(a) \neq f(b)$. We may assume that $a \nleq b$. Let F be a prime filter in L such that $a \in F$, $b \notin F$. Let $(G_s)_{s\in S}$ be the family of distinct maximal filters containing F. Note that $|S| \leq n$. For each $s \in S$, let $f_s: L \to \mathbf{2}$ be defined by $f_s(x) = \mathbf{1}$ if $x \in G_s$ and $f_s(x) = \mathbf{0}$ if $x \notin G_s$. By Theorem 3.3, $f_s \in [L, \mathbf{2}]_{\mathscr{B}_\omega}$ for each $s \in S$. Let $g \in [L, \mathbf{2}^S]_{\mathscr{B}_\omega}$ be defined by $g = \times_{s\in S} f_s$. Next define $h: L \to \mathbf{2}^S \oplus \mathbf{1}$ by $h(x) = \mathbf{1}$ if $x \in F$ and $h(x) = g(x)$ if $x \notin F$. Since $h(a) = \mathbf{1}$ and $h(b) \neq \mathbf{1}$, we have that $h(a) \neq h(b)$.

We now show that $h \in [L, \mathbf{2}^S \oplus \mathbf{1}]_{\mathscr{B}_\omega}$. If $u \leq v$ and $v \in F$ then $h(v) = 1$ and if $v \notin F$ then $h(u) = g(u) \leq g(v) = h(v)$. Thus h preserves order. For "+" we must show $h(u + v) \leq h(u) + h(v)$. If $u \in F$ then $h(u) = 1$ and we are done. Similarly for $v \in F$. If $u, v \notin F$ then $u + v \notin F$ and thus $h(u + v) = g(u + v) = g(u) + g(v) = h(u) + h(v)$. For products, if $uv \in F$, then $1 = h(uv) \geq h(u)h(v)$. If $uv \notin F$, then, say $u \notin F$. If also $v \notin F$, then $h(uv) = g(uv) = g(u)g(v) = h(u)h(v)$. The remaining case is then $uv \notin F$, $u \notin F$, $v \in F$. But then $f_s(v) = 1$ for each $s \in S$, so $g(v) = 1_{2^S}$ and thus $h(u)h(v) = g(u)1 = g(u) = g(u)1_{2^S} = g(u)g(v) = g(uv) = h(uv)$. Since $0 \notin F$ we have $h(0) = g(0) = 0$. Finally, we must show that $(h(u))^* = h(u^*)$. Since h preserves products, we have $h(u)h(u^*) = h(uu^*) = h(0) = 0$ so $h(u^*) \leq (h(u))^*$. To prove $(h(u))^* \leq h(u^*)$. We may assume $u^* \notin F$. Now, if $u \in F$, then $h(u) = 1$ and thus $(h(u))^* = 0$ and we are done. The last possibility is $u^* \notin F$, $u \notin F$. If $[\{u\} \cup F) = L$, then there exists $f \in F$ with $0 = uf$, so $f \leq u^*$ and thus $u^* \in F$, a contradiction. Hence $[\{u\} \cup F) \subset L$, and so there exists $s_o \in S$ such that $G_{s_o} \supseteq [\{u\} \cup F)$ and we have that $(g(u))(s_o) = f_{s_o}(u) = 1$ and therefore $g(u) \neq 0$ since we may assume that $L \neq \mathbf{1}$ and thus $\mathbf{2}^S \neq \mathbf{1}$. Now $h(u) = g(u)$, and $h(u^*) = g(u^*)$. So $(h(u))^* = (g(u))^*$. But since $g(u) \neq 0$ and $g \in [L, \mathbf{2}^S]_{\mathcal{B}_\omega}$ we have $(g(u))^* = g(u^*)$ and thus $(h(u))^* = h(u^*)$. Now $h[L]$ is a $\mathbf{B}_\omega$-subalgebra of $\mathbf{2}^S \oplus \mathbf{1}$ and therefore by Theorem 5.3, $h[L] = L_1 \oplus \mathbf{1}$ for some $L_1 \in \mathbf{B}$. But $|S| \leq n$ so $L_1 = \mathbf{2}^k$ for some $k \leq n$ and so h—with codomain restricted to $\mathbf{2}^k \oplus \mathbf{1}$—is the required homomorphism.

Note that if we drop the condition that $|S| \leq n$ in Lemma 1, the result is that L is a subdirect product of algebras of the form $L_1 \oplus \mathbf{1}$, $L_1 \in \mathbf{B}$. Thus, the lemma provides us with another proof of the "only if" part of Theorem 5.1.

Theorem 2. (K. B. Lee [1]) *For an algebra $L \in \mathbf{B}_\omega$ and $n > 0$, the following are equivalent:*

(i) $(\mathbf{x}_0 \cdots \mathbf{x}_{n-1})^* + \sum_{i<n} (\mathbf{x}_0 \cdots \mathbf{x}_{i-1}\mathbf{x}_i^*\mathbf{x}_{i+1} \cdots \mathbf{x}_{n-1})^* = \mathbf{1}$ *is an identity in* L.

(ii) *Every prime filter is contained in at most n maximal filters.*

(iii) $L \in \mathbf{B}_n$.

Proof. (i) ⇒ (ii) Suppose F is a prime filter and $G_0, \cdots, G_n$ are $n + 1$ distinct maximal filters such that $F \subseteq G_i$ for $i = 0, \cdots, n$. If there exists $i \leq n$ such that $\bigcap_{j \neq i} G_j \subseteq G_i$ then, by Exercise III.3.13, $G_j \subseteq G_i$ for some $j \neq i$ but then the maximality of G_j implies the contradiction $i = j$. For each $i \leq n$, select $a_i \in (\bigcap_{j \neq i} G_j) \sim G_i$. By hypothesis, $a_0, \cdots, a_{n-1}$ satisfy (i). Thus either

(1) $(a_0 \cdots a_{n-1})^* \in F \subseteq G_n$ or

(2) $(a_0 \cdots a_{i-1}a_i^*a_{i+1} \cdots a_{n-1})^* \in F \subseteq G_i$ for some $i \leq n - 1$.

Now for $i \leq n - 1$, $a_i \in \bigcap_{j \neq i} G_j \subseteq G_n$ and so $a_0 \cdots a_{n-1} \in G_n$. This contradicts (1), since G_n is a proper filter. On the other hand, for $i \leq n - 1$, $a_0 \cdots a_{i-1}a_{i+1} \cdots a_{n-1} \in G_i$ and $a_i \notin G_i$. But G_i is maximal, so by Theorem 2.2, $a_i^* \in G_i$ thus $a_0 \cdots a_{i-1}a_i^*a_{i+1} \cdots a_{n-1} \in G_i$ which again is incompatible with (2), since G_i is proper.

(ii) ⇒ (iii) This part follows immediately from Lemma 1 and Exercise 3.6.

(iii) ⇒ (i) By Corollary I.11.7 it suffices to prove that $\mathbf{2}^n \oplus \mathbf{1}$ satisfies the identity. Clearly $\mathbf{2}^n \oplus \mathbf{1}$ has exactly n distinct maximal filters—those generated by the atoms. We will contradict this fact under the assumption that (i) is false. Thus, suppose there exist $a_0, \cdots, a_{n-1}$ in $\mathbf{2}^n \oplus \mathbf{1}$ such that

$$(a_0 \cdots a_{n-1})^* + \sum_{i<n} (a_0 \cdots a_{i-1}a_i^*a_{i+1} \cdots a_{n-1})^* < 1.$$

Let $b_n = a_0 \cdots a_{n-1}$ and $b_i = a_0 \cdots a_{i-1}a_i^*a_{i+1} \cdots a_{n-1}$ for $i = 0, 1, \cdots, n - 1$. There exists a prime filter F which does not contain $b_n^* + \sum_{i<n} b_i^*$ and hence none of $b_0^*, \cdots, b_{n-1}^*, b_n^*$. Let $F_i = [F \cup \{b_i\})$ for $i = 0, 1, \cdots, n$. These $n + 1$ filters are proper, for if $0 \in [F \cup \{b_i\})$, then there would exist $f \in F$ such that $0 = fb_i$ and thus $f \leq b_i^*$ which would imply the contradiction $b_i^* \in F$. Hence there are maximal filters $G_0, \cdots, G_n$ such that $F_i \subseteq G_i$ for $i = 0, \cdots, n$. We will show that these are distinct. Suppose $G_i = G_j$, $i \neq j$ then we may assume that $j \neq n$. If $b_i \in G_j$ then $a_j \in G_j$. But also $b_j \in F_j \subseteq G_j$ so $a_j^* \in G_j$ implying the contradiction $0 = a_j a_j^* \in G_j$. It follows that $b_i \notin G_j$ and thus $b_i \in G_i \sim G_j$.

Corollary 3. *The subdirectly irreducible algebras in* $\mathbf{B_n}$, $n \geq 0$, *are* $\{\mathbf{2}^k \oplus \mathbf{1} : k = 0, 1, \cdots n\}$.

Proof. If $L = \mathbf{2}^k \oplus \mathbf{1}$ for $k \leq n$, then by Theorem 5.1, L is subdirectly irreducible and by Exercise 3.6, $L \in \mathbf{B_n}$. Conversely, if $L \in \mathbf{B_n}$, $n > 0$, and L is subdirectly irreducible then by Theorem 2 and Lemma 1, L is a subdirect product of algebras $\mathbf{2}^k \oplus \mathbf{1}$, $k \leq n$ and hence $L \cong \mathbf{2}^k \oplus \mathbf{1}$ for some $k \leq n$. Finally if $L \in \mathbf{B_0} = \mathbf{B}$ and L is subdirectly irreducible, then $L = L_1 \oplus \mathbf{1}$, where $L_1 \in \mathbf{B}$ thus $L_1 = \mathbf{1}$ and $L = \mathbf{1} \oplus \mathbf{1}$.

Exercise 4. Let $L \in \mathbf{B_\omega}$. Prove that the following are equivalent: (i) $L \in \mathbf{B_n}$, $n > 0$. (ii) Each prime filter is contained in at most n maximal filters. (iii) Each prime ideal contains at most n minimal prime ideals (i.e. minimal in $\mathfrak{P}(L)$). (iv) If $I_1, \cdots, I_{n+1}$ are distinct minimal prime ideals then $\sum^{\mathfrak{I}(L)}_{1 \leq j \leq n+1} I_j = L$.

Exercise 5. Let $L \in \mathbf{B_\omega}$, $n > 0$. Show that the following are equivalent:

(i) $L \in \mathbf{B_n}$.

(ii) If $|S| = n + 1$ and S is a disjoint set, then $\sum \{s^* : s \in S\} = 1$ (G. Grätzer and H. Lakser [3]).

(iii) $(\mathbf{x_0} \cdots \mathbf{x_n})^* = \sum_{\mathbf{i=0}}^{\mathbf{n}} (\mathbf{x_1} \cdots \mathbf{x_{i-1}} \mathbf{x_{i+1}} \cdots \mathbf{x_n})^*$ is an identity in L (J. Berman).

Theorem 6. (K. B. Lee [1]) *The equational subclasses of* $\mathbf{B_\omega}$ *are* $\mathbf{B_{-1}} \subset \mathbf{B_0} \subset \mathbf{B_1} \subset \cdots \subset \mathbf{B_\omega}$.

Proof. Since every equational class contains the one element algebras, $\mathbf{B_{-1}} \subseteq \mathbf{B_0}$. But $\mathbf{2} \in \mathbf{B_0}$ so $\mathbf{B_{-1}} \subset \mathbf{B_0}$. For $n \geq 0$, $\mathbf{2}^n \oplus \mathbf{1}$ is a subalgebra of $\mathbf{2}^{n+1} \oplus \mathbf{1}$ so $\mathbf{2}^n \oplus \mathbf{1} \in \mathbf{B_{n+1}}$. But $\mathbf{2}^n \oplus \mathbf{1}$ generates $\mathbf{B_n}$ so $\mathbf{B_n} \subseteq \mathbf{B_{n+1}}$ for all $\mathbf{n} < \omega$. Now $\mathbf{1}$ is a prime filter in $\mathbf{2}^{n+1} \oplus \mathbf{1}$ and is contained in all $n + 1$ maximal filters of $\mathbf{2}^{n+1} \oplus \mathbf{1}$. So by Theorem 2, $\mathbf{2}^{n+1} \oplus \mathbf{1} \notin \mathbf{B_n}$, hence $\mathbf{B_n} \subset \mathbf{B_{n+1}}$ for all n. Finally, $\mathbf{B_n} \subset \mathbf{B_\omega}$ for all n, for if $\mathbf{B_{n_0}} = \mathbf{B_\omega}$ for some $\mathbf{n_0} < \omega$ then $\mathbf{B_{n_0+1}} \subseteq \mathbf{B_\omega} = \mathbf{B_{n_0}} \subset \mathbf{B_{n_0+1}}$. Now suppose $\mathbf{K}$ is an equational subclass of $\mathbf{B_\omega}$. Assume first that $B_\mathbf{n} \subseteq \mathbf{K}$ for $\mathbf{n} < \omega$. We will show that $\mathbf{B_\omega} = \mathbf{K}$. It suffices to show that, if $L \in \mathbf{B_\omega}$ and L is subdirectly irreducible, then $L \in \mathbf{K}$. But if L is subdirectly irreducible, then $L = L_1 \oplus \mathbf{1}$ for some Boolean algebra L_1. Now $L_1 \oplus \mathbf{1} \in \mathbf{K}$ if and only if each identity that holds for $\mathbf{K}$ also holds for $L_1 \oplus \mathbf{1}$. But an identity holds for $L_1 \oplus \mathbf{1}$ if and only if it holds for all finitely generated subalgebras of $L_1 \oplus \mathbf{1}$. By Theorem 5.3 a finitely generated subalgebra of $L_1 \oplus \mathbf{1}$ is of the form $\mathbf{2}^n \oplus \mathbf{1}$ and since $\mathbf{2}^n \oplus \mathbf{1} \in \mathbf{B_n} \subseteq \mathbf{K}$ we have that $L_1 \oplus \mathbf{1} \in \mathbf{K}$. It follows, in this case, that $L \in \mathbf{K}$.

Next, assume that $\mathbf{B_n} \nsubseteq \mathbf{K}$ for some n. Let n_o be the greatest integer for which $\mathbf{B_n} \subseteq \mathbf{K}$. Thus $\mathbf{B_{n_o}} \subseteq \mathbf{K}$ but $\mathbf{B_n} \nsubseteq \mathbf{K}$ for $n > n_o$. We will show that $\mathbf{K} = \mathbf{B_{n_o}}$. As before, to show that $\mathbf{K} \subseteq \mathbf{B_{n_o}}$, it suffices to show that the subdirectly irreducible algebras of $\mathbf{K}$ are in $\mathbf{B_{n_o}}$. If L is subdirectly irreducible in $\mathbf{K}$ then $L = L_1 \oplus \mathbf{1}$ for some Boolean algebra L_1. If $\mathbf{2}^{n_o+1} \oplus \mathbf{1}$ is a subalgebra of $L_1 \oplus \mathbf{1}$ then $\mathbf{2}^{n_o+1} \oplus \mathbf{1}$ would be in $\mathbf{K}$ and hence $\mathbf{B_{n_o+1}} \subseteq \mathbf{K}$, a contradiction. Hence $\mathbf{2}^{n_o+1}$ is not a subalgebra of L_1. This means that L_1 is finite and $L_1 = \mathbf{2}^k$ for some $k \leq n_o$. Therefore $L = L_1 \oplus \mathbf{1} = \mathbf{2}^k \oplus \mathbf{1} \in \mathbf{B_k} \subseteq \mathbf{B_{n_o}}$.

7. Stone algebras

The smallest equational class $\mathbf{B_{-1}}$ of pseudocomplemented distributive lattices is of course, trivial, and the second, $\mathbf{B_0}$, is the class of Boolean algebras. The next class, $\mathbf{B_1}$, has been studied extensively and is the subject of this section. By Theorem 6.2, $\mathbf{B_1}$ is characterized by the identity $\mathbf{x^* + x^{**} = 1}$. We will continue to refer to the members of $\mathbf{B_1}$ as *Stone algebras*. The various characterizations of Stone algebras and the observation that they form an equational class led to the eventual characterization of all equational classes of $\mathbf{B_\omega}$.

Theorem 1. *For a pseudocomplemented distributive lattice L the following are equivalent.*

(i) *L is a Stone algebra.*

(ii) $\mathbf{x^* + x^{**} = 1}$ *is an identity in L.*

(iii) $\mathbf{(xy)^* = x^* + y^*}$ *is an identity in L.*

(iv) $\mathbf{x^{**} + y^{**} = (x + y)^{**}}$ *is an identity in L.*

(v) $\mathscr{Rg}(L)$ *is a subalgebra and therefore a retract of L.*

(vi) *L is a subdirect product of copies of the algebras* **2** *and* **3**.

(vii) *Each prime ideal in L contains at most one minimal prime ideal.*

(viii) *If I and J are distinct minimal prime ideals in L then $I + J = L$ (the sum here is taken in $\mathfrak{I}(L)$).*

Proof. (i) $\Leftrightarrow$ (ii) By definition of Stone algebras and by Theorem 6.2.

(ii) $\Rightarrow$ (iii) We show directly that for $a, b \in L$, $a^* + b^*$ is the pseudocomplement of $(ab)^*$. First, $ab(a^* + b^*) = 0$. Next, suppose $xab = 0$ for $x \in L$, then by Theorem 2.1, $xa^{**}b = 0$ and hence $xa^{**} \leq b^*$. So $x = x1 = x(a^* + a^{**}) = xa^* + xa^{**} \leq a^* + b^*$.

(iii) $\Rightarrow$ (iv) For $a, b \in L$ we have, by Theorem 2.1: $(a + b)^{**} = (a^*b^*)^* = a^{**} + b^{**}$.

(iv) $\Rightarrow$ (v) For $a, b \in \mathscr{Rg}(L)$ we have by Theorem 4.3: $a \overset{\mathscr{Rg}(L)}{+} b = (a + b)^{**} = a^{**} + b^{**} = a + b$. Thus $\mathscr{Rg}(L)$ is a subalgebra and r_L is a $\mathbf{B_\omega}$-homomorphism.

Also $\mathscr{Rg}(L)$ is a retract of L.

(v) $\Rightarrow$ (ii) For $a \in L$, we have $a^{**} \in \mathscr{Rg}(L)$ and thus $1 = a^{**} \overset{\mathscr{Rg}(L)}{+} a^{***} = a^{**} + a^{*}$.

(i) $\Leftrightarrow$ (vi) Immediate from Corollary 6.3 and Birkhoff's subdirect product theorem.

(i) $\Leftrightarrow$ (vii) $\Leftrightarrow$ (viii) By Exercise 6.4.

In view of the characterization of Stone algebras by Theorem 1(viii), it is natural to ask for a characterization of distributive lattices (with 0(1)) which have that property. W. Cornish [1] has shown that a distributive lattice L with 0 satisfies the condition of Theorem 1(viii) if and only if for $x, y \in L$, $(x + y]^* = (x]^* + (y]^*$ in $\mathfrak{J}(L)$. See also M. Mandelker [1], J. Varlet [2], W. Cornish [2], and B. Davey [3].

Non-Boolean examples of Stone algebras are:

Example 2. Chains with 0, 1. For such a chain $a^* = 0$ for $a \neq 0$ and $0^* = 1$.

Example 3. If L is a dense pseudocomplemented distributive lattice, then $0^* + 0^{**} = 1$ and for $x \neq 0$, $x^* = 0$ so $x^{**} = 1$; thus L is a Stone algebra.

Example 4. The lattice $\mathfrak{J}(L)$ where L is a complete Boolean algebra (cf. Example 1.4). Indeed for $I \in \mathfrak{J}(L)$ and $a = \overline{\sum I}$ then $I^* = (a]$ so $I^* + I^{**} = (a] + (\bar{a}] = L$.

Example 5. Let L be a Boolean algebra and let $L^{[2]} = \{(a, b) \in L \times L : a \leq b\}$. The pseudocomplement of $(a, b) \in L^{[2]}$ is $(\bar{b}, \bar{b})$. This example plays a crucial role in the characterization of injectives Stone algebras (Theorem 8.8).

Exercise 6. The lattice of open sets of an extremally disconnected space (i.e., $\mathrm{Cl}\ U$ is open if U is open) (cf. Example 1.5).

Exercise 7. Prove that if $L \in \mathbf{B}_\omega$, then L is a Stone algebra if and only if $\mathscr{C}(L) = \mathscr{Rg}(L)$.

Exercise 8. Let L be a Stone algebra. Let I_0 be a minimal prime ideal and I_1 and I_2 prime ideals in L such that $I_0 \subseteq I_1 \subseteq I_2$. Let $f: L \to \mathbf{4}$ be defined by $f(x) = \mathbf{0}$ if $x \in I_0$, $f(x) = \mathbf{1}$ if $x \in I_1 \sim I_0$, $f(x) = \mathbf{2}$ if $x \in I_2 \sim I_1$ and $f(x) = \mathbf{3}$ if $x \notin I_2$. Show that f is a $\mathbf{B}_\omega$-homomorphism, and that f is onto $\mathbf{4}$ if $I_0 \subset I_1 \subset I_2$.

Example 4 leads to a different kind of characterization of $\mathbf{B}_1$:

Theorem 9. (G. Grätzer [1]; see also G. Bruns [2] and G. Grätzer [3]) *Every Stone algebra L is isomorphic to a subalgebra of the Stone algebra of all ideals of the Boolean algebra of all subsets of a set X.*

Remark. See the more general theorem of this kind concerning $\mathbf{B}_\omega$ in Exercise 5.7.

Proof. If $L = \mathbf{2}$, let X be a singleton. If $L = \mathbf{3}$, let X be a set such that $|X| = \aleph_0$. Let I be the ideal of finite subsets of X. Note that $I^* = \{\varnothing\}$. The required embedding is given by $\mathbf{0} \to \{\varnothing\}$, $\mathbf{1} \to I$, $\mathbf{2} \to \mathbf{2}^X$. Let $\mathbf{K}$ be the class of Stone algebras which can be represented as in the statement of the theorem, then $\mathbf{2} \in \mathbf{K}$ and $\mathbf{3} \in \mathbf{K}$. Obviously, $\mathbf{K}$ is closed under subalgebras. We will show that $\mathbf{K}$ is also closed under direct products. Indeed, let $(L_s)_{s \in S}$ be a family in $\mathbf{K}$. Then for each L_s there exists a set X_s and a $\mathbf{B}_\omega$-embedding $g_s\colon L_s \to \mathfrak{I}(\mathbf{2}^{X_s})$ and therefore
$g = \times_{s \in S} g_s\colon \times_{s \in S} L_s \to \times_{s \in S} \mathfrak{I}(\mathbf{2}^{X_s})$ is a $\mathbf{B}_\omega$-embedding. But the map
$f\colon \times_{s \in S} \mathfrak{I}(\mathbf{2}^{X_s}) \to \mathfrak{I}(\times_{s \in S} \mathbf{2}^{X_s})$ as defined in Exercise II.9.12, is by Exercise 3.7, a $\mathbf{B}_\omega$-embedding and thus $f \circ g$ is a $\mathbf{B}_\omega$-embedding. It follows that $\times_{s \in S} L_s \in \mathbf{K}$. We can now conclude that $\mathbf{K}$ is the class of all Stone algebras.

The representation space of a Stone algebra is characterized by:

Theorem 10. *Let L be a distributive lattice. Then L is a Stone algebra if and only if $\mathscr{S}(L)$ is a bounded Stone space and the closure of every compact open subset of $\mathscr{S}(L)$ is open.*

Proof. ($\Rightarrow$) Any open compact subset of $\mathscr{S}(L)$ is of the form $\hat{a}$, where $a \in L$. Now $a^* + a^{**} = 1$ and $\widehat{a^*a^{**}} = 0$. Thus $\widehat{a^*} \cup \widehat{a^{**}} = \mathscr{S}(L)$ and $\widehat{a^*} \cap \widehat{a^{**}} = \varnothing$, so $\widehat{a^{**}} = \widetilde{\widehat{a^*}}$. But by Theorem 2.3, $\widehat{a^*} = \widetilde{\mathrm{Cl}\,\hat{a}}$, hence $\widehat{a^{**}} = \mathrm{Cl}\,\hat{a}$ and so $\mathrm{Cl}\,\hat{a}$ is open.

($\Leftarrow$) Let $a \in L$. Then $\mathrm{Cl}\,\hat{a}$ is open so $\widetilde{\mathrm{Cl}\,\hat{a}}$ is closed and hence compact so by Theorem 2.3, $L \in \mathbf{B}_\omega$. It remains to show that L is a Stone algebra. Again, if $a \in L$, then $\widehat{a^*} = \widetilde{\mathrm{Cl}\,\hat{a}}$, so $\widehat{a^{**}} = \widetilde{\mathrm{Cl}(\widetilde{\mathrm{Cl}\,\hat{a}})}$. But $\widetilde{\mathrm{Cl}\,\hat{a}}$ is closed so we obtain $\widehat{a^{**}} = \mathrm{Cl}\,\hat{a}$. It follows that $\widehat{a^*} \cup \widehat{a^{**}} = \mathscr{S}(L)$ and hence $a^* + a^{**} = 1$.

Exercise 11. (cf. Exercise IV.2.8) Let $(L_i)_{i=1,\cdots,n}$ be a family of objects in $\mathscr{B}_1$. Prove that $\mathscr{S}(\times_{i=1}^{n} L_i)$ is homeomorphic to the topological (disjoint) sum of the $\mathscr{S}(L_i)$, $i = 1, \cdots, n$.

C. C. Chen and G. Grätzer [1], [2] developed a structure theorem for Stone algebras by associating with each such algebra L, a triple $(\mathscr{R}g(L), \mathscr{D}s(L), \varphi^L)$ where φ^L is a certain $\mathbf{D}_{01}$-homomorphism from $\mathscr{R}g(L)$ to the lattice of filters of $\mathscr{D}s(L)$. Conversely, they show that if $B \in \mathbf{B}$, $D \in \mathbf{D}_1$ and φ is a $\mathbf{D}_{01}$-homomorphism from B into the filters of D, then there is a Stone algebra L with triple (B, D, φ). T. Katriňák [4], [10] has extended the triple characterization of C. C. Chen and G. Grätzer to all algebras in $\mathbf{B}_\omega$. In addition, he has also given another triple characterization of members of $\mathbf{B}_\omega$ which is useful in obtaining certain kinds of results. This method has also been used by Katriňák to characterize Heyting algebras [5]. See also H. Lakser [1] and T. Katriňák [8], [9].

The connection between Stone algebras and distributive lattices in which each interval has the "Stone property" is described in the next series of results.

Definition 12. A distributive lattice L with 1 is a *relative Stone algebra* provided that each interval $[a, b]$ of L is a Stone algebra.

Theorem 13. *If L is a distributive lattice with 1 and $[a)$ is a Stone algebra for each $a \in L$, then L is a relative Stone algebra.*

Proof. Let $a \leq b$ be members of L and $x \in [a, b]$. We will demonstrate that $x^{*[a,b]} = x^{*[a)}b + a$. Let $u = x^{*[a)}b + a$.

First, $u \in [a, b]$ since $a \leq x^{*[a)}b + a \leq b$. Also, $xu = x(x^{*[a)}b + a) = xx^{*[a)}b + xa = ab + xa = a$. Now suppose $y \in [a, b]$ and $xy = a$. Then $y \leq x^{*[a)}$ so $y \leq x^{*[a)}b + a = u$. Hence u is the pseudocomplement of x in $[a, b]$. Finally, we show that $u + v = b$ where $v = u^{*[a,b]}$. Indeed, since $[a)$ is Stone:

$$\begin{aligned} u + v &= (x^{*[a)}b + a) + (x^{*[a)}b + a)^{*[a)}b + a \\ &= x^{*[a)}b + a + ((x^{*[a)})^{*[a)} + b^{*[a)})a^{*[a)}b + a \\ &= x^{*[a)}b + a + (x^{*[a)})^{*[a)}b + b^{*[a)}b + a \\ &= x^{*[a)}b + a + (x^{*[a)})^{*[a)}b = b + a = b. \end{aligned}$$

Theorem 14. (C. C. Chen and G. Grätzer [2]) *Suppose L is a Stone algebra. Then L is a relative Stone algebra if and only if $\mathscr{D}_\delta(L)$ is a relative Stone algebra.*

Proof. ($\Rightarrow$) It suffices to prove that $\mathscr{D}_\delta(L)$ is convex in L. But $u, v \in \mathscr{D}_\delta(L)$, $u \leq x \leq v$ implies $v^* \leq x^* \leq u^* = 0$ so $x \in \mathscr{D}_\delta(L)$.

($\Leftarrow$) By Theorem 13, we need only prove that $[y)$ is Stone for each $y \in L$.

For each $x \geq y$ let x^a be the pseudocomplement of $x + y^*$ in the interval $[y + y^*) \subseteq \mathscr{D}_\delta(L)$ (So x^a is the greatest $u \in [y + y^*)$ such that $u(x + y^*) = y + y^*$.). We will prove that the element $x^b = x^a(x^* + y^{**})$ is the required pseudocomplement of x in $[y)$ and that $x^b + x^{bb} = 1$.

Now $y \leq y + y^* \leq x^a$ and $y \leq x^* + y^{**}$ so x^b is indeed in $[y)$. Also $xx^b = xx^a(x^* + y^{**}) = xx^a y^{**} = x^a y^{**}(x + y^*) = (y + y^*)y^{**} = y$. Suppose $u \in [y)$ and $ux = y$. We will first show $u \leq x^a$. Now $uxy^* = 0$ so $u \leq (xy^*)^* = x^* + y^{**}$. But also $u + y^* \in [y + y^*)$ and $(u + y^*)(x + y^*) = ux + y^* = y + y^*$ so $u + y^* \leq x^a$ and hence $u \leq x^a$. Hence $u \leq x^a(x^* + y^{**}) = x^b$.

Finally, to verify $x^b + x^{bb} = 1$, we first show

(1) $x^b + y^* = x^a$.

Indeed, since $x^a x^* \leq x^* \leq y^*$ and L is a Stone algebra, $x^b + y^* = x^a(x^* + y^{**}) + y^* = x^a x^* + x^a y^{**} + y^* = x^a y^{**} + y^* = (x^a + y^*)(y^{**} + y^*) = x^a + y^* = x^a$.

Now x^{ba} is the pseudocomplement of $x^b + y^*$ in $[y + y^*)$ and hence by (1) it is the pseudocomplement of $x^a = x^a + y^*$ in $[y + y^*)$; that is, $x^{ba} = x^{aa}$.

By definition, x^a is the pseudocomplement of $x + y^*$ in $[y + y^*)$ and x^{aa} is the pseudocomplement of $x^a + y^* = x^a$ in $[y + y^*)$. But since $\mathscr{D}_\delta(L)$ is a relative Stone algebra, $[y + y^*)$ is Stone so $x^{aa} + x^a = 1$.

Combining the above with $x^a \geq y^* \geq x^*$ we have

$$\begin{aligned}
x^{bb} + x^b &= x^{ba}(x^{b*} + y^{**}) + x^a(x^* + y^{**}) \\
&= x^{aa}((x^a(x^* + y^{**}))^* + y^{**}) + x^a(x^* + y^{**}) \\
&= x^{aa}(x^{a*} + x^{**}y^* + y^{**}) + x^a(x^* + y^{**}) \\
&= (x^{aa} + x^a)y^{**} + x^{aa}x^{a*} + x^{aa}x^{**}y^* + x^a x^* \\
&= y^{**} + x^{aa}x^{a*} + x^{aa}x^{**}y^* + x^a x^* \\
&= y^{**} + x^{aa}x^{a*} + x^{**}y^* + x^a x^* \qquad (\text{since } x^{aa} \geq y^*) \\
&\geq y^{**} + x^{**}y^* + x^a x^* \\
&= (y^{**} + x^{**})(y^{**} + y^*) + x^a x^* \\
&= y^{**} + x^{**} + x^a x^* \\
&= x^{**} + x^* = 1.
\end{aligned}$$

8. Injective Stone algebras

Definition 1. A *double Stone algebra* is a Stone algebra whose dual is also a Stone algebra.

We denote the dual pseudocomplement of a by a^+. Thus a^+ is the least element x for which $a + x = 1$.

Exercise 2. Prove that in a double Stone algebra L

(i) $\mathcal{Rg}(L) = \{a \in L : a = a^{++}\}$.

(ii) $(x^*)^{++} = x^*$ for each $x \in L$.

Exercise 3. Let L be a Boolean algebra. Prove that $L^{[2]}$ is a double Stone algebra and that for $(a, b) \in L^{[2]}$, $(a, b)^+ = (\bar{a}, \bar{a})$. Moreover $\mathcal{Rg}(L^{[2]}) = \{(x, x) : x \in L\} \cong L$ (see Example 7.5 for the definition).

In accordance with our established notation, the category associated with $\mathbf{B}_1$ will be denoted by $\mathcal{B}_1$. Note that $\mathcal{B}_1$ is a full subcategory of $\mathcal{B}_\omega$.

Lemma 4. *In $\mathcal{B}_1$, $\mathbf{3}$ is injective.*

Proof. Let L be a subalgebra of L_1; $L, L_1 \in \text{Ob}\,\mathcal{B}_1$ and let $f \in [L, \mathbf{3}]_{\mathcal{B}_1}$. Let $I_0 = \{x \in L : f(x) = \mathbf{0}\}$ and $I_1 = \{x \in L : f(x) \leq \mathbf{1}\}$. Then I_0 and I_1 are prime ideals in L and I_0 is a minimal prime ideal. Indeed, if $a \in I_0$ then $f(a) = \mathbf{0}$ and thus $f(a^{**}) = \mathbf{0}$, so $a^{**} \in I_0$. In order to show that f can be extended to L_1 it suffices to show (cf. Exercise 7.8) that there exist a minimal prime ideal J_0 in L_1 and a prime ideal J_1 in L_1 such that $J_0 \subseteq J_1$, $L \cap J_0 = I_0$ and $L \cap J_1 = I_1$. By Theorem III.6.5 there exists a prime ideal J_1 in L_1 such that $J_1 \cap L = I_1$. Let F be a maximal filter in L_1 containing

$L_1 \sim J_1$ and let $J_0 = L_1 \sim F$. Then J_0 is a minimal prime ideal in L_1 and $J_0 \subseteq J_1$. So $J_0 \cap L \subseteq J_1 \cap L = I_1$ and also $J_0 \cap L$ is a prime ideal in L. If $a \in J_0 \cap L$ then $a \in J_0$ so $a^{**} \in J_0$, by Theorem 2.2. Also $a^{**} \in L$ and hence $a^{**} \in J_0 \cap L$ and it follows again from Theorem 2.2 that $J_0 \cap L$ is a minimal prime ideal. But $J_0 \cap L$ and I_0 are both contained in I_1 and since I_0 is minimal it follows from Theorem 7.1 that $J_0 \cap L = I_0$ completing the proof.

***Corollary* 5.** *In $\mathscr{B}_1$, **2** is injective.*

Proof. Note that $\mathbf{2} = \mathscr{R}g(\mathbf{3})$ and hence by Theorem 7.1, **2** is a retract of **3**, It follows from Theorem 4 and from Theorem I.16.2 that **2** is injective.

***Corollary* 6.** *$\mathscr{B}_1$ has enough injectives, and therefore $\mathscr{B}_1$ has the congruence extension property, the amalgamation property and each algebra in $\mathbf{B}_1$ has an injective hull.*

Proof. Since all of the subdirectly irreducible algebras in $\mathbf{B}_1$ are injective, the result is immediate from Birkhoff's subdirect product theorem.

***Lemma* 7.** *If L is an injective Stone algebra, then*

(i) *L, $\mathscr{R}g(L)$ and $\mathscr{D}s(L)$ are complete.*

(ii) *$\mathscr{D}s(L)$ is a Boolean algebra.*

(iii) *L is double Stone algebra.*

(iv) *If $a, b \in L$ and $a^* = b^*$, $a^+ = b^+$ then $a = b$.*

Proof. Obviously **2** and **3** satisfy (i), (ii), and (iii) and by componentwise inspection, so does $\mathbf{2}^\alpha \times \mathbf{3}^\beta$ for arbitrary cardinals α, β. By Theorem 7.1(vi) every Stone algebra is a subdirect product of copies of $\mathbf{2}^\alpha$ and $\mathbf{3}^\beta$ for suitable α, β. So if L is injective, then it is a retract of $\mathbf{2}^\alpha \times \mathbf{3}^\beta$. We will now show that (i), (ii), and (iii) are preserved under retracts. Let $h: L' \to L$ define a retract where $L' = \mathbf{2}^\alpha \times \mathbf{3}^\beta$. For (i): L is complete, since L' is complete and L is a retract of L'. It also easily follows that $h[\mathscr{R}g(L')] = \mathscr{R}g(L)$ and so $\mathscr{R}g(L)$ is a retract of $\mathscr{R}g(L')$ and thus $\mathscr{R}g(L)$ is complete. Similarly, $\mathscr{D}s(L)$ is complete. For (ii): $\mathscr{D}s(L')$ is a Boolean algebra, and $\mathscr{D}s(L)$ is a retract of $\mathscr{D}s(L')$ (and thus a homomorphic image), so $\mathscr{D}s(L)$ is a Boolean algebra. For (iii): We must show that for $x \in L$, x^{+L} exists and that $x^{+L} \cdot x^{+L+L} = 0$. We claim that $x^{+L} = h(x^{+L'})$. Indeed, $x + h(x^{+L'}) = h(x) + h(x^{+L'}) = h(x + x^{+L'}) = h(1) = 1$. Again, if $x + y = 1$ then $x^{+L'} \leq y$ so $h(x^{+L'}) \leq h(y) = y$, and we conclude $x^{+L} = h(x^{+L'})$. Finally, for $x \in L$ since L' is a double Stone algebra we have $h(x^{+L'})h((h(x^{+L'}))^{+L'}) = h(x^{+L'}(h(x^{+L'}))^{+L'}) = h((x + h(x^{+L'}))^{+L'}) = h((h(x) + h(x^{+L'}))^{+L'}) = h((h(x + x^{+L'}))^{+L'}) = h((h(1))^{+L'}) = h(1^{+L'}) = h(0) = 0$. But $x^{+L}x^{+L+L} = h(x^{+L'})h((h(x^{+L'}))^{+L'})$, so $x^{+L}x^{+L+L} = 0$.

It remains to show that (iv) holds. Suppose $a^* = b^*$, $a^+ = b^+$, and $b \nleq a$. Then there exists a prime ideal I in L, such that $a \in I$, $b \notin I$, and so $b^{**} \notin I$. But

$a^{**} = b^{**}$, thus also $a^{**} \notin I$, and it follows from Theorem 2.2 that I is not minimal. Hence there exists a minimal prime ideal I_0 such that $I_0 \subset I$. But L is a double Stone algebra, and hence by a dual argument, there exists a (maximal) prime ideal J such that $I \subset J$. By Exercise 7.8 there exists $f \in [L, \mathbf{4}]_{\mathscr{B}\omega}$ such that $f(x) = \mathbf{0}$ if $x \in I_0$, $f(x) = \mathbf{1}$ if $x \in I \sim I_0, f(x) = \mathbf{2}$ if $x \in J \sim I$ and $f(x) = \mathbf{3}$ if $x \notin J$. Now I_0 is minimal and I is not minimal, so by Theorem 2.2 there exists $x \in I \cap \mathscr{D}_\delta(L)$ such that $x \notin I_0$. Let $y \in J$ such that $y > x, y \notin I$. Define $g: \mathbf{4} \to L$ by $g(\mathbf{0}) = 0$, $g(\mathbf{1}) = x$, $g(\mathbf{2}) = y$ and $g(\mathbf{3}) = 1$. Then g is a $\mathbf{D}$-homomorphism. But $x^* = 0$ and $y \geq x$ thus $y^* \leq x^* = 0$ and so g preserves $*$. g also preserves 0 and is therefore a $\mathbf{B}_\omega$-homomorphism. Also $f \circ g$ is the identity on $\mathbf{4}$ and thus $\mathbf{4}$ is a retract of L. It follows that $\mathbf{4}$ is injective. But $\mathscr{D}_\delta(\mathbf{4}) = \{\mathbf{1}, \mathbf{2}, \mathbf{3}\}$ and this is not a Boolean algebra, contradicting (ii). We conclude $b \leq a$. Similarly $a \leq b$ and thus $a = b$.

Theorem 8. (R. Balbes, G. Grätzer [1]) *A Stone algebra is injective if and only if it has the form $L_0 \times L_1^{[2]}$, where L_0 and L_1 are complete Boolean algebras.*

Proof. ($\Leftarrow$) Suppose L_0 and L_1 are complete Boolean algebras. Now L_0 and L_1 are retracts of $\mathbf{2}^\alpha$ for a suitable α, and since $\mathbf{2}$ is injective so is $\mathbf{2}^\alpha$ and hence also L_0 and L_1 (by Theorems I.16.3 and I.16.2). To see that $L_1^{[2]}$ is also injective, first represent $L_1^{[2]}$ as a subalgebra of $\mathbf{2}^\alpha \times \mathbf{3}^\beta$ for suitable α, β (Theorem 7.1(vi)). Since $\mathscr{R}g(L_1^{[2]})$ is a subalgebra of $\mathscr{R}g(\mathbf{2}^\alpha \times \mathbf{3}^\beta)$ and $\mathscr{R}g(L_1^{[2]}) \cong L_1$ (Exercise 3) which is injective, there exists a retract f: $\mathscr{R}g(\mathbf{2}^\alpha \times \mathbf{3}^\beta) \to \mathscr{R}g(L_1^{[2]})$. Since $a^{++} \leq a \leq a^{**}$ we can define $g: \mathbf{2}^\alpha \times \mathbf{3}^\beta \to L_1^{[2]}$ by $g(a) = (p_1(f(a^{++})), p_2(f(a^{**})))$ (where $p_i(u)$ is the ith coordinate of u for $i = 1, 2$). Using Exercise 2(ii), it is easy to verify that g defines a retract. Hence $L_1^{[2]}$ is injective. Therefore, $L_0 \times L_1^{[2]}$ is also injective.

($\Rightarrow$) Thus suppose L is injective. By Lemma 7(i) there exists a smallest dense element d. Setting $a = d^{++}$ we obtain $L = (a] \times (a^*]$ since $a \in \mathscr{R}g(L)$; also $(a], (a^*]$ are both complete since L is complete. Recall, from Theorem 4.1 and its proof, that $(a]$ is a pseudocomplemented distributive lattice with pseudocomplement $b^o = ab^*$ for $b \in (a]$. In fact $(a]$ is a Boolean algebra: $b \leq a$ implies $b^o + b = ab^* + b$ $= (a + b)(b^* + b) = a(b^* + b)$ and $b^* + b \in \mathscr{D}_\delta(L)$ so $a = d^{++} \leq d \leq b + b^*$. The proof will be completed by showing that $(a^*] \cong (\mathscr{R}g((a^*]))^{[2]}$. First, observe that the dual pseudocomplement of $x \in (a^*]$ is $x^\pm = a^*x^+$. Indeed, if $a^*x^+ \leq y$, $y \in (a^*]$ then $a^* = a^*(x^+ + x) = a^*x^+ + a^*x \leq y + x$ and conversely, if $x + y$ $= a^*$ for $y \leq a^*$ then $x + y + a^{**} = a^* + a^{**} = 1$ so $y \geq (x + a^{**})^+ = x^+a^{**+}$ $= x^+a^*$. Now denote the pseudocomplement of $x \in (a^*]$ by x^o (i.e. $x^o = x^*a^*$). Define a map f: $(a^*] \to (\mathscr{R}g((a^*]))^{[2]}$ by $f(x) = (x^{\pm\pm}, x^{oo})$. Then f is an embedding since $a^* = (d^{++})^* = d^{+++} = a^+$ implies that $x = y$ if $x^\pm = y^\pm$ and $x^o = y^o$ (Lemma 7(iv)). It remains to show that for $b, c \in \mathscr{R}g((a^*]), b \leq c$, there exists $x \in (a^*]$ such that $x^{\pm\pm} = b, x^{oo} = c$. Indeed, take $x = c(b + d)$ then since $d^{++} = a$ and $d^* = 0$, we have $x^{\pm\pm} = c^{\pm\pm}(b^{\pm\pm} + d^{\pm\pm}) = c^{++}a^*(b^{++}a^* + d^{++}a^*)$ $= a^*b^{++}c^{++} = a^*b^{++} = b^{\pm\pm} = b$ and $x^{oo} = c^{oo}(b^{oo} + d^{oo})$ $= c^{**}a^*(b^{**}a^* + d^{**}a^*) = c^{**}a^* = c^{oo} = c$.

Exercise 9. Prove that a finite Stone algebra is injective if and only if it has the form $\mathbf{2}^\alpha \times \mathbf{3}^\beta$ for some finite cardinals α, β.

Exercise 10. Prove that a finite Stone algebra L is weakly projective if and only if $L = \times_{i=1}^{n} L_i$ where $L_1 = \mathbf{1} \oplus L_1' \oplus \mathbf{1}$ and L_i' is a finite distributive lattice in which the product of join irreducibles is join irreducible.

A. Day [2] has shown that $\mathbf{B}_0$, $\mathbf{B}_1$, and $\mathbf{B}_2$ are the only non-trivial equational subclasses of $\mathbf{B}_\omega$ which have enough injectives. In each of the other equational subclasses, an algebra is injective if and only if it is a complete Boolean algebra. H. Lakser [1] has used the 'triple' method of Stone algebras (see Section 7) to describe the essential extensions and injective hulls of Stone algebras.

G. Grätzer and H. Lakser [3] define a *weak injective A*, in an equational category, to be an algebra with the property that if $h: B \to C$ is a one-one homomorphism and $f: B \to A$ is an *onto* homomorphism, then there is a homomorphism $g: C \to A$ such that $g \circ h = f$. They show that for $\mathscr{B}_1$ and $\mathscr{B}_2$, weak injectivity is equivalent to injectivity and characterize the weak injectives in $\mathscr{B}_\mathbf{n}$, $n \geq 1$ thus greatly extending Theorem 8.

Since $\mathscr{B}_1$, $\mathscr{B}_2$ have enough injectives, the congruence extension property and the amalgamation property hold in these classes. Moreover, it is shown in G. Grätzer and H. Lakser [2] that the congruence extension property holds in $\mathbf{B}_\omega$ (and hence $\mathscr{B}_\mathbf{n}$, $\mathbf{n} \leq \boldsymbol{\omega}$) and that the amalgamation property holds in $\mathscr{B}_\mathbf{n}$ only for $\mathbf{n} \leq \mathbf{2}$ or $\mathbf{n} = \omega$.

9. Coproducts of Stone algebras. Free Stone algebras

We have seen in Theorem VII.1.2 that the coproduct of a family of Boolean algebras in $\mathscr{D}_{01}$ is also the coproduct of this family in $\mathscr{B}$. A similar situation exists for the category $\mathscr{B}_1$, which, as we have seen in Section 3, can be considered as a subcategory of $\mathscr{D}_{01}$. In other words, we will show that the coproduct of a family of Stone algebras in $\mathscr{D}_{01}$ is also the coproduct in $\mathscr{B}_1$. We first prove a preliminary result.

Theorem 1. *The topological product of a family of representation spaces of Stone algebras is the representation space of a Stone algebra.*

Proof. Let $(X_s)_{s \in S}$ be a family of representation spaces of Stone algebras and let Y be its topological product. By Theorem VII.7.2, Y is a bounded Stone space. Thus, by Theorem 7.10 we only need to show that if U is a compact open subset of Y then $\mathrm{Cl}\, U$ is open. It follows from the proof of Theorem VII.7.2 that U is a finite union of sets of the type $\times_{s \in S} U_s$ where for each $s \in S$, U_s is a compact open subset of X_s and such that $U_s = X_s$ for all but a finite number of $s \in S$. Hence, it suffices to show that $\mathrm{Cl} \times_{s \in S} U_s$ is open. But since $U_s = X_s$ for all but a finite number of $s \in S$, it follows from the hypothesis that $\mathrm{Cl} \times_{s \in S} U_s = \times_{s \in S} \mathrm{Cl}\, U_s$ is open.

Theorem 2. (G. Grätzer [4]) *Suppose $(L_s)_{s \in S}$ is a family of objects in $\mathscr{B}_1$ and suppose $(j_s: L_s \to L)$ is a coproduct of this family in $\mathscr{D}_{01}$. Then it is also a coproduct in $\mathscr{B}_1$.*

Proof. It follows from Theorem VII.7.3 that the representation space of L is $\bigtimes_{s\in S}\mathscr{S}(L_s)$ and $p_s = \mathscr{S}(j_s)$ for each $s \in S$ where $p_s\colon \mathscr{S}(L) \to \mathscr{S}(L_s)$ are the projections. By Theorem 1, L is a Stone algebra and it follows from Exercise 3.8 that each j_s is a $\mathbf{B}_1$-homomorphism. It remains to show that the extension property for homomorphisms holds. Thus, let $(f_s\colon L_s \to L_1)$ be a family of $\mathbf{B}_1$-homomorphisms. Then there exists a unique $h \in [L, L_1]_{\mathscr{D}_{01}}$ such that $h \circ j_s = f_s$ for each $s \in S$. That h preserves the operation * follows easily from the fact that h is a $\mathbf{D}_{01}$-homomorphism and from Theorems 2.1(vii) and 7.1(iii).

Corollary 3. *The free Stone algebra* $\mathscr{F}_{\mathbf{B}_1}(\alpha)$ *on* α *free generators is the coproduct in* $\mathscr{D}_{01}$ *of* α *copies of* $\mathscr{F}_{\mathbf{B}_1}(\mathbf{1})$.

The following theorem is due to R. Balbes and A. Horn [3]; the proof is due to G. Grätzer and H. Lakser:

Theorem 4. (see G. Grätzer [4]) *Let n be a non-negative integer. Then the free Stone algebra on n free generators is* $\bigtimes_{S\subseteq \mathbf{n}}\mathscr{F}_{\mathbf{D}_{01}}(S)$.

Proof. Obviously $\mathscr{F}_{\mathbf{B}_1}(\mathbf{1}) \cong \mathbf{2} \times \mathbf{3}$ and thus by Exercise 7.11, $\mathscr{S}(\mathscr{F}_{\mathbf{B}_1}(\mathbf{1}))$ is the topological union of $\mathscr{S}(\mathbf{2})$ and $\mathscr{S}(\mathbf{3})$. By Corollary 3, $\mathscr{F}_{\mathbf{B}_1}(\mathbf{n})$ is the coproduct in $\mathscr{D}_{01}$ of n copies of $\mathscr{F}_{\mathbf{B}_1}(\mathbf{1})$ and thus by Theorem VII.7.3, $\mathscr{S}(\mathscr{F}_{\mathbf{B}_1}(\mathbf{n}))$ is the topological product of n copies of the topological union of $\mathscr{S}(\mathbf{2})$ and $\mathscr{S}(\mathbf{3})$. But $\mathscr{S}(\mathbf{2})$ is the one-point space, so $\mathscr{S}(\mathscr{F}_{\mathbf{B}_1}(\mathbf{n}))$ is the topological union of the one-point space and the union of $\binom{n}{k}$ topological products of k copies of $\mathscr{S}(\mathbf{3})$, where $k = 1, 2, \cdots, n$. But $\mathscr{S}(\mathbf{3}) \cong \mathscr{S}(\mathscr{F}_{\mathbf{D}_{01}}(\mathbf{1}))$, and thus again by Exercise IV.2.8, and Theorem VII.7.3. $\mathscr{F}_{\mathbf{B}_1}(\mathbf{n})$ is isomorphic to $\bigtimes_{S\subseteq \mathbf{n}}\mathscr{F}_{\mathbf{D}_{01}}(S)$.

Exercise 5. Obtain the lattice diagram of $\mathscr{F}_{\mathbf{B}_1}(\mathbf{n})$ for $n = 2, 3$.

A complete description, although somewhat complicated, has been given by A. Urquhart [2] for $\mathscr{F}_{\mathbf{B}_\mathbf{m}}(\mathbf{n})$ for each $\mathbf{n} < \aleph_0$, $\mathbf{m} \leq \omega$.

IX

Heyting Algebras

1. Introduction

We have seen in Section II.6 that the classical propositional calculus gives rise, in a natural way, to the study of Boolean algebras. By exactly the same method, we can associate with other logics (e.g. non-classical logics, n-valued logics) various classes of lattices.

In this chapter we will investigate a class of lattices which arises from the intuitionistic logic; that is, from the logic obtained from the classical 2-valued logic by deleting from the tautologies of the latter: "$a + \neg a$". It turns out that for formulas x, y, the identity $[x \to y] = [\bar{x}] + [y]$, which holds in the lattices associated with classical 2-valued logic (Boolean algebras) is no longer true in the lattices which are associated with intuitionistic logic; but rather $[x \to y]$ is the largest $[u]$ in the lattice for which $[x][u] \leq [y]$.

We will call the class of algebras arising in this way Heyting algebras, since it was A. Heyting [1] who in 1930 "formalized" the propositional and predicate calculus for the intuitionistic view of mathematics. The lattice theoretic approach is due to G. Birkhoff (cf. [6, 1948 edition]) and the study of these algebras was carried out in the classic work of J. C. C. McKinsey and A. Tarski [1], [2]. The use of the term Heyting algebra is by no means standard as other authors have called them pseudo-Boolean algebras or relatively pseudocomplemented distributive lattices with 0; J. C. C. McKinsey and A. Tarski [2] call the algebraic duals of Heyting algebras, Brouwerian algebras for L. E. J. Brouwer who was the leading exponent of the intuitionistic school of mathematics.

The papers of J. C. C. McKinsey and A. Tarski [1], [2] together with the books of H. Rasiowa [3], and H. Rasiowa and R. Sikorski [1], clearly show the applications of Heyting algebras to topology and logic. A selection of papers which deal with intuitionistic logical systems and the lattice theoretic counterparts include A. Horn [3], [6], [7]; J. C. C. McKinsey and A. Tarski [3]; L. Rieger [2].

2. Definitions

Definition 1. Suppose that L is a lattice and $a, b \in L$. If there is a largest $x \in L$ such that $ax \leq b$, then this element is denoted by $a \to b$ and is called the *relative pseudocomplement of a with respect to b*.

The definition of relative pseudocomplement is equivalent to the existence of an element $a \to b$ such that

(1) $ax \leq b \Leftrightarrow x \leq a \to b$.

Definition 2. A *Heyting algebra* is a lattice with 0 in which $a \to b$ exists for each $a, b \in L$.

Heyting algebras are not only distributive, but they satisfy the following infinite distributive law (cf. Theorem II.6.9).

(2) If L is a Heyting algebra and $\sum S$ exists for some $S \subseteq L$ then for each $a \in L$, $\sum\{as : s \in S\}$ exists and $a \sum S = \sum\{as : s \in S\}$.

Indeed, if $\sum S$ exists, then $as \leq a\sum S$ for each $s \in S$ and if $as \leq x$ for each $s \in S$, then $s \leq a \to x$ and consequently $a \sum S \leq x$. Note that L has a greatest element and by (1), $x \to x = 1$ for each $x \in L$. Evidently $a \to 0$ is the pseudocomplement of a and will therefore be denoted by a^*. In order to simplify formulas, we rank the operation $\to$ above $+$ but below $\cdot$, so that $a + b \to c$ is written in place of $a + (b \to c)$ but $ab \to c$ means $(ab) \to c$.

Considered as lattices, Heyting algebras are pseudocomplemented distributive lattices. Caution should be used however because, considered as algebras, pseudocomplemented distributive lattices and Heyting algebras are not of the same similarity type (see Section 4). But we can conclude that all of the formulas of Theorem VIII.2.1 and VIII.2.2 hold and that the set $\mathscr{D}\!s(L)$ of dense elements and the set $\mathscr{R}g(L)$ of regular elements are defined for every Heyting algebra L.

In addition to the above mentioned formulas, we have:

Theorem 3. *For elements x, y, z in a Heyting algebra:*

(i) $x(x \to y) \leq y$.

(ii) $xy \leq z \Leftrightarrow y \leq x \to z$.

(iii) $x \leq y \Leftrightarrow x \to y = 1$.

(iv) $y \leq x \to y$.

(v) $x \leq y \Rightarrow z \to x \leq z \to y$ and $y \to z \leq x \to z$.

(vi) $x \to (y \to z) = xy \to z$.

(vii) $x(y \to z) = x(xy \to xz)$.

(viii) $x(x \to y) = xy$.

(ix) $(x + y) \to z = (x \to z)(y \to z)$.

(x) $x \to yz = (x \to y)(x \to z)$.

(xi) $(x \to y)^* = x^{**}y^*$.

Proof. (i) Obvious from the definition.

(ii) ($\Rightarrow$) By definition.

($\Leftarrow$) $xy \leq x(x \to z) \leq z$.

(iii) $x \to y = 1 \Leftrightarrow 1 \leq x \to y \Leftrightarrow x \cdot 1 \leq y$.

(iv) $xy \leq y \Rightarrow y \leq x \to y$.

(v) $z(z \to x) \leq x \leq y$ so $z \to x \leq z \to y$; $x(y \to z) \leq y(y \to z) \leq z$ so $y \to z \leq x \to z$.

(vi) $xy(x \to (y \to z)) = y(x(x \to (y \to z))) \le y(y \to z) \le z$ so $x \to (y \to z) \le xy \to z$. Conversely, $xy(xy \to z) \le z$ so $x(xy \to z) \le y \to z$ and hence $xy \to z \le x \to (y \to z)$.

(vii) $x(x \to y) \le x$ and $(xy)x(y \to z) \le xz$ so $x(y \to z) \le xy \to xz$. Hence $x(y \to z) \le x(xy \to xz)$, Conversely, $x(xy \to xz) \le x$ and $yx(xy \to xz) \le xz \le z$ so $x(xy \to xz) \le y \to z$. Hence $x(xy \to xz) \le x(y \to z)$.

(viii) Clearly $x(x \to y) \le x, y$; also $xy \le x, x \to y$.

(ix) $x, y \le x + y$ implies $(x + y) \to z \le x \to z, y \to z$. Conversely, $(x + y)(x \to z)(y \to z) \le x(x \to z) + y(y \to z) \le z + z = z$ so $(x \to z)(y \to z) \le (x + y) \to z$.

(x) $yz \le y, z \Rightarrow x \to yz \le x \to y, x \to z$. Also $x(x \to y)(x \to z) \le xy(x \to z) \le yz$ so $(x \to y)(x \to z) \le x \to yz$.

(xi) $y \le x \to y$ implies $(x \to y)^* \le y^*$ and $x^* = x \to 0 \le x \to y$ implies $(x \to y)^* \le x^{**}$ so $(x \to y)^* \le x^{**}y^*$. Conversely $x^{**}y^*(x \to y) = x^{**}y^*(xy^* \to yy^*) = x^{**}y^*(xy^* \to 0) = x^{**}y^*(xy^* \to 0y^*) = x^{**}y^*(x \to 0) = x^{**}y^*x^* = 0$ so $x^{**}y^* \le (x \to y)^*$.

Exercise 4. Prove that in a free distributive lattice, $a \to b$ exists whenever $a \nleq b$.

Exercise 5. Prove that the free algebras in $\mathbf{D_{01}}$ are all Heyting algebras.

Exercise 6. In a Heyting algebra, define $[x_1] = x_1$ and for $n \geq 1, [x_1, \cdots, x_{n+1}] = [x_1, \cdots, x_n] \to x_{n+1}$. Show that for each $i \in \{1, \cdots, n\}$

$$x[x_1, \cdots, x_n] = x[x_1, \cdots, xx_i, \cdots, x_n].$$

Exercise 7. Prove that a Heyting algebra L is a Boolean algebra if and only if $\mathscr{D}_\mathscr{s}(L) = \{1\}$.

We now point out some of the relationships between pseudocomplemented distributive lattices and Heyting algebras (see also W. C. Nemitz [2] and T. Katrinák [3]).

Theorem 8. *Suppose that L is a distributive lattice with* 0, 1. *The following are equivalent:*

(i) *L is a Heyting algebra.*

(ii) *Every interval* $[a, b]$ *in L is pseudocomplemented.*

Moreover, if L is a Heyting algebra, then every interval $[a, b]$ *in L is a Heyting algebra.*

Proof. Assume (i). We will prove that $[a, b]$ is a Heyting algebra. Let $c, d \in [a, b]$. We will verify that $c \xrightarrow{[a,b]} d = (c \to d)b$. First, observe that $a \le (c \to d)b \le b$ so $(c \to d)b \in [a, b]$. Also $c((c \to d)b) = c(c \to d)b = cdb \le d$. If $x \in [a, b]$ and $cx \le d$ then $x \le c \to d$; but $x \le b$ so $x \le (c \to d)b$.

Next, suppose (ii) holds. We will prove $a \to b = a^{*[ab)}$. Indeed, $a^{*[ab)}$ is the largest $x \in [ab)$ such that $ax = ab$ so $a(a^{*[ab)}) = ab \leq b$. Suppose $ax \leq b$. Clearly, $ab \leq x + ab \leq 1$ and $a(x + ab) = ax + ab = ab$, so $x + ab \leq a^{*[ab)}$. Hence $x \leq a^{*[ab)}$.

Lemma 9. *If L is a Heyting algebra and*

$$\mathbf{x \to y + y \to x = 1}$$

is an identity in L then it is an identity in each interval of L.

Proof. Let $a, b \in [c, d]$. By Theorem 8, $a \xrightarrow{[c,d]} b + b \xrightarrow{[c,d]} a$ $= (a \to b)d + (b \to a)d = ((a \to b) + (b \to a))d = 1d = d = 1_{[c,d]}$.

Theorem 10. *Let L be a distributive lattice with* 0, 1. *Then L is a relative Stone algebra if and only if L is a Heyting algebra in which*

$$\mathbf{x \to y + y \to x = 1}$$

is an identity in L.

Proof. $(\Rightarrow)$ If L is a relative Stone algebra, then each interval is a Stone algebra and, in particular, is pseudocomplemented. By Theorem 8, L is a Heyting algebra. By the proof of Theorem 8, $a \to b + b \to a = a^{*[ab)} + b^{*[ab)} = (ab)^{*[ab)} = 1$.

$(\Leftarrow)$ By Theorem VIII.7.13 it suffices to prove $[a)$ is a Stone algebra for all $a \in L$. By Theorem 8, $[a)$ is a Heyting algebra. By hypothesis and Lemma 9, $\mathbf{x \to y + y \to x = 1}$ is an identity in $[a)$. So we must show that if L is a Heyting algebra that satisfies $\mathbf{x \to y + y \to x = 1}$ then L is a Stone algebra. Setting $y = x \to 0$ we have $1 = x \to (x \to 0) + (x \to 0) \to x = x \to 0 + (x \to 0) \to 0$ $= x^* + x^{**}$ since $(x \to 0) \to x = (x \to 0) \to 0$.

Exercise 11. Prove the following strengthened version of Theorem 10: Let L be a distributive lattice with 1. Then L is a relative Stone algebra if and only if $x \to y$ exists for each $x, y \in L$ and

$$\mathbf{x \to y + y \to x = 1}$$

is an identity in L.

From Exercise 11, it is seen that relative Stone algebras can be defined as the class $\mathbf{K}$ of algebras of the form $(L, (+, \cdot, \to))$ which satisfy identities that define distributive lattices and the identity $\mathbf{x \to y + y \to x = 1}$.

Let $\mathbf{K_n}$ be the equational subclass of $\mathbf{K}$ generated by $\mathbf{n}$. T. Hecht and T. Katrinák [1] have characterized all equational subclasses of $\mathbf{K}$. They are $\mathbf{K_1} \subset \mathbf{K_2}$ $\subset \cdots \subset \mathbf{K}$. Moreover, for $n \geq 2$, $L \in \mathbf{K_n}$ if and only if L satisfies the identity

$$\mathbf{x_1 \to x_2 + x_2 \to x_3 + \cdots + x_n \to x_{n+1} = 1}.$$

Closely related to $\mathbf{K}$ is the class of L-algebras. This is the equational subclass of Heyting algebras which satisfy the identity $\mathbf{x \to y + y \to x = 1}$ (i.e. they are relative Stone algebras with 0). Using the same argument as for **K**, T. Hecht and T. Katrinák [1] determined all equational subclasses of L-algebras. In [6], A. Horn studies these algebras with particular interest in their relation to logic. The free L-algebras are completely characterized by A. Horn in [7]. See also A. S. Troelstra [1].

3. Examples

There are many interesting examples of Heyting algebras in various fields of mathematics.

Example 1. Boolean algebras: $a \to b = \bar{a} + b$.

Example 2. Chains with 0, 1: $a \to b = \begin{cases} 1 \text{ if } a \leq b \\ b \text{ if } b < a. \end{cases}$

Example 3. As pointed out in the introduction, the algebraic system associated with the intuitionistic propositional calculus is a Heyting algebra.

Example 4. Let $\mathfrak{O}$ be the ring of open sets of a topological space X, with $0 = \varnothing$, $1 = X$. For $O_1, O_2 \in \mathfrak{O}$, $O_1 \to O_2 = \text{Int}\,(\widetilde{O_1} \cup O_2)$, since $O_1 \cap (\text{Int}\,(\widetilde{O_1} \cup O_2)) \subseteq O_1 \cap (\widetilde{O_1} \cup O_2) \subseteq O_2$, and if O is an open set such that $O \cap O_1 \subseteq O_2$, then $O \subseteq \widetilde{O_1} \cup O_2$ so $O = \text{Int}\ O \subseteq \text{Int}\,(\widetilde{O_1} \cup O_2)$ (see also Section 5).

Example 5. For a distributive lattice with 0, we have seen that $\mathfrak{J}(L)$ is in $\mathbf{B}_\omega$. It is, in fact, a Heyting algebra with $I \to J = \{x \in L : xi \in J \text{ for each } i \in I\}$. Indeed, first verify that $K = \{x \in L : xi \in J \text{ for each } i \in I\}$ is an ideal then note that for $k \in I \cap K$, $k = k \cdot k \in J$ so $I \cap K \subseteq J$. Finally, if $K_1 \in \mathfrak{J}(L)$, $I \cap K_1 \subseteq J$ and $k_1 \in K_1$ then for every $i \in I$, $k_1 i \in I \cap K_1 \subseteq J$ so $k_1 \in K$ and hence $K_1 \subseteq K$.

4. The class H

Let **H** be the class of algebras of the form $(L, (+, \cdot, 0, \to))$ which satisfy the following identities:

(1) A set of identities which define lattices with 0.

(2) $\mathbf{x(x \to y) = xy}$.

(3) $\mathbf{x(y \to z) = x(xy \to xz)}$.

(4) $\mathbf{z(xy \to x) = z}$.

Theorem 1. *The equational class* **H** *is exactly the class of Heyting algebras.*

Proof. We have seen that every Heyting algebra L, considered as an algebra with operations $+, \cdot, 0, \to$ satisfies (1)–(4). Conversely, suppose $(L, (+, \cdot, 0, \to))$ satisfies these identities. If $a, b \in L$, then $a(a \to b) = ab \leq b$, and if $ac \leq b$, then

$c(a \to b) = c(ca \to cb) = c(abc \to bc) = c$ so $c \le a \to b$. Thus, L is a Heyting algebra.

Remark. Recall that L is necessarily distributive.

Theorem 2. *Let L be a Heyting algebra. The map $x \mapsto (x]$ is an* **H**-*embedding of L into the complete Heyting algebra $\mathfrak{I}(L)$.*

Proof. Corollary II.9.4 shows that $x \mapsto (x]$ preserves $+$, $\cdot$, and 0 and is one-one. Also, since the map preserves order, $(x \to y] \subseteq (x] \to (y]$. If $I \in \mathfrak{I}(L)$, $(x] \cap I \subseteq (y]$ and $i \in I$, then $xi \in (x] \cap I \subseteq (y]$ so $i \le x \to y$ and thus $i \in (x \to y]$. We have, then, $I \subseteq (x \to y]$ so that $(x \to y] = (x] \to (y]$.

As in the proof of Theorem 2, we will often use the fact that a **D**-homomorphism f for which $f(0) = 0$ and $f(x) \to f(y) \le f(x \to y)$ holds for all x, y, is an **H**-homomorphism.

Exercise 3. Show that the notions of order isomorphism and **H**-isomorphism are the same in **H**.

The equational category associated with **H** is $\mathscr{H}$.

It follows from Example 3.1 that we can assign to every Boolean algebra $(L, (+, \cdot, {}^-, 0, 1))$ a well-defined Heyting algebra $(L, (+, \cdot, 0, \to))$ by disregarding the operations ${}^-$, 1 and adding the operation $\to$. By assigning to each morphism in $\mathscr{B}$ the same morphism in $\mathscr{H}$, we have therefore a functor from $\mathscr{B}$ to $\mathscr{H}$. It is clear that this functor is one-one on the objects. Using a slight extension of the procedure described in Section I.19, we will therefore often consider the category $\mathscr{B}$ as a subcategory of $\mathscr{H}$.

Lemma 4. *Let L be a Heyting algebra and F a filter in L. Then $\boldsymbol{\theta}(F)$ is an* **H**-*congruence relation. Moreover $\boldsymbol{\theta}$ establishes an isomorphism between the filters of L and $\mathfrak{C}(L)$.*

Proof. $\boldsymbol{\theta}(F)$ is a **D**-congruence relation. Suppose $(x, x'), (y, y') \in \boldsymbol{\theta}(F)$. Then there exist $u, v \in F$ such that $xu = x'u$, $yv = y'v$. Thus $uv(x \to y) = uv(xuv \to yuv) = uv(xu \to yv) = uv(x'u \to y'v) = uv(x' \to y')$ and $uv \in F$ so $(x \to y, x' \to y') \in \boldsymbol{\theta}(F)$.

Clearly, $F \subseteq G \Rightarrow \boldsymbol{\theta}(F) \subseteq \boldsymbol{\theta}(G)$. Suppose $\boldsymbol{\theta}(F) \subseteq \boldsymbol{\theta}(G)$. Let $x \in F$. Then $(x, 1) \in \boldsymbol{\theta}(F)$, since $x \cdot x = 1 \cdot x$ so $(x, 1) \in \boldsymbol{\theta}(G)$. There exists $g \in G$ such that $xg = g$. Hence $g \le x$, $g \in G$ implies $x \in G$.

Let θ be an **H**-congruence relation on L. Set $F = \{x \in L : (x, 1) \in \theta\}$. Then F is a filter and we will show that $\boldsymbol{\theta}(F) = \theta$. Now $(u, v) \in \boldsymbol{\theta}(F)$ implies $ux = vx$, $x \in F$ so $(x, 1) \in \theta$ and $(xu, u) \in \theta$, $(xv, v) \in \theta$. Hence $(u, v) \in \theta$. Conversely, $(u, v) \in \theta$ implies $(u \to v, 1) = (u \to v, v \to v) \in \theta$. So $u \to v \in F$. Similarly $v \to u \in F$ and hence $u((u \to v)(v \to u)) = uv = v((u \to v)(v \to u))$ proving that $(u, v) \in \boldsymbol{\theta}(F)$.

Theorem 4 shows that a homomorphism f is one-one if and only if $f(x) = 1$ implies $x = 1$, a fact which can also be obtained directly.

Theorem 5. An algebra L is subdirectly irreducible in **H** *if and only if* $L = L_1 \oplus \mathbf{1}$ *where* $L_1 \in \mathbf{H}$.

Remark. The fact that there are so many subdirectly irreducibles in **H** implies that relatively little information can be obtained from Birkhoff's subdirect product theorem.

Proof of Theorem 5. $\Rightarrow$ By Lemma 4, L has a smallest filter $F \neq [1)$, and therefore $|F| = 2$. So $F = [a)$ for some $a \neq 1$. If $x \in L$ and $x \neq 1$, then $[x) \supseteq [a)$ so $x \leq a$. By Theorem 2.8, $(a] \in \mathbf{H}$ and thus $L = L_1 \oplus \mathbf{1}$ for some $L_1 \in \mathbf{H}$.
($\Leftarrow$) Immediate from Lemma 4.

Exercise 6. Prove that **H** has the congruence extension property.

The characterization of the equational subclasses of an equational class is, in general, a difficult problem. Very useful tools—besides Birkhoff's subdirect product theorem—are the results contained in a paper by B. Jónsson [2] and which can be applied when the lattice of congruence relations, of each algebra in the class, is distributive. Although a detailed discussion of these methods is beyond the scope of this book, we note that the class **H** satisfies the condition of Jónsson. This fact has been exploited by A. Day [3] to obtain interesting results concerning some of the equational subclasses of **H**.

5. Some fundamental theorems

The purpose of this section is to extend some of the basic results of pseudo-complemented distributive lattices to Heyting algebras, and at the same time to develop some tools intrinsic to **H**.

Theorem 1. *Let* $L \in \mathbf{H}$ *and* $f: L \to \mathbf{2}$ *a function. The set* $\{x \in L : f(x) = 1\}$ *is a maximal filter if and only if f is an* **H**-*homomorphism.*

Proof. ($\Leftarrow$) Since $f(x^*) = f(x \to 0) = f(x) \to f(0) = f(x) \to \mathbf{0} = (f(x))^*$, f is a $\mathbf{B}_\omega$-homomorphism, so by Theorem VIII.3.3, $\{x \in L : f(x) = \mathbf{0}\}$ is a minimal prime ideal. Thus, by Theorem VIII.2.2, $\{x \in L : f(x) = \mathbf{1}\} = L \sim \{x \in L : f(x) = \mathbf{0}\}$ is a maximal filter.
($\Rightarrow$) Since $\{x \in L : f(x) = \mathbf{1}\}$ is maximal and hence prime, f is a $\mathbf{D}_{01}$-homomorphism, so we need only prove that $f(a) \to f(b) \leq f(a \to b)$. If $f(a) \to f(b) = \mathbf{1}$ then $f(a) \leq f(b)$. Suppose $f(a) = \mathbf{0}$. Then $a \notin \{x \in L : f(x) = \mathbf{1}\}$ which is maximal so by Theorem VIII.2.2, $a^* \in \{x \in L : f(x) = \mathbf{1}\}$. Hence $a^* \leq a \to b$ implies $f(a \to b) = \mathbf{1}$. On the other hand, if $f(a) = \mathbf{1}$ then also $f(b) = \mathbf{1}$ so $b \leq a \to b$ implies $f(a \to b) = \mathbf{1}$.

Exercise 2. Let $L \in \mathbf{H}$ and $f: L \to \mathbf{3}$ a function. Set $F_1 = \{x \in L : f(x) \geq \mathbf{1}\}$ and $F_2 = \{x \in L : f(x) = \mathbf{2}\}$. Prove that f is an **H**-homomorphism if and only if

(i) $F_2 \subseteq F_1$.

(ii) F_1 is a maximal filter.

(iii) F_2 is a prime filter.

(iv) If $x \notin F_1$ then $x^* \in F_2$.

(v) If $x, y \in F_1 \sim F_2$ then $x \to y \in F_2$.

Theorem 3. *If $L \in \mathbf{H}$, then $\mathscr{Rg}(L)$ is a Boolean algebra under the operations induced by the partial ordering of L. The operations are given by*

$$u \overset{\mathscr{Rg}(L)}{+} v = (u + v)^{**}$$

$$u \overset{\mathscr{Rg}(L)}{\cdot} v = uv$$

$$0_{\mathscr{Rg}(L)} = 0$$

$$u \xrightarrow{\mathscr{Rg}(L)} v = (u^* + v)^{**}$$

The map $r_L : L \to \mathscr{Rg}(L)$ defined by $x \mapsto x^{**}$ is an $\mathbf{H}$-homomorphism of L onto $\mathscr{Rg}(L)$. Furthermore $\mathscr{Ds}(L)$ is a filter and $\mathscr{Rg}(L) \cong L/\mathscr{Ds}(L)$.

Proof. By Theorem VIII.4.3 we only need to prove that r_L preserves $\to$. Indeed, $r_L(x) \xrightarrow{\mathscr{Rg}(L)} r_L(y) = x^{**} \xrightarrow{\mathscr{Rg}(L)} y^{**} = (x^{***} + y^{**})^{**} = (x^* + y^{**})^{**}$ $= (x^{**}y^*)^* = (x \to y)^{**} = r_L(x \to y)$ by Theorem 2.3(xi).

Corollary 4. *$\mathscr{B}$ is a reflective subcategory of $\mathscr{H}$ and the reflector preserves monomorphisms.*

Proof. From Theorem 3, the assignments $L \mapsto \mathscr{Rg}(L)$ and $\Phi_{\mathscr{Rg}}(L)$ $= r_L : L \to \mathscr{Rg}(L)$ have the properties $\mathscr{Rg}(L) \in \mathbf{B}$ and $\Phi_{\mathscr{Rg}}(L) \in [L, \mathscr{Rg}(L)]_{\mathscr{H}}$. To prove the extension property, let $f \in [L, L_1]_{\mathscr{H}}$, $L_1 \in \mathrm{Ob}\, \mathscr{B}$. We have seen from Theorem VIII.4.3 that $f|\mathscr{Rg}(L)$ preserves $+, \cdot, 0$. But it also preserves $\to$ for if $u, v \in \mathscr{Rg}(L)$ then $f|\mathscr{Rg}(L)(u \to v) = f(\bar{u}^{\mathscr{Rg}(L)} \overset{\mathscr{Rg}(L)}{+} v) = f((u^* + v)^{**})$ $= ((f(u))^* + f(v))^{**} = \overline{f(u)} + f(v) = f(u) \to f(v) = f|\mathscr{Rg}(L)(u) \to f|\mathscr{Rg}(L)(v)$. The remainder of the proof is essentially the same as the proof of Theorem VIII.4.9.

Theorem 5. (R. Balbes and A. Horn [4]) *L is injective in $\mathscr{H}$ if and only if L is a complete Boolean algebra.*

Proof. ($\Leftarrow$) Complete Boolean algebras are injective in $\mathscr{B}$ which, by Corollary 4, is a reflective subcategory of $\mathscr{H}$ and such that the reflector preserves monomorphisms. Hence Theorem I.18.6 yields the result.

($\Rightarrow$) Theorem 4.2 shows that there exists an embedding of any Heyting algebra into a complete Heyting algebra. Now if L is injective in $\mathscr{H}$, then L is a retract of a complete Heyting algebra. Thus L is itself complete.

To prove that L is a Boolean algebra it suffices to show $x^* = 0$ only for $x = 1$. Assume there exists $a \in L$ such that $a^* = 0$, $a < 1$. Hence $\{0, a, 1\}$ is a subalgebra of L. Let C be a chain in L which is maximal subject to the conditions of being a subalgebra of L and containing $\{0, a, 1\}$. Let C_1 be the chain $\mathbf{2} \oplus (C \sim \{0\})$. Then $1_{C,L}$ can be extended to an **H**-homomorphism $h: C_1 \to L$. By the maximality of C, $h(\mathbf{1}) \in C$. Now for each $x \in C \sim \{0\}$, $x \to h(\mathbf{1}) = h(x \to \mathbf{1}) = h(\mathbf{1}) \leq h(a) = a < 1$. So $x \nleq h(\mathbf{1})$. Therefore $h(\mathbf{1}) < x$ for all $x \in C \sim \{0\}$ and since $h(\mathbf{1}) \in C$, we have $h(\mathbf{1}) = 0$. But this is impossible, since $(h(\mathbf{1}))^* = h(\mathbf{1}^*) = h(\mathbf{0}) = 0$.

The category $\mathscr{H}$ does not have enough injectives since the existence of a one-one **H**-homomorphism $f: L_1 \to L_2$, with L_2 a Boolean algebra, implies that L_1 is also a Boolean algebra.

Concerning the equational subclasses of **H**, A. Day [4] has shown that if **K** is a non-trivial equational subclass of **H**, then **K** has enough injectives if and only if **K** is generated by $\{\mathbf{2}\}$ or **K** contains a non-Boolean injective. Moreover, if **K** does not have enough injectives, then the only injectives in **K** are the complete Boolean algebras. In fact, it is shown that the only equational subclasses with enough injectives are those generated by $\{\mathbf{2}\}$, $\{\mathbf{3}\}$ and $\{\mathbf{2}^2 \oplus \mathbf{1}\}$. A. Day also shows that the amalgamation property holds in very few equational subclasses of **H** but in fact a stronger form of the property holds in $\mathscr{H}$. For other recent results on the equational subclasses of **H**, see K. A. Baker [1] and A. Day [3].

We close this section with a representation theorem for Heyting algebras. First, recall that the open subsets of a topological space form a Heyting algebra. We will need the notion of a closure algebra (see Section II.4).

Definition 6. A Boolean algebra L with an additive closure operator ^{c} in which $0^c = 0$ is called a *closure algebra.* An element $a \in L$ is called *open* if $\bar{a} \in L^c$. The set of open elements is denoted by L^o.

Theorem 7. *If L is a closure algebra, then L^o is a Heyting algebra, under the partial ordering of L and is also a* $\mathbf{D}_{01}$*-subalgebra of L.*

Proof. Since $0^c = 0$ and $1^c = 1$, $\{0, 1\} \subseteq L^c$ so $\{0, 1\} \subseteq L^o$. Suppose $x, y \in L^o$. Then $\bar{x}, \bar{y} \in L^c$. Now $(\bar{x}\bar{y})^c = \bar{x}\bar{y}$ so $\overline{x + y} = \bar{x}\bar{y} \in L^c$ and hence $x + y \in L^o$. Since ^{c} is additive, $(\bar{x} + \bar{y})^c = \bar{x}^c + \bar{y}^c = \bar{x} + \bar{y}$ so $\overline{xy} = \bar{x} + \bar{y} \in L^c$ and hence $xy \in L^o$. Finally, let $x, y \in L^o$. We will verify that $x \xrightarrow{L^o} y$ exists and equals $\overline{(x\bar{y})^c}$. Indeed, by definition, $\overline{(x\bar{y})^c} \in L^o$. Also $x(\overline{(x\bar{y})^c}) \leq x(\overline{x\bar{y}}) = x(\bar{x} + y) \leq y$. Suppose $u \in L^o$ and $xu \leq y$. Then $u \leq \bar{x} + y \Rightarrow x\bar{y} \leq \bar{u} \Rightarrow (x\bar{y})^c \leq \bar{u}^c = \bar{u}$ since $u \in L_0$.

More remarkable is the fact that every Heyting algebra can be represented in this way.

Lemma 8. *Let $L \in \mathbf{H}$ and let $A = \{\bar{x} \in B_L : x \in L\}$. Then for each $a \in B_L$, $[a)_{B_L} \cap A$ has a smallest element and A is a $\mathbf{D}_{01}$-subalgebra of B_L.*

Proof. That A is a $\mathbf{D}_{01}$-subalgebra of B_L is immediate. Now let $a \in B_L$, then $\bar{a} = \prod_{i=1}^{n} (\bar{x}_i + y_i)$ for some $x_i, y_i \in L$. We will prove that $u = \overline{\prod_{i=1}^{n} (x_i \xrightarrow{L} y_i)}$ is the smallest member of $[a)_{B_L} \cap A$. Clearly, $u \in A$. Now $x_i \xrightarrow{L} y_i \leq \bar{x}_i + y_i$ for $1 \leq i \leq n$ so

$$u = \overline{\prod_{i=1}^{n} (x_i \xrightarrow{L} y_i)} \geq \overline{\prod_{i=1}^{n} (\bar{x}_i + y_i)} = a.$$

Next, suppose $y \in [a)_{B_L} \cap A$. Then $y \geq a \Rightarrow \bar{y} \leq \bar{a} \Rightarrow \bar{y} \leq \bar{x}_i + y_i$ for $1 \leq i \leq n$ $\Rightarrow \bar{y}x_i \leq y_i$ for $1 \leq i \leq n \Rightarrow \bar{y} \leq x_i \xrightarrow{L} y_i$ for $1 \leq i \leq n$ (note $\bar{y} \in L$) $\Rightarrow \bar{y} \leq \prod_{i=1}^{n} (x_i \xrightarrow{L} y_i) = \bar{u}$, thus $y \geq u$.

Theorem 9. (J. C. C. McKinsey and A. Tarski [2]) *Every Heyting algebra is isomorphic to the algebra of open elements of some closure algebra.*

Proof. By Lemma 8 and Theorem II.4.11 there exists a closure operator ^{c} on B_L such that $(B_L)^c = \{\bar{x} \in B_L : x \in L\}$, so $(B_L)^o = L$.

The reader interested in pursuing the topic of closure algebras is referred to J. C. C. McKinsey and A. Tarski [1], [2], H. Rasiowa [3], H. Rasiowa and R. Sikorski [1], R. Sikorski [4], [6], [7] and P. Wilker [1].

Exercise 10. Prove that coproducts exist in $\mathscr{H}$.

6. Free Heyting algebras

Although very little is known about the structure of free Heyting algebras with more than one free generator, the structure of $\mathscr{F}_{\mathbf{H}}(\mathbf{1})$ has been completely determined by I. Nishimura [1].

Theorem 1. *The free Heyting algebra on one free generator is the sublattice*

$$L = \left[\left(\bigcup_{n=1}^{\infty} \{(n+1, p) : n \leq p \leq n+3\}\right) \cup (\{(1,1), (1,2), (1,3)\})\right] \oplus \mathbf{1}$$

of $(N \times N) \oplus \mathbf{1}$ (see diagram).

Proof. To simplify notation, let $s_n = (n, n)$, $t_n = (n, n+1)$, $r_n = (n+1, n)$ and $u_n = (n, n+2)$. It is evident that L is a $\mathbf{D}_{01}$-subalgebra of $(N \times N) \oplus \mathbf{1}$. Although it is clear from the diagram of L that $u \to v$ exists for all $u, v \in L$, we provide the following table for later reference.

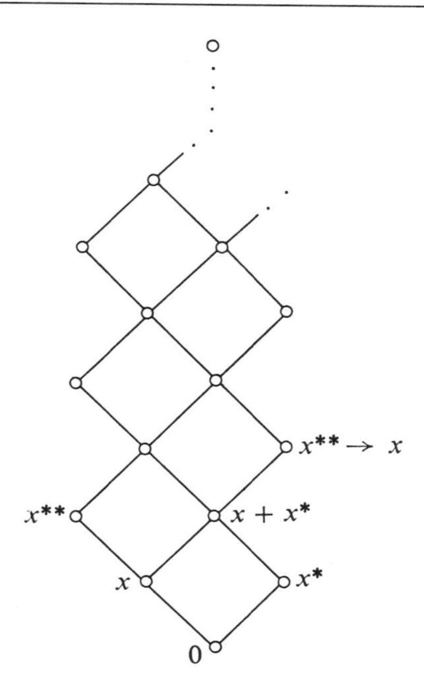

Table (1)

$\to$	s_p	t_p	r_p	u_p
s_n	1 if $p \geq n$ s_p if $p < n$	1 if $p \geq n$ u_p if $p = n - 1$ t_p if $p < n - 1$	1 if $p \geq n$ r_p if $p < n$	1 if $p \geq n$ u_p if $p < n$
t_n	1 if $p \geq n + 1$ r_p if $p = n$ s_p if $p < n$	1 if $p \geq n$ t_p if $p < n$	1 if $p \geq n + 1$ r_p if $p < n + 1$	1 if $p \geq n$ u_p if $p < n$
r_n	1 if $p \geq n + 1$ u_p if $p = n$ s_p if $p < n$	1 if $p \geq n + 1$ u_p if $p = n$ or $p = n - 1$ t_p if $p < n - 1$	1 if $p \geq n$ r_p if $p < n$	1 if $p \geq n + 1$ u_p if $p < n + 1$
u_n	1 if $p \geq n + 2$ r_p if $p = n + 1$ or $p = n$ s_p if $p < n$	1 if $p \geq n + 1$ r_{p+1} if $p = n$ t_p if $p < n$	1 if $p \geq n + 2$ r_p if $p < n + 2$	1 if $p \geq n$ u_p if $p < n$

From Table (1) we extract the following formulas, which show that $\{t_1\}$ **H**-generates L.

$$r_1 = t_1^*$$

$$s_n = t_{n-1} + r_{n-1} \qquad (n > 1)$$

$$t_n = u_{n-1} + s_n \qquad (n > 1)$$

$$r_n = u_{n-1} \to t_{n-1} \qquad (n > 1)$$

$$u_n = r_n \to s_n \qquad (n > 1)$$

For the extension property, let L_1 be a Heyting algebra and $x \in L_1$. Define a function $f: L \to L_1$ by $f(0) = 0, f(1) = 1, f(t_1) = x, f(r_1) = x^*, f(u_1) = x^{**}$ and for $n > 1: f(s_n) = f(t_{n-1}) + f(r_{n-1}), f(r_n) = f(u_{n-1}) \to f(t_{n-1}), f(t_n) = f(u_{n-1}) + f(s_n)$, $f(u_n) = f(r_n) \to f(s_n)$.

In order to show that f preserves order, it suffices to prove $f(u) \leq f(v)$ when v covers u. This can easily be verified by considering the cases $u = s_n$, $u = r_n$, $u = t_n$ and $u = u_n$ for each n. The computations are left to the reader.

For preservation of $+$ and $\cdot$, let

$$P_1 = \{(t_1, r_1), (r_1, u_1), (u_1, s_2), (u_1, r_2)\}$$

and

$$P_n = \{(s_n, u_{n-1}), (r_n, u_{n-1}), (t_n, r_n), (u_n, s_{n+1}), (r_n, u_n), (u_n, r_{n+1})\}$$

for $n \geq 2$. Now, since f preserves order, we need only verify that it preserves sums and products of these pairs. This is easy for P_1, so proceeding by induction, assume that f preserves sums and products of all pairs in P_{n-1}. Then f has this property for the first two pairs in P_n by hypothesis. The sums of the second two pairs are preserved by definition, and since $t_n \leq u_n, f(r_n) + f(u_n) = f(r_n) + f(t_n) + f(u_n)$ $= f(s_{n+1}) + f(u_n) = f(t_{n+1}) = f(r_n + u_n)$. Similarly, $f(u_n) + f(r_{n+1})$ $= f(u_n + r_{n+1})$. The only difficult remaining product is verified as follows:

$$\begin{aligned}
f(t_n)f(r_n) &= (f(u_{n-1}) + f(s_n))(f(_{n-1}) \to f(t_{n-1})) \\
&= f(u_{n-1})(f(u_{n-1}) \to f(t_{n-1})) + f(s_n)(f(u_{n-1}) \to f(t_{n-1})) \\
&= f(u_{n-1})f(t_{n-1}) + f(s_n)(f(s_n)f(u_{n-1}) \to f(s_n)f(t_{n-1})) \\
&= f(u_{n-1}t_{n-1}) + f(s_n)(f(s_n u_{n-1}) \to f(s_n t_{n-1})) \\
&= f(t_{n-1}) + f(s_n)(f(t_{n-1}) \to f(t_{n-1})) \\
&= f(t_{n-1}) + f(s_n) = f(s_n) = f(t_n r_n).
\end{aligned}$$

The verification that f preserves $\to$ will be accomplished by showing that Table (2)—obtained from Table (1) by replacing each element of L by its image under f—is valid. Now, the verification of this for the first entry in each box is trivial, since f preserves order and $f(1) = 1$. For what remains we use induction on n. The case $n = 1$ is easily verified by showing

$$x^{**} \to x = x^{**} \to (x + x^*)$$

Now, assume the induction hypothesis. If $c \in L$, then

$$\begin{aligned}
(3)\ f(s_n) \to f(c) &= f(t_{n-1} + f(r_{n-1})) \to f(c) \\
&= (f(t_{n-1}) \to f(c))(f(r_{n-1}) \to f(c)) \\
&= f(t_{n-1} \to c) f(r_{n-1} \to c) \\
&= f((t_{n-1} + r_{n-1}) \to c) \\
&= f(s_n \to c).
\end{aligned}$$

Similarly,

$$(4)\ f(t_n) \to f(c) = f(t_n \to c)$$

By (3) and (4), the first two rows of Table (2) are verified. The remaining formulas in the third row require separate treatment. As an example, we verify that $f(r_n) \to f(r_p) = f(r_p)$ for $p < n$. Indeed,

$$\begin{aligned} f(u_{p-1})(f(r_n) \to f(r_p)) &= f(u_{p-1})(f(u_{p-1}r_n) = f(u_{p-1}r_p)) \\ &= f(u_{p-1})(f(u_{p-1}) = f(t_{p-1})) \\ &= f(u_{p-1})f(t_{p-1}) = f(t_{p-1}). \end{aligned}$$

So by induction

$$f(r_n) \to f(r_p) \leq f(u_{p-1}) \to f(t_{p-1}) = f(r_p).$$

Finally, using the symmetry of the lattice diagram, the proofs of the formulas in the fourth row of Table (2) are patterned after those in the third row. This completes the construction.

A recent paper of A. Urquhart [1] gives a description—though less transparent that that just given for $\mathscr{F}_{\mathbf{H}}(\mathbf{1})$—for $\mathscr{F}_{\mathbf{H}}(\mathbf{n})$, $n < \aleph_0$. See also P. Köhler [1].

Exercise 2. (R. Balbes and A. Horn [4]) Show that the subalgebra of $\mathscr{F}_{\mathbf{H}}(\{x, y\})$ generated by $\{x^{**}y^{*}, x^{**}y^{**}\}$ is infinite. (Hint: Let S be an infinite set and let $(a_n)_{n=0,1,\ldots}$, $(b_n)_{n=0,1,\ldots}$, $(c_n)_{n=0,1,\ldots}$ be non-empty sets satisfying

$$a_i \cap b_j = a_i \cap c_j = b_i \cap c_j = \varnothing \text{ for all } i, j$$

$$a_i \cap a_j = b_i \cap b_j = c_i \cap c_j = \varnothing \text{ for all } i \neq j,$$

and

$$S = \bigcup_{n=0}^{\infty} (a_n \cup b_n \cup c_n).$$

Set

$$s_n = \bigcup_{i=0}^{n-1} (a_i \cup b_i \cup c_i)\,,\; x_0 = a_0,\; y_0 = b_0,\; z_0 = c_0$$

$$x_{n+1} = a_{n+1} \cup b_n \cup c_n \cup s_n,$$

$$y_{n+1} = a_n \cup b_{n+1} \cup c_n \cup s_n$$

and

$$z_{n+1} = a_n \cup b_n \cup c_{n+1} \cup s_n.$$

First prove that

$$\{x_n : n = 0, 1, \cdots\} \cup \{y_n : n = 0, 1, \cdots\} \cup \{z_n : n = 0, 1, \cdots\}$$

form a basis for a topology on S. Next prove that in the algebra L of open sets of S that $x_n \to (y_n \cup z_n) = x_{n+1}$, $y_n \to (x_n \cup z_n) = y_{n+1}$, $z_n \to (x_n \cup y_n) = z_{n+1}$, and that L is generated by $\{x_0, y_0, z_0\}$. Then observe that $x_0 = y_0^{*}z_0^{*}$, $y_0 = x_1^{**}y_1^{*}$, $z_0 = x_1^{**}y_1^{**}$ and therefore $\{x_1^{**}y_1^{**}, x_1^{**}y_1^{*}\}$ generate L.)

7. Finite weakly projective Heyting algebras

In order to describe the finite weakly projective Heyting algebras, it is convenient to be able to combine Heyting algebras in the following way.

Definition 1. Let $n > 0$ be a fixed integer. For each r, $0 \leq r < n$ suppose L_r is a Heyting algebra. Define $L_0 \dagger \cdots \dagger L_{n-1}$ to be the lattice with nodes $0 = a_0 < \cdots < a_n = 1$ and such that $[a_i, a_{i+1}] \cong L_i$ for $i = 0, \cdots, n-1$.

In practice we identify L_i with $[a_i, a_{i+1}]$ so that $a_i = 0_{L_i}$ and $a_{i+1} = 1_{L_i}$ for $i = 0, \cdots, n-1$. Thus, for example, $L \oplus \mathbf{1} = L \dagger \mathbf{2}$; see also Figure 1.

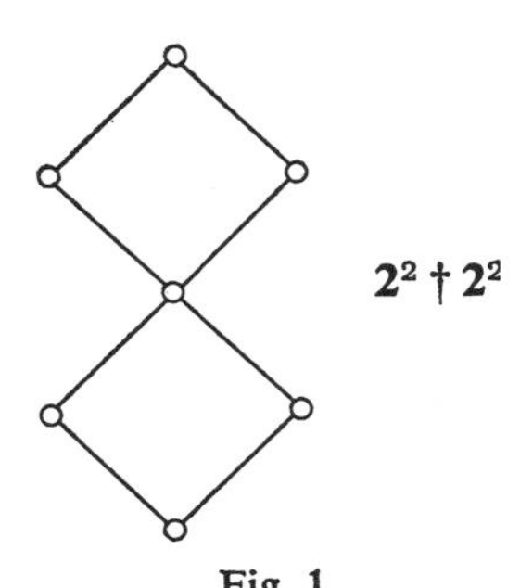

Fig. 1

Clearly, $L = L_0 \dagger \cdots \dagger L_{n-1}$ has a 1 and, if $x \nleq y$ in L, then

$$x \xrightarrow{L} y = \begin{cases} y & \text{if } x \in L_r, y \in L_s, r > s. \\ x \xrightarrow{L_r} y & \text{if } x, y \in L_r \end{cases}$$

Exercise 2. Show that if $f_1 : L_1 \to L_1'$ is an **H**-isomorphism and $f_2 : L_2 \to L_2'$ is an **H**-homomorphism, then the function $f: L_1 \dagger L_2 \to L_1' \dagger L_2'$ defined by $f | L_i = f_i$ is an **H**-homomorphism.

All of the results in this section are from R. Balbes and A. Horn [4].

Lemma 3. *Let L be a Heyting algebra with a node a such that $[a) \cong \mathbf{2}$ or $[a) \cong \mathbf{2}^2$. Suppose L_1 is a Heyting algebra and $f\colon (a] \to L_1$ is a* **D***-homomorphism such that $f(x) \to f(y) = f(x \to y)$ when $x \nleq y$. Then for any $a_1 \in L_1$ there is an extension g of f to L such that g is a* **D***-homomorphism and $g(x \to y) = g(x) \to g(y)$ whenever $x \nleq y$. Moreover if $[a) \cong \mathbf{2}$ the extension can be defined so that $g(1) = a_1 + a_1 \to f(a)$ and if $[a) \cong \mathbf{2}^2$, the extension is defined so that if b, c are the atoms of $\mathbf{2}^2$,* then *$g(b) = (a_1 \to f(a)) \to f(a)$ and $g(c) = a_1 \to f(a)$.*

Proof. First, suppose $[a) \cong \mathbf{2}$. Then g is a **D**-homomorphism, since $g(1) = a_1 + a_1 \to f(a) \geq f(a)$. It remains to show that $g(1) \to g(u) = g(1 \to u)$ for $u \in (a]$:

$$\begin{aligned} g(1) \to g(u) &= (a_1 + a_1 \to f(a)) \to f(u) \\ &= (a_1 \to f(u))((a_1 \to f(a)) \to f(u)) \\ &= (a_1 \to f(u))((a_1 \to f(u))(a_1 \to f(a)) \to f(u)) \\ &= (a_1 \to f(u))((a_1 \to f(u)) \to f(u)) \\ &= (a_1 \to f(u)) f(u) = f(u) = g(u) = g(1 \to u). \end{aligned}$$

Next, suppose $[a) \cong \mathbf{2}^2$. Define $g(1) = g(b) + g(c)$. Since $g(b)g(c) = f(a) = g(a) = g(bc)$, g is a **D**-homomorphism. Now if $b \nleq u$ then either $u = c$ or $u \in (a]$. In the first case

$$\begin{aligned} g(b) \to g(c) &= ((a_1 \to f(a)) \to f(a)) \to (a_1 \to f(a)) \\ &= a_1((a_1 \to f(a)) \to f(a)) \to f(a) \\ &= a_1 \to f(a) = g(c) = g(b \to c). \end{aligned}$$

In the other case $u \leq a$ implies,

$$\begin{aligned} &[a_1 \to f(a)][((a_1 \to f(a)) \to f(a)) \to f(u)] \\ &\quad = [a_1 \to f(a)][(a_1 \to f(a))(a_1 \to f(a) \to f(a)) \to f(u)] \\ &\quad = (a_1 \to f(a))(f(a) \to f(u)) = (a_1 \to f(a))(f(a \to u)) \\ &\quad = (a_1 \to f(a))\,(f(u)) \leq f(u) \end{aligned}$$

so

$$((a_1 \to f(a)) \to f(a)) \to f(u) \leq (a_1 \to f(a)) \to f(u).$$

Similarly, $((a_1 \to f(a)) \to f(a)) \to f(u) \leq a_1 \to f(a)$ so

$$\begin{aligned} g(b) \to g(u) &= ((a_1 \to f(a)) \to f(a)) \to f(u) \\ &\leq [a_1 \to f(a)][(a_1 \to f(a)) \to f(u)] \\ &\leq f(u) = g(u) = g(b \to u). \end{aligned}$$

The fact that the reverse inequality holds, follows as usual, from the preservation of products. Similarly, $g(c) \to g(u) = g(c \to u)$ for $c \nleq u$. Also,

$$\begin{aligned} g(c) \to g(b) &= (a_1 \to f(a)) \to ((a_1 \to f(a)) \to f(a)) \\ &= (a_1 \to f(a)) \to f(a) = g(b) = g(c \to b). \end{aligned}$$

Finally,

$$\begin{aligned} g(1) \to g(b) &= (g(b) + g(c)) \to g(b) \\ &= (g(b) \to g(b))(g(c) \to g(b)) \\ &= g(c) \to g(b) = g(c \to b) \\ &= g(b) = g(1 \to b), \end{aligned}$$

and similarly, $g(1) \to g(c) = g(1 \to c)$. Suppose, $u \in (a]$, then

$$\begin{aligned} g(1) \to g(u) &= (g(b) + g(c)) \to g(u) \\ &= (g(b) \to g(u))(g(c) \to g(u)) \\ &= g((b \to u)(c \to u)) \\ &= g((b + c) \to u) = g(1 \to u). \end{aligned}$$

Theorem 4. *Let $n > 0$ be an integer and $(L_i)_{i=0,\cdots,n-1}$ a family of Heyting algebras such that $L_i \cong \mathbf{2}$ or $L_i \cong \mathbf{2}^2$ for $0 \le i \le n-1$. Then $L = (L_0 \dagger \cdots \dagger L_{n-1}) \oplus \mathbf{1}$ is weakly projective.*

Proof. Let $0 = a_0 < a_1 < \cdots < a_n < 1$ be the nodes of L and let $J = \{j : L_j \cong \mathbf{2}^2\}$. We will prove that L is a retract of the free algebra

$\mathscr{F}_{\mathbf{H}}(X)$ where $X = \{x_i : i = 0, \cdots, n\} \cup \{y_j : j \in J\}$.

Let $f: \mathscr{F}_{\mathbf{H}}(X) \to L$ be the onto $\mathbf{H}$-homomorphism such that $f(x_i) = a_i$, $0 \le i \le n$ and $f(y_j)$ is one of the atoms of L_j for each $j \in J$. Define $g_0: (a_0] \to \mathscr{F}_{\mathbf{H}}(X)$ by $g_0(0) = 0$. Suppose $g_i: (a_i] \to \mathscr{F}_{\mathbf{H}}(X)$ has been defined and preserves $+$, $\cdot$, 0 and $x \to y$ whenever $x \nleq y$, and $f \circ g_i = 1_{(a_i],L}$. By Lemma 3, g_i can be extended to a function $g_{i+1}: (a_{i+1}] \to \mathscr{F}_{\mathbf{H}}(X)$ preserving $+$, $\cdot$, 0 and $x \to y$ whenever $x \nleq y$. If $[a_i, a_{i+1}] \cong \mathbf{2}$, Lemma 3 also shows that g_{i+1} can be defined so that $g_{i+1}(a_{i+1}) = x_{i+1} + x_{i+1} \to g_i(a_i)$. In this case,

$$\begin{aligned} f \circ g_{i+1}(a_{i+1}) &= f(x_{i+1}) + f(x_{i+1}) \to f(g_i(a_i)) \\ &= a_{i+1} + a_{i+1} \to a_i = a_{i+1}; \end{aligned}$$

and we have $f \circ g_{i+1} = 1_{(a_{i+1}],L}$. If $[a_i, a_{i+1}] \cong \mathbf{2}^2$ we can define g_{i+1} so that $g_{i+1}(f(y_i)) = (y_i \to g_i(a_i)) \to g_i(a_i)$ and $g_{i+1}(f(y_i) \to a_i) = y_i \to g(a_i)$ since $f(y_i) \to a_i$ is the other atom of $[a_i, a_{i+1}]$. But here also we obtain $f \circ g_{i+1} = 1_{(a_{i+1}],L}$ since

$$\begin{aligned} f \circ g_{i+1}(f(y_i)) &= (f(y_i) \to f(g_i(a_i))) \to f \circ g_i(a_i) \\ &= (f(y_i) \to a_i) \to a_i = f(y_i), \end{aligned}$$

$$f \circ g_{i+1}(f(y_i) \to a_i) = f(y_i) \to f \circ g_i(a_i) = f(y_i) \to a_i,$$

and

$$\begin{aligned} f \circ g_{i+1}(a_{i+1}) &= f \circ g_{i+1}(f(y_i) + f(y_i) \to a_i) \\ &= f \circ g_{i+1}(f(y_i)) + f \circ g_{i+1}(f(y_i) \to a_i) \\ &= f(y_i) + f(y_i) \to a_i = a_{i+1}. \end{aligned}$$

Finally, extend $g_n: (a_n] \to \mathscr{F}_{\mathbf{H}}(X)$ to $g: L \to \mathscr{F}_{\mathbf{H}}(X)$ by setting $g(1) = 1$. It is easy to see that g is an $\mathbf{H}$-homomorphism and that $f \circ g = 1_L$. Hence L, being a retract of a free algebra, is projective.

Lemma 5. *Suppose $h \in [L_1, L_2]_{\mathscr{H}}$, $f \in [L_3, L_2]_{\mathscr{H}}$ and*

(i) f is onto.

(ii) $f(u) = 0 \Rightarrow u = 0$.

If there exists an algebra L_0 such that $L_0 \dagger L_1$ is weakly projective then there is a morphism $g \in [L_1, L_3]$ such that $f \circ g = h$.

Proof. By Exercise 2 there are homomorphisms $f': L_0 \dagger L_3 \to L_0 \dagger L_2$ and $h': L_0 \dagger L_1 \to L_0 \dagger L_2$ such that $f'|L_3 = f$, $h'|L_1 = h$, $h'|L_0 = 1_{L_0}$ and $f'|L_0 = 1_{L_0}$.

Since f' is onto and $L_0 \dagger L_1$ is weakly projective there exists $g': L_0 \dagger L_1 \to L_0 \dagger L_3$ in $\mathscr{H}$ such that $f' \circ g' = h'$. Now $= f' \circ g'(0_{L_1}) = h'(0_{L_1}) = 0_{L_2}$ so by (ii), $g'(0_{L_1}) = 0_{L_3}$. Hence $g'|L_1 : L_1 \to L_3$ defines a morphism in $\mathscr{H}$ such that $f \circ g = h$.

The next lemma enables us to give a complete description of finite weakly projective Heyting algebras.

Lemma 6. *Let L be a projective Heyting algebra. Then*

(i) 1 *is join irreducible.*

(ii) *If L is finite and $a \in L$ is a node then there exist at most two elements which cover a.*

(iii) *If L has a node a such that a_1, a_2 are distinct elements which cover a then $a_1 + a_2$ is a node.*

Proof. (i) Define a homomorphism $f: L \oplus \mathbf{1} \to L$ by $f|L = 1_L$ and $f(1) = 1$. Since f is onto, there exists $g: L \to L \oplus \mathbf{1}$ in $\mathscr{H}$ such that $f \circ g = 1$. Now if $x, y \in L$ and $x + y = 1$ then $g(x) + g(y) = g(x + y) = 1$ is join irreducible in $L \oplus \mathbf{1}$ so, say $g(x) = 1$. Hence $x = f \circ g(x) = f(1) = 1$. (ii) Consider the free Heyting algebra $\mathscr{F}_{\mathbf{H}}(x, y)$. Now $b_1 = x^{**}y^{**}$, $b_2 = x^{**}y^*$, $b_3 = x^*y^{**}$, $b_4 = x^*y^*$ are the atoms of $\mathscr{Rg}(\mathscr{F}_{\mathbf{H}}(x, y))$. Note that the canonical homomorphism r of $\mathscr{F}_{\mathbf{H}}(x, y)$ onto $\mathscr{Rg}(\mathscr{F}_{\mathbf{H}}(x, y))$ is such that $r(u) = 0 \Rightarrow u = 0$. Next, suppose that there are three distinct atoms a_1, a_2, a_3 in $[a)$. Let $a_4 = a_3$. Since $[a_i)$ is a maximal filter in $[a)$, there is a homomorphism f: $[a) \to \mathbf{2} \times \mathbf{2} \times \mathbf{2}$ such that $f(u)(i) = f_{[a_i)}(u)$. Equivalently, there exists a homomorphism f: $[a) \to \mathscr{Rg}(\mathscr{F}_{\mathbf{H}}(x, y))$ such that $f(u) = \sum \{b_i : a_i \leq u\}$. Since $(a] \dagger [a) = L$ is weakly projective, Lemma 5 implies the existence of a homomorphism g: $[a) \to \mathscr{F}_{\mathbf{H}}(x, y)$ such that $r \circ g = f$. Now

$$g((a_1 \to a) \to a) = (g(a_1)^{**} = r \circ g(a_1) = f(a_1) = b_1$$

and similarly $g((a_2 \to a) \to a) = b_2$. By Exercise 6.2, $g[[a)]$ is infinite. It follows that L is also infinite. (iii) Suppose $a_1 + a_2$ is not a node. Then there exists $z \in L$ which is incomparable with $a_1 + a_2$. But a is a node, so $a < z$. Since a_1 covers a, we may assume, without loss of generality, that $a_1 < z$ and $a_2 \nleq z$. Let y be an element such that $y \leq z$ and which covers a_1 (see Figure 1).

We first verify the hypothesis of Exercise 5.2 in order to show that there exists a homomorphism h_1: $[a) \to \mathbf{3}$ such that $h_1(a_1) = \mathbf{1}$. Indeed, let $F_1 = [a_1)$, $F_2 = [y)$. Then (i) $F_2 \subseteq F_1$ and (ii) F_1 is a maximal filter since a_1 is an atom in $[a)$. For (iii) suppose $u, v \in [a)$ and $y \leq u + v$ but $y \nleq u, y \nleq v$. Now $a_2 \leq yu$ or $a_1 \leq yu$ or $yu = 0$. The first case yields the contradiction $a_2 \leq y \leq z$, and if $a_1 \leq yu < y$ then $a_1 < y$ implies $a_1 = yu$ so $yu \leq a_1$. Similarly, $yv \leq a_1$ so $y = yu + yv \leq a_1$, a contradiction. (iv) Assume $u \in [a) \sim F_1$. Then $a_1 \nleq uy$ and $a_2 \nleq uy$ so $uy = 0$, i.e. $y \leq u^*$. (v) If $u, v \in F_1 \sim F_2$ then, as before, $yu \leq a_1$ so $yu \leq a_1 \leq v$ and $y \leq u \to v$. Define a homomorphism f from $\mathscr{F}_{\mathbf{H}}\{x\}$ into $\mathbf{3} \times \mathbf{2}$ by $f(x) = (h_1(a_1), \mathbf{0})$. Since $\{f(x)\}$ generates $\mathbf{3} \times \mathbf{2}$, f is onto. It follows from Figure 1 that $f(u) = (\mathbf{0}, \mathbf{0})$ only for $u = 0$. By Lemma 5, there exist $g \in [[a), \mathscr{F}_{\mathbf{H}}\{x\}]_{\mathscr{H}}$ and $f \circ g = h$. Now $f \circ g(a_1) = h(a_1)$

$= (h_1(a_1), 0)$. But, again by the figure, we see that $g(a_1)$ must be x. Finally, $\{x\}$ generates an infinite algebra $\mathscr{F}_{\mathbf{H}}\{x\}$ so $[a)$ must be infinite, a contradiction.

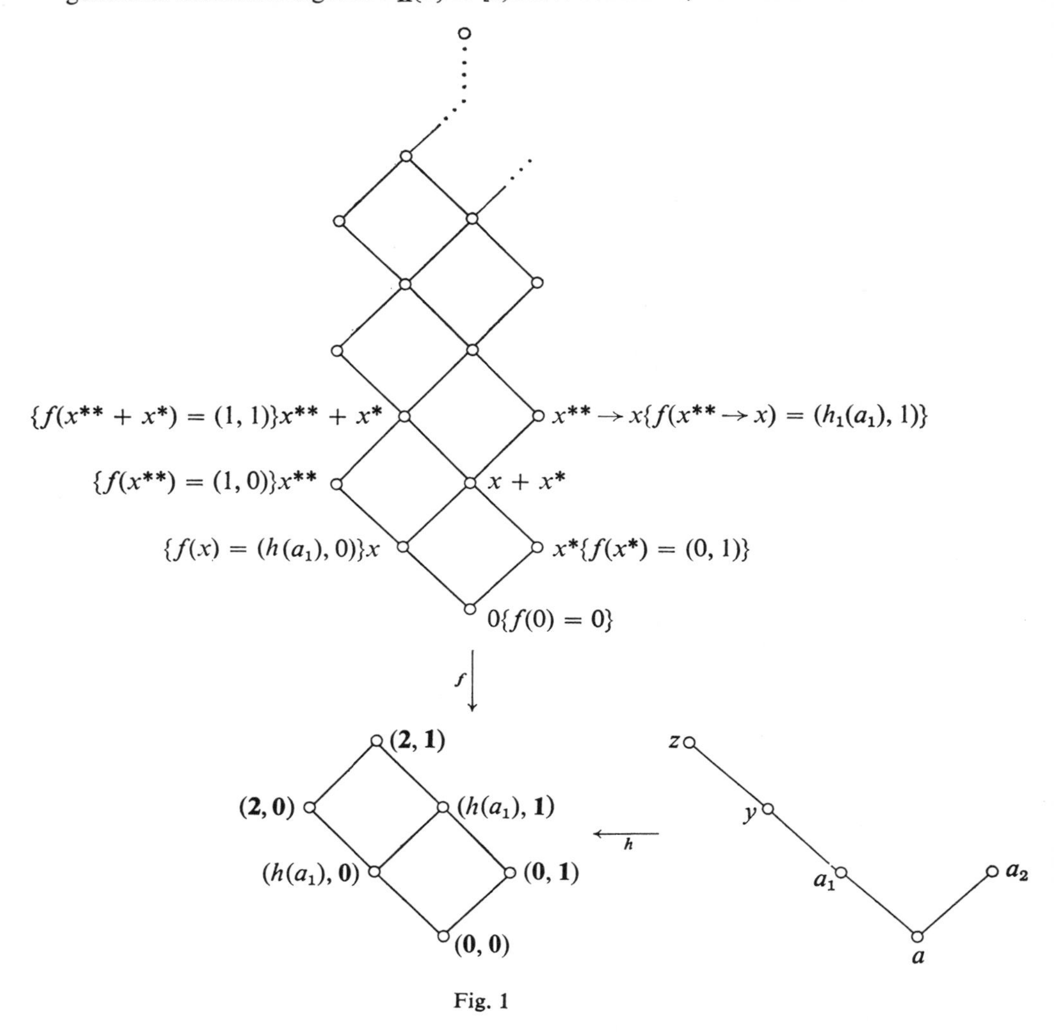

Fig. 1

Theorem 7. (R. Balbes and A. Horn [4]) *A finite Heyting algebra L is weakly projective if and only if $L \cong L_0 \dagger \cdots \dagger L_n$ where $L_n \cong \mathbf{2}$ and for each $0 \leq i \leq n - 1$, $L_i = \mathbf{2}$ or $L_i = \mathbf{2}^2$.*

Proof. The conditions are sufficient by Theorem 4. Conversely, let $0 = a_0 < a_1 < \cdots < a_{n+1} = 1$ be the nodes of L and set $L_i = [a_i, a_{i+1}]$ for $i = 0, \cdots, n$. By Lemma 6 each a_i has at most two covers. If a_i has one cover a then $a = a_{i+1}$. In this case $[a_i, a_{i+1}] \cong \mathbf{2}$. If a_i has two covers b_1, b_2 then $b_1 + b_2$ is a node so $b_1 + b_2 = a_{i+1}$, so here $[a_i, a_{i+1}] \cong \mathbf{2}^2$. Finally, by Lemma 6(i), 1 is join irreducible so $[a_n, 1] \cong \mathbf{2}$.

X

Post Algebras

1. Introduction

In the next two chapters we will discuss distributive lattices which are the algebras corresponding to n-valued classical propositional logic for $n \geq 2$, just as Boolean algebras are the algebras corresponding to 2-valued logic. These logics were developed for $n \geq 2$ in 1921 by E. L. Post [1] and by J. Lukasiewicz [1]. The subject of this chapter will be Post algebras, whereas Lukasiewicz algebras will be discussed in Chapter XI. A set of axioms for Post algebras was formulated first in 1942 by P. C. Rosenbloom [1]. This system of axioms was improved considerably in 1960, by G. Epstein [1] who also developed a representation theory for Post algebras. This work was further developed by T. Traczyk e.g. [1]. Since then many other workers have expanded the field. In addition, in recent years generalizations of Post algebras have been suggested and examined: C. C. Chang and A. Horn [1]; T. Traczyk [3]; Ph. Dwinger [5], [6]; G. Rousseau [1]; R. Balbes and Ph. Dwinger [2], [3]; T. P. Speed [4]; H. Sawicka [1]; G. Epstein and A. Horn [1]; T. Katrinák and A. Mitschke [1]. For the logical aspects of Post algebras, see H. Rasiowa [3].

Perhaps one of the most striking features of Post algebras is its representation theory, since it provides a very simple way to describe these algebras as the lattices of continuous functions from Boolean spaces to finite discrete chains. In this respect they can also be considered as natural generalizations of Boolean algebras which indeed can be defined as the lattices of continuous functions from Boolean spaces to the discrete two elements chain (cf. Section VII.4). Another interesting feature of Post algebras of order $n \geq 2$ is that they are coproducts (in the category $\mathscr{D}_{01}$) of Boolean algebras and n-element chains. Thus, much of Sections 4 and 6 of Chapter VII will be used in this chapter.

As with other classes of algebras we have seen, Post algebras can be studied from both a categorical point of view and from an algebraic point of view. It turns out to be practical to start our study by considering the category $\mathscr{P}$ of Post algebras and suitably defined morphisms, as a subcategory of the category $\mathscr{D}_{01}$. Thus at this point we do not yet give Post algebras the status of an "equational class" but rather consider them as special objects in $\mathscr{D}_{01}$. The setting will be the category $\mathscr{D}_{01}$.

Later in this chapter we will consider the class $\mathbf{P_n}$ of Post algebras of order $n \geq 2$, where n is now fixed, and show that $\mathbf{P_n}$ can be considered as an equational class of algebras with which we will associate, in the usual way, the equational category $\mathscr{P}_{\mathbf{n}}$. At the end of the chapter some remarks will be made on generalized Post algebras and we will consider, in particular, Stone algebras of order n.

2. Definitions and basic properties

There are several ways to define Post algebras. We will give one of these definitions and then state some theorems, which at the same time provide equivalent definitions.

Definition 1. Let $L \in \mathbf{D}_{01}$. L is a *Post algebra* if the following conditions are satisfied:

(i) L has a subset $C = \{0 = c_0 \leq c_1 \leq \cdots \leq c_{n-1} = 1\}$ where $n \geq 2$ and such that each $x \in L$ can be written as $x = \sum_{i=1}^{n-1} a_i c_i$, $a_i \in \mathscr{C}(L)$ for $1 \leq i \leq n-1$.

(ii) If $a \in \mathscr{C}(L)$ and $ac_i \leq c_{i-1}$ for some i, $1 \leq i \leq n-1$ then $a = 0$.

The class of Post algebras will be denoted by $\mathbf{P}$ and we note that $\mathbf{1} \in \mathbf{P}$. We also note that if $L \neq \mathbf{1}$ then the c_i are necessarily distinct. Indeed $c_i = c_{i-1}$, for some $1 \leq i \leq n-1$, contradicts (ii) since $0 \neq 1$.

Exercise 2. Prove that if $L \neq \mathbf{1}$, then $c_i \notin \mathscr{C}(L)$ for $1 \leq i \leq n-2$.

The chain C in Definition 1 will be called a *chain of constants* of L and we will prove that C is uniquely determined. For this, we first prove a theorem that will also provide us with a useful equivalent definition of Post algebras. We will adopt here the coproduct convention as established in Section VII.2.

Theorem 3. *Let L be a Post algebra. Then $L \doteq \mathscr{C}(L) + C$ where C is a chain of constants of L. Conversely, if $L \doteq B + C$, $B \in \mathbf{B}$, $C \in \mathbf{C}_{01}$, C finite, then L is a Post algebra whose center is B and with C as a chain of constants.*

Proof. First, assume that L satisfies Definition 1. Certainly, $[\mathscr{C}(L) \cup C]_{\mathbf{D}_{01}} = L$. To show that the condition (3) of VII.4 is satisfied, assume $ac_i \leq c_j$ for $a \in \mathscr{C}(L)$, $\{c_i, c_j\} \subseteq C$. If $c_i \nleq c_j$ then $c_j < c_i$ and then $i \geq 1$ and $c_j \leq c_{i-1}$. It follows that $ac_i \leq c_{i-1}$ and thus $a = 0$. Next, assume $L \doteq B + C$. If $C = \mathbf{1}$, then $L = \mathbf{1}$. Let $C = \{0 = c_0 < c_1 < \cdots < c_{n-1} = 1\}$, $n \geq 2$. By Theorem VII.3.2, B is the center of L, and it follows from Theorem VII.4.1 that (i) of Definition 1 is satisfied. For (ii) note that $ac_i \leq c_{i-1}$ for $a \in B$ and some i, $1 \leq i \leq n-1$ implies $a = 0$ by condition (3) of VII.4, since $c_{i-1} < c_i$.

Theorem 3 has an important implication. If L is a Post algebra with a chain of constants $C = \{0 = c_0 \leq c_1 \leq \cdots \leq c_{n-1} = 1\}$, then it follows from Corollary VII.6.3 that C is uniquely determined (see also Exercise 8 for a direct proof). Thus for $L \neq \mathbf{1}$, the integer n is called the *order* of L. Notice that $n \geq 2$. For reasons of convenience we will say that $L = \mathbf{1}$ has *order* n for each $n \geq 2$. We can, therefore, talk about *the* order and *the* chain of constants of a Post algebra L ($L \neq \mathbf{1}$). It also follows that if L is a Post algebra whose center is B and whose chain of constants is C, then L is uniquely determined (up to isomorphism) by B and C, and we will often write $L = \langle B, C \rangle$. Note that with this notation the Post algebra $\mathbf{1}$ can be denoted by

$\langle \mathbf{1}, \mathbf{1} \rangle$. Conversely, if $B \in \mathbf{B}$, $C \in \mathbf{C}_{01}$, C finite and $B \neq \mathbf{1}$, $C \neq \mathbf{1}$, then there is a unique Post algebra L of order $n \geq 2$, where $n = |C|$ and such that $L = \langle B, C \rangle$. This correspondence will be investigated in more detail in Theorem 3.1. The class of Post algebras of order $n \geq 2$ will be denoted by $\mathbf{P_n}$. The characterization of Post algebras by means of Theorem 3 enables us to apply the results on coproducts of Boolean algebras and chains obtained in Chapter VII (particularly Theorem VII.4.1). We therefore have the following theorem, which also gives a very useful characterization of Post algebras.

Theorem 4. (G. Epstein [1], T. Traczyk [1]) *Suppose* $L \in \mathbf{D}_{01}$, $L \neq \mathbf{1}$. *Then* $L \in \mathbf{P_n}$, $n \geq 2$ *if and only if* L *has a subset* $C = \{0 = c_0 < c_1 < \cdots < c_{n-1} = 1\}$ *such that for each* $x \in L$, x *has a unique representation* $x = \sum_{i=1}^{n-1} a_i c_i$ *where* $\{a_1, a_2, \cdots, a_{n-1}\} \subseteq \mathscr{C}(L)$, $a_1 \geq a_2 \geq \cdots \geq a_{n-1}$. *Moreover, if these conditions are satisfied then* C *is the chain of constants of* L.

The representation of an element x in the sense of Theorem 4 is called the *monotonic representation of* x.

Proof. ($\Rightarrow$) Suppose $L = \langle B, C \rangle$ where $C = \{0 = c_0 < c_1 < \cdots < c_{n-1} = 1\}$, $n \geq 2$. By Theorem 3, $L \doteq B + C$ and thus by Theorem VII.4.1 we have for $x \in L$ that

(1) $x = \sum_{k=1}^{m} a_{j_k} c_{j_k}$, $k \geq 1$, where $\{c_{j_1}, \cdots, c_{j_m}\} \subseteq C$, $\{a_{j_1}, \cdots, a_{j_m}\} \subseteq B$, $0 = c_{j_1} < c_{j_2} < \cdots < c_{j_m}$, $1 = a_{j_1}, a_{j_2} > \cdots > a_{j_m} > 0$.

Now if $x \neq 0$, then $m \geq 2$ and we can write

(2) $x = \sum_{i=1}^{n-1} a_i c_i$ where $a_i = a_{j_2}$ for $1 \leq i \leq j_2$, $a_i = a_{j_{k+1}}$ for $j_k < i \leq j_{k+1}$ and $2 \leq k \leq m - 1$ and $a_i = 0$ for $i > j_m$. If $x = 0$ then x can also be expressed by taking $a_i = 0$ for $1 \leq i \leq n - 1$ and in either case $a_1 \geq a_2 \geq \cdots \geq a_{n-1}$.

It remains to show that the expression (2) is unique. Suppose $x = \sum_{i=1}^{n-1} a_i' c_i$, $a_1' \geq a_2' \geq \cdots \geq a_{n-1}'$ then $x = \sum_{i=0}^{n-1} a_i' c_i$, where $a_0' = 1$. If $a_1' \neq 1$, then by dropping all but the maximal members of $\{a_i' c_i, i = 0, 1, \cdots, n - 1\}$ (cf. the argument in the second half of the proof of Theorem VII.4.1) we get an expression for x in the form (1). But this expression is unique and it follows that $a_i = a_i'$ for $1 \leq i \leq n - 1$. If $a_1' = 0$ then $x = 1 \cdot 0 + 0c_1 + \cdots + 0c_{n-1}$ and it follows again by uniqueness of (1) that $a_i = a_i' = 0$ for $1 \leq i \leq n - 1$.

($\Leftarrow$) It suffices to show that Definition 1(ii) is satisfied. Thus, suppose $ac_i \leq c_{i-1}$, $a \in \mathscr{C}(L)$ for some i, $1 \leq i \leq n - 1$. Then $ac_{i-1} = ac_i$. But then the hypothesis of the uniqueness of the monotonic representation implies $a = 0$. The remainder of the theorem is obvious.

Exercise 5. Show that in the monotonic representation $x = \sum_{i=1}^{n-1} a_i c_i$, $x \in \mathscr{C}(L) \Leftrightarrow a_1 = a_2 = \cdots = a_{n-1}$.

The following theorem illustrates the significance of the monotonic representation.

Theorem 6. *Let* $L \in \mathbf{P_n}$, $n \geq 2$, $L \neq \mathbf{1}$, $C = \{0 = c_0 < c_1 < \cdots < c_{n-1} = 1\}$, $L = \langle B, C \rangle$. *Suppose* $x, y \in L$ *have monotonic representations* $x = \sum_{i=1}^{n-1} a_i c_i$, $y = \sum_{i=1}^{n-1} a_i' c_i$. *Then* $x \leq y$ *if and only if* $a_i \leq a_i'$ *for* $1 \leq i \leq n-1$ *and the monotonic representations of* $x + y$ *and* xy *are* $x + y = \sum_{i=1}^{n-1} (a_i + a_i')c_i$, $xy = \sum_{i=1}^{n-1} (a_i a_i')c_i$.

Proof. It is immediate by the distributive law that $x + y = \sum_{i=1}^{n-1} (a_i + a_i')c_i$. If $x \leq y$, then $x + y = y$, thus $\sum_{i=1}^{n-1} (a_i + a_i')c_i = \sum_{i=1}^{n-1} a_i'c_i$. We have monotonic representations on either side and therefore by uniqueness $a_i \leq a_i'$ for $1 \leq i \leq n-1$. Conversely, if $a_i \leq a_i'$ for $1 \leq i \leq n-1$ then it follows by a similar argument that $x \leq y$. Finally, observe that $\sum_{i=1}^{n-1} (a_i a_i')c_i \leq x, y$. Suppose $u = \sum_{i=1}^{n-1} b_i c_i$ (monotonic representation) and $u \leq x$, $u \leq y$. Then $b_i \leq a_i$, $b_i \leq a_i'$ so $b_i \leq a_i a_i'$. Hence $u \leq \sum_{i=1}^{n-1} (a_i a_i')c_i'$.

Theorem 7. (G. Epstein [1]) *L Let* $L = \langle B, C \rangle$. *is complete if and only if B is complete.*

The proof is left to the reader as an exercise.

Exercise 8. (G. Rousseau [1]) Prove the uniqueness of the chain of constants in a Post algebra by proving that c_i is the least x for which $ax \leq c_{i-1}$, $a \in \mathscr{C}(L) \Rightarrow a = 0$ $(i = 1, \cdots, n)$.

Exercise 9. Prove that a Post algebra is a Stone algebra. (In section 9 we will see that every Post algebra is a relative Stone algebra and therefore by Exercise IX.2.11, a Heyting algebra.) (Hint: Consider a Post algebra as a coproduct of a Boolean algebra and a chain.)

G. Rousseau [1] has introduced the notion of pseudo-Post algebra. We will restrict ourselves to a few remarks and refer the reader to the literature for more details.

A *pseudo-Post algebra* of order $n \geq 2$ may be defined as the coproduct of a Heyting algebra and a chain of length n. If $L \doteq L_1 + C$, where L_1 is a Heyting algebra and C a chain of length $n \geq 2$, then Theorem 4 still holds after suitable changes have been made.

Theorem 10. *Let* $L \in \mathbf{D_{01}}$, $L \neq \mathbf{1}$. *L is a pseudo-Post algebra of order* $n \geq 2$ *if and only if L contains a Heyting algebra* L_1 *which is a* $\mathbf{D_{01}}$*-subalgebra of L and a* $\mathbf{D_{01}}$*-subalgebra* $C = \{0 = c_0 < c_1 < \cdots < c_{n-1} = 1\}$ *such that for each* $x \in L$, *x has a unique representation* $x = \sum_{i=1}^{n-1} a_i c_i$ *where* $\{a_1, \cdots, a_{n-1}\} \subseteq L_1$ *and* $a_1 \geq a_2 \geq \cdots \geq a_{n-1}$.

We note that the order of a pseudo-Post algebra and its chain of constants are in general not uniquely determined. Indeed if C_1 and C_2 are chains of n and m

elements respectively and $n \neq m$, then the coproduct of C_1 and C_2 is a pseudo-Post algebra with a chain of constants of n elements, but at the same time a pseudo-Post algebra with a chain of constants of m elements (cf. Exercise VII.6.8). On the other hand, if $L \doteq L_1 + C$ and $L \doteq L_1 + C'$ where L_1 is a Heyting algebra and C and C' are finite chains, then $C = C'$ (Exercise VII.6.10).

3. The category of Post algebras

We will denote by $\mathscr{P}$, the subcategory of $\mathscr{D}_{01}$ such that Ob $\mathscr{P} = \mathbf{P}$ and if $L, L_1 \in \mathbf{P}$, $L = \langle B, C\rangle$, $L_1 = \langle B_1, C_1\rangle$ then $[L, L_1]_{\mathscr{P}} = \{h \in [L, L_1]_{\mathscr{D}_{01}} : h[C] \subseteq C_1\}$. In order to avoid cumbersome formulations we introduce some additional notation: $\mathscr{P}_1$ is the full subcategory of $\mathscr{P}$ whose objects have at least two elements; $\mathscr{B}_1$ is the full subcategory of $\mathscr{B}$ whose objects have at least two elements; $\mathscr{C}_1$ is the full subcategory of $\mathscr{C}_{01}$ whose objects are finite and have at least two elements.

It is not difficult to see that monomorphisms and epimorphisms in $\mathscr{B}_1$ and $\mathscr{C}_1$ coincide with one-one and onto morphisms, respectively. Indeed, if $h \in [L, L_1]_{\mathscr{B}_1}$ is monic (epic), then h is monic (epic) in $\mathscr{B}$, since there is exactly one morphism in $\mathscr{B}$ from L_1 to $\mathbf{1}$ and there are no morphisms in $\mathscr{B}$ from $\mathbf{1}$ to L. Conversely, one-one (onto) in $\mathscr{B}_1$ implies monic (epic) in $\mathscr{B}_1$. If $h \in [C, C_1]_{\mathscr{C}_1}$ is monic and $h(c) = h(c')$ for $c \neq c'$, let $f, g \in [\mathbf{3}, C_1]_{\mathscr{C}_1}$ be defined by $f(\mathbf{1}) = c$ and $g(\mathbf{1}) = c'$. Then $h \circ f = h \circ g$ and $f \neq g$, a contradiction. If $h \in [C, C_1]_{\mathscr{C}_1}$ is epic and there exists $y \in C_1$, $y \notin \operatorname{Im} h$ (note $y \neq 0, 1$), let $f, g \in [C_1, \mathbf{2}]_{\mathscr{C}_1}$ be defined by $f(x) = \mathbf{0} \Leftrightarrow x \leq y$ and $g(x) = \mathbf{0} \Leftrightarrow x < y$. Then $f \circ h = g \circ h$ and $f \neq g$, a contradiction. Finally, one-one (onto) in $\mathscr{C}_1$ implies monic (epic) in $\mathscr{C}_1$.

We also note that $\mathscr{B}_1(\mathscr{C}_1)$ contains all $\mathscr{B}(\mathscr{C}_{01})$-isomorphisms. Moreover if $L, L_1 \in$ Ob $\mathscr{P}_1$ and $h \in [L, L_1]_{\mathscr{D}_{01}}$ is an isomorphism then $h \in [L, L_1]_{\mathscr{P}}$ since the chain of constants is uniquely determined.

Theorem 1. *The categories $\mathscr{P}_1$ and $\mathscr{B}_1 \times \mathscr{C}_1$ are equivalent.*

Proof. Define a functor $\mathscr{F}: \mathscr{P}_1 \to \mathscr{B}_1 \times \mathscr{C}_1$ as follows. For $L \in \mathscr{P}_1$, $L = \langle B, C\rangle$ set $\mathscr{F}(L) = (B, C)$. For $h \in [L, L_1]_{\mathscr{P}_1}$, $L = \langle B, C\rangle$, $L_1 = \langle B_1, C_1\rangle$ set $\mathscr{F}(h) = (h|B, h|C)$. Obviously $\mathscr{F}$ is a functor. Next, let $(B, C) \in$ Ob $\mathscr{B}_1 \times \mathscr{C}_1$ and let $L = \langle B, C\rangle$. Then $\mathscr{F}(L) = (B, C)$. Now suppose $L = \langle B, C\rangle$, $L_1 = \langle B_1, C_1\rangle$. We will show that the function from $[L, L_1]_{\mathscr{P}_1}$ to $[\mathscr{F}(L), \mathscr{F}(L_1)]_{\mathscr{B}_1 \times \mathscr{C}_1}$ induced by $\mathscr{F}$, is one-one and onto. Indeed, let $g, h \in [L, L_1]_{\mathscr{P}_1}$ and suppose $\mathscr{F}(g) = \mathscr{F}(h)$. Then $g|B = h|B$ and $g|C = h|C$ so g and h agree on $L \doteq B + C$ thus $g = h$. Finally, if $k = (k_1, k_2) \in [(B, C), (B_1, C_1)]_{\mathscr{B}_1 \times \mathscr{C}_1}$ then $k_1 \in [B, B_1]_{\mathscr{D}_{01}}$ and $k_2 \in [C, C_1]_{\mathscr{D}_{01}}$ so there exists $h: L \to L_1$ such that $h|B = k_1$, $h|C = k_2$ since $L \doteq B + C$ and $L_1 \doteq B_1 + C_1$ so $\mathscr{F}(h) = k$.

Theorem 2. *A morphism in $\mathscr{P}$ is monic (epic) if and only if it is one-one (onto).*

Proof. It suffices to prove the theorem for the category $\mathscr{P}_1$. Let $h \in [L, L_1]_{\mathscr{P}_1}$, $L = \langle B, C\rangle$, $L_1 = \langle B_1, C_1\rangle$, $h_1 = h|B$, $h_2 = h|C$. If h is monic (epic) in $\mathscr{P}_1$, then by

Theorem 1, h is monic (epic) in $\mathscr{B}_1 \times \mathscr{C}_1$ thus h_1 is monic (epic) in $\mathscr{B}_1$ and h_2 is monic (epic) in $\mathscr{C}_1$. But then h_1 and h_2 are one-one (onto). To show h is one-one, let $x, y \in L$, and let $x = \sum_{i=1}^{n-1} a_i c_i$, $y = \sum_{i=1}^{n-1} a_i' c_i$ be the monotonic representations. Suppose $h(x) = h(y)$, then $\sum_{i=1}^{n-1} h_1(a_i)h_2(c_i) = \sum_{i=1}^{n-1} h_1(a_i')h_2(c_i)$. Now the same members of C_1 appear and are strictly increasing on both sides of the expression. Hence $h_1(a_i) = h_1(a_i')$ and thus $a_i = a_i'$ for $1 \leq i \leq n-1$. Therefore $x = y$ (To prove that h is onto if h_1 and h_2 are onto, observe that $h_1[B] \cup h_2[C] = B_1 \cup C_1$. But $[B_1 \cup C_1]_{\mathbf{D_{01}}} = L_1$ hence h is onto.). The converse is immediate.

We will now use Theorem 1 to characterize the injectives in $\mathscr{P}$. First, we prove a lemma concerning $\mathscr{C}_1$.

Lemma 3. *Every object in $\mathscr{C}_1$ is both injective and projective.*

Proof. Let $h \in [C_1, C]_{\mathscr{C}_1}$, $g \in [C_1, C_2]_{\mathscr{C}_1}$ and suppose g is monic. Define $f \in [C_2, C]_{\mathscr{C}_1}$ by $f(x) = \sum \{h(u) : u \in C_1, g(u) \leq x\}$. It follows from the fact that g is monic that $f \circ g = h$. Next, suppose $h \in [C, C_1]_{\mathscr{C}_1}$, $g \in [C_2, C_1]_{\mathscr{C}_1}$ and g epic. Define $f \in [C, C_2]_{\mathscr{C}_1}$ by $f(x) = \sum \{u \in C_2 : g(u) = h(x)\}$. Then $g \circ f = h$.

Since an object (B, C) in $\mathscr{B}_1 \times \mathscr{C}_1$ is injective if and only if B is injective and C is injective and since the injectives in $\mathscr{B}$ and thus in $\mathscr{B}_1$ are the complete Boolean algebras, we infer from Theorems 2.7 and 1, the following theorem.

Theorem 4. *If $L \in \mathrm{Ob}\,\mathscr{P}$, then L is injective if and only if $\mathscr{C}(L)$ is complete.*

There is little known about the projectives in $\mathscr{B}$, but we know (cf. Corollary V.7.6) that all countable Boolean algebras are projective. Thus, if $L \in \mathrm{Ob}\,\mathscr{P}$ and L is countable, then L is projective. Of course, it is a trivial observation that L is projective if and only if its center is a retract of a free Boolean algebra.

Exercise 5. Let $L \in \mathrm{Ob}\,\mathscr{P}$, $L = \langle B, C \rangle$. Show that the subobjects of L are the $\mathbf{D_{01}}$-subalgebras of L which are $\mathbf{D_{01}}$-generated by the union of a $\mathbf{B}$-subalgebra of B and a $\mathbf{D_{01}}$-subchain of C.

Exercise 6. Show that the image of a $\mathscr{P}$-morphism is a subobject.

Exercise 7. (T. Traczyk [1]) Let $L = \langle B, C \rangle$ be a Post algebra of order $n \geq 2$. $C = \{0 = c_0 < \cdots < c_{n-1} = 1\}$. Call a subset I of L a *Post ideal* if I is an ideal in L, and if there exists an integer k, $1 \leq k \leq n-1$ and an ideal I' in B such that an element $x = \sum_{i=1}^{n-1} a_i c_i$ is a member of I (monotonic representation) if and only if $a_i \in I'$ for $i \geq k$. Show that $I \cap B = I'$ and that I is proper if and only if I' is proper. If I is proper, then k is called the *order* of I. Show that the order of $I = \min\{k : c_k \notin I\}$ and that the order of $(0]$ is 1. Show that if I' is a proper ideal in B, then for every k, $1 \leq k \leq n-1$, there exists precisely one Post ideal I in L of order k such that $I \cap B = I'$.

Exercise 8. (Ph. Dwinger [3]) Prove that a proper ideal I in a Post algebra $L = \langle B, C\rangle$ is a Post ideal if and only if there exists an $L_1 \in \mathrm{Ob}\,\mathscr{P}$ and $h \in [L, L_1]_{\mathscr{P}}$ such that $I = \{x \in L : h(x) = 0\}$.

Exercise 9. (Ph. Dwinger [3]) Prove that a prime ideal of a Post algebra is a Post ideal.

Exercise 10. Suppose L is a Post algebra of order $n \geq 2$ and that I is a prime ideal of L of order k, $1 \leq k \leq n - 1$. Show that for every m, $2 \leq m \leq n - k + 1$ there exists an $h \in [L, L_1]_{\mathscr{P}}$ where $L_1 = \langle \mathbf{2}, \mathbf{m}\rangle$ such that $I = \{x \in L : h(x) = 0\}$.

4. Representation theory

We have seen in Theorem 2.3 that Post algebras are coproducts of Boolean algebras and finite chains. Therefore, the results of Section VII.4 apply. We first restate Theorem VII.4.2 for this special situation.

Theorem 1. (G. Epstein [1]) *Let $L = \langle B, C\rangle$ be a Post algebra. Then L is a $\mathbf{D_{01}}$-subdirect product of C^S for some set S. Moreover, L is a $\mathbf{D_{01}}$-subalgebra of C^S and is $\mathbf{D_{01}}$-generated by the union of a $\mathbf{B}$-subalgebra of C^S and the set D of diagonal elements of C^S. Conversely, let C be a finite chain and S any set. Any $\mathbf{D_{01}}$-subalgebra of C^S which is $\mathbf{D_{01}}$-generated by the union of a $\mathbf{B}$-subalgebra B of C^S and the set D of diagonal elements of C^S is a Post algebra whose center is B and whose chain of constants is D.*

Remark. The first part of Theorem 1 implies, in particular, that every Post algebra of order n belongs to the class $\mathbf{C_n}$ (cf. Chapter VI).

Exercise 2. Prove the first part of Theorem 1, using the results of Chapter VI (Hint: Every prime ideal of a Post algebra of order $n \geq 2$ is contained in a basic n-chain of prime ideals.).

Exercise 3. Let $L = \langle B, C\rangle$, $L \neq \mathbf{1}$, be a Post algebra of order $n \geq 2$. Consider L as a $\mathbf{D_{01}}$-subalgebra of C^S for some set S. For every $x \in L$ and every i, $1 \leq i \leq n - 1$, let $x_i \in C^S$ be defined by $x_i(s) = 1$ if $x(s) = c_i$, and $x_i(s) = 0$ if $x(s) \neq c_i$ for $s \in S$. Prove that $B = \{x_i : x \in L, 1 \leq i \leq n - 1\}$. Also prove that for $x \in L$, the monotonic representation of x is $x = \sum_{i=1}^{n-1} x_i d_i$, where for every i, $1 \leq i \leq n - 1$, d_i is the diagonal element defined by $d_i(s) = c_i$ for each $s \in S$.

We restate Theorem VII.4.3, which also applies to Post algebras.

Theorem 4. *A Post algebra $\langle B, C\rangle$ is isomorphic to C^A for some set A if and only if B is a complete field of sets.*

The topological formulation of the representation theorem for Post algebras is obtained by Theorem VII.4.5.

Theorem 5. (G. Epstein [1]) *Let $L = \langle B, C\rangle$ be a Post algebra. Then L is isomorphic to $\langle \mathcal{S}_{\mathcal{B}}(B), C\rangle$. Conversely, let X be a Boolean space and C a finite discrete chain, where if $X = \varnothing$, $C \neq \mathbf{1}$, then $L = \langle X, C\rangle$ is a Post algebra and X is homeomorphic to the representation space of $\mathscr{C}(L)$.*

The following corollary is an immediate consequence of Theorem 4.

Corollary 6. (P. C. Rosenbloom [1]) *Let B be a finite Boolean algebra, $B = \mathbf{2}^m$, $m \geq 1$. If $L = \langle B, C\rangle$, then L is isomorphic to C^m. Thus, if L is a finite Post algebra of order $n \geq 2$ and m is the number of atoms in its center then $|L| = n^m$.*

5. The partially ordered set of prime ideals of a Post algebra

It follows from Theorem 2.3 and Theorem VII.2.1 that for a Post algebra L of order $n \geq 2$, $\mathfrak{P}(L)$ is isomorphic to $S \times \mathbf{n-1}$ where S is the totally unordered poset of prime ideals of $\mathscr{C}(L)$. The question arises as to whether the converse holds: if $L \in \mathrm{Ob}\, \mathscr{D}_{01}$ and $\mathfrak{P}(L) \simeq S \times \mathbf{n-1}$, $n \geq 2$, where S is a totally unordered poset then is L a Post algebra of order n? The answer to this question is negative. Before we present a counterexample we make some more observations concerning $\mathfrak{P}(L)$, where $L \in \mathbf{P_n}$. Let $(I_s)_{s\in S}$ be the family of all prime ideals of $\mathscr{C}(L)$. We have seen that $\mathfrak{P}(L) \simeq S \times \mathbf{n-1}$. If $C = \{0 = c_0 < c_1 < \cdots < c_{n-1} = 1\}$ is the chain of constants of L, then for $(s, \mathbf{i}) \in S \times \mathbf{n-1}$, let $I_{s,i}$ be the corresponding prime ideal of L. (See Theorem VII.2.1.) We then have

(1) $I_{s,i} = (I_s \cup \{c_i\}] = \{x \in L : a_{i+1} \in I_s\}$, where for each $x \in L$, a_{i+1} is the coefficient of c_{i+1} in the monotonic representation of x: We first show $(I_s \cup \{c_i\}] = \{x \in L : a_{i+1} \in I_s\}$. Indeed, $x \in (I_s \cup \{c_i\}] \Leftrightarrow x \leq a + c_i$ for some $a \in I_s \Leftrightarrow a_{i+1} \leq a$ for some $a \in I_s \Leftrightarrow a_{i+1} \in I_s$. To show $I_{s,i} = (I_s \cup \{c_i\}]$ note that $(I_s \cup \{c_i\}] \subseteq I_{s,i}$. If $x, y \in L$ with monotonic representations $x = \sum_{i=1}^{n-1} a_i c_i$ and $y = \sum_{i=1}^{n-1} b_i c_i$ and $xy \in (I_s \cup \{c_i\}]$ then $a_{i+1} b_{i+1} \in I_s$ so either a_{i+1} or $b_{i+1} \in I_s$, and therefore either x or $y \in (I_s \cup \{c_i\}]$ and it follows that $(I_s \cup \{c_i\}]$ is a prime ideal in L. Finally, for $x = \sum_{i=1}^{n-1} a_i c_i$ (monotonic representation) in L, $x \in (I_s \cup \{c_i\}] \cap \mathscr{C}(L) \Leftrightarrow a_{i+1} \in I_s$ and $x \in \mathscr{C}(L) \Leftrightarrow x \in I_s$; hence $(I_s \cup \{c_i\}] \cap \mathscr{C}(L) = I_s$. Also $x \in (I_s \cup \{c_i\}] \cap C \Leftrightarrow a_{i+1} \in I_s$ and $x \in C \Leftrightarrow a_{i+1} = 0$ and $x \in C \Leftrightarrow x \in \{c_0, \cdots, c_i\}$, hence $(I_s \cup \{c_i\}] \cap C = \{c_0, \cdots, c_i\}$. It follows from the second part of Theorem VII.1.7 that $I_{s,i} = (I_s \cup \{c_i\}]$.

The following property also holds:

(2) $\bigcap_{s\in S} I_{s,i} \sim \bigcup_{s\in S} I_{s,i-1} = \{c_i\}$ for $1 \leq i \leq n-1$, where for each $s \in S$, $I_{s,n-1} = L$: Indeed, it follows immediately from (1) that for $1 \leq i \leq n-2$, $c_i \in \bigcap_{s\in S} I_{s,i} \sim \bigcup_{s\in S} I_{s,i-1}$. If x belongs to $\bigcap_{s\in S} I_{s,i} \sim \bigcup_{s\in S} I_{s,i-1}$ for some i, $1 \leq i \leq n-2$ and $x = \sum_{i=1}^{n-1} a_i c_i$ (monotonic representation), then by (1) $a_{i+1} \in \bigcap_{s\in S} I_s$, so $a_{i+1} = 0$. If $a_i < 1$, then there exists $s_0 \in S$ such that $a_i \in I_{s_0}$, so again by (1), $x \in I_{s_0,i-1}$ and so $x \in \bigcup_{s\in S} I_{s,i-1}$, a contradiction. So $a_i = 1$ and thus $x = c_i$. For the proof of (2) for $i = n-1$, it is immediate that $1 \in L \sim \bigcup_{s\in S} I_{s,n-2}$. Conversely if $x \in L \sim \bigcup_{s\in S} I_{s,n-2}$, $x = \sum_{i=1}^{n-1} a_i c_i$ (monotonic representation), then

$x \notin I_{s,n-2}$ for each $s \in S$ so $a_{n-1} \notin I_s$ for each $s \in S$, but then $a_{n-1} = 1$ so $x = c_{n-1} = 1$.

We will now give an example, which is due to C. C. Chang and A. Horn [1], of an object L of $\mathcal{D}_{01}$ such that $\mathfrak{P}(L) \cong S \times \mathbf{2}$ where S is a totally unordered set but such that L is not a Post algebra of order 3.

Let S be an infinite countable set. For the purpose of our example it will be useful to index the elements of the set S by the members of the ordinal $\omega \oplus \mathbf{1}$, so $S = \{s_\alpha : \alpha \in \omega \oplus \mathbf{1}\}$. Let $L = \mathbf{3}^S$. Then L is an object of $\mathcal{D}_{01}$ and is, in fact, a Post algebra of order 3 whose chain of constants is the chain of diagonal elements (= constant functions) of L (cf. Theorem 4.4). We will define a $\mathbf{D}_{01}$-subalgebra L_1 of L and show that $\mathfrak{P}(L_1) \cong S \times \mathbf{2}$, but that L_1 is not a Post algebra.

Let L_1 be the set of all those $f \in L$ for which either $f(s_\mathbf{n}) = \mathbf{2}$ for all $\mathbf{n} \in \omega$ such that $\mathbf{n} \geq \mathbf{n_o}$ for some $\mathbf{n_o} \in \omega$ and $f(s_\omega) = \mathbf{1}$ or $\mathbf{2}$, or $f(s_\mathbf{n}) = \mathbf{0}$ for all $\mathbf{n} \in \omega$ such that $\mathbf{n} \geq \mathbf{n_o}$ for some $\mathbf{n_o} \in \omega$ and $f(s_\omega) = \mathbf{0}$.

Obviously, L_1 is a $\mathbf{D}_{01}$-subalgebra of L. It is also easy to see that for $\alpha \in \omega \oplus \mathbf{1}$ and $\mathbf{k} \in \mathbf{2}$, $I_{\alpha,\mathbf{k}} = \{f \in L : f(s_\alpha) \leq \mathbf{k}\}$ is a prime ideal of L and therefore $J_{\alpha,\mathbf{k}} = I_{\alpha,\mathbf{k}} \cap L_1$ is a prime ideal of L_1. We will first show that all the prime ideals of L_1 are obtained in this way. That is, we will show that $\mathfrak{P}(L_1) = \{J_{\alpha,\mathbf{k}} : \alpha \in \omega \oplus \mathbf{1}, \mathbf{k} \in \mathbf{2}\}$. Let $I \in \mathfrak{P}(L_1)$. Suppose that for each $\alpha \in \omega \oplus \mathbf{1}$ there exists $f_\alpha \in I$ and $g_\alpha \in L_1 \sim I$ such that $g_\alpha(s_\alpha) \leq f_\alpha(s_\alpha)$. It would follow from this and from the definition of the members of L_1 that there exists $\mathbf{n_o} \in \omega$ such that $g_\omega(s_\mathbf{n}) \leq f_\omega(s_\mathbf{n})$ for $\mathbf{n} \geq \mathbf{n_o}$ and $\mathbf{n} \in \omega$. Assume that $\mathbf{n_o}$ is the first member of ω for which this inequality holds. Let $f = f_0 + f_1 + \cdots + f_{\mathbf{n_o}-1} + f_\omega$ and $g = g_0 g_1 \cdots g_{\mathbf{n_o}-1} g_\omega$ (if $\mathbf{n_o} = \mathbf{0}$ then $f = f_\omega$ and $g = g_\omega$). For $\mathbf{m} \in \omega$, $\mathbf{m} \geq \mathbf{n_o}$, $g(s_\mathbf{m}) \leq g_\omega(s_\mathbf{m}) \leq f_\omega(s_\mathbf{m}) \leq f(s_\mathbf{m})$. For $\mathbf{m} \in \omega$, $\mathbf{m} < \mathbf{n_o}$, $g(s_\mathbf{m}) \leq g_\mathbf{m}(s_\mathbf{m}) \leq f_\mathbf{m}(s_\mathbf{m}) \leq f(s_\mathbf{m})$ and $g(s_\omega) \leq g_\omega(s_\omega) \leq f_\omega(s_\omega) \leq f(s_\omega)$. It follows that $g \leq f$. But $f \in I$ and therefore $g \in I$ and so there exists $\alpha \in \omega \oplus \mathbf{1}$ such that $g_\alpha \in I$, a contradiction. So there exists $\alpha_o \in \omega \oplus \mathbf{1}$ such that if $f \in I$, $g \in L_1$ and $g(s_{\alpha_o}) \leq f(s_{\alpha_o})$ then $g \in I$. Let $\mathbf{k} = \max\{f(s_{\alpha_o}) : f \in I\}$ then we have $I = \{f \in L_1 : f(s_{\alpha_o}) \leq \mathbf{k}\}$. Note that $\mathbf{k} \neq \mathbf{2}$ (otherwise $I = L_1$) and it follows that $I = J_{\alpha_o,\mathbf{k}}$. The next step is to show that $\mathfrak{P}(L_1) \cong S \times \mathbf{2}$ (where S is considered as a totally unordered set). For this, we will show that the map $(s_\alpha, \mathbf{k}) \mapsto J_{\alpha,\mathbf{k}}$ for $\alpha \in \omega \oplus \mathbf{1}$, $\mathbf{k} \in \mathbf{2}$ defines an isomorphism between $S \times \mathbf{2}$ and $\mathfrak{P}(L_1)$. Indeed, suppose first $(s_\alpha, \mathbf{k}) \leq (s_{\alpha'}, \mathbf{k}')$ for $\alpha, \alpha' \in \omega \oplus \mathbf{1}$, $\mathbf{k}, \mathbf{k}' \in \mathbf{2}$. But then $\alpha = \alpha'$ and $\mathbf{k} \subseteq \mathbf{k}'$. If $f \in J_{\alpha,\mathbf{k}}$ then $f(s_\alpha) \subseteq \mathbf{k}$ so $f(s_\alpha) \subseteq \mathbf{k}'$ and thus $f \in J_{\alpha,\mathbf{k}'}$. Conversely, suppose $J_{\alpha,\mathbf{k}} \subseteq J_{\alpha',\mathbf{k}'}$ for $\alpha, \alpha' \in \omega \oplus \mathbf{1}$, $\mathbf{k}, \mathbf{k}' \in \mathbf{2}$. Then $\alpha = \alpha'$. Indeed, if $\alpha \neq \alpha'$ let $f \in L_1$ be such that $f(s_\alpha) = \mathbf{k}$ and $f(s_{\alpha'}) = \mathbf{2}$. But then $f \in J_{\alpha,\mathbf{k}}$ and $f \notin J_{\alpha',\mathbf{k}'}$, a contradiction. Hence $\alpha = \alpha'$. Now, if $\mathbf{k} \nsubseteq \mathbf{k}'$, then $\mathbf{k}' \subset \mathbf{k}$, so $\mathbf{k}' = \mathbf{0}$ and $\mathbf{k} = \mathbf{1}$. Let $f \in L_1$ be such that $f(s_\alpha) = \mathbf{1}$, then $f \in J_{\alpha,\mathbf{k}}$ but $f \notin J_{\alpha',\mathbf{k}'}$, a contradiction. It follows that $(\alpha, \mathbf{k}) \leq (a', \mathbf{k}')$. Since, for every $I \in \mathfrak{P}(L_1)$, $I = J_{\alpha,\mathbf{k}}$ for some $\alpha \in \omega \oplus \mathbf{1}$ and $\mathbf{k} \in \mathbf{2}$, it follows that $\mathfrak{P}(L_1) \cong S \times \mathbf{2}$.

It remains to show that L_1 is not a Post algebra. Let $f_1 \in L$ be the constant function defined by $f_1(s_\alpha) = \mathbf{1}$ for all $\alpha \in \omega \oplus \mathbf{1}$. It is easy to see that $\bigcap_{\alpha\in\omega\oplus\mathbf{1}} I_{\alpha,\mathbf{1}} \sim \bigcup_{\alpha\in\omega\oplus\mathbf{1}} I_{\alpha,\mathbf{0}} = \{f_1\}$. But also $\bigcap_{\alpha\in\omega\oplus\mathbf{1}} J_{\alpha,\mathbf{1}} \sim \bigcup_{\alpha\in\omega\oplus\mathbf{1}} J_{\alpha,\mathbf{0}}$

$\subseteq \bigcap_{\alpha\in\omega\oplus 1} I_{\alpha,1} \sim \bigcup_{\alpha\in\omega\oplus 1} I_{\alpha,0}$. However, $f_1 \notin L_1$ so $\bigcap_{\alpha\in\omega\oplus 1} J_{\alpha,1} \sim \bigcup_{\alpha\in\oplus 1} J_{\alpha,0} = \varnothing$. It follows from (2) that L_1 is not a Post algebra.

It has been proven by C. C. Chang and A. Horn [1] (also cf. T. Katrinák and A. Mitschke [1]) that if $L \in \text{Ob}\,\mathcal{D}_{01}$ and $\mathfrak{P}(L) \cong S \times \mathbf{n-1}$, $n \geq 2$ and where S is a totally unordered set, then L is Post algebra of order n, if in addition the following condition holds (cf. (2)): $\bigcap_{s\in S} I_{s,i} \sim \bigcup_{s\in S} I_{s,i-1} \neq \varnothing$ for $1 \leq i \leq n-1$ where for $(s, 1) \in S \times \mathbf{n-1}$, $I_{s,i}$ is the corresponding prime ideal of L and where $I_{s,n-1} = L$ for each $s \in S$.

Exercise 1. Prove that every Post algebra of order $n \geq 2$ belongs to $\mathbf{C_n}$ (Hint: use the results of this section and Theorem VI.4.2).

6. Post algebras as equational classes of algebras

In the previous sections we have considered the class $\mathbf{P}$ as a subclass of the class $\mathbf{D_{01}}$ and for each $n \geq 2$ the class $\mathbf{P_n}$ was considered as a subclass of $\mathbf{P}$, and thus as a subclass of $\mathbf{D_{01}}$. Clearly, $\mathbf{P_n}$ is not an equational subclass of $\mathbf{D_{01}}$. It is the purpose of this section to define Post algebras of order n, $n \geq 2$ as an equational class of algebras. This is done by adding additional operations and identities to those which define the class $\mathbf{D_{01}}$. We shall also use the symbol $\mathbf{P_n}$ to denote these equational classes for each $n \geq 2$.

Theorem 1. (T. Traczyk [2]) *Consider the class of algebras of the form* $(L, (+, \cdot, C, D_1, \cdots, D_{n-1}, c_0, \cdots, c_{n-1}))$, $n \geq 2$ *where* $+$ *and* $\cdot$ *are binary operations* $C, D_1, \cdots, D_{n-1}$ *are unary operations and* $c_0, c_1, \cdots, c_{n-1}$ *are nullary operations. The following identities characterize the class of Post algebras of order n.*

P_1: *A set of identities in* $+, \cdot$ *that define distributive lattices.*
P_2: $\mathbf{c_0 + x = x}$, $\mathbf{c_i c_j = c_i}$ *for* $1 \leq i \leq j \leq n-1$, $\mathbf{x c_{n-1} = x}$.
P_3: $\mathbf{C(x)D_1(x) = c_0}$, $\mathbf{C(x) + D_1(x) = c_{n-1}}$.
P_4: $\mathbf{D_i(x)D_j(x) = D_j(x)}$ *for* $1 \leq i \leq j \leq n-1$.
P_5: $\mathbf{D_i(x + y) = D_i(x) + D_i(y)}$, $\mathbf{D_i(xy) = D_i(x)D_i(y)}$, *for* $1 \leq i \leq n-1$.
P_6: $\mathbf{D_i(D_j(x)) = D_j(x)}$ *for* $1 \leq i, j \leq n-1$.
P_7: $\mathbf{D_i(c_j) = c_{n-1}}$ *for* $1 \leq i \leq j \leq n-1$, $\mathbf{D_i(c_j) = c_0}$ *for* $0 \leq j < i \leq n-1$.
P_8: $\mathbf{x = D_1(x)c_1 + \cdots + D_{n-1}(x)c_{n-1}}$.

Proof. First assume that $P_1, \cdots, P_8$ hold in an algebra L of the type under consideration. By P_2, $c_0 = 0$ and $c_{n-1} = 1$ and $c_0 \leq c_1 \leq \cdots \leq c_{n-1}$. Thus L is a distributive lattice with 0 and 1. By P_3, $C(x)$ and $D_1(x)$ are in $\mathscr{C}(L)$ for each $x \in L$. By P_6, $D_1(D_j(x)) = D_j(x)$ for $1 \leq j \leq n-1$, thus $D_j(x) \in \mathscr{C}(L)$ for $1 \leq j \leq n-1$. Since P_8 holds, condition (i) of Definition 2.1 is satisfied. It remains to show that (ii) of that condition holds. Suppose $ac_i \leq c_{i-1}$ for $a \in \mathscr{C}(L)$ and for some i, $1 \leq i \leq n-1$. Then $ac_i + c_{i-1} = c_{i-1}$. By P_5, $D_j(a)D_j(c_i) + D_j(c_{i-1}) = D_j(c_{i-1})$ for $1 \leq j \leq n-1$. Taking $j = i$ we obtain $D_i(a)D_i(c_i) + D_i(c_{i-1}) = D_i(c_{i-1})$. By P_7, $D_i(c_i) = 1$ and $D_i(c_{i-1}) = 0$ thus $D_i(a) = 0$. But then by P_4, $D_{n-1}(a) = 0$. By P_5 and

P_7, $D_{n-1}(a) + D_{n-1}(\bar{a}) = D_{n-1}(1) = D_{n-1}(c_{n-1}) = c_{n-1} = 1$. But $D_{n-1}(a) = 0$ thus $D_{n-1}(\bar{a}) = 1$. By P_8, $D_{n-1}(\bar{a})c_{n-1} \leq \bar{a}$, thus $\bar{a} = 1$ and so $a = 0$. For the converse, suppose $L = \langle B, C\rangle$, $C = \{0 = c_0, c_1, \cdots, c_{n-1} = 1\}$, where $n \geq 2$. For $x \in L$, let $x = \sum_{i=1}^{n-1} a_i c_i$ be the monotonic representation of x. Putting $D_i(x) = a_i$ for $1 \leq i \leq n-1$ and $C(x) = \overline{D_1(x)}$, it is easy to verify that the conditions $P_1, \cdots, P_8$ are satisfied.

Corollary 2. *The class* $\mathbf{P_n}$ *of Post algebras of order* $n \geq 2$ *is equational.*

It is interesting to note that unless the Post algebra is the one-element algebra, the nullary operations are always distinct.

The class of Post algebras of order $n \geq 2$, as defined in this section as an equational class of algebras can be considered as a subclass of the class $\mathbf{D_{01}}$ (cf. p. 10 and 11 of Chapter I). From now on we will also use the symbol $\mathbf{P_n}$ to denote this equational class and use the symbol $\mathscr{P}_{\mathbf{n}}$, in accordance with our convention, to denote the associated equational category. Recall that, if L is a Post algebra of order n, $n \geq 2$, then the chain of constants is uniquely determined. Likewise, because of the uniqueness of the monotonic representation and of the uniqueness of complementation, the operations $D_1, \cdots, D_{n-1}, C$ (cf. second half of proof of Theorem 1) are also uniquely determined. Therefore, the functor defined in Section I.19 (page 31) from $\mathscr{P}_{\mathbf{n}}$ to $\mathscr{D}_{\mathbf{01}}$ is an embedding which is one-one on the objects. Thus, in accordance with the convention established in Section I.19 (page 31) we will often consider $\mathscr{P}_{\mathbf{n}}$ as a subcategory of $\mathscr{D}_{\mathbf{01}}$. Note, however, that $\mathscr{P}_{\mathbf{n}}$ is not a full subcategory of $\mathscr{D}_{\mathbf{01}}$.

It also follows from the uniqueness of the chain of constants that order-isomorphisms and $\mathbf{P_n}$-isomorphisms are the same in $\mathbf{P_n}$. Thus, there will be no confusion if we identify the Post algebra determined by $\langle \mathbf{2}, \mathbf{n}\rangle$ with $\mathbf{n}$. That is, $\mathbf{n}$ is the Post algebra of order n in which $C(\mathbf{0}) = \mathbf{n-1}$, $C(\mathbf{m}) = \mathbf{0}$ for $\mathbf{m} \neq \mathbf{0}$ and

$$D_i(\mathbf{m}) = \begin{cases} \mathbf{n-1} \text{ if } i \leq m \\ \mathbf{0} \quad \text{ if } i > m \end{cases} \text{for } i = 1, \cdots, n-1.$$

The nullary operations are, of course, $\mathbf{0}, \mathbf{1}, \cdots, \mathbf{n-1}$.

The following result is also immediate.

Theorem 3. *Suppose* $L_1, L_2 \in \mathrm{Ob}\,\mathscr{P}_{\mathbf{n}}$, $n \geq 2$, $L_1 = \langle B_1, C_1\rangle$, $L_2 = \langle B_2, C_2\rangle$. *If* $h \in [L_1, L_2]_{\mathscr{D}_{01}}$ *and* $h[C_1] = C_2$ *then* $h \in [L_1, L_2]_{\mathscr{P}_{\mathbf{n}}}$.

Theorem 4. *Let* $L \in \mathbf{P_n}$, $n \geq 2$, $L = \langle B, C\rangle$. *Then* $\mathfrak{C}(L) \simeq \mathfrak{C}(B)$.

Proof. Consider the map $\theta \mapsto \theta \cap B^2$ for $\theta \in \mathfrak{C}(L)$. We will show that this map defines an isomorphism between $\mathfrak{C}(L)$ and $\mathfrak{C}(B)$. Obviously, $\theta_1 \subseteq \theta_2 \Rightarrow \theta_1 \cap B^2 \leq \theta_2 \cap B^2$. Conversely, if $\theta_1 \cap B^2 \subseteq \theta_2 \cap B^2$ for $\theta_1, \theta_2 \in \mathfrak{C}(L)$ and $(x, y) \in \theta_1$ for $x, y \in L$, then it follows from Theorem 1 ($P8$) that $(x, y) \in \theta_2$. Finally, let $\theta \in \mathfrak{C}(B)$. Define $\theta' \subseteq L^2$ by $(x, y) \in \theta' \Leftrightarrow (D_i(x), D_i(y))) \in \theta$ for $1 \leq i \leq n-1$. It follows again from Theorem 1 that $\theta' \in \mathfrak{C}(L)$, and it is easy to see that $\theta = \theta' \cap B^2$ (cf. Exercise 2.5).

The next theorem is an immediate consequence of Theorem 4 and the fact that the only subdirectly irreducible in **B** is the algebra **2**.

Theorem 5. *Let $L \in \mathbf{P_n}$, $n \geq 2$. Then L is subdirectly irreducible if and only if $L \cong \mathbf{n}$. Therefore L is a subdirect product of copies of* **n**.

Note that the last part of Theorem 5 also follows from Theorem 4.1 and Theorem 3.

Exercise 6. (G. Rousseau [1]) Prove that the class of pseudo-Post algebras is equational (see Section 2).

If $L = \langle B, C\rangle$ is a Post algebra of order $n \geq 2$ and L' is a $\mathbf{P_n}$-subalgebra of L, then $\mathscr{C}(L')$ is obviously a **B**-subalgebra of $\mathscr{C}(B)$ and the chain C' of constants of L', is C. In particular, $L' = [\mathscr{C}(L') \cup C]_{\mathbf{D_{01}}}$, in fact L' is the coproduct (in $\mathscr{D}_{01}$) of $\mathscr{C}(L')$ and C. The following theorem is also useful.

Theorem 7. *Let $L \in \text{Ob}\,\mathscr{P}_{\mathbf{n}}$, $n \geq 2$, $L = \langle B, C\rangle$ and $S \subseteq L$. Let $D = \{D_i(x) : 1 \leq i \leq n-1, x \in S\}$ and let B' be the* **B***-subalgebra of* **B** *generated by D. Then the $\mathbf{P_n}$-subalgebra of L generated by S is the $\mathbf{D_{01}}$-subalgebra of L which is $\mathbf{D_{01}}$-generated by $B' \cup C$.*

The proof is immediate.

We close this section with the following theorem, which establishes some additional facts about the category $\mathscr{P}_{\mathbf{n}}$.

Theorem 8. (cf. G. Georgescu and C. Vraciu [3]) (i) *The categories $\mathscr{B}$ and $\mathscr{P}_{\mathbf{n}}$ are equivalent for each $n \geq 2$.*

(ii) *A morphism in $\mathscr{P}_{\mathbf{n}}$ is monic* (*epic*) *if and only if it is one-one* (*onto*).

(iii) *An object in $\mathscr{P}_{\mathbf{n}}$ is injective if and only if it is complete.*

Proof. (i) The proof is analogous to the proof of Theorem 3.1. (ii) This follows from (i) and from the fact that the statement holds for the category $\mathscr{B}$. (iii) This follows from (i), Theorem V.9.2, and Theorem 2.7.

7. Coproducts in $\mathscr{P}_{\mathbf{n}}$ and free Post algebras

Since $\mathbf{P_n}$ is an equational category for $n \geq 2$ and satisfies the condition of Theorem I.20.8, $\mathscr{P}_{\mathbf{n}}$ has coproducts and it also has free algebras. We will first characterize coproducts in $\mathscr{P}_{\mathbf{n}}$.

Theorem 1. (R. Cignoli [3], Ph. Dwinger [7]) *Let $(L_s)_{s\in S}$ be a family of objects in $\mathscr{P}_{\mathbf{n}}$, $n \geq 2$ and $L_s = \langle B_s, C_s\rangle$ for each $s \in S$. Then a coproduct of this family in $\mathscr{P}_{\mathbf{n}}$ is the Post algebra $L = \langle B, C\rangle$ where B is a coproduct in $\mathscr{B}$ of the family $(B_s)_{s\in S}$ and where C is a chain of length n.*

Proof. We can consider each L_s as an $\mathbf{P_n}$-subalgebra of L, and it is immediate that $[\bigcup_{s\in S} L_s]_{\mathbf{P_n}} = L$. Let $(h_s : L_s \to L')_{s\in S}$ be a family of morphisms in $\mathscr{P}_{\mathbf{n}}$ and where $L' = \langle B', C'\rangle$. Then the family $(h_s | B_s)_{s\in S}$ can be extended to a morphism $h \in [B, B']_{\mathscr{D}_{01}}$. But L is the $\mathscr{D}_{01}$-coproduct of B and C, hence h can be extended to a morphism $h^* \in [L, L']_{\mathscr{D}_{01}}$ and obviously $h^* \in [L, L']_{\mathscr{P}_{\mathbf{n}}}$.

Next, we characterize the free objects of $\mathscr{P}_{\mathbf{n}}$, $n \geq 2$. Since the free Post algebras of order n are the coproducts of free Post algebras of order n on one generator we will characterize these first.

Theorem 2. *$\mathscr{F}_{\mathbf{P_n}}(\mathbf{1})$, is $\langle \mathbf{2}^n, \mathbf{n}\rangle$ for $n \geq 2$.*

In $\mathbf{2}^n$, choose a sequence of elements $1 > a_1 > a_2 > \cdots > a_{n-1} > 0$. Let $y = \sum_{i=1}^{n-1} a_i\mathbf{i}$. Since the set $\{a_i : 1 \leq i \leq n - 1\}$ $\mathbf{B}$-generates $\mathbf{2}^n$ it follows from Theorem 6.7 that $\{y\}$ $\mathbf{P_n}$-generates the Post algebra $\langle \mathbf{2}^n, \mathbf{n}\rangle$. Now, let $h: \{y\} \to L'$ be a function where $L' = \langle B', \mathbf{n}\rangle$ is an object of $\mathscr{P}_{\mathbf{n}}$. Let $h(y) = \sum_{i=1}^{n-1} a_i'\mathbf{i}$ (monotonic representation). Let $C = \{a_{n-1} < a_{n-2} < \cdots < a_1\}$. Define $f: C \to B'$ by $f(a_i) = a_i'$ for $1 \leq i \leq n - 1$. Since $\mathbf{2}^n$ is obviously a free Boolean extension of $C \cup \{0, 1\}$, f can be extended to a morphism $g \in [\mathbf{2}^n, B']_{\mathscr{B}}$. But $\langle \mathbf{2}^n, \mathbf{n}\rangle$ is a $\mathscr{D}_{01}$-coproduct of $\mathbf{2}^n$ and $\mathbf{n}$ and thus g can be extended to a morphism $h^* \in [\langle \mathbf{2}^n, \mathbf{n}\rangle, L']$ and $h^*(y) = h(y)$.

We can now characterize $\mathscr{F}_{\mathbf{P_n}}(S)$.

Corollary 3. *Suppose $n \geq 2$ and S is a set. Let B_S denote the $\mathscr{B}$-coproduct of $|S|$ copies of $\mathbf{2}^n$. Then $\mathscr{F}_{\mathbf{P_n}}(S)$ is the Post algebra $\langle B_S, \mathbf{n}\rangle$.*

Proof. Immediate from Theorems 1 and 2.

Corollary 4. (R. Cignoli [1], Ph. Dwinger [7], also cf. A. L. Foster [1]) (i) *$\mathscr{F}_{\mathbf{P_n}}(S)$, $n \geq 2$ is isomorphic to $\langle X, \mathbf{n}\rangle$ where X is the topological product of $|S|$ copies of the discrete space consisting of n points.* (ii) *$\mathscr{F}_{\mathbf{P_n}}(\mathbf{m})$, $n \geq 2$, is isomorphic to the product of n^m copies of $\mathbf{n}$ and therefore $|\mathscr{F}_{\mathbf{P_n}}(\mathbf{m})| = n^{n^m}$.* (In particular, $|\mathscr{F}_{\mathbf{B}}(\mathbf{m})| = 2^{2^m}$.)

Proof. (i) Follows from Corollary 3, Theorem 4.5, and Theorem VII.7.3. (ii) Immediate from (i) since, in this case, X is discrete and has n^m points.

J. Berman [3] has developed an alternate method which can be used for various classes of algebras, to determine $\mathscr{F}_{\mathbf{P_n}}(\mathbf{m})$.

8. Generalized Post algebras

In recent years Post algebras have been generalized in different directions. We will sketch some of these generalizations in this section, but we must refer the reader to the literature for more details. A generalization that we will describe in greater detail—Stone algebras of order n—will be the subject of the next section.

In Theorems 2.3 and 4.5 we have seen that a Post algebra is a coproduct (in $\mathscr{D}_{01}$) of its center and of its chain of constants and that a Post algebra can also be represented as a lattice $\langle X, C\rangle$ of continuous functions on a Boolean space X to a finite discrete chain C. From this viewpoint, it is natural to define a generalized Post algebra as a coproduct (in $\mathscr{D}_{01}$) of a Boolean algebra and a chain with 0, 1 or equivalently (cf. Theorem VII.4.5), as a lattice of continuous functions on a Boolean space to an arbitrary chain (with 0, 1) the latter endowed with the discrete topology (C. C. Chang and A. Horn [1]). It is not difficult to prove that if L is a generalized Post algebra in this sense, that L satisfies the following conditions (G. Epstein and A. Horn [1]) (also cf. Exercise 2.9).

(i) L is a Heyting algebra in which the identity $\mathbf{x} \rightarrow \mathbf{y} + \mathbf{y} \rightarrow \mathbf{x} = \mathbf{1}$ holds.

(ii) For each $a, b \in L$, there is a largest element in $\mathscr{C}(L)$ denoted by $a \Rightarrow b$ such $a(a \Rightarrow b) \leq c$ and the identity $\mathbf{x} \Rightarrow \mathbf{y} + \mathbf{y} \Rightarrow \mathbf{x} = \mathbf{1}$ holds in L.

Exercise 1. Verify (i) and (ii).

Note that it follows from Theorem IX.2.10 that L is a relative Stone algebra. It is also easy to see that if L is a generalized Post algebra in the above sense, then so is $\breve{L}$, and thus $\breve{L}$ satisfies the duals $\breve{\text{(i)}}$, $\breve{\text{(ii)}}$ of the conditions (i) and (ii) respectively.

In G. Epstein and A. Horn [1], P-algebras are investigated. P-algebras are objects L of $\mathscr{D}_{01}$ such that L satisfies (i) and (ii) and $\breve{L}$ satisfies (i) and (ii). Further abstractions of Post algebras have been studied by G. Epstein and A. Horn in [2].

In considering generalized Post algebras as lattices $\langle X, C\rangle$ of continuous functions on a Boolean space X to a discrete chain C with 0, 1, one can go a step further and drop the condition that C has a 0 and 1. Obviously, these lattices are not necessarily objects of $\mathscr{D}_{01}$ and they are not coproducts of Boolean algebras and chains. However, many of the properties as described in Section 2 still hold for this class of "generalized Post algebras"—although in a more general fashion—as has been shown by W. J. Blok [1].

A generalization of Post algebras in a direction which is based on Theorem 2.4 has been proposed by T. Traczyk [4] and which we will briefly describe. Let $B \in \mathrm{Ob}\,\mathscr{B}$ and let $T = [0, 1]$ be the unit interval on the real line. Let L be the lattice of all maps $x \in B^T$ such that $t \leq t' \Rightarrow x(t) \geq x(t')$ for $x \in B^T$ and $t, t' \in T$. It is easy to see that $L \in \mathrm{Ob}\,\mathscr{D}_{01}$.

If we define, for each $t \in T$, $c_t \in L$ by

(1) $c_t(s) = 1$ for $s \leq t$ and $c_t(s) = 0$ for $s > t$, then the family $(c_t)_{t\in T}$ is a chain in L which is isomorphic to T, since $t \leq t' \Leftrightarrow c_t \leq c_{t'}$. Furthermore, $c_0 = 0$ and $c_1 = 1$. It also follows that B is isomorphic to $\mathscr{C}(L)$. Indeed, for $a \in B$, let $x_a \in L$ be defined by $x_a(t) = a$ for all $t \in T$. Then the map $a \mapsto x_a$ establishes an isomorphism between B and $\mathscr{C}(L)$. If we define unary operations D_t, $t \in T$, on L by the assignment $t' \mapsto x(t)$ then for $x \in L$ we have:

(2) $D_0(x) = 1$, $D_t(x) \in \mathscr{C}(L)$ for each $t \in T$;
$D_t(x) \geq D_{t'}(x)$ for $t \leq t'$, $t, t' \in T$

(3) $x = \sum_{t \in T} D_t(x)c_t$ for each $x \in L$.

(4) If $(a_t)_{t \in T}$ is a family of elements in $\mathscr{C}(L)$ such that $a_t \geq a_{t'}$ for $t \leq t'$, $t, t' \in T$, then there exists exactly one element $x \in L$ such that $x(t) = a_t = D_t(x)$ for each $t \in T$ and $x = \sum_{t \in T} a_t c_t$.

Conversely, if $L \in \text{Ob}\, \mathscr{D}_{01}$ and L has a chain of elements $(c_t)_{t \in T}$, where $T = [0, 1]$ and a family $(D_t)_{t \in T}$ of unary operations which satisfy (1), (2), (3) and (4) then L is a generalized Post algebra in the above sense.

Exercise 2. Verify the above statements in detail.

For further generalizations in this direction, see also Ph. Dwinger [5].

Generalized Post algebras with an "infinite chain of constants" which are on one hand a generalization of the generalized Post algebras introduced by T. Traczyk [4] and, on the other hand, a special case of the generalized Post algebras as introduced by Ph. Dwinger [5], have been examined by H. Sawicka [1].

Pseudo-Post algebras where introduced by G. Rousseau [1] and have already been mentioned in Section 2.

A definition of generalized Post algebras, where the "chain of constants" has been replaced by an arbitrary distributive lattice has been given by T. P. Speed [4].

B. Weglorz [1] has considered Post-like algebras as a certain equational class of algebras and N. C. Ho [1] considers a certain type of "generalized Post algebras" which relate to semi-lattices and which have applications to many valued logic with infinitely long formulas.

Finally, we mention that T. Traczyk [3] has examined the α-representability of α-complete Post algebras (α-representable, α-complete distributive lattices will be discussed in Chapter XII) and has generalized the Loomis-Sikorski theorem (XII.5.21) to σ-complete Post algebras. See also H. Sawicka [2].

9. Stone algebras of order n

In this section we will discuss a special class of Stone algebras: the Stone algebras of order $n \geq 1$. They can also be considered as a generalization of Post algebras and were introduced by T. Katrinák and A. Mitschke [1]. Since Post algebras of order n, considered as algebras (cf. Section 6), and Stone algebras are of different similarity types, it will be practical to consider—in this section—Post algebras and Stone algebras as subclasses of the class $\mathbf{D_{01}}$. Thus, a Post algebra is a member of $\mathbf{D_{01}}$ satisfying the conditions of Definition 2.1 and a Stone algebra is a member of $\mathbf{D_{01}}$ which is pseudocomplemented and for which the identity $\mathbf{x^* + x^{**} = 1}$ holds. (Note, however, that a $\mathbf{D_{01}}$-isomorphism between Stone algebras is also a $\mathbf{B_1}$-isomorphism.) The results in this section are due to T. Katrinák and A. Mitschke [1].

In this section, we will always denote the chain of constants of a Post algebra $L \neq \mathbf{1}$ of order $n \geq 2$, by $0 = c_0 < c_1 < \cdots < c_{n-1} = 1$, and if we write $a = a_1c_1 + \cdots + a_{n-1}c_{n-1}$ for $a \in L$, we will assume that this is the monotonic representation of a.

We recall (Exercise 2.9) that every Post algebra is a Stone algebra. Indeed, if $L \neq \mathbf{1}$ is a Post algebra of order $n \geq 2$ and $a = \sum_{i=1}^{n-1} a_i c_i$, $a \in L$, then it is easy to see that $a^* = \bar{a}_1$ and thus $a^{**} = a_1$ and so $a^{**} + a^* = 1$. Of course, the fact that L is a Stone algebra also follows from Theorem 2.3 and VIII.9.2. From this it follows that $\mathscr{D}_\mathscr{d}(L) = \{a \in L : a = c_1 + a_2c_2 + \cdots + a_{n-1}c_{n-1}\}$. Obviously, $\mathscr{D}_\mathscr{d}(L)$ is again a Post algebra of order $n - 1$, whose zero element is c_1 and $\mathscr{D}_\mathscr{d}(L) = [c_1)$. Again, $\mathscr{D}_\mathscr{d}(L)$ is a Stone algebra and $\mathscr{D}_\mathscr{d}(\mathscr{D}_\mathscr{d}(L)) = [c_2)$. Writing $\mathscr{D}_\mathscr{d}^i(L)$ for $\mathscr{D}_\mathscr{d}(\mathscr{D}_\mathscr{d}^{i-1}(L))$, for $i = 2, \cdots, n - 1$, we see that $\mathscr{D}_\mathscr{d}^i(L) = [c_i)$ is a Post algebra of order $n - i$ for $i \leq n - 2$ and is also a Stone algebra. In particular, $\mathscr{D}_\mathscr{d}^{n-2}(L)$ is a Boolean algebra and $\mathscr{D}_\mathscr{d}^{n-1}(L) = \mathbf{1}$.

Theorem 1. *Let L be a Post algebra of order n, $n \geq 2$. Then L is a relative Stone algebra.*

Proof. $\mathscr{D}_\mathscr{d}^{n-1}(L) = \mathbf{1}$, which is a relative Stone algebra. The theorem follows then from Theorem VIII.7.14 and by induction.

Remark. It also follows from Exercise IX.2.11 that L is a Heyting algebra in which the identity $\mathbf{x} \rightarrow \mathbf{y} + \mathbf{y} \rightarrow \mathbf{x} = \mathbf{1}$ holds.

Exercise 2. Let L be a Post algebra of order $n \geq 2$. Let $a = \sum_{i=1}^{n-1} a_i c_i$, $b = \sum_{i=1}^{n-1} b_i c_i$, $a, b \in L$. Prove that $a \rightarrow b = \sum_{i=1}^{n-1} (a \rightarrow b)_i c_i$ where $(a \rightarrow b)_i = (\bar{a}_1 + b_1) \cdots (\bar{a}_i + b_i)$ for $i = 1, \cdots, n - 1$. Also verify directly that $a \rightarrow b + b \rightarrow a = 1$.

Exercise 3. Let L be a Post algebra of order $n \geq 2$. Let $a = \sum_{i=1}^{n-1} a_i c_i$, $b = \sum_{i=1}^{n-1} b_i c_i$, $x = \sum_{i=1}^{n-1} x_i c_i$, $a \leq x \leq b$, $a, x, b \in L$. Prove that if $x^{*[a,b]} = \sum_{i=1}^{n-1} (x^{*[a,b]})_i c_i$ then $(x^{*[a,b]})_i = (\bar{x}_1 + a_1)(\bar{x}_2 + a_2) \cdots (\bar{x}_i + a_i) b_i$ for $i = 1, \cdots, n - 1$.

Before giving the definition of Stone algebras of order n, we recall (Theorem VIII.7.1 and Exercise VIII.7.7) that if L is a Stone algebra, then $\mathscr{C}(L) = \mathscr{R}_\mathscr{g}(L)$ is a retract of L.

Definition 4. Let L be a Stone algebra. L is a *Stone algebra of order 1*, if $L = \mathbf{1}$. L is a *Stone algebra of order n*, $n \geq 2$, if $L \neq \mathbf{1}$ and $\mathscr{D}_\mathscr{d}(L)$ is a Stone algebra of order $n - 1$.

If L is a Stone algebra of order $n \geq 2$ we will again write $\mathscr{D}_\mathscr{d}^i(L)$ for $\mathscr{D}_\mathscr{d}(\mathscr{D}_\mathscr{d}^{i-1}(L))$ for $i = 2, \cdots, n - 1$ and also, for convenience, use the symbols $\mathscr{D}_\mathscr{d}^0(L)$ and $\mathscr{D}_\mathscr{d}^1(L)$ to denote L and $\mathscr{D}_\mathscr{d}(L)$ respectively and thus $L = \mathscr{D}_\mathscr{d}^0(L)$

$\supseteq \mathscr{D}\!\mathscr{s}^1(L) \supseteq \mathscr{D}\!\mathscr{s}^2(L) \supseteq \cdots \supseteq \mathscr{D}\!\mathscr{s}^{n-1}(L) = \mathbf{1}$ are Stone algebras. We will denote the smallest element of $\mathscr{D}\!\mathscr{s}^i(L)$ by c_i for $i = 0, 1, \cdots, n-1$, thus $\mathscr{D}\!\mathscr{s}^i(L) = [c_i)$ for $i = 0, 1, \cdots, n-1$ and the chain $c_1 < \cdots < c_{n-1} = 1$ is called the *chain of smallest dense elements* of L. We will also denote the element 0 by c_0. It follows from the previous discussion that if $L \neq \mathbf{1}$ is a Post algebra of order $n \geq 2$ then L is a Stone algebra of order n and the chain of constants of L (0 deleted) is the chain of smallest dense elements of L. Note that this again yields the result (Section 2) that the chain of constants of a Post algebra is unique determined.

Theorem 5. *A Stone algebra of order n is relative Stone, and is therefore a Heyting algebra in which the identity* $\mathbf{x} \rightarrow \mathbf{y} + \mathbf{y} \rightarrow \mathbf{x} = \mathbf{1}$ *holds.*

Proof. Similar to proof of Theorem 1.

Before we can state the representation theorem for Stone algebras of order n, we first define a certain class of lattices, *P-lattices*, which can be considered as a generalization of Post algebras.

First we make the following observation. If L is a Post algebra of order $n \geq 2$ and $B = \mathscr{C}(L)$, then it follows immediately from Theorem 2.4 that L is isomorphic to a $\mathbf{D_{01}}$-subalgebra L' of $\times_{i=1}^{n-1} B_i$, where $B_i = B$ for i $= 1, \cdots, n-1$ and where L' is defined by $L' = \{a \in \times_{i=1}^{n-1} B_i : a_{i+1} \leq a_i$ for $i = 1, \cdots, n-2\}$ (we write a_i instead of $a(i)$). This observation suggests the following generalization. Suppose $B_1, B_2, \cdots, B_{n-1}, n \geq 2$ are Boolean algebras and $h_i\colon B_i \rightarrow B_{i+1}$ are **B**-homomorphisms for $i = 1, \cdots, n-2$. Let $B = \times_{i=1}^{n-1} B_i$. Define L by

$$L = \left\{a \in \mathop{\times}_{i=1}^{n-1} B_i : a_{i+1} \leq h_i(a_i) \text{ for } i = 1, \cdots, n-2\right\}$$

(we write again a_i instead of $a(i)$). It is easy to see that $L \in \mathbf{D_{01}}$. In fact L is a $\mathbf{D_{01}}$-subalgebra of B. But L is also a Stone algebra of order n. Indeed, if $a \in L$, $a = (a_1, \cdots, a_{n-1})$ then it is easy to see that

$$a^* = (\bar{a}_1, h_1(\bar{a}_1), (h_2 \circ h_1)(\bar{a}_1), \cdots, (h_{n-2} \circ \cdots \circ h_1)(\bar{a}_1)).$$

Note that

$$a \in \mathscr{C}(L) = \mathscr{R}\!\mathscr{g}(L) \Leftrightarrow a^{**} = a \Leftrightarrow$$

$$a = (a_1, h_1(a_1), (h_2 \circ h_1)(a_1), \cdots, (h_{n-2} \circ \cdots \circ h_1)(a_1)).$$

It follows that $\mathscr{C}(L) \cong B_1$. Moreover for $a \in L$, we have $a \in \mathscr{D}\!\mathscr{s}(L) \Leftrightarrow a^* = 0 \Leftrightarrow a_1 = 1$. So $\mathscr{D}\!\mathscr{s}(L) = \{a \in L : a_1 = 1\}$. The smallest element c_1 of $\mathscr{D}\!\mathscr{s}(L)$ is obviously defined by $c_1 = (1, 0, 0, \cdots, 0)$ and $\mathscr{D}\!\mathscr{s}(L) = [c_1)$. Again, it easily follows that $\mathscr{D}\!\mathscr{s}(L)$ is a Stone algebra and if $a \in \mathscr{D}\!\mathscr{s}(L)$, $a = (1, a_2, \cdots, a_{n-1})$ then

$$a^{*\mathscr{D}\!\mathscr{s}(L)} = a^{*[c_1)} = (1, \bar{a}_2, h_2(\bar{a}_2), \cdots, (h_{n-2} \circ \cdots \circ h_2)(\bar{a}_2))$$

and

$$a \in \mathscr{C}(\mathscr{D}\!\mathscr{s}(L)) \Leftrightarrow a = (1, a_2, h_2(a_2), \cdots, (h_{n-2} \circ \cdots \circ h_2)(a_2)).$$

So $\mathscr{C}(\mathscr{D}\delta(L)) = \mathscr{C}([c_1]) \cong B_2$. Finally the map $h_1': \mathscr{C}(L) \to \mathscr{C}(\mathscr{D}\delta(L))$ defined by $h_1'(a) = a + c_1$ is obviously a **B**-homomorphism and the diagram

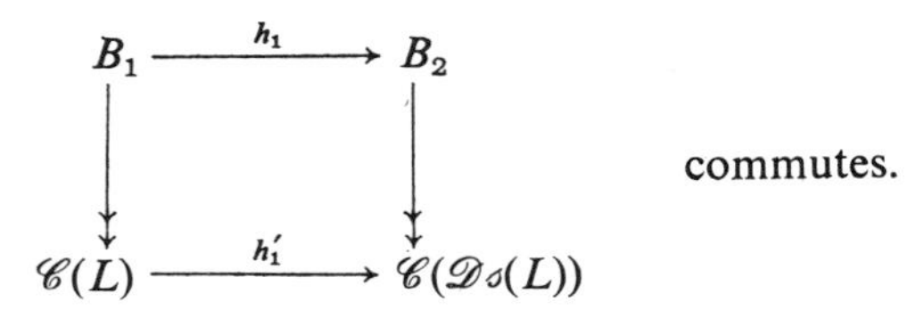

commutes.

Proceeding in this way, we see that L is a Stone algebra of order n and thus by Theorem 5, L is a relative Stone algebra.

If $c_1 < c_2 < \cdots < c_{n-1} = 1$ is the chain of smallest dense elements then

$$c_1 = (1, 0, 0 \cdots 0),\ c_2 = (1, 1, 0, \cdots 0), \cdots, c_{n-1} = (1, 1, \cdots 1)$$

whereas

$$\mathscr{D}\delta^i(L) = \{a \in L : a_1 = a_2 = \cdots = a_i = 1\},$$

for $i = 1, \cdots, n - 1$; and we have

$$\mathscr{C}(\mathscr{D}\delta^i(L)) = \{(1, 1, \cdots, 1, a_{i+1}, h_{i+1}(a_{i+1}), \cdots, (h_{n-2} \circ \cdots \circ h_{i+1})(a_{i+1})) : a \in L\}$$

for $i = 1, 2, \cdots, n - 1$. It is easy to see that

(1) $\mathscr{C}(\mathscr{D}\delta^i(L)) \cong B_{i+1}$ for $i = 0, 1, \cdots, n - 2$.

Finally, defining $h_i': \mathscr{C}(\mathscr{D}\delta^{i-1}(L)) \to \mathscr{C}(\mathscr{D}\delta^i(L))$ by

$$h_i'(a) = a + c_i, \text{ for } i = 1, \cdots, n - 2$$

we have that h_i' is a **B**-homomorphism for $i = 1, \cdots, n - 2$ and the diagram

$$(2)\qquad \begin{array}{ccc} B_i & \xrightarrow{h_i} & B_{i+1} \\ \downarrow & & \downarrow \\ \mathscr{C}(\mathscr{D}\delta^{i-1}(L)) & \xrightarrow{h_i'} & \mathscr{C}(\mathscr{D}\delta^i(L)) \end{array}$$

commutes for $i = 1, \cdots, n - 2$.

Exercise 6. Verify (1) and (2). Note that if all h_i are isomorphisms then L is of course a Post algebra of order n, whose chain of constants is $0 = c_0 < c_1 < \cdots < c_{n-1} = 1$.

Definition 7. Suppose $B_1, B_2, \cdots, B_{n-1}, n \geq 2$ are Boolean algebras and $h_i: B_i \to B_{i+1}$ are **B**-homomorphisms for $i = 1, \cdots, n - 2$. If $L = \{a \in \times_{i=1}^{n-1} B_i : a_{i+1} \leq h(a_i) \text{ for } i = 1, \cdots, n - 2\}$ then L is called a *P-lattice.*

Note that it follows immediately from (1) and (2) that if L is a P-lattice, then the B_i and h_i are uniquely determined (up to isomorphisms).

The representation theorem now states that every Stone algebra of order $n \geq 2$ is isomorphic to a P-lattice. We need first some more notation and theorems.

If L is a Stone algebra of order $n \geq 2$ whose chain of smallest dense elements is $c_1 < c_2 < \cdots < c_{n-1} = 1$, then we will also denote $\mathscr{C}(\mathscr{D}d^i(L))$ by B_{i+1} for $i = 0, 1, \cdots, n-2$.

If $a \in \mathscr{D}d^i(L)$, $i = 1, \cdots, n-1$ then we will denote the pseudocomplement $a^{*[c_i)}$ of a in $\mathscr{D}d^i(L)$ by a^{*_i}. For the pseudocomplement a^* of a in L, we will also use the symbol a^{*_0}.

Theorem 8. *Let L be a Stone algebra of order $n \geq 2$ with chain of smallest dense elements $c_1 < c_2 < \cdots < c_{n-1} = 1$. If $x \in L$, then x has a unique representation of the form* (i) $x = x_1 \cdots x_{n-1}$, *where $x_i \in B_i$ for $i = 1, \cdots, n-1$ satisfies* (ii) $x_i + c_i \geq x_{i+1}$ *for* $i = 1, \cdots, n-2$.

Proof. We first prove that x can be represented as in (i). The case $n = 2$ is obvious, so assume $n > 2$. If $x \in L$, then $x = x^{**}(x + c_1)$. Indeed, $x^{**}(x + c_1) = x + x^{**}c_1$. But $x + x^* \in \mathscr{D}d(L)$, so $x + x^* \geq c_1$ and thus $x = x^{**}(x + x^*) \geq x^{**}c_1$. Thus, letting $x_1 = x^{**}$ and $d_1 = x + c_1$, we can write $x = x_1 d_1$, where $x_1 \in B_1$, and $d_1 \in \mathscr{D}d(L)$. Again, applying the same procedure to d_1, we can write $d_1 = d_1^{*_1*_1}(d_1 + c_2) = (x + c_1)^{*_1*_1}(x + c_2)$ or $d_1 = x_2(x + c_2)$, where $x_2 = (x + c_1)^{*_1*_1}$. Proceeding in this way we find: $x = x_1 x_2 \cdots x_{n-1}$ where

(3) $x_i = (x + c_{i-1})^{*_{i-1}*_{i-1}}$ for $i = 1, 2, \cdots, n-1$ (and where, as usual, $c_0 = 0$)

Next, we show that the x_i satisfy (ii). Indeed $x_i \geq x$ and thus $x_i + c_i \geq x + c_i$ for $i = 1, \cdots, n-2$. But $x_i \in B_i = \mathscr{C}[c_{i-1}, 1]$ and thus $x_i + c_i \in B_{i+1} = \mathscr{C}[c_i, 1]$ so

$x_i + c_i = (x_i + c_i)^{*_i*_i} \geq (x + c_i)^{*_i*_i}$ or, by (3),

$x_i + c_i \geq x_{i+1}$ for $i = 1, \cdots, n-2$.

Finally, we prove the uniqueness. If $n = 2$, then L is a Boolean algebra and the uniqueness of the expression (i), subject to the condition (ii) for $x \in L$, is obvious. We assume therefore that the uniqueness has been proven for Stone algebras of order $< n$, where $n > 2$ and we prove the uniqueness for the case n. Suppose $x \in L$ and $x = x_1 \cdots x_{n-1} = y_1 \cdots y_{n-1}$, $x_i, y_i \in B_i$ for $i = 1, \cdots, n-1$ and $x_i + c_i \geq x_{i+1}$, $y_i + c_i \geq y_{i+1}$ for $i = 1, \cdots, n-2$. Hence $x^{**} = x_1^{**} \cdots x^{**}_{n-1}$ but $x_i \in \mathscr{D}d(L)$ for $i = 2, \cdots, n-2$ so $x^{**} = x_1^{**}$ and similarly $x^{**} = y_1^{**}$. But x_1 and $y_1 \in B_1$ thus $x^{**} = x_1 = y_1$. Let $d_1 = x_2 \cdots x_{n-1}$, $d_1' = y_2 \cdots y_{n-2}$, thus $x = x_1 d_1 = y_1 d_1'$ and $d_1, d_1' \geq c_1$. By hypothesis $x_1 + c_1 \geq x_2$, $y_1 + c_1 \geq y_2$ but $x_1 = y_1$, so $x_1 + c_1 \geq x_2, y_2$. It follows that $x_1 + c_1 \geq d_1, d_1'$. Thus $x + c_1 = x_1 d_1 + c_1 = (x + c_1)(d_1 + c_1) = d_1$ and similarly $x + c_1 = d_1'$. Hence $d_1 = d_1'$. But d_1 and d_1' are elements of $[c_1)$ which is a Stone algebra of order $n - 1$ and it follows from the induction hypothesis that $x_i = y_i$ for $i = 2, \cdots, n-1$.

From now on we will call the representation $x = x_1 \cdots x_{n-1}$ of an element x of a Stone algebra of order $n \geq 2$ which satisfies (i) and (ii) of Theorem 8, the *canonical representation* of x.

Theorem 9. *Let L be a Stone algebra of order $n \geq 2$. Suppose $x, y \in L$ with canonical representations $x = x_1 \cdots x_{n-1}$, $y = y_1 \cdots y_{n-1}$. Then $x \leq y \Leftrightarrow x_i \leq y_i$ for $i = 1, \cdots, n-1$. Moreover $xy = (x_1 y_1) \cdots (x_{n-1} y_{n-1})$* and *$x + y = (x_1 + y_1) \cdots (x_{n-1} + y_{n-1})$ are canonical representations.*

Proof. Obviously $x_i y_i \in B_i$ for $i = 1, \cdots, n-1$ and $x_i y_i + c_i = (x_i + c_i)(y_i + c_i) \geq x_{i+1} y_{i+1}$ for $i = 1, \cdots, n-2$. It follows that $xy = (x_1 y_1) \cdots (x_{n-1} y_{n-1})$ is the canonical representation of xy. If $x \leq y$ then $x = xy$, so by Theorem 8, $x_i = x_i y_i$ or $x_i \leq y_i$ for $i = 1, \cdots, n-1$. Conversely, if $x_i \leq y_i$ for $i = 1, \cdots, n-1$ then $x_i y_i = x_i$ for $i = 1, \cdots, n-1$ and thus $xy = x$ or $x \leq y$. Finally, $x_i + y_i \in B_i$ for $i = 1, \cdots, n-1$ and $x_i + y_i + c_i = x_i + c_i + y_i + c_i \geq x_{i+1} + y_{i+1}$ for $i = 1, \cdots, n-2$. So letting $z = (x_1 + y_1) \cdots (x_{n-1} + y_{n-1})$ we obtain the canonical representation of z. Since $x_i + y_i \geq x_i, y_i$ for $i = 1, \cdots, n-1$, $z \geq x, y$. If $u \geq x, y$, $u \in L$ and $u = u_1 \cdots u_{n-1}$ is the canonical representation of u then $u_i \geq x_i, y_i$ so $u_i \geq x_i + y_i$ for $i = 1, \cdots, n-1$ and thus $u \geq z$. It follows that $z = x + y$.

We can now prove the representation theorem.

Theorem 10. (T. Katrinák and A. Mitschke [1]) *Every Stone algebra of order $n \geq 2$ is isomorphic to a P-lattice.*

Proof. Let L be a Stone algebra of order n, $n \geq 2$. Define $h_i: B_i \rightarrow B_{i+1}$ by $h_i(x) = x + c_i$, for $i = 1, \cdots, n-2$. Obviously, $h_i \in [B_i, B_{i+1}]_{\mathscr{B}}$. Let L' be the P-lattice defined by

$$L' = \left\{ a \in \bigtimes_{i=1}^{n-1} B_i : a_{i+1} \leq h_i(a_i) \text{ for } i = 1, \cdots, n-2 \right\}$$

(where again we write a_i for $a(i)$, for $i = 1, \cdots, n-1$). Define $f: L \rightarrow L'$ by $f(x)(i) = x_i$ for $x \in L$ and where $x = x_1 \cdots x_{n-1}$ is the canonical representation of x. Since, by Theorem 8(ii), $x_{i+1} \leq h_i(x_i)$ for $i = 1, \cdots, n-2$ we have indeed that $f(x) \in L'$. That f is an isomorphism is an immediate consequence of Theorem 9.

In this section we have considered Stone algebras of order n as special members of $\mathbf{D}_{01}$. One can however also define Stone algebras of order n as an equational class of algebras of a certain similarity type. Obviously the operations will be $+, \cdot, ^*, c_0, \cdots c_n$. We will however not go into this but refer the reader to the literature (T. Katrinák and A. Mitschke [1]).

XI

De Morgan Algebras and Lukasiewicz Algebras

1. Introduction

We have already observed in Section X.1 that besides Post algebras, the Lukasiewicz algebras of order $n \geq 2$ are also algebras corresponding to higher valued propositional logic. Lukasiewicz algebras of order 3, were introduced first by G. C. Moisil [2], [3], and investigated further by several authors such as A. Monteiro [3], [5], L. Monteiro [1], L. Monteiro and L. Coppola [1], J. Varlet [3], R. Cignoli [1], [4] and C. O. Sicoe [1], [2], [3], [4] (also cf. A. Tarski [6]). Before we start our discussion of Lukasiewicz algebras we will first discuss de Morgan algebras which are related to constructive logic with strong negation (A. A. Markov [1], D. Nelson [1]), and N. N. Vorobev [1]). As in the case of Lukasiewicz algebras, de Morgan algebras are also interesting from the algebraic point of view. They also form a natural introduction to Lukasiewicz algebras. De Morgan algebras were studied by J. A. Kalman [1] (as *i*-lattices), G. C. Moisil [1], A. Bialynicki-Birula and H. Rasiowa [1], A. Bialynicki-Birula [1] (under the name of quasi-Boolean algebras), A. Monteiro [2], R. Cignoli [5], and others.

In contrast to our treatment of Post algebras, which were defined as a special subclass of $\mathbf{D_{01}}$ and later algebraically, we will define Lukasiewicz algebras immediately as algebras. In a sense, Lukasiewicz algebras of order $n \geq 2$ can be obtained by deleting the nullary operations (except 0 and 1) from Post algebras of order n. In fact, we will see that a Post algebra of order n can be considered as a special case of a Lukasiewicz algebra of the same order, and that every Lukasiewicz algebra of order $n \geq 2$ can be embedded in a Post algebra of order n. This situation is completely described by the fact that the category $\mathscr{P}_\mathbf{n}$ of Post algebras of order $n \geq 2$ is a reflective subcategory of the category $\mathscr{L}_\mathbf{n}$ of Lukasiewicz algebras of order n. We will also determine the subdirectly irreducibles in the class $\mathbf{L_n}$ of Lukasiewicz algebras of order n, and it will not be surprising that the results obtained here have a certain resemblance to the results obtained for Post algebras. For example, we have seen (Theorem X.6.5) that the subdirectly irreducibles in $\mathbf{P_n}$ are the chains $\mathbf{n}$. The subdirectly irreducibles in $\mathbf{L_n}$ are the $\mathbf{L_n}$-subalgebras of $\mathbf{n}$ (Theorem 7.1).

We also mention that generalizations of Lukasiewicz algebras of order n (and in connection with these, Post algebras of order n) have been investigated recently (e.g. G. Georgescu and C. Vraciu [4]).

We close this introduction by mentioning that another class of algebras—the Nelson algebras—have been studied in connection with logic. We refer the interested reader to the literature (H. Rasiowa [1], A. Monteiro [4], D. Brignole and A. Monteiro [1]).

2. De Morgan algebras. Basic properties and representation theory

Definition 1. A *de Morgan algebra* is an algebra of the form $(L, (+, \cdot, ^-, 0, 1))$ where $+$ and $\cdot$ are binary operations, $^-$ is a unary operation and 0, 1 are nullary operations, satisfying:

M_1: A set of identities in $+$, $\cdot$, and with a 0 and 1 which define distributive lattices with 0, 1.

M_2: $\overline{\mathbf{x}+\mathbf{y}} = \bar{\mathbf{x}}\bar{\mathbf{y}}$; $\overline{\mathbf{xy}} = \bar{\mathbf{x}} + \bar{\mathbf{y}}$ is an identity.

M_3: $\bar{\bar{\mathbf{x}}} = \mathbf{x}$ is an identity.

We note that it follows from the axioms that $\bar{0} = 1$ and $\bar{1} = 0$. The assignment $x \mapsto \bar{x}$ satisfies $x \leq y \Leftrightarrow \bar{x} \geq \bar{y}$ for all $x, y \in L$ and is onto L. Thus, the mapping $x \mapsto \bar{x}$ is an antiautomorphism of L. If X is a set, then a function $\varphi: X \to X$ which satisfies $\varphi \circ \varphi = 1_X$ is called an *involution of* X. Thus, the assignment $x \mapsto \bar{x}$ is an involution. Conversely, if $\varphi: L \to L$ is an involution of L, $L \in \mathbf{D}_{01}$ which is at the same time an antiautomorphism of L, then L can be made into a de Morgan algebra by defining $\bar{x} = \varphi(x)$ for $x \in L$. More generally, if X is a set and φ is an involution of X, then the ring of all subsets of X can be made into a de Morgan algebra, by defining $\bar{A} = X \sim \varphi[A]$ for each $A \subseteq X$. Conversely, if L is a de Morgan algebra and L consists of all subsets of a set X, where $\cup$ and $\cap$ are the binary operations and $\varnothing$ and X are the nullary operations, then we can associate with L an involution of X as follows. If $x \in X$, then $\{x\}$ is an atom of L and thus $\overline{\{x\}}$ is a dual atom of L and so $X \sim \overline{\{x\}}$ is a singleton. We define $\varphi(x)$ so that $\{\varphi(x)\} = X \sim \overline{\{x\}}$. Now $\{\varphi(x)\} \nsubseteq \overline{\{x\}} \Rightarrow \overline{\{\varphi(x)\}} \nsupseteq \{x\}$. But $\overline{\{\varphi(x)\}}$ is a dual atom of L, so $\{x\} = X \sim \overline{\{\varphi(x)\}}$ and it follows that $\varphi(\varphi(x)) = x$ and therefore φ is an involution. This leads us to the following definition.

Definition 2. Let L be a de Morgan algebra. If L consists of all the subsets of a set X with $\cup$ and $\cap$ as binary operations and $\varnothing$ and X as nullary operations, then L is called a *complete de Morgan field of sets*. If L is a subalgebra of a complete de Morgan field of sets, then L is called a *de Morgan field of sets*.

The reader should note that the operation $^-$ is, in general, not set theoretic complementation.

Exercise 3. (A. Bialynicki-Birula and H. Rasiowa [1]) Prove that there is a one-one correspondence between all complete de Morgan fields of subsets of a set X and the involutions of X.

We will denote the equational class of de Morgan algebras by $\mathbf{M}$ and the associated equational category by $\mathscr{M}$. Note that the classes $\mathbf{M}$ and $\mathbf{B}$ have the same similarity type and moreover the category $\mathscr{B}$ of Boolean algebras is a full subcategory of $\mathscr{M}$. Another less trivial example of a class of algebras which can be considered as de Morgan algebras are the Post algebras. Indeed, suppose L is a Post algebra of order $n \geq 2$. Define a unary operation on L, by $\bar{x} = \sum_{i=1}^{n-1} \bar{a}_{n-i}c_i$, where $x = \sum_{i=1}^{n-1} a_i c_i$ is the monotonic representation of x, $x \in L$, and $\bar{a}_{n-i}$ is the complement of a_{n-i} in $\mathscr{C}(L)$. It is not difficult to see that under the operations $+, \cdot, ^-, 0$ and 1, L becomes a de Morgan algebra.

Unlike any of the classes we have considered, order-isomorphisms and **M**-isomorphisms are not the same in **M**. Indeed, the Boolean algebra $\mathbf{2}^2$ is not **M**-isomorphic with the de Morgan algebra obtained from $\mathbf{2}^2$ by redefining $\overline{(\mathbf{0}, \mathbf{0})} = (\mathbf{1}, \mathbf{1})$, $\overline{(\mathbf{0}, \mathbf{1})} = (\mathbf{0}, \mathbf{1})$, $\overline{(\mathbf{1}, \mathbf{0})} = (\mathbf{1}, \mathbf{0})$, and $\overline{(\mathbf{1}, \mathbf{1})} = (\mathbf{0}, \mathbf{0})$.

Our first objective will be to determine the subdirectly irreducibles in **M**.

For the purpose of our discussion we will introduce the following notation. If $L \in \mathbf{M}$, then we will denote the member of $\mathbf{D}_{01}$ which is obtained from L by deleting the operation $^-$, by L^*. Also, if $L \in \mathbf{M}$ and $a, b \in L$, $a \leq b$, $\boldsymbol{\theta}(a, b)$ will denote the smallest $\theta \in \mathfrak{C}(L)$ for which $(a, b) \in \theta$, and $\boldsymbol{\theta}^*(a, b)$ will denote the smallest $\theta \in \mathfrak{C}(L^*)$ for which $(a, b) \in \theta$ (cf. Exercise II.9.24). Note that for $a, b \in L$, $a \leq b$, $L \in \mathbf{M}$, $\boldsymbol{\theta}^*(a, b) \subseteq \boldsymbol{\theta}(a, b)$. Also, products in $\mathfrak{C}(L)$ and $\mathfrak{C}(L^*)$ are the same (intersections), but sums need not be the same. If θ_1 and θ_2 are in $\mathfrak{C}(L^*)$, $L \in \mathbf{M}$, then we will denote the sum of θ_1 and θ_2 in $\mathfrak{C}(L^*)$ by $\theta_1 \overset{\mathfrak{C}(L^*)}{+} \theta_2$.

Lemma 4. *Let $L \in \mathbf{M}$ and let $a, b \in L$, $a \leq b$. Then $\boldsymbol{\theta}(a, b) = \boldsymbol{\theta}^*(a, b) \overset{\mathfrak{C}(L^*)}{+} \boldsymbol{\theta}^*(\bar{b}, \bar{a})$.*

Proof. We have $\boldsymbol{\theta}(a, b) \geq \boldsymbol{\theta}^*(a, b)$. But $(\bar{b}, \bar{a}) \in \boldsymbol{\theta}(a, b)$ so also $\boldsymbol{\theta}(a, b) \geq \boldsymbol{\theta}^*(\bar{b}, \bar{a})$, hence $\boldsymbol{\theta}(a, b) \geq \boldsymbol{\theta}^*(a, b) \overset{\mathfrak{C}(L^*)}{+} \boldsymbol{\theta}^*(\bar{b}, \bar{a})$. But since $(a, b) \in \boldsymbol{\theta}^*(a, b)$, we have $(a, b) \in \boldsymbol{\theta}(a, b) \overset{\mathfrak{C}(L^*)}{+} \boldsymbol{\theta}(\bar{b}, \bar{a})$. To complete the proof, it suffices to show that $\boldsymbol{\theta}^*(a, b) \overset{\mathfrak{C}(L^*)}{+} \boldsymbol{\theta}^*(\bar{b}, \bar{a}) \in \mathfrak{C}(L)$. For this we only need to show that $(x, y) \in \boldsymbol{\theta}^*(a, b) \overset{\mathfrak{C}(L^*)}{+} \boldsymbol{\theta}^*(\bar{b}, \bar{a})$ implies $(\bar{x}, \bar{y}) \in \boldsymbol{\theta}^*(a, b) \overset{[\mathfrak{C}(L^*)}{+} \boldsymbol{\theta}^*(\bar{b}, \bar{a})$ for $x, y \in L$. Now it follows from II.9.14 and II.9.24 that there exist $x = z_0, z_1, \cdots, z_{n-1}, z_n = y$ such that $(x, z_1) \in \boldsymbol{\theta}^*(a, b)$, $(z_1, z_2) \in \boldsymbol{\theta}^*(\bar{b}, \bar{a}), \cdots, (z_{n-1}, y) \in \boldsymbol{\theta}^*(a, b)$ and $ax = az_1$, $z_1\bar{b} = z_2\bar{b}, \cdots, z_{n-1}a = ya$, $b + x = b + z_1$, $z_1 + \bar{a} = z_2 + \bar{a}, \cdots, z_{n-1} + b = y + b$. By applying the $^-$ operation to all these equalities and letting $w_i = \bar{z}_i$ for $i = 1, \cdots, n - 1$ we have $(\bar{x}, w_1) \in \boldsymbol{\theta}^*(\bar{b}, \bar{a})$, $(w_1, w_2) \in \boldsymbol{\theta}^*(a, b), \cdots, (w_{n-1}, \bar{y}) \in \boldsymbol{\theta}^*(\bar{b}, \bar{a})$ so $(\bar{x}, \bar{y}) \in \boldsymbol{\theta}^*(a, b) \overset{\mathfrak{C}(L^*)}{+} \boldsymbol{\theta}^*(\bar{b}, \bar{a})$.

Lemma 5. *Let $L \in \mathbf{M}$. If there exists $a \in L \sim \{0, 1, \bar{a}\}$ then L is not subdirectly irreducible.*

Proof. We will construct two non-trivial congruence relations θ_1 and θ_2, such that $\theta_1\theta_2 = 0$.

Case 1. *a and $\bar{a}$ not comparable.* Then $a < a + \bar{a}$ and $\bar{a} < a + \bar{a}$. Let $\theta_1 = \boldsymbol{\theta}(a, a + \bar{a})$ and $\theta_2 = \boldsymbol{\theta}(\bar{a}, a + \bar{a})$. Obviously, $\theta_1 \neq 0$ and $\theta_2 \neq 0$. By Lemma 4,

$\theta_1 = \boldsymbol{\theta}^*(a, a + \bar{a}) \overset{\mathfrak{C}(L^*)}{+} \boldsymbol{\theta}^*(a\bar{a}, \bar{a})$. Now $(a\bar{a}, \bar{a}) = (a\bar{a}, \bar{a}(a + \bar{a})) \in \boldsymbol{\theta}^*(a, a + \bar{a})$ $\Rightarrow \boldsymbol{\theta}^*(a, a + \bar{a}) \geq \boldsymbol{\theta}^*(a\bar{a}, \bar{a})$ thus $\theta_1 = \boldsymbol{\theta}^*(a, a + \bar{a})$. Again,

$$\theta_2 = \boldsymbol{\theta}^*(\bar{a}, a + \bar{a}) \overset{\mathfrak{C}(L^*)}{+} \boldsymbol{\theta}^*(a\bar{a}, a).$$

Now $(a\bar{a}, a) \in \boldsymbol{\theta}^*(a\bar{a}, a) \Rightarrow (\bar{a}, a + \bar{a}) \in \boldsymbol{\theta}^*(a\bar{a}, a) \Rightarrow \boldsymbol{\theta}^*(\bar{a}, a + \bar{a}) \leq \boldsymbol{\theta}^*(a\bar{a}, a)$ so $\theta_2 = \boldsymbol{\theta}^*(a\bar{a}, a)$. Suppose $(x, y) \in \theta_1\theta_2$. Then $(x, y) \in \theta_1 = \boldsymbol{\theta}^*(a, a + \bar{a})$ so $xa = ya$. Also $(x, y) \in \theta_2$ so $x + a = y + a$. By distributivity, $x = y$, and it follows that $\theta_1\theta_2 = 0$.

Case 2. $a < \bar{a}$. Let $\theta_1 = \boldsymbol{\theta}(0, a)$, $\theta_2 = \boldsymbol{\theta}(a, \bar{a})$. Clearly, θ_1 and θ_2 are non-trivial. Note that $\theta_2 = \boldsymbol{\theta}^*(a, \bar{a})$ since $(x, y) \in \boldsymbol{\theta}^*(a, \bar{a}) \Rightarrow (\bar{x}, \bar{y}) \in \boldsymbol{\theta}^*(a, \bar{a})$. By Lemma 4 and by Theorem II.9.15 we have $\theta_1\theta_2 = \boldsymbol{\theta}^*(0, a)\boldsymbol{\theta}^*(a, \bar{a}) \overset{\mathfrak{C}(L^*)}{+} \boldsymbol{\theta}^*(\bar{a}, 1)\boldsymbol{\theta}^*(a, \bar{a})$. But $xa = ya$ and $x + a = y + a$ imply $x = y$, so $\boldsymbol{\theta}^*(0, a)\boldsymbol{\theta}^*(a, \bar{a}) = 0$. Similarly, $\boldsymbol{\theta}^*(\bar{a}, 1)\boldsymbol{\theta}^*(a, \bar{a}) = 0$ so $\theta_1\theta_2 = 0$.

Case 3. $\bar{a} < a$. Similar to Case 2.

We are now able to characterize the subdirectly irreducibles in $\mathbf{M}$. We introduce the following notation. Let M_0, M_1, and M_2 be the de Morgan algebras defined by $M_0 = \mathbf{2}$, $M_1 = \mathbf{3}$ with $\bar{\mathbf{1}} = \mathbf{1}$, and

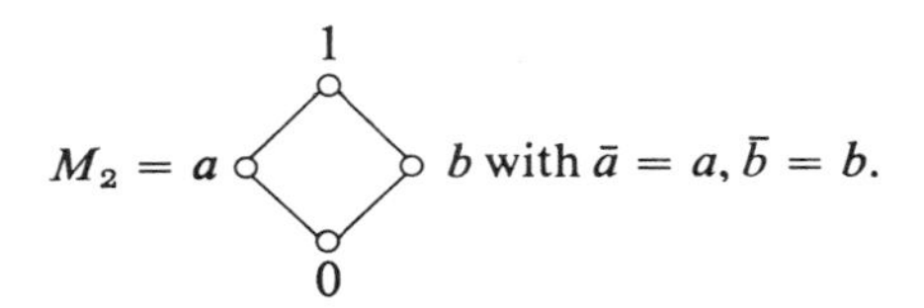

Theorem 6. (J. A. Kalman [1]) *The subdirectly irreducibles in* $\mathbf{M}$ *are* M_0, M_1 *and* M_2.

Proof. It is immediate that M_0, M_1, and M_2 are subdirectly irreducible. Conversely, suppose $L \in \mathbf{M}$ and L is subdirectly irreducible and L is neither isomorphic to M_0 nor to M_1. Then $|L| > 3$. If $|L| = 4$, then there exist $a, b \in L$, $a \neq b$, $a, b \neq 0, 1$. So by Lemma 5, $a = \bar{a}$ and $b = \bar{b}$. If $a + b \neq 1$, then $\bar{a}\bar{b} \neq 0$ so again by Lemma 5, $a + b = \overline{a + b} = \bar{a}\bar{b} = ab$ thus $a = b$, a contradiction. Hence $a + b = 1$ and $ab = 0$ and therefore $L \cong M_2$. If $|L| > 4$ then L has 5 distinct elements: $0, a, b, c, 1$, so by the preceding argument, $a + c = 1$, $a + b = 1$, $ac = 0$, and $ab = 0$. By distributivity $b = c$, a contradiction.

Remark. M_0 is isomorphic to a subalgebra of M_1, and M_1 is isomorphic to a subalgebra of M_2 as can be easily seen.

Corollary 7. *Let* $L \in \mathbf{M}$. *Then* L *is a subdirect product of copies of* M_0, M_1, *and* M_2.

Corollary 8. (A. Bialynicki-Birula and H. Rasiowa [1]) *Let* $L \in \mathbf{M}$. *Then* L *is isomorphic to a de Morgan field of sets.*

Proof. It follows from Corollary 7 and from the remark following the proof of Theorem 6, that L is isomorphic to a subalgebra of a product L_1 of copies of M_2. But $M_2 \simeq \mathbf{2} \times \mathbf{2}$ and so L_1^* is isomorphic to the ring of all subsets of a set. (Note that these last two isomorphisms are $\mathbf{D}_{01}$-isomorphisms.) Therefore L_1 is isomorphic to a complete de Morgan field of sets and L is isomorphic to a de Morgan field of sets.

3. Equational subclasses of M

We will denote the trivial equational subclass of $\mathbf{M}$, consisting of all one element algebras, by $\mathbf{M}_{-1}$. $\mathbf{M}_0$, $\mathbf{M}_1$, and $\mathbf{M}_2$ will denote the equational subclasses of $\mathbf{M}$ generated by M_0, M_1, and M_2, respectively.

Theorem 1. (cf. J. A. Kalman [1]) *The equational subclasses of* $\mathbf{M}$ *are* $\mathbf{M}_{-1} \subset \mathbf{M}_0 \subset \mathbf{M}_1 \subset \mathbf{M}_2 = \mathbf{M}$. *Moreover for* $L \in \mathbf{M}$, $L \in \mathbf{M}_0$ *if and only if*

(1) $\mathbf{x\bar{x} = 0}$ *is an identity in* L

and $L \in \mathbf{M}_1$ *if and only if*

(2) $\mathbf{(x + \bar{x})y\bar{y} + (y + \bar{y})x\bar{x} = y\bar{y} + x\bar{x}}$ *is an identity in* L.

Proof. Obviously, $\mathbf{M}_0$ is the class of Boolean algebras, $\mathbf{M}_{-1} \subset \mathbf{M}_0$, and $\mathbf{M}_1$ is characterized by (1). Since M_0 is a subalgebra of M_1, and M_1 is a subalgebra of M_2, we have $\mathbf{M}_0 \subseteq \mathbf{M}_1 \subseteq \mathbf{M}_2$ and by Theorem 2.6, $\mathbf{M}_2 = \mathbf{M}$. Since $\mathbf{x\bar{x} = 0}$ is an identity in M_0 but not in M_1, we have $\mathbf{M}_0 \subset \mathbf{M}_1$. An examination of the possible values in M_1 shows that (2) is an identity in M_1 but (2) does not hold in M_2, when $x = a$ and $y = b$. So $\mathbf{M}_1 \subset \mathbf{M}_2$. It easily follows that $\mathbf{M}_{-1}$, $\mathbf{M}_0$, $\mathbf{M}_1$, and $\mathbf{M}_2$ are all the equational subclasses of $\mathbf{M}$. Indeed, suppose $\mathbf{K}$ is an equational subclass of $\mathbf{M}$. If $\mathbf{K} \neq \mathbf{M}$, then $M_2 \notin \mathbf{K}$. Thus, if $\mathbf{K} \neq \mathbf{M}$, then for $L \in \mathbf{K}$, L is a subdirect product of copies of M_0 and M_1 only. So $\mathbf{K} \subseteq \mathbf{M}_1$. Again, if $\mathbf{K} \subset \mathbf{M}_1$ then $M_1 \notin \mathbf{K}$ so $\mathbf{K} \subseteq M_0$. If $\mathbf{K} \subset \mathbf{M}_0$ then $M_0 \notin \mathbf{K}$ so $\mathbf{K} = \mathbf{M}_{-1}$. Finally, suppose $L \in \mathbf{M}$ and (2) is an identity in L. If $L \in \mathbf{M} \sim \mathbf{M}_1$ then M_2 is a homomorphic image of L but (2) does not hold in $\mathbf{M}_2$, so $L \in \mathbf{M}_1$.

Exercise 2. (J. A. Kalman [1]) Let $L \in \mathbf{M}$. Prove that $L \in \mathbf{M}_1 \Leftrightarrow x\bar{x} \leq y + \bar{y}$ for all $x, y \in L$.

Exercise 3. If $L \in \mathbf{M}$ and L is a chain, then $L \in \mathbf{M}_1$.

Exercise 4. Let L be a Post algebra of order $n \geq 2$. Prove that L, considered as a de Morgan algebra (cf. Section 2), belongs to $\mathbf{M}_1$.

Remark. A distributive lattice in which (2) holds (or equivalently, the inequality of Exercise 2) is also called a normal *i*-lattice (J. A. Kalman [1]) or a Kleene algebra (D. Brignole and A. Monteiro [1] and S. C. Kleene [1] and [2]).

Exercise 5. (cf. A. Bialynicki-Birula and H. Rasiowa [1]) Let $L \in \text{Ob}\,\mathcal{M}$. Define $\varphi: \mathcal{S}(L) \to \mathcal{S}(L)$ by $\varphi(I) = L \sim I^-$ for $I \in \mathcal{S}(L)$ and where $I^- = \{\bar{a} : a \in I\}$. Prove that φ is an involution of $\mathcal{S}(L)$ and that $\hat{\bar{a}} = \mathcal{S}(L) \sim \varphi[\hat{a}]$ for $a \in L$.

Exercise 6. (cf. I. Petrescu [1]) Let L_1 and L_2 be objects of $\mathbf{M}$ and let the corresponding involutions on $\mathcal{S}(L_1)$ and $\mathcal{S}(L_2)$ as defined in Exercise 5, be φ_1 and φ_2, respectively. Suppose $f \in [L_1, L_2]_{\mathcal{D}_{01}}$. Prove that $f \in [L_1, L_2]_{\mathcal{M}}$ if and only if $\mathcal{S}(f) \circ \varphi_2 = \varphi_1 \circ \mathcal{S}(f)$.

4. Coproducts of de Morgan algebras. Free de Morgan algebras

We will use the symbol L^*, as in Section 2, to denote the member of $\mathbf{D_{01}}$ which is obtained from $L \in \mathbf{M}$ by deleting the unary operation on L. In addition, we will use the following notations. Recall that for a lattice L, the elements of $\breve{L}$ are the same as the elements of L but $a \leq b$ in $\breve{L}$ if and only if $a \geq b$ in L. The map $h_L: L \to \breve{L}$ is defined by $h_L(a) = a$ for $a \in L$. Note that $a \leq b \Leftrightarrow h_L(a) \geq h_L(b)$ for $a, b \in L$. If $f: L_1 \to L_2$ is a homomorphism between lattices then $\breve{f}: \breve{L}_1 \to \breve{L}_2$ is defined by $\breve{f}(h_{L_1}(a)) = h_{L_2}(f(a))$ for $a \in L_1$. Obviously, $\breve{f}$ is a lattice homomorphism and the diagram

(1) 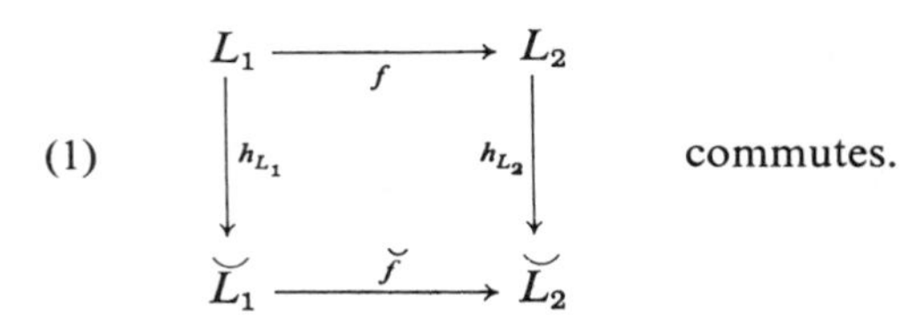 commutes.

Note that if $f_1: L_1 \to L_2, f_2: L_2 \to L_3$ are lattice homomorphisms, then $\widetilde{f_2 \circ f_1} = \breve{f}_2 \circ \breve{f}_1$. If $L \in \mathbf{M}$ then $\breve{L}$ can be made into a de Morgan lattice by defining $\overline{h_L(x)} = h_L(\bar{x})$. We will also use the symbol $\breve{L}$ to denote this de Morgan algebra. More generally, if $L \in \mathbf{D_{01}}$ and if $^-$ is a unary operation on L which makes L into a de Morgan algebra then we will (where there is no danger of confusion), also use the symbol L to denote this de Morgan algebra. Note that if $f \in [L_1, L_2]_{\mathcal{M}}$ then $\breve{f} \in [\breve{L}_1, \breve{L}_2]_{\mathcal{M}}$.

Lemma 1. *Let $(L_s)_{s \in S}$ be a family of objects in $\mathcal{D}_{01}$ and let $(j_s: L_s \to L)$ be a coproduct of this family. Then $(\breve{j}_s: \breve{L}_s \to L)$ is a coproduct of $(\breve{L}_s)_{s \in S}$.*

Proof. Suppose $(f_s: \breve{L}_s \to L_1)_{s \in S}$ is a family of morphisms in $\mathcal{D}_{01}$ then there exists a (unique) morphism $g \in [L, \breve{L}_1]$ with $g \circ j_s = \breve{f}_s$ for each $s \in S$, hence $\breve{g} \circ \breve{j}_s = f_s$ for each $s \in S$.

Theorem 2. (J. Berman and Ph. Dwinger [1]) *Let $(L_s)_{s \in S}$ be a family of objects in $\mathcal{M}$ and let $(j_s: L_s^* \to L)_{s \in S}$ be a coproduct of $(L_s^*)_{s \in S}$ (in $\mathcal{D}_{01}$). Then L admits a (unique) unary operation $^-$ which makes L into a de Morgan algebra such that $(j_s: L_s \to L)$ is a coproduct of $(L_s)_{s \in S}$ (in $\mathcal{M}$).*

Proof. By Lemma 1, $(\breve{j}_s: \breve{L}_s^* \to \breve{L})_{s\in S}$ is a coproduct of $(\breve{L}_s^*)_{s\in S}$ (in $\mathscr{D}_{01}$). Define $f_s: L_s^* \to \breve{L}_s^*$ for each $s \in S$ by

(2) $f_s(a) = \overline{h_{L_s^*}(a)} = h_{L_s^*}(\bar{a})$ for each $a \in L_s$.

Note that f_s is a $\mathbf{D_{01}}$-isomorphism.

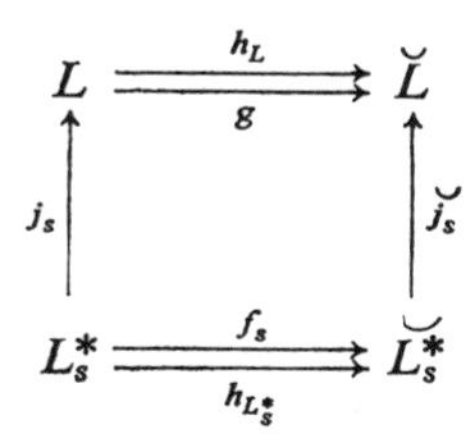

Therefore, by Lemma 1 there exists an isomorphism $g: L \to \breve{L}$ such that

(3) $g \circ j_s = \breve{j}_s \circ f_s$ for all $s \in S$.

Define a unary operation $^-$ on L by

(4) $\bar{a} = (h_L^{-1} \circ g)(a)$ for $a \in L$.

Obviously for $a, b \in L$, $a \leq b \Rightarrow \bar{a} \geq \bar{b}$. In order to show that $^-$ makes L into a de Morgan algebra (also denoted by L), it suffices to show that $\bar{\bar{a}} = a$ for all $a \in L$. By (4) it suffices therefore to show that

(5) $h_L^{-1} \circ g \circ h_L^{-1} \circ g = 1_L$.

It follows from (1), (2) and (3) that for $a \in L_s$, $(h_L^{-1} \circ g \circ h_L^{-1} \circ g \circ j_s)(a) = j_s(a)$ for all $s \in S$. By uniqueness, (5) follows. We now show that $(j_s: L_s \to L)_{s\in S}$ is a coproduct of $(L_s)_{s\in S}$ (in $\mathscr{M}$). It follows easily from (1), (2), (3), (4), and (5) that $j_s(\bar{a}) = \overline{j_s(a)}$ for each $a \in L_s$ and each $s \in S$. Hence j_s is a morphism in $\mathscr{M}$ for each $s \in S$. Finally, let $(k_s: L_s \to L_1)_{s\in S}$ be a family of morphisms in $\mathscr{M}$. Then by (1) we have

(6) $\breve{k}_s \circ h_{L_s^*} = h_{L_1^*} \circ k_s$ for each $s \in S$.

There exists a unique $k \in [L, L_1^*]_{\mathscr{D}_{01}}$ such that

(7) $k \circ j_s = k_s$ for each $s \in S$.

It remains to show:

(8) $k(\bar{a}) = \overline{k(a)}$ for each $a \in L$.

Define $f \in [L_1^*, \breve{L}_1^*]_{\mathscr{D}_{01}}$ by

(9) $f(a) = \overline{h_{L_1^*}(a)} = h_{L_1^*}(\bar{a})$.

It follows from the fact that k_s preserves the operation $^-$, for each $s \in S$, and from (2), (6) and (9) that

(10) $\breve{k}_s \circ f_s = f \circ k_s$ for each $s \in S$.

By (1) we have

(11) $\check{k} \circ h_L = h_{L_1^\bullet} \circ k$.

It follows from (4), (11), and (9) that $k(\bar{a}) = (h_{L_1^\bullet}^{-1} \circ \check{k} \circ g)(a)$ and that $\overline{k(a)} = (h_{L_1^\bullet}^{-1} \circ f \circ k)(a)$, for $a \in L$. Hence, in order to show (8) it suffices to show that

(12) $\check{k} \circ g = f \circ k$.

Now it follows from (3), (7), and (10) that $\check{k} \circ g \circ j_s = f \circ k \circ j_s$ for each $s \in S$, and (12) follows by uniqueness.

In order to characterize the free de Morgan algebra $\mathscr{F}_{\mathbf{M}}(S)$ on a set S of free generators, we note that the free de Morgan algebra $\mathscr{F}_{\mathbf{M}}(\mathbf{1})$ on one free generator is given by the following diagram:

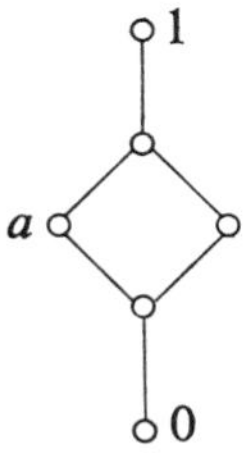

1

a $\bar{a}$

0

hence $(\mathscr{F}_{\mathbf{M}}(\mathbf{1}))^* = \mathscr{F}_{\mathbf{D01}}(\mathbf{2})$.

Theorem 3. (J. Berman and Ph. Dwinger [1]) $(\mathscr{F}_{\mathbf{M}}(S))^* \cong \mathscr{F}_{\mathbf{D01}}(S_1 \cup S_2)$, *where S_1 and S_2 are disjoint sets with $|S_1| = |S_2| = |S|$. In particular, $(\mathscr{F}_{\mathbf{M}}(\mathbf{n}))^* = \mathscr{F}_{\mathbf{D01}}(\mathbf{2n})$ and thus $|\mathscr{F}_{\mathbf{M}}(\mathbf{n})| = |\mathscr{F}_{\mathbf{D01}}(\mathbf{2n})|$.*

Proof. $\mathscr{F}_{\mathbf{M}}(S)$ is the coproduct of $|S|$ copies of $\mathscr{F}_{\mathbf{M}}(\mathbf{1})$ and thus by Theorem 2, $(\mathscr{F}_{\mathbf{M}}(S))^*$ is a coproduct (in $\mathscr{D}_{01}$) of $|S|$ copies of $(\mathscr{F}_{\mathbf{M}}(\mathbf{1}))^*$. Hence $(\mathscr{F}_{\mathbf{M}}(S))^*$ is isomorphic to a coproduct of $|S|$ copies of $\mathscr{F}_{\mathbf{D01}}(\mathbf{2})$ and this again is isomorphic to $\mathscr{F}_{\mathbf{D01}}(S_1 \cup S_2)$.

Exercise 4. Give a direct proof of Theorem 3 by extending a one-one correspondence between S_1 and S_2 to a unary operation on $\mathscr{F}_{\mathbf{D01}}(S_1 \cup S_2)$.

The injective de Morgan algebras have been characterized by R. Cignoli [5].

5. Lukasiewicz algebras. Definitions and basic theorems

Definition 1. A *Lukasiewicz algebra* of order $n \geq 2$, is an algebra $(L, (+, \cdot, ^-, D_1, D_2, \cdots, D_{n-1}, 0, 1))$ where $+$, $\cdot$, are binary operations, $^-$, $D_1, D_2, \cdots, D_{n-1}$ are unary operations, and 0 and 1 are nullary operations, satisfying:

L_1: L is de Morgan algebra under $+, \cdot, ^-, 0, 1$.
L_2: $\mathbf{D_i(x + y) = D_i(x) + D_i(y); D_i(xy) = D_i(x)D_i(y)}$, $1 \leq i \leq n - 1$.
L_3: $\mathbf{D_i(x)D_j(x) = D_j(x)}$, $1 \leq i \leq j \leq n - 1$.
L_4: $\mathbf{D_i(x) + \overline{D_i(x)} = 1; D_i(x)\overline{D_i(x)} = 0}$, $1 \leq i \leq n - 1$.
L_5: $\mathbf{D_i(\bar{x}) = \overline{D_{n-i}(x)}}$, $1 \leq i \leq n - 1$.
L_6: $\mathbf{D_i(D_j(x)) = D_j(x)}$, $1 \leq i \leq n - 1, 1 \leq j \leq n - 1$.
L_7: $\mathbf{x + D_1(x) = D_1(x); xD_{n-1}(x) = D_{n-1}(x)}$.
L_8: $\mathbf{D_i(0) = 0; D_i(1) = 1}$, $1 \leq i \leq n - 1$.
L_9: $\mathbf{\bar{x}D_{n-1}(x) = 0; \bar{x} + D_1(x) = 1}$.
L_{10}: $\mathbf{y(x + \overline{D_i(x)} + D_{i+1}(y)) = y}$, $1 \leq i \leq n - 2$.

We note that these axioms are not independent.

Before discussing the relationship between Lukasiewicz algebras and Post algebras we first establish some elementary properties. We will denote the equational class of Lukasiewicz algebras of order $n \geq 2$, by $\mathbf{L_n}$ and the corresponding equational category by $\mathscr{L}_{\mathbf{n}}$. Note that it is immediate from L_4, that if $L \in \mathbf{L_n}$, then $D_i(x) \in \mathscr{C}(L)$ for each $x \in L$ and $1 \leq i \leq n - 1$.

Exercise 2. Prove that the following identity holds in $\mathbf{L_n}$, $n \geq 2$: $\mathbf{x\overline{D_{i+1}(x)}D_i(y) + y = y}$, $1 \leq i \leq n - 2$. (Hint: Use Definition 2.1. M_2, L_{10}, and L_5.)

Lemma 3. The following identities hold in L_n, $n \geq 2$:

$$\mathbf{x\prod_{i=1}^{k} (\overline{D_i(x)} + D_{i+1}(x)) = xD_{k+1}(x)}, \quad 1 \leq k \leq n - 2.$$

Proof. The proof goes by induction on k. It is easily verified that the identity holds for $k = 1$. Suppose the identity holds for $1 \leq k \leq m$. Then by L_3 and L_4

$$x\prod_{i=1}^{m+1} (\overline{D_i(x)} + D_{i+1}(x)) = xD_{m+1}(x)(\overline{D_{m+1}(x)} + D_{m+2}(x)) = xD_{m+2}(x).$$

Theorem 4. *Let* $L \in \mathbf{L_n}$, $n \geq 2$. *For* $x, y \in L$, $x = y \Leftrightarrow D_i(x) = D_i(y)$, $1 \leq i \leq n - 1$.

Proof. ($\Rightarrow$) Trivial.

($\Leftarrow$) We have, by L_7, L_{10} and by the hypothesis:

$$y = y(x + \overline{D_{n-2}(x)} + D_{n-1}(y)) = y(x + \overline{D_{n-2}(x)} + D_{n-1}(x))$$
$$= y(x + \overline{D_{n-2}(x)}) = y(x + \overline{D_{n-2}(y)}).$$

Again, by L_{10}, L_1, the hypothesis and Lemma 3,

$$y = \prod_{i=1}^{n-3} (yx + y\overline{D_i(x)} + yD_{i+1}(y)) = yx + y\prod_{i=1}^{n-3} (\overline{D_i(x)} + D_{i+1}(y))$$
$$= yx + y\prod_{i=1}^{n-3} (\overline{D_i(y)} + D_{i+1}(y)) = yx + yD_{n-2}(y) = y(x + D_{n-2}(y)).$$

Hence $y = y(x + \overline{D_{n-2}(y)})(x + D_{n-2}(y)) = yx$, thus $y \le x$. Similarly, $x \le y$ and thus $y = x$.

Corollary 5. *Let $L \in \mathbf{L_n}$, $n \ge 2$. For $x, y \in L$, $x \le y \Leftrightarrow D_i(x) \le D_i(y)$ for $1 \le i \le n - 1$.*

Proof. By Theorem 4 and L_2.

Exercise 6. Suppose $h_1, h_2 \in [L, L_1]_{\mathscr{L}\mathbf{n}}$. Prove that $h_1 = h_2$ if and only if $\mathbf{I}(\mathrm{Ker}(h_1|\mathscr{C}(L))) = \mathbf{I}(\mathrm{Ker}(h_2|\mathscr{C}(L)))$.

Theorem 7. *Let $L \in \mathbf{L_n}$, $n \ge 2$. If $x \in \mathscr{C}(L)$, then $\bar{x}$ is the complement of x in $\mathscr{C}(L)$.*

Proof. Let x' be the complement of x in $\mathscr{C}(L)$. By L_2 and L_8, $D_i(x) + D_i(x') = 1$ and $D_i(x)D_i(x') = 0$, $1 \le i \le n - 1$. Thus, by L_4, $D_i(x') = \overline{D_i(x)}$ and thus by L_5, $D_i(x') = D_{n-i}(\bar{x})$. It follows from L_3 that $D_1(x') \ge D_2(x') \ge \cdots \ge D_{n-1}(x') = D_1(\bar{x}) \ge D_2(\bar{x}) \ge \cdots \ge D_{n-1}(\bar{x}) = D_1(x')$ and thus by Theorem 4, $x' = \bar{x}$.

Theorem 8. *Let $L \in \mathbf{L_n}$, $n \ge 2$ and let $x \in L$. The following are equivalent:*

(i) $x \in \mathscr{C}(L)$.

(ii) $D_1(x) = D_2(x) = \cdots = D_{n-1}(x)$.

(iii) $x = D_1(x)$.

In particular, $x = 0(1) \Leftrightarrow D_i(x) = 0(1)$ for $1 \le i \le n - 1$.

Proof. (i) $\Rightarrow$ (ii) Immediate from the proof of Theorem 7.

(ii) $\Rightarrow$ (iii) By L_9 and the hypothesis, $\bar{x}D_1(x) = 0$ and $\bar{x} + D_1(x) = 1$ so $\bar{x} \in \mathscr{C}(L)$ and $D_1(x)$ is the complement of $\bar{x}$ in $\mathscr{C}(L)$ which is, by Theorem 7, $\bar{\bar{x}}$. By L_1 and by Definition 2.1, M_3, $\bar{\bar{x}} = x$ and thus $D_1(x) = x$.

(iii) $\Rightarrow$ (i) By hypothesis and L_6, $D_{n-1}(x) = D_{n-1}(D_1(x)) = D_1(x)$. Thus, by L_9, $\bar{x} \in \mathscr{C}(L)$. Hence by Theorem 7 and by L_1, $x \in \mathscr{C}(L)$.

Exercise 9. (A. Monteiro [5]) Suppose $L \in \mathbf{L_3}$. Prove that for $x \in L$, $D_1(\overline{D_1(\bar{x})}) \le x$.

Exercise 10. (A. Monteiro [5]) Let $L \in \mathbf{L_3}$. Prove that for $x \in L$: (i) $D_1(x) + D_1(\bar{x}) = 1$. (ii) $D_2(x)D_2(\bar{x}) = 0$.

Exercise 11. Let L be a Lukasiewicz algebra of order $n \ge 2$. Prove that L, considered as a de Morgan algebra, belongs to $\mathbf{M_1}$.

6. Post algebras as special cases of Lukasiewicz algebras

Suppose $(L, (+, \cdot, C, D_1, \cdots, D_{n-1}, c_0, \cdots, c_{n-1}))$ is a Post algebra of order $n \geq 2$ (cf. Section X.6). It is not difficult to see that

$$(L, (+, \cdot, D_1, \cdots, D_{n-1}, {}^-, c_0, \cdots, c_{n-1}))$$

is a Lukasiewicz algebra of order n, where the unary operation $^-$ is defined by

(1) $\bar{x} = \sum_{i=1}^{n-1} \overline{D_{n-i}(x)} c_i$

and where $\overline{D_{n-i}(x)} = C(D_{n-i}(x))$ is the complement of $D_{n-1}(x)$ in $\mathscr{C}(L)$. We have already noted in Section 2 that $(L, (+, \cdot, {}^-, 0, 1))$ is a de Morgan algebra, so L_1 holds. The axioms $L_1, \cdots, L_{10}$ of Definition 5.1 also hold as can easily be seen. We will check L_{10}. For $x, y \in L$ and $1 \leq i \leq n - 2$,

$$\bar{x} D_i(x) \overline{D_{i+1}(y)} = \left(\sum_{j=1}^{n-1} \overline{D_{n-j}(x)} c_j \right) D_i(x) \overline{D_{i+1}(y)}$$

$$= \left(\sum_{j=1}^{n-i-1} \overline{D_{n-j}(x)} c_j \right) D_i(x) \overline{D_{i+1}(y)}$$

$$\leq \left(\sum_{j=1}^{n-i-1} \overline{D_{n-1}(x)} c_j \right) D_i(x) \overline{D_{i+1}(y)} \leq c_{n-i-1} \overline{D_{i+1}(y)} \leq \bar{y}.$$

Hence $x + \overline{D_i(x)} + D_{i+1}(y) \geq y$. We also note that if $L, L_1 \in \mathbf{P_n}$ and $h: L \to L_1$ is a homomorphism, then h preserves the operation $^-$. Indeed, for $x \in L$,

$$h(\bar{x}) = h\left(\sum_{i=1}^{n-1} \overline{D_{n-i}(x)} c_i \right) = \sum_{i=1}^{n-1} h(\overline{D_{n-i}(x)}) c_i$$

$$= \sum_{i=1}^{n-1} \overline{h(D_{n-i}(x))} c_i = \sum_{i=1}^{n-1} \overline{D_{n-i}(h(x))} c_i = \overline{h(x)}.$$

Conversely, suppose $(L, (+, \cdot, {}^-, D_1, \cdots, D_{n-1}, 0, 1))$ is a Lukasiewicz algebra of order $n \geq 2$ and suppose L has elements $c_0, c_1, \cdots, c_{n-1}$ such that

(2) $D_i(c_j) = 1$ for $1 \leq i \leq j \leq n - 1$

$D_i(c_j) = 0$ for $0 \leq j < i \leq n - 1$.

then $(L, (+, \cdot, C, D_1, \cdots, D_{n-1}, c_0, \cdots, c_{n-1}))$ is a Post algebra of order n, where the unary operation C is defined by $C(x) = \overline{D_1(x)}$ for $x \in L$. The verification of the axioms of Theorem X.6.1 is left to the reader as an exercise.

It now follows from the previous discussion that we can assign to every Post algebra of order n, a well-defined Lukasiewicz algebra of the same order by disregarding the operations $C, c_1, \cdots, c_{n-2}$ and by adding the operation $^-$. Furthermore, every $\mathbf{P_n}$-homomorphism is also an $\mathbf{L_n}$-homomorphism. We therefore have a functor from $\mathscr{P}_\mathbf{n}$ into $\mathscr{L}_\mathbf{n}$ and it also follows from our previous observations that this functor is an embedding which is one-one on the objects. Using a slight extension of the procedure described in Section I.19, we will often therefore consider the category $\mathscr{P}_\mathbf{n}$ as a subcategory of the category $\mathscr{L}_\mathbf{n}$.

That $\mathscr{P}_\mathbf{n}$ is a full subcategory of $\mathscr{L}_\mathbf{n}$ follows from the following slightly more general result.

Theorem 1. *Suppose* $h \in [L, L_1]_{\mathscr{L}_\mathbf{n}}$,

(i) *If* $L_1 \in \mathrm{Ob}\,\mathscr{P}_\mathbf{n}$ *and* $c \in L$ *exists such that for some* j, $1 \leq j \leq n-1$, $D_i(c) = 1$ *for* $1 \leq i \leq j$, $D_i(c) = 0$ *for* $j < i \leq n-1$, *then* $h(c) = c_j$.

(ii) *If* $L \in \mathrm{Ob}\,\mathscr{P}_\mathbf{n}$, *then* $L_1 \in \mathrm{Ob}\,\mathscr{P}_\mathbf{n}$ *and* $h \in [L, L_1]_{\mathscr{P}_\mathbf{n}}$. *It follows that* $\mathscr{P}_\mathbf{n}$ *is a full subcategory of* $\mathscr{L}_\mathbf{n}$.

The proof is immediate.

We note that if $L \in \mathrm{Ob}\,\mathbf{P_n}$, $n \geq 2$, then L can of course have $\mathbf{L_n}$-subalgebras which are not $\mathbf{P_n}$-subalgebras. For instance, consider the Post algebra $\mathbf{n}$ of order $n \geq 2$. This Post algebra obviously has no proper $\mathbf{P_n}$-subalgebras, but it has proper $\mathbf{L_n}$-subalgebras. It is not difficult to determine these subalgebras. Indeed, as an $\mathbf{L_n}$-algebra, $\mathbf{n}$ has the operations $+$, $\cdot$, 0, 1 together with $D_1, \cdots, D_{n-1}, {}^-$ defined by

$$D_i(\mathbf{m}) = \begin{cases} \mathbf{n-1} & \text{if } i \leq m, \\ \mathbf{0} & \text{if } i > m \end{cases} \quad \mathbf{m} \in \mathbf{n},\ 1 \leq i \leq n-1$$

and $\bar{\mathbf{m}} = \mathbf{n - 1 - m}$.

Now if L is an $\mathbf{L_n}$-subalgebra of $\mathbf{n}$ then certainly $\mathbf{0}$ and $\mathbf{n-1} \in L$ so every subset of $\mathbf{n}$ containing $\mathbf{0}$ and $\mathbf{n-1}$ and which is closed under the operation $^-$ is an $\mathbf{L_n}$-subalgebra of $\mathbf{n}$. So, it is easy, for example, to check that the $\mathbf{L_5}$-subalgebras of $\mathbf{5}$ are: $\{\mathbf{0, 4}\}$, $\{\mathbf{0, 1, 2, 3, 4}\}$, $\{\mathbf{0, 1, 3, 4}\}$ and $\{\mathbf{0, 2, 4}\}$.

Exercise 2. (R. Cignoli [1]) Prove that the number of $\mathbf{L_n}$-subalgebras of $\mathbf{n}$, $n \geq 2$ is $2^{(n-2)/2}$ for n even, and $2^{(n-1)/2}$ for n odd.

We have seen in Theorem X.6.8 that in the category $\mathscr{P}_\mathbf{n}$, the monomorphisms are the one-one morphisms, and that the epimorphisms are the onto morphisms. Obviously, in the category $\mathscr{L}_\mathbf{n}$ the monomorphisms are also the one-one morphisms, and onto morphisms are epic; but an epimorphism in $\mathscr{L}_\mathbf{n}$ need not be onto. In order to provide a large class of such examples we first make the following observation. If L is a Boolean algebra then L can be made into a Lukasiewicz algebra of order $n \geq 2$, by defining $D_i(x) = x$ for $x \in L$ and $1 \leq i \leq n-1$. Rather than using a different symbol for each n, we will simply say that we "consider L as a Lukasiewicz algebra of order n". Note that if $L \neq \mathbf{1}$ and $n > 2$ then L, considered as a Lukasiewicz algebra of order n, is not a Post algebra of order n. If $n = 2$, then L considered as a Lukasiewicz algebra of order 2 is also a Post algebra of order 2. The following is an example of an epimorphism in $\mathscr{L}_\mathbf{n}$ which is not onto. Let $L \in \mathrm{Ob}\,\mathscr{L}_\mathbf{n}$, $n \geq 2$ such that $\mathscr{C}(L) \neq L$ (for example $L = \mathbf{n}$, $n > 2$) and consider $\mathscr{C}(L)$ as a Lukasiewicz algebra of order n, then the inclusion map $j = 1_{\mathscr{C}(L),L} \in \mathbf{L_n}$ and is not onto.

However, j is an epimorphism. Indeed, suppose $h, k \in [L, L_1]_{\mathscr{L}_\mathbf{n}}$ and $h \circ j = k \circ j$. If $x \in L$, then $D_i(x) \in \mathscr{C}(L)$ for $1 \leq i \leq n-1$, and thus $h(D_i(x)) = k(D_i(x))$ for $1 \leq i \leq n-1$ and so $D_i(h(x)) = D_i(k(x))$ for $1 \leq i \leq n-1$. It follows from Theorem 5.4 that $h(x) = k(x)$ and so $h = k$.

7. A representation theory for Lukasiewicz algebras

The main purpose of this section is to characterize the subdirectly irreducibles in $\mathbf{L_n}$, which in turn, by applying Birkhoff's subdirect product theorem, yields a representation theorem for the class $\mathbf{L_n}$. First note that it follows from Theorem X.6.5 that $\mathbf{n}$ is subdirectly irreducible. Indeed, if $L \in \mathbf{P_n}$, and θ is an $\mathbf{L_n}$-congruence relation, then by Theorem 6.1(ii) θ is also a $\mathbf{P_n}$-congruence relation. So if θ is minimal as a $\mathbf{P_n}$-congruence relation, it is also minimal as an $\mathbf{L_n}$-congruence relation. The following theorem characterizes all the subdirectly irreducibles in $\mathbf{L_n}$ and also provides additional information on $\mathbf{L_n}$.

Theorem 1. (R. Cignoli [1]) *Let $L \in \mathbf{L_n}$, $|L| > 1$, $n \geq 2$. The following are equivalent:*

(i) $\mathscr{C}(L) = \mathbf{2}$.

(ii) *L is a chain.*

(iii) *L is an $\mathbf{L_n}$-subalgebra of $\mathbf{n}$.*

(iv) *L is subdirectly irreducible.*

It follows, in particular, that every chain in $\mathbf{L_n}$ is finite.

Proof. (i) $\Rightarrow$ (ii) Suppose $x, y \in L$. Then $\{D_i(x), D_i(y)\} \subseteq \{0, 1\}$ for $0 \leq i \leq n - 1$. It follows from L_3 (Definition 5.1) and from Corollary 5.5 that $x \leq y$ or $y \leq x$. Hence L is a chain.

(ii) $\Rightarrow$ (iii) First, note that for $x \in L$, $D_i(x) = 0$ or 1, for $1 \leq i \leq n - 1$. We define a map $h: L \to \mathbf{n}$ by $D_i(h(x)) = \mathbf{0}$ if $D_i(x) = 0$ and $D_i(h(x)) = \mathbf{n-1}$ if $D_i(x) = 1$ for $1 \leq i \leq n - 1$. It is obvious, that this map is well defined and it follows from Corollary 5.5 and from Theorem X.6.1 that $x \leq y \Leftrightarrow h(x) \leq h(y)$ and that $h(0) = \mathbf{0}$ and $h(1) = \mathbf{n-1}$. Thus h is one-one and h preserves the operations $+$, $\cdot$, 0 and 1. It remains to show that $h(D_i(x)) = D_i(h(x))$ for $x \in L$ and $1 \leq i \leq n - 1$, and that $h(\bar{x}) = \overline{h(x)}$ for $x \in L$. For $x \in L$ and i fixed, $1 \leq i \leq n - 1$, it follows from Theorem 5.8 and from the definition of h that $h(D_i(x)) = \mathbf{0} \Leftrightarrow D_j(h(D_i(x))) = \mathbf{0}$ for $1 \leq j \leq n - 1 \Leftrightarrow D_j(D_i(x)) = 0$ for $1 \leq j \leq n - 1 \Leftrightarrow D_i(x) = 0 \Leftrightarrow D_i(h(x)) = \mathbf{0}$. Hence $h(D_i(x)) = D_i(h(x))$. Again, for $x \in L$, it follows from the definition of h and from L_5 (Definition 5.1) that $D_i(h(\bar{x})) = \mathbf{0} \Leftrightarrow D_i(\bar{x}) = 0 \Leftrightarrow \overline{D_{n-i}(x)} = 0 \Leftrightarrow D_{n-i}(x) = 1$ $\Leftrightarrow D_{n-i}(h(x)) = \mathbf{n-1} \Leftrightarrow \overline{D_{n-i}(h(x))} = \mathbf{0} \Leftrightarrow D_i(\overline{h(x)}) = \mathbf{0}$ for $1 \leq i \leq n - 1$. Hence by Theorem 5.4, $h(\bar{x}) = \overline{h(x)}$.

(iii) $\Rightarrow$ (iv) Since $|L| > 1$, L has a non-trivial congruence relation θ. Thus there exist $x, y \in L$, $x \neq y$ such that $(x, y) \in \theta$ and thus $\nu_\theta(x) = \nu_\theta(y)$. This implies $D_i(\nu_\theta(x)) = D_i(\nu_\theta(y))$ or $\nu_\theta(D_i(x)) = \nu_\theta(D_i(y))$ for $1 \leq i \leq n - 1$. Suppose $\theta \neq 1$. But L/θ is a chain and since $\theta \neq 1$, $\mathscr{C}(L/\theta) \cong \mathbf{2}$. But then ν_θ restricted to $\mathscr{C}(L)$ is one-one and onto $\mathscr{C}(L/\theta)$ and we obtain $D_i(x) = D_i(y)$ for $1 \leq i \leq n - 1$ and thus $x = y$, a contradiction. It follows that $\theta = 1$ and thus L is subdirectly irreducible.

(iv) $\Rightarrow$ (i) Since $|L| > 1$, $|\mathscr{C}(L)| \geq 2$. Suppose $|\mathscr{C}(L)| > 2$. Then $\mathscr{C}(L)$, considered as a member of $\mathbf{B}$ has two non-trivial $\mathbf{B}$-congruence relation θ_1, θ_2 with

$\theta_1 \cdot \theta_2 = 0$. Define binary relations θ_1', θ_2' on L by $(x, y) \in \theta_i' \Leftrightarrow (D_j(x), D_j(y)) \in \theta_i$ for $1 \leq j \leq n-1$, $i = 1, 2$. It is not difficult to see that θ_1' and θ_2' are non-trivial $\mathbf{L_n}$-congruence relations on L. But $(x, y) \in \theta_1' \cdot \theta_2' \Rightarrow (D_i(x), D_i(y)) \in \theta_1, \theta_2, 1 \leq i \leq n-1 \Rightarrow D_i(x) = D_i(y), 1 \leq i \leq n-1 \Rightarrow x = y$. Hence $\theta_1' \cdot \theta_2' = 0$, which contradicts the fact that L is subdirectly irreducible.

We now have the following representation theorem.

Corollary 2. (G. C. Moisil [3]; R. Cignoli [1] and [4]) *If $L \in \mathbf{L_n}$ then L is a subdirect product of subalgebras of* $\mathbf{n}$.

Exercise 3. (R. Cignoli [1]) Let I be an ideal of L, $L \in \mathbf{L_n}$. Prove that I is generated by an ideal of $\mathscr{C}(L)$ if and only if $x \in I$ implies $D_1(x) \in I$.

R. Cignoli [4] has given a characterization of the representation spaces of Lukasiewicz and Post algebras.

8. The partially ordered set of prime ideals of a Lukasiewicz algebra

We have seen in Section X.5 that if L is a Post algebra of order $n \geq 2$, then $\mathfrak{P}(L)$ is isomorphic to $S \times \mathbf{n-1}$, where S is the totally unordered set of prime ideals of $\mathscr{C}(L)$. We will prove that a similar theorem, although in a somewhat weaker form, holds for Lukasiewicz algebras.

Theorem 1. (R. Cignoli [1]) *Let $L \in \mathbf{L_n}$. Then each prime ideal of L is contained in exactly one maximal chain of at most $n-1$ prime ideals of L.*

Proof. We first introduce the following notation. Let J be a prime ideal of $\mathscr{C}(L)$. Then for each k, $1 \leq k \leq n-1$, let $J_k = \{x \in L : D_k(x) \in J\}$. It easily follows that J_k is a prime ideal in L for $1 \leq k \leq n-1$; also $J_1 \subseteq J_2 \subseteq \cdots \subseteq J_{n-1}$. We will now show that if I is a prime ideal of L, then I is a member of a chain of this type. Let $J = I \cap \mathscr{C}(L)$, then J is a prime ideal in $\mathscr{C}(L)$. We have by L_7 that for $x \in L$, $x \in J_1 \Rightarrow D_1(x) \in J \Rightarrow D_1(x) \in I \Rightarrow x \in I$. So $J_1 \subseteq I$. Again, by L_7, $x \in I \Rightarrow D_{n-1}(x) \in I \Rightarrow D_{n-1}(x) \in J \Rightarrow x \in J_{n-1} \Rightarrow I \subseteq J_{n-1}$. Let $k_0 = \max\{k : J_k \subseteq I\}$. Suppose $I \nsubseteq J_{k_0}$. Then there exists $x \in I$, $x \notin J_{k_0}$. Thus $D_{k_0}(x) \notin J$ and therefore $\overline{D_{k_0}(x)} \in I$. Since $J_{k_0+1} \nsubseteq I$, there exists $y \in J_{k_0+1} \sim I$, so $D_{k_0+1}(y) \in J$ and thus $D_{k_0+1}(y) \in I$. It follows that $x + \overline{D_{k_0}(x)} + D_{k_0+1}(y) \in I$ and therefore by L_{10}, $y \in I$, a contradiction. Thus $I \subseteq J_{k_0}$. But also $J_{k_0} \subseteq I$, so $I = J_{k_0}$. It follows that each prime ideal I of L is contained in a maximal chain of the form $J_1 \subseteq J_2 \subseteq \cdots \subseteq J_{n-1}$ of at most $n-1$ prime ideals of L, where $J = I \cap \mathscr{C}(L)$. It remains to show that if I, I' are prime ideals of L and $I = J_k$, $I' = J'_{k'}$, where $J = I \cap \mathscr{C}(L)$, $J' = I' \cap \mathscr{C}(L)$, $1 \leq k \leq n-1$, $1 \leq k' \leq n-1$ and $J \neq J'$ then $I \nsubseteq I'$. But $I \subseteq I'$ implies $J \subseteq J'$, thus $J = J'$, a contradiction.

9. Embedding of Lukasiewicz algebras in Post algebras. Injective objects in $\mathscr{L}_n$

Lemma 1. *Let L be an $\mathbf{L_n}$-subalgebra of L_1, $L_1 \in \mathbf{P_n}$, $n \geq 2$, such that $[L]_{\mathbf{P_n}} = L_1$. Then $\mathscr{C}(L) = \mathscr{C}(L_1)$.*

Proof. Certainly $\mathscr{C}(L) \subseteq \mathscr{C}(L_1)$. For the converse, note that if $D = \{D_i(x) : 1 \leq i \leq n-1, x \in L\}$, then $[D]_{\mathbf{B}} = D = \mathscr{C}(L)$. Hence by Theorem X.6.7, $L_1 = [\mathscr{C}(L) \cup \{c_0, \cdots, c_{n-1}\}]_{\mathbf{D_{01}}}$. Suppose $x \in \mathscr{C}(L_1)$, then $x = a_1c_1 + \cdots + a_{n-1}c_{n-1}$, $a_i \in \mathscr{C}(L)$, $1 \leq i \leq n-1$ and therefore by Theorem 5.8 and Theorem X.6.1 (P_7), $x = D_1(x) = a_1 + \cdots + a_{n-1} \in \mathscr{C}(L)$.

Theorem 2. *Let L be an $\mathbf{L_n}$-subalgebra of L_1, $L_1 \in \mathbf{P_n}$, $n \geq 2$ and suppose $[L]_{\mathbf{P_n}} = L_1$. If $f \in [L, L_2]_{\mathscr{L}_\mathbf{n}}$, $L_2 \in \mathbf{P_n}$, then there exists a unique $g \in [L_1, L_2]_{\mathscr{P}_\mathbf{n}}$ with $g|L = f$.*

Proof. By Lemma 1, $\mathscr{C}(L) = \mathscr{C}(L_1)$. We define g as follows. For $x \in L_1$, let $g(x) = f(D_1(x))c_1 + \cdots + f(D_{n-1}(x))c_{n-1}$. Since $f \in [L, L_2]_{\mathscr{L}_\mathbf{n}}$, we have $f(D_i(x)) \geq f(D_j(x))$ for $i \leq j$ and it follows from Theorem X.2.4 that $f(D_i(x)) = D_i(g(x))$ for $1 \leq i \leq n-1$. From this and from Theorem X.6.1 (P_5 and P_7) it follows that g preserves joins and meets and that $g(c_i) = c_i$ for $1 \leq i \leq n-1$. Next, we show that $g(\bar{x}) = \overline{g(x)}$ for $x \in L_1$. Indeed $D_i(g(\bar{x})) = f(D_i(\bar{x})) = f(\overline{D_{n-i}(x)}) = \overline{f(D_{n-i}(x))} = \overline{D_{n-i}(g(x))} = D_i(\overline{g(x)})$ for $1 \leq i \leq n-1$, so by Theorem 5.4, $g(\bar{x}) = \overline{g(x)}$. Now, if $x \in L$, then $g(x) = f(D_1(x))c_1 + \cdots + f(D_{n-1}(x))c_{n-1} = D_1(f(x))c_1 + \cdots + D_{n-1}(f(x))c_{n-1} = f(x)$ and therefore $g|L = f$. Finally, for $x \in L_1$, $D_i(g(x)) = f(D_i(x)) = g(D_i(x))$ for $1 \leq i \leq n-1$.
The uniqueness of g is obvious.

Theorem 3. (cf. R. Cignoli [4]) *$\mathscr{P}_\mathbf{n}$ is a full reflective subcategory of $\mathscr{L}_\mathbf{n}$, $n \geq 2$, and the corresponding reflector preserves and reflects monomorphisms.*

Proof. It follows from Corollary 7.2 that if $L \in \text{Ob}\, \mathscr{L}_\mathbf{n}$ then L can be embedded as an $\mathbf{L_n}$-subalgebra of an object of $\mathscr{P}_\mathbf{n}$. Therefore we may assume that L is an $\mathbf{L_n}$-subalgebra of L_1, $L_1 \in \text{Ob}\, \mathscr{P}_\mathbf{n}$ and that $[L]_{\mathbf{P_n}} = L_1$. This observation in combination with Theorem 2, yields the result that $\mathscr{L}_\mathbf{n}$ is a reflective subcategory of $\mathscr{P}_\mathbf{n}$. That this subcategory is full, follows from Theorem 6.1. For the last part of the theorem, suppose $f \in [L, L']_{\mathscr{L}_\mathbf{n}}$. Let $L_1, L_1' \in \text{Ob}\, \mathscr{P}_\mathbf{n}$ such that $[L]_{\mathbf{P_n}} = L_1$ and $[L']_{\mathbf{P_n}} = L_1'$ and let $g \in [L_1, L_1']_{\mathscr{P}_\mathbf{n}}$ with $g|L = f$. If f is monic, and $g(x) = g(y)$, $x, y \in L$ then $g(D_i(x)) = g(D_i(y))$ for $1 \leq i \leq n-1$. But by Lemma 1, $\mathscr{C}(L) = \mathscr{C}(L_1)$ hence $f(D_i(x)) = f(D_i(y))$ for $1 \leq i \leq n-1$. We infer from Theorem 5.4 that $f(x) = f(y)$ and thus $x = y$. It follows that g is monic. Conversely, if g is monic, then g is one-one, hence $f = g|L$ is one-one and thus f is monic.

Exercise 4. Prove that if $[L]_{\mathbf{P_n}} = L_1$, $L \in \text{Ob}\, \mathscr{L}_\mathbf{n}$, $L_1 \in \text{Ob}\, \mathscr{P}_\mathbf{n}$ then the inclusion $1_{L,L_1}$ is epic. (Hint: Use Theorem 6.1.)

We are now able to characterize the injectives in $\mathscr{L}_\mathbf{n}$.

Theorem 5. *Let $L \in \text{Ob}\, \mathscr{L}_\mathbf{n}$, $n \geq 2$. The following are equivalent:*

(i) *L is injective.*

(ii) *$L \in \text{Ob}\, \mathscr{P}_\mathbf{n}$ and L is complete.*

Proof. (i) $\Rightarrow$ (ii) We may assume that L is an $\mathbf{L_n}$-subalgebra of L_1, where $L_1 \in \mathrm{Ob}\,\mathscr{P}_\mathbf{n}$. By hypothesis, there exists an $f \in [L_1, L]_{\mathscr{L}_\mathbf{n}}$ with $f|L = 1_L$. By Theorem 6.1, $L \in \mathrm{Ob}\,\mathscr{P}_\mathbf{n}$. But then L is injective in $\mathscr{P}_\mathbf{n}$ and thus, by Theorem X.6.8, L is complete.

(ii) $\Rightarrow$ (i) By Theorem X.6.8, L is injective in $\mathscr{P}_\mathbf{n}$. It follows from Theorem I.18.6 and from Theorem 3 that L is injective in $\mathscr{L}_\mathbf{n}$.

10. Free Lukasiewicz algebras

In this section we characterize the free Lukasiewicz algebras of order $n \geq 2$ and, in particular, the finite members of this class.

Theorem 1. *$\mathscr{F}_{\mathbf{L_n}}(S)$ is the $\mathbf{L_n}$-subalgebra of $\mathscr{F}_{\mathbf{P_n}}(S)$ generated by S. Moreover, $\mathscr{C}(\mathscr{F}_{\mathbf{L_n}}(\mathbf{m})) \cong \mathbf{2}^{n^m}$.*

Proof. Let L be the $\mathbf{L_n}$-subalgebra of $\mathscr{F}_{\mathbf{P_n}}(S)$ generated by S, and let $f\colon S \to L'$ be a map, $L' \in \mathbf{L'_n}$. We may assume that L' is an $\mathbf{L_n}$-subalgebra of a Post algebra L'_1 of order n. There exists $g \in [\mathscr{F}_{\mathbf{P_n}}(S), L'_1]_{\mathscr{P}_\mathbf{n}}$ with $g|S = f$. Let $f_1 = f|L$, then $f_1 \in [L, L'_1]_{\mathbf{L_n}}$. If L_2 is the $\mathbf{L_n}$-subalgebra of L' generated by $f[S]$, then $g^{-1}[L_2] \supseteq L$ and it follows that $f_1[L] \subseteq L'$. Hence $f_1 \in [L, L']_{\mathscr{L}_\mathbf{n}}$ and $f_1|S = f$. Therefore $L_1 = \mathscr{F}_{\mathbf{L_n}}(S)$. The last part of the theorem follows from Lemma 9.1 and Corollary X.7.4.

In order to determine the structure of finite free Lukasiewicz algebras, we first prove two lemmas.

If $L \in \mathbf{L_n}$ and $a \in \mathscr{C}(L)$ then for $x \in (a]$, $D_i(x) \leq D_i(a) = a$, so $D_i(x) \in (a]$ for $1 \leq i \leq n-1$. If we further define a unary operation $\tilde{}$ on $(a]$ by $\tilde{x} = \bar{x}a$ for $x \in (a]$ then we have the following lemma.

Lemma 2. *Suppose $L \in \mathbf{L_n}$ and $a \in \mathscr{C}(L)$*

(i) $((a], (+, \cdot, \tilde{}, D_1, \cdots, D_{n-1}, 0, a)) \in \mathbf{L_n}$.

(ii) *If a is an atom of $\mathscr{C}(L)$ then $((a], (+, \cdot, \tilde{}, D_1, \cdots, D_{n-1}, 0, a))$ is isomorphic to an $\mathbf{L_n}$-subalgebra of $\mathbf{n}$.*

Proof. (i) $L_1, \cdots, L_{10}$ are easily verified. (ii) If $x \in (a]$, then $D_i(x) = 0$ or a for $1 \leq i \leq n-1$. It follows from L_3 and Corollary 5.5 that $(a]$ is a chain and the result then follows from Theorem 7.1(iii).

For our further discussion we will denote by $(a]$, the algebra $((a], +, \cdot, \tilde{}, D_1, \cdots, D_{n-1}, 0, a))$ for $a \in \mathscr{C}(L)$.

Lemma 3. *Let $L \in \mathrm{Ob}\,\mathscr{L}_\mathbf{n}$ and let $a \in \mathscr{C}(L)$. If $h\colon L \to (a]$ is defined by $h(x) = xa$ for $x \in L$, then $h \in [L, (a]]_{\mathscr{L}_\mathbf{n}}$ and $\mathbf{I}(\mathrm{Ker}(h|\mathscr{C}(L))) = (\bar{a}]_{\mathscr{C}(L)}$.*

Proof. Obviously, $h \in [L, (a]]_{\mathscr{D}_{01}}$. For $x \in L$, $h(D_i(x)) = D_i(x)a = D_i(x)D_i(a) = D_i(xa) = D_i(h(a))$ for $1 \leq i \leq n - 1$. Again $h(\bar{x}) = \bar{x}a = \bar{x}a + \bar{a}a = (\bar{x} + \bar{a})a = \overline{xa} \cdot a = \overline{h(x)}a = \widetilde{h(x)}$. Hence $h \in [L, (a]]_{\mathscr{L}_{\mathbf{n}}}$. Finally, for $x \in \mathscr{C}(L)$, $h(x) = 0 \Leftrightarrow xa = 0 \Leftrightarrow x \in (\bar{a}]_{\mathscr{C}(L)}$.

Theorem 4. (R. Cignoli [1]) *Let S be a finite set,* $|S| = m$. *Let* $\{f_j : 1 \leq j \leq n^m\}$ *be the set of all maps from S to* **n** *and for each j,* $1 \leq j \leq n^m$ *let* $L_j = [f_j[S]]_{\mathbf{L_n}}$. *Then* $\mathscr{F}_{\mathbf{L_n}}(S) \cong \times_{j=1}^{n^m} L_j$.

Proof. Let $a_1, \cdots, a_k$ be the atoms of $\mathscr{C}(\mathscr{F}_{\mathbf{L_n}}(S))$ and for each j, $1 \leq j \leq k$, let $h_j \in [\mathscr{F}_{\mathbf{L_n}}(S), (a_j]]_{\mathscr{L}_{\mathbf{n}}}$ be defined by $h_j(x) = xa_j$ for each $x \in \mathscr{F}_{\mathbf{L_n}}(S)$. Define $h: \mathscr{F}_{\mathbf{L_n}}(S) \to (a_1] \times \cdots \times (a_k]$ by $h(x)(j) = h_j(x)$ for $1 \leq j \leq k$. Then, since $a_1 + \cdots + a_k = 1$, we have that h is an isomorphism (cf. Theorem II.6.12). Therefore we may assume by Lemma 2 that $\mathscr{F}_{\mathbf{L_n}}(S) = L_1 \times \cdots \times L_k$, where L_j is an $\mathbf{L_n}$-subalgebra of **n**. Also if $h_j: \mathscr{F}_{\mathbf{L_n}}(S) \to \mathbf{n}$ is defined by $h_j(x) = x(j)$ for each j, $1 \leq j \leq k$, then by Lemma 3, $\mathbf{I}(\mathrm{Ker}(h_j|\mathscr{C}(\mathscr{F}_{\mathbf{L_n}}(S)))) = (\bar{a}_j]_{\mathscr{C}(\mathscr{F}_{\mathbf{L_n}}(S))}$ for $1 \leq j \leq k$. Now for $1 \leq j \leq k$ let $f_j = h_j|S$ then for $j \neq j'$, $1 \leq j \leq k$, $1 \leq j' \leq k$, we have $f_j \neq f_{j'}$. Indeed, $f_j = f_{j'} \Rightarrow h_j = h_{j'} \Rightarrow \mathbf{I}(\mathrm{Ker}(h_j|\mathscr{C}(\mathscr{F}_{\mathbf{L_n}}(S)))) = \mathbf{I}(\mathrm{Ker}(h_{j'}|\mathscr{C}(\mathscr{F}_{\mathbf{L_n}}(S)))) \Rightarrow (\bar{a}_j]_{\mathscr{C}(\mathscr{F}_{\mathbf{L_n}}(S))} = (\bar{a}_{j'}]_{\mathscr{C}(\mathscr{F}_{\mathbf{L_n}}(S))} \Rightarrow \bar{a}_j = \bar{a}_{j'} \Rightarrow j = j'$.

Conversely, suppose $f: S \to \mathbf{n}$ is a map. Let $f^* \in [\mathscr{F}_{\mathbf{L_n}}(S), \mathbf{n}]_{\mathscr{L}_{\mathbf{n}}}$ be such that $f^*|S = f$. It follows from Section 6 that distinct subalgebras of **n** are non-isomorphic. Hence there exist $j \in \{1, \cdots, k\}$ such that $\mathbf{I}(\mathrm{Ker}(f^*|\mathscr{C}(\mathscr{F}_{\mathbf{L_n}}(S)))) = (\bar{a}_j]$ and so by Exercise 5.6, $f^* = h_j$ and thus $f = f_j$. It follows that $\{f_j : 1 \leq j \leq k\}$ is the set of all maps from S to **n** and thus $k = n^m$. Finally, note that for $1 \leq j \leq k$, $L_j = [f_j[S]]_{\mathbf{L_n}}$ and thus $\mathscr{F}_{\mathbf{L_n}}(S) \cong \times_{j=1}^{n^m} L_j$.

Remark. Taking $n = 2$, it follows again from Theorem 4 that $\mathscr{F}_{\mathbf{B}}(\mathbf{n}) = \mathbf{3}^{2^m}$.

Exercise 5. Prove that $\mathscr{F}_{\mathbf{L_3}}(\mathbf{1}) \cong \mathbf{2} \times \mathbf{3} \times \mathbf{2}$ and that $\mathscr{F}_{\mathbf{L_3}}(\mathbf{2}) = \mathbf{2}^4 \times \mathbf{3}^5$.

XII

Complete and α-Complete Distributive Lattices

1. Introduction. Definitions and some theorems

In this chapter we will discuss lattices which satisfy a higher degree of distributivity, and in connection with this, complete distributive lattices and completions of special types of distributive lattices. Finally, we will discuss α-representability of distributive lattices. We will first have to extend the notions of order preserving maps between posets, order embeddings, and homomorphisms between lattices. Throughout this chapter $\alpha, \beta, \cdots$ will denote *infinite* cardinal numbers.

Definition 1. Let P_1 and P_2 be posets. A map $f: P_1 \to P_2$ is called an *α-complete map* (or simply an *α-map*) provided that whenever $\varnothing \neq S \subseteq P_1$, $|S| \leq \alpha$, and $\sum S$ exists, then $\sum f[S]$ exists and $f(\sum S) = \sum f[S]$ and dually for products. f is called a *complete map* if f is an α-map for each α.

Note that an α-map $f: P_1 \to P_2$ always preserves order and preserves finite sums and products, whenever they exist. In particular, if P_1 and P_2 are lattices then an α-map $f: P_1 \to P_2$ is a (lattice) homomorphism. In this case we also say, that f is an *α-complete homomorphism*, or simply an *α-homomorphism* and we say that f is a *complete homomorphism* if f is a complete map.

Definition 2. Let P_1 and P_2 be posets. P_1 can be *α-regularly embedded* into P_2 if there exists an α-map $f: P_1 \to P_2$ which is also an order embedding. P_1 can be *regularly embedded* into P_2 if there exists a complete map $f: P_1 \to P_2$ which is an order embedding.

Definition 3. Let L_1 be a sublattice of a lattice L, then L_1 is called an *α-regular sublattice* of L, or simply an *α-sublattice* of L, if the inclusion is an α-homomorphism. L_1 is called a *regular sublattice* of L if the inclusion map is a complete homomorphism.

Definition 4. Let L be a lattice. L is called *α-complete* or simply an *α-lattice*, if $\sum S$ exists for all $\varnothing \neq S \subseteq L$, $|S| \leq \alpha$ and dually for products. Thus L is *complete* if L is an α-lattice for each α.

It may be helpful to the reader at this point to make some observations in regard to the terminology introduced in the previous definitions. First, note that if L_1 is an α-sublattice of a lattice L, then this means that if $\varnothing \neq S \subseteq L_1$, $|S| \leq \alpha$ and $\sum^{L_1} S$ exist, then $\sum^{L} S$ exists and $\sum^{L_1} S = \sum^{L} S$ and dually for products. Also note that if L is an α-lattice (complete lattice) and L_1 is an α-sublattice (regular sublattice) of L, then L_1 is not necessarily an α-lattice (complete lattice). Simple example: the rationals form a regular sublattice of the reals under the ordering of the reals, but the first is not complete, whereas the latter is complete (also see Theorem 7). On the other hand, L_1 may be an α-lattice (complete lattice) without being an α-sublattice (regular sublattice) of L. Simple example: if L is a complete distributive lattice, then $\hat{L}$ is a complete sublattice of the power set of $\mathfrak{P}(L)$ but $\hat{L}$ is, in general, not a regular sublattice of this power set. Finally, if L is a complete lattice and $\varnothing \neq S \subseteq L$, then there exists a smallest complete lattice L_1 which is a regular sublattice of L containing S. Indeed, L_1 is simply the intersection of all complete lattices which are regular sublattices of L which contain S.

Definition 5. An *α-complete ring of sets* (*α-complete field of sets*) is a ring (field) of subsets of a set X which is both α-complete and an α-sublattice of the power set of X. A *complete ring of sets* (*complete field of sets*) is a ring (field) which is α-complete for each α.

Thus a ring R is α-complete if and only if for $\varnothing \neq S \subseteq R$, $|S| \leq \alpha$, $\cup S \in R$ and $\cap S \in R$. We will again talk about α-rings of sets (α-fields of sets) instead of α-complete rings of sets (α-complete fields of sets).

One of the questions that we will discuss in this chapter will be the conditions under which a distributive lattice can be regularly embedded in a complete distributive lattice or in a complete Boolean algebra, and the properties of such embeddings. Another question is that of determining when a distributive lattice is an α-homomorphic image (complete homomorphic image) of an α-ring (complete ring) of sets, or an α-homomorphic image (complete homomorphic image) of an α-field (complete field) of sets. It will turn out that in most of these cases the lattice has to satisfy some "higher degree of distributivity".

We will start out our investigation with a discussion of lattices L which satisfy one or both of the following conditions of distributivity.

Let L be a lattice:

D_1: If $(x_s)_{s \in S}$ is family of elements in L such that $\sum_{s \in S} x_s$ exist then for each $y \in L$, $\sum_{s \in S} y x_s$ exists and $y \sum_{s \in S} x_s = \sum_{s \in S} y x_s$.

$\breve{D}_1$: the dual of D_1.

Obviously if L satisfies D_1 or $\breve{D}_1$ then $L \in \mathbf{D}$.

We have already seen (cf. Exercise II.6.9(iv)) that D_1 and $\breve{D}_1$ hold in any Boolean algebra. More generally we have

Theorem 6. (N. Funayama [2]) *Every relatively complemented distributive lattice satisfies D_1 and $\breve{D}_1$.*

Proof. We prove D_1. Pick an $s_o \in S$. Then $yx_{s_0} \leq x_s \dotplus x_{s_0} \leq (\sum_{s \in S} x_s) + y$. But $[yx_{s_0}, (\sum_{s \in S} x_s) + y]$ is a Boolean algebra and so $y \sum_{s \in S} x_s = y \sum_{s \in S} (x_s + x_{s_0})$ $= \sum_{s \in S} y(x_s + x_{s_0}) = \sum_{s \in S} yx_s$.

Theorem 7. (C. C. Chang and A. Horn [2]) *If L is a lattice that satisfies D_1 and $\breve{D}_1$ then L is a regular sublattice of its free Boolean extension B_L.*

Proof. Suppose $(x_s)_{s \in S} \subseteq L$ and $\prod^L_{s \in S} x_s$ exist. Certainly $\prod^L_{s \in S} x_s \leq x_s$ for every $s \in S$. Now suppose $y \leq x_s$ for every $s \in S$ and for $y \in B_L$. y can be written as $y = \sum_{i=1}^n u_i \bar{v}_i$ where $u_i, v_i \in L \cup \{0_{B_L}, 1_{B_L}\}$ for $1 \leq i \leq n$. Now $y \leq x_s \Rightarrow u_i \bar{v}_i \leq x_s \Rightarrow u_i \leq x_s + v_i$ for all $s \in S$ and $1 \leq i \leq n$. By $\breve{D}_1$, $v_i + \prod^L_{s \in S} x_s = \prod^L_{s \in S} (v_i + x_s)$ for each i and thus $v_i + \prod^L_{s \in S} x_s \geq u_i$ for each i. Therefore $u_i \bar{v}_i \leq \prod^L_{s \in S} x_s$ for each i and it follows that $y \leq \prod^L_{s \in S} x_s$. We conclude that $\prod^{B_L}_{s \in S} x_s$ exists and equals $\prod^L_{s \in S} x_s$. Dually for sums.

We will see later that not every distributive lattice can be regularly embedded in a complete distributive lattice. However, either one of the conditions D_1 or $\breve{D}_1$ is sufficient as we will now see.

Exercise 8. Prove that every Post algebra satisfies D_1 and $\breve{D}_1$.

Exercise 9. Let $P = \langle B, C \rangle$. Then B is a regular sublattice of P.

Definition 10. An ideal I of a lattice L is called α-*complete* (or simply an α-*ideal*) provided that if $\sum^L S$ exists and $\varnothing \neq S \subseteq I$, $|S| \leq \alpha$, then $\sum^L S \in I$. I is *complete* if it is α-complete for each α. The definition of α-*complete* and *complete filter* is dual.

Exercise 11. Prove that for each element a of a lattice L, $(a]$ is a complete ideal.

Exercise 12. Prove that if $\mathfrak{U}$ is a collection of α-ideals (complete ideals) of a lattice L and $\cap \mathfrak{U} \neq \varnothing$, then $\cap \mathfrak{U}$ is an α-ideal (complete ideal).

For a lattice L, let $I_{\mathfrak{C}}(L)$ denote the set of complete ideals—together with the void set in case L has no 0. $I_{\mathfrak{C}}(L)$, partially ordered by set inclusion, becomes a complete lattice in which products are set-theoretic intersections (Exercise 12). There is a natural map $\varphi: L \to I_{\mathfrak{C}}(L)$ defined by $\varphi(a) = (a]$, and this map is a regular embedding as can be easily seen. Indeed, it is obvious that φ is order preserving and one-one. Next, we show $\varphi(\prod_{s \in S} a_s) \supseteq \prod_{s \in S} \varphi(a_s) = \bigcap_{s \in S} \varphi(a_s)$ if the product on the left side exists. But $(\prod_{s \in S} a_s] \supseteq \bigcap_{s \in S} (a_s]$. Finally, suppose $\sum_{s \in S} a_s$ exists in L. We show $\varphi(\sum_{s \in S} a_s) \subseteq \sum_{s \in S} \varphi(a_s)$ or equivalently $(\sum_{s \in S} a_s] \subseteq \sum_{s \in S} (a_s]$. But for each $s \in S$, $a_s \in (a_s]$ so $a_s \in \sum_{s \in S} (a_s]$. Now $\sum_{s \in S} (a_s]$ is a complete ideal so $\sum_{s \in S} a_s \in \sum_{s \in S} (a_s]$. It follows that φ is complete.

Lemma 13. *Suppose L is a lattice that satisfies* D_1. *Let* $(I_s)_{s\in S}$ *be a family in* $I_{\mathfrak{C}}(L)$. *Then* $\sum_{s\in S}^{I_{\mathfrak{C}}(L)} I_s$ *consists of all elements of the form* $\sum (\bigcup_{s\in S} J_s)$ *where* $J_s \subseteq I_s$ *for each* $s \in S$ *and whenever these sums exist.*

Proof. Let U be the set of all elements of the type described. It is obvious that U is closed under sums in L, whenever they exist. Again, if $y \in U$ and $y = \sum(\bigcup_{s\in S} J_s)$, $J_s \subseteq I_s$ for $s \in S$, and $x \leq y$ then by D_1, $x = xy = \sum \bigcup_{s\in S} \{xu : u \in J_s\}$; and thus $x \in U$. Hence U is a complete ideal and U is obviously the smallest complete ideal containing each I_s.

Theorem 14. (cf. P. Crawley [1]) *Suppose L is a lattice that satisfies* D_1. *Then* $I_{\mathfrak{C}}(L)$ *is distributive and thus L can be regularly embedded in the complete distributive lattice* $I_{\mathfrak{C}}(L)$.

Proof. Let I_1, I_2 and $I_3 \in I_{\mathfrak{C}}(L)$. To show $I_1(I_2 + I_3) \subseteq I_1I_2 + I_1I_3$. Suppose $x \in I_1(I_2 + I_3)$. By Lemma 13 there exists $J_2 \subseteq I_2$, $J_3 \subseteq I_3$ such that $x = \sum (J_2 \cup J_3)$. By D_1, $x = x \sum (J_2 \cup J_3) = \sum \{xu : u \in J_2 \cup J_3\}$ which is a member of $I_1I_2 + I_1I_3$ since $\{xu : u \in J_2\} \subseteq I_1 \cap I_2$ and $\{xu : u \in I_3\} \subseteq I_1 \cap I_3$.

Remark. If L satisfies $\breve{D}_1$ then L can also be regularly embedded in a complete distributive lattice, namely the complete distributive lattice of complete filters. The argument is of course dual to that of the proof of Theorem 14.

We will also be concerned with lattices which are (α, β)-*meet distributive*; that is, lattices L which satisfy the following conditions:

D_2: If $(x_{st})_{s\in S, t\in T}$ is a family of elements in L satisfying

(i) $0 < |S| \leq \alpha$, $0 < |T| \leq \beta$,

(ii) $\sum_{t\in T} x_{st}$ exists for each $s \in S$,

(iii) $\prod_{s\in S} \sum_{t\in T} x_{st}$ exists,

(iv) $\prod_{s\in S} x_{s\varphi(s)}$ exists for each $\varphi \in T^S$,

then $\sum_{\varphi\in T^s} \prod_{s\in S} x_{s\varphi(s)}$ exists and $\prod_{s\in S} \sum_{t\in T} x_{st} = \sum_{\varphi\in T^s} \prod_{s\in S} x_{s\varphi(s)}$.

A lattice L will be called (α, β)-*join distributive* if it satisfies $\breve{D}_2$: the dual of D_2.

A lattice L is (α, ∞)-*meet distributive* ((∞, β)-*meet distributive*) if L is (α, β)-meet distributive for each β (for each α). (α, ∞) and (∞, β)-*join distributivity* are defined similarly. Note that a lattice satisfies D_1 if and only if it is $(2, \infty)$-meet distributive and satisfies $\breve{D}_1$ if and only if it is $(2, \infty)$-join distributive.

Remark. It is easy to see that Theorem 7 can be strengthened as follows: If L is a lattice and L is $(2, \alpha)$-meet and $(2, \alpha)$-join distributive then L is an α-sublattice of B_L.

Finally, a lattice L is called *completely meet distributive* if L is (∞, ∞)-meet distributive, i.e. (α, β)-meet distributive for each α and β. The definition for *com-*

pletely join distributive is, of course, dual. L is called *completely distributive* if it is both completely meet and completely join distributive. However, we will see (Theorem 17) that for a complete lattice L the conditions of complete meet- and complete join distributivity are equivalent.

It is sometimes convenient to write the condition for complete meet distributivity in the following equivalent form:

$$D_2': \prod_{s\in S} \sum_{t\in T_s} x_{st} = \sum_{\substack{\varphi\in X T_s \\ s\in S}} \prod_{s\in S} x_{s\varphi(s)}$$

for each family $(x_{st})_{s\in S, t\in T_s}$, where the usual assumptions on the existence of the sums and products on the left side, and products on the right side are made. Similarly for complete join distributivity:

$\breve{D}_2'$: the dual of D_2'.

Exercise 15. Prove that both definitions, as given above, for complete meet (join) distributivity are equivalent.

Theorem 16. *Every chain is completely distributive.*

Proof. We prove that D_2 holds for arbitrary α and β (The proof of $\breve{D}_2$ is then dual.). Thus, let $(x_{st})_{s\in S, t\in T}$ be a set of elements of the chain such that $\sum_{t\in T} x_{st}$ exists for each $s \in S$, $\prod_{s\in S} \sum_{t\in T} x_{st} = y$ exists and $\prod_{s\in S} x_{s\varphi(s)}$ exists for each $\varphi \in T^S$. We must show that $y = \sum_{\varphi\in T^S} \prod_{s\in S} x_{s\varphi(s)}$. Certainly, $y \geq \prod_{s\in S} x_{s\varphi(s)}$ for each $\varphi \in T^S$. Suppose $u \geq \prod_{s\in S} x_{s\varphi(s)}$ for each $\varphi \in T^S$. To show $u \geq y$ let us suppose $u < y$. Hence $u < \sum_{t\in T} x_{st}$ for each $s \in S$. Thus there exists $\varphi_u \in T^S$ with $u < x_{s\varphi_u(s)}$ for each $s \in S$. Thus $u \leq \prod_{s\in S} x_{s\varphi_u(s)}$. Hence $u = \prod_{s\in S} x_{s\varphi_u(s)}$. It follows that y covers u. Indeed if $u \leq u' < y$, then by a similar argument $u' = \prod_{s\in S} x_{s\varphi_{u'}(s)}$ but then $u' \leq u$ and thus $u = u'$. Now $u < x_{s\varphi_u(s)}$, thus $y \leq x_{s\varphi_u(s)}$ for each $s \in S$ hence $y \leq \prod_{s\in S} s_{s\varphi_u(s)} = u$, a contradiction.

Theorem 17. (G. N. Raney [1]) *For complete lattices, complete meet distributivity and complete join distributivity are equivalent.*

Proof. (The present proof is due to Ph. Dwinger and can also be obtained from Theorem 14.) It suffices to show that $D_2 \Rightarrow \breve{D}_2$ for sets S and T of arbitrary cardinality. Thus we assume that for every family $(x_{st})_{s\in S, t\in T}$ of elements

$$\prod_{s\in S} \sum_{t\in T} x_{st} = \sum_{\varphi\in T^S} \prod_{s\in S} x_{s\varphi(s)}.$$

We will show

$$\sum_{s\in S} \prod_{t\in T} x_{st} = \prod_{\varphi\in T^S} \sum_{s\in S} x_{s\varphi(s)}.$$

For each $s \in S$ and $\varphi \in T^S$ let $y_{\varphi s} = x_{s\varphi(s)}$. Then by hypothesis

$$\prod_{\varphi\in T^S} \sum_{s\in S} x_{s\varphi(s)} = \prod_{\varphi\in T^S} \sum_{s\in S} y_{\varphi s} = \sum_{\psi\in S^{(T^S)}} \prod_{\varphi\in T^S} y_{\varphi\psi(\varphi)}.$$

Now it is easy to see that for every $\psi \in S^{(T^S)}$, the family $(y_{\varphi\psi(\varphi)})_{\varphi\in T^S}$ contains a subfamily $(x_{st})_{t\in T}$ for some $s \in S$. (Otherwise there exists $\varphi_0 \in T^S$ and $\psi_0 \in S^{(T^S)}$ such that $x_{s,\varphi_0(s)} \notin (y_{\varphi\psi_0(\varphi)})_{\varphi\in T^S}$ for all $s \in S$. But then for $s = s_0 = \psi_0(\varphi_0)$, $y_{\varphi_0\psi_0(\varphi_0)} = x_{s_0,\varphi_0(s_0)}$.) For each $s \in S$ let $\psi_s \in S^{(T^S)}$ be defined by $\psi_s(\varphi) = s$ for each $\varphi \in T^S$. Then for each $s \in S$, $(y_{\varphi\psi_s(\varphi)})_{\varphi\in T^S} = (x_{st})_{t\in T}$. Hence

$$\sum_{\psi\in S^{(T^S)}} \prod_{\varphi\in T^S} y_{\varphi\psi(\varphi)} = \sum_{s\in S}\prod_{t\in T} x_{st} + \sum_{\psi\in S^{(T^S)}} \prod_{\varphi\in T^S} y_{\varphi\psi(\varphi)} = \sum_{s\in S}\prod_{t\in T} x_{st}$$

which completes the proof.

Exercise 18. Prove that if a lattice L is (β^α, α)-meet distributive and complete, then L is (α, β)-join distributive.

2. Normal completion of lattices. The special case of distributive lattices

The question as to whether every distributive lattice can be regularly embedded in a complete distributive lattice has to be answered negatively. We will furnish below an example of a distributive lattice which fails to have this property. On the other hand, every Boolean algebra can be regularly embedded in a complete Boolean algebra. In fact, we will see that the category of complete Boolean algebras and complete homomorphisms is a reflective subcategory of the category of Boolean algebras and complete homomorphisms. Thus it follows from Theorem 1.7 that every lattice which is $(2, \infty)$-meet and join distributive is regularly embeddable in a complete Boolean algebra. Also note (Theorem 1.14) that if L is $(2, \infty)$-meet or join distributive then L can be regularly embedded in a complete distributive lattice.

We first present an example of a distributive lattice which cannot be regularly embedded in a complete distributive lattice and which is due to P. Crawley [1].

Let X_1, X_2, and X_3 be infinite disjoint sets and let $X = X_1 \cup X_2 \cup X_3$. Let $A_1, A_2, \cdots, B_1, B_2, \cdots$ be subsets of X_3 such that

$$X_3 \supset A_1 \supset A_2 \supset \cdots$$

$$X_3 \supset B_1 \supset B_2 \supset \cdots$$

and such that $B_j \supset X_3 \sim (\bigcap_{i=1}^\infty A_i)$ for $j = 1, 2, \cdots$. That such subsets exist is easy to see. Indeed, first pick the subsets $A_1, A_2, \cdots$ such that $X_3 \supset A_1 \supset A_2 \supset \cdots$ and such that, in addition, $\bigcap_{i=1}^\infty A_i$ is infinite. Then pick sets, $C_1, C_2, \cdots$ such that $\varnothing \subset C_1 \subset C_2 \subset \cdots$ and such that $C_j \subset \bigcap_{i=1}^\infty A_i$ for $j = 1, 2, \cdots$. Then take $B_i = X_3 \sim C_i$ for $i = 1, 2, \cdots$. Note that $X_3 = A_i \cup B_j$ for $i = 1, 2, \cdots$, $j = 1, 2, \cdots$. Let

$$L_1 = \{X, \varnothing, X_1 \cup X_3, X_2 \cup X_3, X_3, A_i, B_j, X_1 \cup A_k, X_2 \cup B_l, A_m \cap B_n : i, j, k, l, m, n \in \{1, 2, \cdots\}\}$$

It is easy to see that L_1 is a ring of subsets of X (containing $\varnothing$ and X). Let

$$L = \{U \cup V : U \in L_1, V \text{ a finite subset of } X_1 \cup X_2\}$$

Again, it is not difficult to check that L is a ring of subsets of X. Note that all finite subsets of $X_1 \cup X_2$ belong to L since $\varnothing \in L_1$ and that $X_3 \in L$. Also $\varnothing$ and X belong to L. Now suppose $f: L \to L_2$ is a regular embedding, where L_2 is a complete distributive lattice. We will show that L_2 cannot, in fact, be distributive.

We proceed by steps.

(1) Let S_1 be the set of all finite subsets of X_1, then $S_1 \subseteq L$. Let S_2 be the set of all finite subsets of X_2, then $S_2 \subseteq L$. Let $S = S_1 \cup S_2$. Suppose $A \in L$ and A contains each member of S. Then A must contain $X_1 \cup X_2$. But the only member of L that contains $X_1 \cup X_2$ is X_3. So we have $X = \sum^L S$. Let $u = \sum^{L_2} f[S_1]$, $v = \sum^{L_2} f[S_2]$ and $z = f(X_3)$. Then by hypothesis

$$f(X) = f(\textstyle\sum^L S) = \sum^{L_2} f[S] = \sum^{L_2} f[S_1] + \sum^{L_2} f[S_2] = u + v.$$

Thus $u + v = f(X)$.

(2) For $V \in S_2$, we have $V \subseteq X_2 \cup B_i$ for $i = 1, 2, \cdots$. So $f(V) \subseteq f(X_2 \cup B_i)$ for $i = 1, 2, \cdots$. Hence $\sum^{L_2} f(V) \leq v \leq f(X_2 \cup B_i)$ for $i = 1, 2, \cdots$. Thus $vz \leq f(X_2 \cup B_i)z = f(X_2 \cup B_i)f(X_3) = f((X_2 \cup B_i) \cap X_3) = f(B_i)$ for $i = 1, 2, \cdots$. It follows that

$$vz \leq \prod_{i=1}^{\infty}{}^{L_2} f(B_i).$$

But it is easy to see that $\prod_{i=1}^{\infty}{}^{L} B_i = \varnothing$ and therefore $\prod_{i=1}^{\infty}{}^{L_2} f(B_i) = f(\varnothing)$ and so we have $vz = f(\varnothing)$.

(3) For $V \in S_1$, we have $V \subseteq X_1 \cup A_i$, for $1 = 1, 2, \cdots$. So $f(V) \subseteq f(X_1 \cup A_i)$ for $i = 1, 2, \cdots$. So $\sum^{L_2} f(V) \subseteq \sum^{L_2} f[S_1] = u \subseteq f(X_1 \cup A_i)$ for $i = 1, 2, \cdots$. Similarly, $v \leq f(X_2 \cup B_j)$, for $j = 1, 2, \cdots$. Thus we have

$$\begin{aligned} u(v + z) &\leq f(X_1 \cup A_i)(f(X_2 \cup B_j) + z) \\ &= f((X_1 \cup A_i)(f(X_2 \cup B_j) + f(X_3)) \\ &= f(X_1 \cup A_i) f(X_2 \cup B_j \cup X_3) \\ &= f(X_1 \cup A_i) f(X_2 \cup X_3) \\ &= f((X_1 \cup A_i) \cap (X_2 \cup X_3)) = f(A_i) \end{aligned}$$

and hence $u(v + z) \leq f(A_i)$ for $i = 1, 2, \cdots$. It follows that $u(v + z) \leq \prod_{i=1}^{\infty}{}^{L_2} f(A_i)$.

But again $\prod_{i=1}^{\infty}{}^{L} A_i = \varnothing$ and therefore $u(v + z) = f(\varnothing)$ and thus also $uv = f(\varnothing)$.

(4) Consider the elements $f(\varnothing), u, v, v + z, f(X)$. We have $u \nleq v$. Indeed suppose $u \leq v$. Then by (3) $uv = u = f(\varnothing)$. So $f(\varnothing) = f(V)$ for each $V \in S_1$. But f is an embedding and thus $\varnothing = V$ for each $V \in S_1$, a contradiction. Similarly $v \nleq u$. We also have $v \neq v + z$. Indeed, if $v = v + z$, then $vz = z$. But by (2) $vz = f(\varnothing)$ and it follows that $f(\varnothing) = f(X_3)$ or $X_3 = \varnothing$, a contradiction.

Finally, we have $u + v + z = f(X)$ and by (3), $u(v + z) = f(\varnothing)$ and also $uv = f(\varnothing)$. We conclude that L_2 has the 5 element sublattice

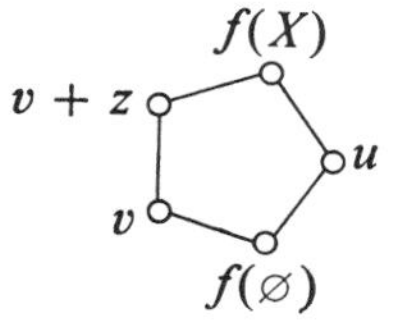

$f(X)$

$v + z$ u

v

$f(\varnothing)$

and therefore L_2 is not distributive.

Although a distributive lattice cannot always be regularly embedded in a complete distributive lattice, we have seen in Section 1 that every lattice can always be regularly embedded in a complete lattice. More generally, every partially ordered set can be regularly embedded in a complete lattice by a procedure which is a generalization of the completion of the rationals by Dedekind cuts (H. M. MacNeille [1]). In view of the fact that several classes of distributive lattices have "a completion by cuts" which is again distributive, we will describe this method here.

We begin by generalizing the notion of ideal.

Definition 1. Let P be a poset. An *ideal* in P is a non-empty subset $I \subseteq P$ satisfying: $x \in I, y \leq x \Rightarrow y \in I$, and if $x, y \in I$ and $x + y$ exists then $x + y \in I$. An ideal I is *closed* if $I = \bigcap_{x \in T} (x]$ for some $T \subseteq P$.

Note that for $T \subseteq P, I = \bigcap_{x \in T} (x]$ is a closed ideal if and only if $I \neq \varnothing$, for if $u, v \in I$ then $u + v \in I$. In particular $(x]$ is a closed ideal for each $x \in P$. Finally, by taking $T = \varnothing$, we see that P itself is a closed ideal.

Exercise 2. Let I be an ideal in P. Show that I is closed if and only if I contains every element which is a lower bound for each upper bound of I.

Let $\bar{P}$ denote the set of closed ideal of P to which is adjoined the void set, if P has no smallest element. Then $\bar{P}$—partially ordered by set inclusion—is a complete lattice, in which products are set-theoretic intersections. We will also use the symbol $\bar{P}$ to denote this complete lattice. It can be shown directly that there exists a regular embedding of P into $\bar{P}$, but in view of subsequent applications we will work in a somewhat more general setting.

For each $A \subseteq P$ we define the set

$$A^c = \cap \{(x] : x \text{ is an upper bound for } A\}.$$

Note that A^c is the set of all lower bounds of the set of all upper bounds of A.

Lemma 3. *The map ^{c} is a closure operator on the power set of P and if $A \subseteq P$ we have that $A \in \bar{P}$ if and only if A is a closed element of the power set of P under the closure operator ^{c}.*

Proof. We must check the conditions of Definition II.4.10. Let $A \subseteq B \subseteq P$ and suppose $u \in A^c$. Now each upper bound of B is an upper bound of A so $u \in B^c$.

Also $A \subseteq A^c$ for if $a \in A$ then a is a lower bound of all upper bounds of A, hence $a \in A^c$. To prove that $A^c = A^{cc}$ we only need to prove that $A^{cc} \subseteq A^c$. Suppose $u \in A^{cc}$. If x is an upper bound of A, then obviously x is an upper bound of A^c so $u \leq x$ and thus $u \in A^c$. This establishes that ^{c} is a closure operator.

Next suppose $A \subseteq P$ and $A \in \bar{P}$. We only need to show that $A^c \subseteq A$, but this follows immediately from Exercise 2. The converse is also immediate.

We now define a map $\nu: P \to \bar{P}$ by $\nu(x) = (x]$ for $x \in P$, and we will show that ν is a regular embedding. Certainly, $x \leq y \Rightarrow \nu(x) \subseteq \nu(y)$. Next, suppose $\sum^P S$ exists for $S \subseteq P$. To show $\nu(\sum^P S) = \sum^{\bar{P}} \nu[S]$ we observe that $\sum^{\bar{P}} \nu[S] = \sum^{\bar{P}}_{s \in S} (s]$ $= (\bigcup_{s \in S} (s])^c$ by Theorem II.4.12 and $\nu(\sum^P S) = (\sum^P S]$. If $y \leq \sum^P S$ then obviously $y \in (\bigcup_{s \in S} (s])^c$ hence $(\sum^P S] \subseteq (\bigcup_{s \in S} (s])^c$ and thus $\nu(\sum^P S) \subseteq \sum^{\bar{P}} \nu[S]$. The reverse inequality follows, since ν is order preserving. That ν preserves products, if they exist, follows from $\nu(\prod^P S) = (\prod^P S] = \bigcap_{s \in S} (s] = \prod^{\bar{P}}_{s \in S} (s] = \prod^{\bar{P}} \nu[S]$. Hence $\nu: P \to \bar{P}$ is a regular embedding. Note that if P has a unit element 1_P then $\nu(1_P)$ $= 1_{\bar{P}}$. Similarly, if P has a zero element 0_P, then $\nu(0_P) = 0_{\bar{P}}$; indeed $\nu(0_P) = \{0_P\}$ is the smallest closed ideal of P and hence the smallest element of $\bar{P}$, thus $\nu(0_P)$ $= 0_{\bar{P}}$.

If P is a partially ordered set and (L, φ) is a pair such that L is a complete lattice and $\varphi: P \to L$ is a regular embedding, then (L, φ) is called a *normal completion* of P, if there exists an isomorphism $\psi: \bar{P} \to L$ such that $\psi \circ \nu = \varphi$. Note that $(\bar{P}, \nu)$ itself is a normal completion of P and is often referred to as the *MacNeille completion* of P. Also note that any two normal completions of P are isomorphic; we will therefore often talk about *the* normal completion of P. If P is a partially ordered subset of a complete lattice L (with the same partial ordering as that of L) and (L, j) is a normal completion of P, where $j: P \to L$ is the inclusion map, then we will simply say that L is a normal completion of P.

Exercise 4. Prove that in the above definition of normal completion it is sufficient to require that φ is an embedding.

Lemma 5. *Let P be a partially ordered set and let $A \in \bar{P}$. Then*

(i) $A = \sum^{\bar{P}} \{\nu(x): \nu(x) \subseteq A, x \in P\}$.

(ii) $A = \prod^{\bar{P}} \{\nu(x): A \subseteq \nu(x), x \in P\}$.

Proof. (i) Since A is an ideal, $A = \bigcup_{a \in A} (a]$ and $\nu(x) \subseteq A \Leftrightarrow x \in A$, we have

$$\sum\nolimits^{\bar{P}} \{\nu(x): \nu(x) \subseteq A, x \in P\} = \sum\nolimits^{\bar{P}} \{(x]: x \in A\} = \Big(\bigcup_{x \in A} (x]\Big)^c = A^c = A.$$

(ii) Now, $\prod^{\bar{P}} \{\nu(x): A \subseteq \nu(x), x \in P\} = \cap \{(x]: A \subseteq (x], x \in P\}$. But $u \in \cap \{(x]: A \subseteq (x], x \in P\} \Leftrightarrow u \leq x$ for each x such that $a \leq x$ for all $a \in A \Leftrightarrow u \in A^c = A$.

Definition 6. Let P be a subset of a partially ordered set Q. P is *join dense* (*meet dense*) in Q if every element of Q is the sum (product) of elements in P. We now have the following theorem.

Theorem 7. (Cf. B. Banaschewski and G. Bruns [1] and [2]) *Let L be a complete lattice and let P be a partially ordered subset of L, with the partial ordering of L. Then L is a normal completion of P if and only if P is join and meet dense in L.*

Proof. ($\Rightarrow$) If $(\bar{P}, \nu)$ is the MacNeille completion of P, then by Lemma 5, $\nu[P]$ is join and meet dense in $\bar{P}$. By the definition of a normal completion, it follows that P is join and meet dense in L.

($\Leftarrow$) Let $(\bar{P}, \nu)$ be the MacNeille completion of P. By the definition of normal completion we must show that there exists an isomorphism $\varphi: L \to \bar{P}$ such that $\varphi|P = \nu$. Define a map $\varphi: L \to \bar{P}$ by $\varphi(a) = (a]_L \cap P$. We first show that $\varphi(a)$ is, in fact, a closed ideal of P. By hypothesis, $a = \sum^L (a]_L \cap P$. Thus for $y \in L$, we have that $y \in P$ and y is an upperbound of $\varphi(a)$ if and only if $y \in [a)_L \cap P$. Again, by hypothesis $a = \prod [a)_L \cap P$. So if $x \in P$ and x is a lower bound of every upper bound of $\varphi(a)$ then $x \leq a$ and thus $x \in \varphi(a)$. It follows from Exercise 2 that $\varphi(a)$ is a closed ideal of P. We now show that for $a, b \in L$, $a \leq b \Leftrightarrow \varphi(a) \subseteq \varphi(b)$. That $a \leq b$ implies $\varphi(a) \subseteq \varphi(b)$ is immediate. Conversely, if $\varphi(a) \subseteq \varphi(b)$, then by hypothesis $a = \sum \varphi(a) \leq \sum \varphi(b) = b$. Next we show that φ is onto $\bar{P}$. Let $A \in \bar{P}$ and let $a = \sum^L A$. We claim that $\varphi(a) = A$. Certainly $\varphi(a) = (a] \cap P \supseteq A$. Suppose $x \in (a] \cap P$, then x is a lower bound in P of all upper bounds of A. But A is a closed ideal, hence $x \in A$. Finally, $\varphi|P = \nu$ is obvious.

Note that Theorem 7 implies that if a partially ordered subset P of a complete lattice L is join and meet dense in L, then P is regularly embedded in L.

We close this general discussion of the normal completion with the following theorem which shows that the normal completion of a partially ordered set is—in a certain sense—a minimal completion.

Theorem 8. (cf. B. Banaschewski and G. Bruns [1] and [2]) *Let P be a poset and let L be a normal conpletion of P. If $h: P \to L_1$ is an order embedding of P in a complete lattice L_1 then there exists an order embedding $h^*: L \to L_1$ which is an extension of h.*

Notice that h^* only satisfies $x \leq y \Rightarrow h^*(x) \leq h^*(y)$ and h^* need not preserve sums or products.

Proof. Define $h^*: L \to L_1$ by $h^*(a) = \sum h[(a]_L \cap P]$ for $a \in L$. Clearly h^* preserves order and $h^*|P = h$. Suppose $h^*(a) \leq h^*(b)$. We have for each $x \in (a]_L \cap P$ and each $y \in [b)_L \cap P$: $h(x) = h^*(x) \leq h^*(a) \leq h^*(b) \leq h^*(y) = h(y)$ and thus $x \leq y$. But by Theorem 7, $a = \sum (a]_L \cap P$ and $b = \prod [b)_L \cap P$ so $a \leq b$.

A simple example that shows that h^* need not preserve sums (products) is the following: Let $P = \omega \oplus \breve{\omega}$ and let $L = \omega \oplus \mathbf{1} \oplus \breve{\omega}$ and consider P as a partially ordered subset of L. By Theorem 7, $\bar{L}$ is a normal completion of P. Let $L_1 = \omega \oplus \mathbf{2} \oplus \breve{\omega}$. There is a natural embedding of P in L_1 which can be extended in two different ways to an embedding of $\bar{L}$ in L_1. But both of these extensions fail to preserve all sums and products. There are distributive lattices whose normal completion is not distributive although they can be embedded regularly in a complete distributive lattice. The following example is due to N. Funayama [1]. Let C_1, C_2, and C_3 be the following chains

$$C_1: a_1 < a_2 < \cdots < \cdots < b_2 < b_1$$

$$C_2: \qquad p < q$$

$$C_3: c_1 < c_2 < \cdots < \cdots < d_2 < d_1$$

Let L be the sublattice of $C_1 \times C_2 \times C_3$ consisting of the following elements

$$A_{ik} = (b_i, q, d_k), \qquad i, k = 1, 2, \cdots$$

$$B_{ik} = (a_i, q, d_k), \qquad i, k = 1, 2, \cdots$$

$$C_{ik} = (a_i, p, d_k), \qquad i, k = 1, 2, \cdots$$

$$D_{ik} = (a_i, p, c_k) \qquad i, k = 1, 2, \cdots$$

It is easy to check that L is indeed a sublattice of $C_1 \times C_2 \times C_3$, and since the latter is distributive so is L. Let $A = \{D_{ik}: i, k = 1, 2, \cdots\}$. Then A, $(B_{11}]_L$ and $(C_{11}]_L$ are closed ideals of L. Now we have:

$$A \cap (B_{11}]_L = \{(a_1, p, c_k): k = 1, 2, \cdots\} = A \cap (C_{11}]_L,$$

$$(A \cup (B_{11}]_L)^c = \{B_{ik}, C_{i'k'}, D_{i''k''}: i, k, i', k', i'', k'' = 1, 2, \cdots\}$$

$$= (A \cup (C_{11}]_L)^c.$$

Also $(B_{11}]_L \supset (C_{11}]_L$. Thus the 5 closed ideals A, $(A \cup (B_{11}]_L)^c = (A \cup (C_{11}]_L)^c$, $(B_{11}]_L$, $(C_{11}]_L$, and $A \cap (B_{11}]_L = A \cap (C_{11}]_L$ form a 5 element non-distributive lattice which is a sublattice of $\bar{L}$, hence $\bar{L}$ is not distributive. It is, however, easy to see that L can be regularly embedded in $\bar{C}_1 \times C_2 \times \bar{C}_3$ which is a complete distributive lattice. (The example of P. Crawley that we have given before in this section is, of course, also an example of a distributive lattice whose normal completion is not distributive.)

Exercise 9. (N. Funayama [1]) Let L be a distributive lattice and let A be a closed ideal of L. Prove that the maps $x \mapsto (x]_L \cap A$ and $x \mapsto ((x]_L \cap A)^c$ are homomorphisms from L into $\bar{L}$.

Exercise 10. Let L be a distributive lattice. Prove that $\bar{L}$ is distributive if and only if the map $x \mapsto ((x]_L \cap A, ((x]_L \cup A)^c)$ is a monomorphism from L to $\bar{L} \times \bar{L}$ for each closed ideal A of L.

Exercise 11. Prove that the normal completion of a Heyting algebra is also a Heyting algebra and hence distributive.

In contrast to the general situation, the normal completion of a Boolean algebra is again distributive and, in fact, is a Boolean algebra. In addition, we shall also show that the normal completion of a Boolean algebra is maximal in a sense which makes the category of complete Boolean algebras and complete homomorphisms a reflective subcategory of the category of Boolean algebras and complete homomorphisms.

Lemma 12. *Suppose that L_1 is a subalgebra of a Boolean algebra L, and that L_1 is dense in L (i.e. for each $a \in L$, $a \neq 0$, there exists $b \in L_1$, $b \neq 0$ such that $b \leq a$). Then L_1 is meet and join dense in L.*

Proof. Suppose $a \in L$, then $a \geq x$ for each $x \in (a]_L \cap L_1$. If $b \in L$, $b \geq x$ for each $x \in (a]_L \cap L_1$ and $b \ngeq a$, then $\bar{b}a \neq 0$. Hence, by hypothesis, there exists $x_0 \in L_1, x_o \neq 0$, $x_o \leq \bar{b}a \leq a$. Therefore $x_o \leq b$. Also $x_o \leq \bar{b}$ so $x_o = 0$, a contradiction. It follows that $a = \sum (a]_L \cap L_1$ and thus L_1 is join dense in L. That L_1 is meet dense in L follows from a dual argument.

Theorem 13. *Let L be a Boolean algebra. Then the normal completion of L is isomorphic to the Boolean algebra of regular open subsets of $\mathscr{S}_{\mathscr{B}}(L)$.*

Proof. By Corollary II.9.4, $\mathfrak{J}(L)$ is a complete lattice and by Exercise VIII.1.4, $\mathfrak{J}(L) \in \mathbf{B}_\omega$. By Theorems VIII.4.3 and VIII.4.4, $\mathscr{R}g(\mathfrak{J}(L))$ is a complete Boolean algebra. Again by Theorem VIII.4.5, $\mathscr{C}(\mathfrak{J}(L))$ is a subalgebra of $\mathscr{R}g(\mathfrak{J}(L))$. But $\mathscr{C}(\mathfrak{J}(L))$ consists (Ex. VIII.4.6) of the principal ideals of L and is therefore dense in $\mathfrak{J}(L)$ and so it is dense in $\mathscr{R}g(\mathfrak{J}(L))$. It follows from Lemma 12 and Theorem 7, that $\mathscr{R}g(\mathfrak{J}(L))$ is a normal completion of $\mathscr{C}(\mathfrak{J}(L))$, but $L \cong \mathscr{C}(\mathfrak{J}(L))$, so $\mathscr{R}g(\mathfrak{J}(L))$ is a normal completion of L. The last part of the theorem follows from the fact that $\mathfrak{J}(L)$, by Exercise IV.1.13, is isomorphic to the lattice L' of open subsets of $\mathscr{S}_{\mathscr{B}}(L)$ and that $\mathscr{R}g(L')$ is the Boolean algebra of regular open sets of $\mathscr{S}_{\mathscr{B}}(L)$ (cf. remark prior to Exercise VIII.1.6).

Remark. If L is a Boolean algebra and ^{c} is the closure operator on the power set of L as defined before, then we have seen that $\bar{L}$ is the lattice of closed ideals of L. It is easy to see that the restriction of ^{c} to the lattice $\mathfrak{J}(L)$ is precisely the map $r_{\mathfrak{J}(L)}$ as defined in Glivenko's Theorem (VIII.4.3).

Exercise 14. Prove the statement in the above remark.

Exercise 15. Let $L \in \mathbf{B}_\omega$ and suppose L is complete with $\mathscr{C}(L)$ dense in L. Then $\mathscr{R}g(L)$ is a normal completion of $\mathscr{C}(L)$. The following result is a corollary of Theorem 13 and of the second ***Remark*** on page 231.

Corollary 16. (N. Funayama [2]) *Let $L \in \mathbf{D}$. L can be embedded as an α-sublattice of a complete Boolean algebra if and only if L is* (2, α)*-meet and join distributive.*

3. The normal completion of a Boolean algebra

We have seen in the previous section that the normal completions of a Boolean algebra L are Boolean algebras and, in fact, they are isomorphic to the Boolean algebra of regular open subsets of $\mathscr{S}_{\mathscr{B}}(L)$. In order to simplify the notation in this section, we will let the symbol $\bar{L}$ denote a specific normal completion of L under the assumption that the corresponding embedding is the inclusion map (the notation $(\bar{L}, \nu)$ has been used in the previous section to denote the MacNeille completion which is, as we recall, one of the normal completions of L, but there is of course no danger of confusion).

We further introduce the following terminology. The category of Boolean algebras and complete homomorphisms will be denoted by $\mathscr{B}_c$ and the full subcategory of $\mathscr{B}_c$ whose objects are the complete Boolean algebras by $\overline{\mathscr{B}_c}$. It is obvious that the monomorphisms in $\mathscr{B}_c$ and in $\overline{\mathscr{B}_c}$ are the one-one homomorphisms (since the free Boolean algebra on one generator is finite).

We will also extend the notion of regular sublattice to that of regular subalgebra.

Definition 1. A subalgebra L_1 of a Boolean algebra L is called a *regular subalgebra* of L if the inclusion map is regular.

Definition 2. Let $L_1 \in \mathrm{Ob}\,\mathscr{B}$ be a regular subalgebra of L, $L \in \mathrm{Ob}\,\overline{\mathscr{B}_c}$. Then L_1 *completely generates* L, if L is the smallest complete regular subalgebra of L containing L_1.

Lemma 3. *Let $L \in \mathrm{Ob}\,\mathscr{B}_c$, $g \in [L, L_1]_{\mathscr{B}_c}$ and $L_1 \in \mathrm{Ob}\,\overline{\mathscr{B}_c}$. Then there exists a unique $h \in [\bar{L}, L_1]_{\mathscr{B}c}$ with $h|L = g$.*

Proof. We proceed in steps:

(1) Let $\{a\} \cup S \subseteq L$ and suppose $a \le \sum^{\bar{L}} S$. Then $g(a) \le \sum g[S]$. Indeed by Theorem 1.6, $a = a \sum^{\bar{L}} S = \sum^{\bar{L}} \{as : s \in S\} = \sum^{L} \{as : s \in S\}$ (we also use here the fact that L is regular in $\bar{L}$ and that $a \in L$). But g is complete, hence $g(a) = g(\sum^{L} \{as : s \in S\}) = g(a) \sum g[S]$. Thus $g(a) \le \sum g[S]$.

(2) Let $S \subseteq L$, $T \subseteq L$ and suppose $\sum^{\bar{L}} S = \sum^{\bar{L}} T$. Then $\sum g[S] = \sum g[T]$. Indeed, for each $s \in S$, $s \le \sum^{\bar{L}} T$, so by (1), $g(s) \le \sum g[T]$ and thus $\sum g[S] \le \sum g[T]$.

(3) Define $h : \bar{L} \dot{\to} L_1$ by $h(a) = \sum g[(a]_{\bar{L}} \cap L]$. It is obvious that $h|L = g$ and that $h(0) = g(0) = 0$ and $h(1) = g(1) = 1$.

(4) We show that h preserves arbitrary sums. Suppose $S \subseteq \bar{L}$. By Theorem 2.7 we have for each $s \in S$, $s = \sum^{\bar{L}} (s]_{\bar{L}} \cap L$. Hence $\sum^{\bar{L}} S = \sum^{\bar{L}}_{s \in S} (\sum^{\bar{L}} (s]_{\bar{L}} \cap L)$. It follows from (2) and (3) that

$$h(\textstyle\sum^{\bar{L}} S) = \sum_{s \in S} (\sum g[(s]_{\bar{L}} \cap L]) = \sum_{s \in S} h(s) = \sum h[S]$$

(5) We show that h preserves complements and arbitrary products. Let $a \in \bar{L}$. Then again by Theorem 2.7 $a = \sum^{\bar{L}} (a]_{\bar{L}} \cap L$ and $\bar{a} = \sum^{\bar{L}} (\bar{a}]_{\bar{L}} \cap L$. Thus, we have $1 = a + \bar{a} = \sum^{\bar{L}}(((a]_{\bar{L}} \cap L) \cup ((\bar{a}]_{\bar{L}} \cap L))$. Hence by (2),

$$\begin{aligned} 1 = h(1) &= \sum g[((a]_{\bar{L}} \cap L) \cup ((\bar{a}]_{\bar{L}} \cap L)] \\ &= \sum g[(a]_{\bar{L}} \cap L] + \sum g[(\bar{a}]_{\bar{L}} \cap L] \\ &= h(a) + h(\bar{a}). \end{aligned}$$

Again, by Theorem 1.6,

$$\begin{aligned} h(a)\, h(\bar{a}) &= \sum g[(a]_{\bar{L}} \cap L] \sum g[(\bar{a}]_{\bar{L}} \cap L] \\ &= \sum \{g(x)\, g(y) : x \in (a]_{\bar{L}} \cap L, y \in (\bar{a}]_{\bar{L}} \cap L\} = 0. \end{aligned}$$

It follows that $h(\bar{a}) = \overline{h(a)}$. That h preserves products follows from the fact that h preserves 0, 1, sums and from the de Morgan laws.

(6) The uniqueness of h follows from the fact that L completely generates $\bar{L}$ (by Theorem 2.7) and from a standard argument about uniqueness of extensions on algebras generated by a set.

Theorem 4. (cf. B. Banaschewski and G. Bruns [1]) *$\overline{\mathscr{B}_c}$ is a reflective subcategory of $\mathscr{B}_c$ and the reflector preserves and reflects monomorphisms.*

Proof. That $\overline{\mathscr{B}_c}$ is a reflective subcategory of $\mathscr{B}_c$ is an immediate consequence of Lemma 3. That the reflector reflects monomorphisms follows from the fact that the inclusion $1_{L,\bar{L}}$, $L \in \mathrm{Ob}\, \mathscr{B}$ is a monomorphism. It remains to show that monomorphisms are preserved. Suppose $g \in [L, L_1]_{\mathscr{B}_c}$ and g monic. Let $g^* \in [\bar{L}, \bar{L}_1]_{\mathscr{B}_c}$ be the unique extension of g. Suppose $g^*(a) = 0$ for $a \in \bar{L}$. But then $0 = g^*(a) = \sum [g(a]_{\bar{L}} \cap L]$ and so $g(x) = 0$ for each $x \in (a]_{\bar{L}} \cap L$. It follows from Theorem 2.7 that $a = 0$.

Corollary 5. *Let L_1 be a subalgebra of a complete Boolean algebra L. Then L is a normal completion of L_1 if and only if L_1 is a regular subalgebra of L which completely generates L.*

Proof. The "only if" part follows from Theorem 2.7. The "if" part follows from the fact that the reflector in Theorem 4, preserves monomorphisms.

Theorem 6. (B. Banaschewski and G. Bruns [1]) *Let $L, L_1 \in \mathrm{Ob}\, \mathscr{B}$. The following are equivalent:*

(i) *L is a normal completion of L_1.*

(ii) *L_1 is a regular subalgebra of L and completely generates L.*

(iii) *L is the injective hull of L_1.*

Proof. (i) $\Rightarrow$ (ii) Immediate from Corollary 5.

(ii) $\Rightarrow$ (iii) Suppose $f \in [L, L_2]_{\mathscr{B}}$ such that $f|L_1$ is monic. By Theorem 2.7, L_1 is join dense in L, hence if $f(a) = 0$, $a \in L$, then $f(x) = 0$ for each $x \in (a]_L \cap L$. Hence $x = 0$ for all such x, so f is monic. Since L is complete, it is injective in $\mathscr{B}$.

(iii) $\Rightarrow$ (i) Suppose L_1 is not join dense in L. Then there exists an element $a \in L$, such that a is not the sum of $(a]_L \cap L_1$. But L is complete so $b = \sum^L (a] \cap L_1$ satisfies $b < a$ and $(b]_L \cap L_1 = (a]_L \cap L_1$. Let $I = (0, a\bar{b}]_L$. Then since $a\bar{b} \neq 0$, $I \neq (0]$, but is easy to see that $I \cap L_1 = (0]$. Thus the map $\nu_I : L \to L/I$ is one-one on L_1 but not on L. It follows that L_1 is join dense in L and thus dense. It follows from Lemma 2.12 that L_1 is meet dense in L and the result follows now from Theorem 2.7.

Although the normal completions of distributive lattices are not necessarily distributive, the injective hulls of objects in $\mathscr{D}$ do exist (there are enough injectives) and are characterized as follows.

Theorem 7. (B. Banaschewski and G. Bruns [2]) *Let* $L, L_1 \in \text{Ob}\,\mathscr{D}$ (*or* $\text{Ob}\,\mathscr{D}_{01}$). *The following are equivalent*:

(i) *L is the normal completion of the free Boolean extension B_{L_1} of L_1.*

(ii) *L is the injective hull of L_1.*

Proof. Consider the diagram

$$L_1 \xrightarrow{\ i\ } B_{L_1} \xrightarrow{\ j\ } L$$

where (B_{L_1}, i) is the free Boolean extension of L_1 and L is the normal completion of B_{L_1}, with embedding j. Now $j \circ i$ is a monomorphism in $\mathscr{D}$ ($\mathscr{D}_{01}$) so it suffices to show that it is essential. Let $f \in [L, L']_{\mathscr{D}}$ ($f \in [L, L']_{\mathscr{D}_{01}}$) and suppose $f \circ j \circ i$ is one-one. Define $f' : L \to f[L]$ by $f'(x) = f(x)$ for all x. Then $f[L] \in \text{Ob}\,\mathscr{B}$ (since $L \in \text{Ob}\,\mathscr{B}$ and so $f' \in [L, f[L]]_{\mathscr{B}}$. Now $f' \circ j \circ i$ is one-one and, by Theorem V.4. 8, i is essential, so $f' \circ j$ is monic. But $f' \circ j \in \mathscr{B}$ and by the previous theorem j is essential in $\mathscr{B}$ so f' is one-one. Hence f is one-one.

In B. Banaschewski and G. Bruns [1], a characterization of the normal completion of posets is given within the framework of the category of posets. This involves replacing monomorphisms with "strict" monomorphisms in the definition of essential extension, and is beyond the scope of this book.

Exercise 8. (cf. G. Epstein [1]) Let L_1 and L_2 be Post algebras. Prove that the following are equivalent:

(i) L_2 is an injective hull of L_1.

(ii) The order of L_1 equals the order of L_2 and the center of L_2 is a normal completion of L_1.

(iii) L_2 is a normal completion of L_1.

Exercise 9. Let $L \in \text{Ob}\,\mathscr{B}$. Prove that $\bar{L}$ is atomic if and only if L is atomic.

Exercise 10. (cf. B. Banaschewski and G. Bruns [2]) Let C be a chain which is dense in itself. Then the normal completion of B_C is isomorphic to the Boolean algebra of regular open subsets of C, if C is endowed with the interval topology.

4. Complete, completely distributive lattices

In this section we will investigate completely distributive lattices. We recall (Theorem 1.17) that for complete lattices, complete join distributivity and complete meet distributivity are equivalent. In order to avoid unnecessarily complicated formulations we will restrict ourselves—in this section—to complete lattices. First, we will deal with complete rings of sets and prove a theorem of V. K. Balachandran [2] and G. N. Raney [1], which characterizes those complete lattices which are complete rings of sets.

The complete lattices and the complete homomorphisms form, of course, a category. However, it is not practical to introduce a notation for this category. Thus we will work in the category $\mathscr{L}$ of lattices and (lattice) homomorphisms and, in particular, with those objects which are complete and those homomorphisms which are complete. We note that the product of a family of complete (and completely distributive) lattices is again complete (and completely distributive) and that the projections are complete homomorphisms—as can be easily seen. We will use the term *regular subdirect product* for a subdirect product of a family of complete lattices for which the embedding map is regular.

In the second part of this section we will deal with complete, completely distributive lattices in general, and prove the theorem of G. N. Raney [2] that a complete lattice is completely distributive if and only if it is a regular subdirect product of complete chains.

Finally, we recall (Theorem 1.16) that every chain and thus in particular, every complete chain is completely distributive. Hence it follows that a product of complete, completely distributive chains is a complete, completely distributive lattice.

We will first characterize the complete lattices which are isomorphic to complete rings of sets.

Exercise 1. Prove that a complete ordinal is isomorphic to a complete ring of sets.

A complete, completely distributive lattice need not be isomorphic to a complete ring of sets. However, for Boolean algebras this is always true.

Theorem 2. (A. Tarski [1]) *Let L be a complete Boolean algebra. The following are equivalent:*

(i) *L is isomorphic to a complete field of sets.*

(ii) *L is completely distributive.*

(iii) *L is atomic.*

(iv) *L is isomorphic to the field of all subsets of a set.*

Proof. (i) $\Rightarrow$ (ii) Obvious.

(ii) $\Rightarrow$ (iii) We have $1 = \prod_{x \in L} (x + \bar{x})$. Using the fact that L is completely distributive (note that we really only need that L is $(\infty, 2)$-meet distributive) it easily follows that 1 is the sum of atoms and by applying Exercise III.1.6 we obtain the result.

(iii) $\Rightarrow$ (iv) If X is the set of atoms of L, for each $a \in L$, let $f(a) = (a] \cap X$. It is easy to see that f is an isomorphism between L and the field of all subsets of X.

(iv) $\Rightarrow$ (i) Trivial.

Theorem 2 states, in particular, that a complete Boolean algebra L is isomorphic to a complete field of sets if and only if L is atomic; that is, if every element of L is the sum of atoms. This last formulation suggests a generalization to complete lattices.

Definition 3. Let L be a complete lattice. An element $x \in L$ is *completely join irreducible* if whenever $x \leq \sum S$ for $S \subseteq L$, then there exists $s \in S$ such that $x \leq s$.

We note that the requirement for x to be completely join irreducible in the sense of Definition 3, is stronger than what would be a natural generalization of the notion of join irreducible in the sense of Definition III.1.1, to complete lattices: x is "completely join irreducible" if whenever $x = \sum S$, $S \subseteq L$, then there exists $s \in S$ such that $x = s$. It is not difficult to see that the two notions are equivalent for a complete lattice L if L is $(2, \infty)$-meet distributive. However, in the representation theorem that we will prove (Theorem 6) we do not assume as a hypothesis anything more than that L is a complete lattice, and in order to prove this theorem we need the stronger notion of completely join irreducible as stated in the definition.

Exercise 4. Let C be a complete chain. An element $a \neq 0$, $a \in C$ is completely join irreducible if and only if a has an immediate predecessor.

Exercise 5. Let L be a complete ring of subsets of a set X. Prove that a non-empty element $A \in L$ is completely join irreducible if and only if there exists a point $p \in X$ such that $A = \bigcap\{B \in L : p \in B\}$.

Theorem 6. (V. K. Balachandran [2], G. N. Raney [1]) *Let L be a complete lattice. L is isomorphic to a complete ring of sets if and only if every element of L is the sum of completely join irreducible elements.*

Proof. ($\Rightarrow$) Suppose $A \in L$. For each $p \in A$, let $A_p = \cap\{B \in L : p \in B\}$. By Exercise 5, A_p is completely join irreducible for each $p \in A$, but also $A_p \subseteq A$ for each $p \in A$, hence $A = \bigcup_{p \in A} A_p$.

($\Leftarrow$) Let X be the set of completely join irreducible elements of L. For each $a \in L$, let $f(a) = (a] \cap X$ (cf. (iii) $\Rightarrow$ (iv) of the proof of Theorem 2). It suffices to show that f is a regular embedding of L in the power set of X. Clearly $a \leq b \Rightarrow f(a) \subseteq f(b)$. Suppose $a \nleq b$, then there exists an $x_0 \in X$ with $x_0 \in (a] \sim (b]$, thus $f(a) \nsubseteq f(b)$. Hence $a \leq b \Leftrightarrow f(a) \subseteq f(b)$. Next, suppose $S \subseteq L$. To show $f(\sum S) \subseteq \bigcup f[S]$, suppose $x_0 \in f(\sum S)$ for some $x_0 \in X$. Then $x_0 \leq \sum S$. But x_0 is completely join irreducible, so there exists an $s \in S$ with $x_0 \leq s_0$, thus $x_0 \in \bigcup f[S]$. For $\bigcap f[S] \subseteq f(\prod S)$, suppose $x_0 \in \bigcap f[S]$ for some $x_0 \in X$; hence $x_0 \leq s$ for each $s \in S$ so $x_0 \leq \prod S$ and it follows that $x_0 \in f(\prod S)$.

Exercise 7. Prove that the unit interval of the real numbers and the unit interval of the rational numbers are not complete rings of sets. Also prove that $\omega + \mathbf{1} + \breve{\omega}$ is a complete ring of sets. (Hint: Use Exercise 4.)

Exercise 8. Solve Exercise 1, using Theorem 6.

We have stated above that a complete lattice is completely distributive if and only if it is a regular subdirect product of complete chains. In the remaining part of this section our main purpose will be to prove this statement. Most of the results stated below—in particular, the main theorem—are due to G. N. Raney [1], [2].

Definition 9. Let L be a lattice. A *semi-ideal* of L is a non-void subset A of L such that $a \in A$, $b \leq a$, $b \in L$ implies $b \in A$.

It is easy to see that the set $\mathfrak{S}(L)$ of semi-ideals of a complete lattice L is a complete lattice with set-theoretic unions and intersections as sums and products, respectively. Therefore $\mathfrak{S}(L)$ is a complete ring of sets.

Theorem 10. (G. N. Raney [1]) *Let L be a complete lattice and let f: $\mathfrak{S}(L) \to L$ be defined by $f(A) = \sum A$ for $A \in \mathfrak{S}(L)$. Then f is onto and, moreover, L is completely distributive if and only if L is a complete homomorphism. Thus L is completely distributive if and only if L is a complete homomorphic image of a complete ring of sets.*

Proof. f is onto since $(a] \in \mathfrak{S}(L)$ and $f((a]) = a$.

($\Rightarrow$) Let $(A_s)_{s \in S} \subseteq \mathfrak{S}(L)$ and for each $s \in S$ let $A_s = (x_{st})_{t \in T_s}$. It is easy to see that $f(\bigcup_{s \in S} A_s) = \sum_{s \in S} f(A_s)$. Next, observe that $x \in \bigcap_{s \in S} A_s \Leftrightarrow x = \prod_{s \in S} x_{s\varphi(s)}$ for some $\varphi \in \bigtimes_{s \in S} T_s$. Hence, we have by D_2'

$$f(\bigcap_{s \in S} A_s) = \sum_{\varphi \in \bigtimes_{s \in S} T_s} \prod_{s \in S} x_{s\varphi(s)}$$

$$= \prod_{s \in S} \sum_{t \in T_s} x_{st} = \prod_{s \in S} f(A_s).$$

Hence f is a complete homomorphism.

($\Leftarrow$) Trivial.

Remark. We observe that in the first part of the proof we only used D_2'. This enables us to give a short proof of Theorem 1.17. Indeed, if L satisfies D_2', then since L is a complete homomorphic image of a complete ring of sets, it follows that L satisfies $\breve{D}_2'$. Conversely, if L satisfies $\breve{D}_2'$, then $\check{L}$ satisfies D_2' and hence $\check{L}$ satisfies $\breve{D}_2'$ and this again implies that L satisfies D_2'.

Exercise 11. Let L be a lattice and $\varnothing \neq S \subseteq L$. The set $\{x \in L : x \leq s \text{ for some } s \in S\}$ is called the *semi-ideal generated* by S. Prove that $\{x \in L : x \leq s \text{ for some } s \in S\}$ is indeed a semi-ideal and is contained in every other semi-ideal that contains S.

Definition 12. (G. N. Raney [1]. Also cf. G. Bruns [1]) Let L be a complete lattice. For $a \in L$, let $(A_s)_{s \in S}$ be the family of all subsets $A_s = (x_{st})_{t \in T_s}$ of L such that $\sum A_s = \sum_{t \in T_s} x_{st} \geq a$. Define $\rho(a)$ to be the semi-ideal of L generated by $\{\prod_{s \in S} x_{s\varphi(s)} : \varphi \in \times_{s \in S} T_s\}$.

Lemma 13. $\sum \rho(a) \leq a$ for each $a \in L$.

Proof. $A_{s_0} = \{a\}$ for some $s_0 \in S$ so $a = x_{s_0\varphi(s_0)} \geq \prod_{s \in S} x_{s\varphi(s)}$ for each $\varphi \in \times_{s \in S} T_s$.

Theorem 14. *Let L be a complete lattice. Then for each $a \in L$, $\rho(a)$ is the intersection of all semi-ideals I for which $\sum I \geq a$.*

Proof. Let $a \in L$ and let $(A_s)_{s \in S}$ be the family of all subsets of L such that if $s \in S$ and $A_s = (x_{st})_{t \in T_s}$, then $\sum A_s = \sum_{t \in T_s} x_{st} \geq a$. Suppose $x \in \rho(a)$. Then there exists $\varphi_o \in \times_{s \in S} T_s$ with $x \leq \prod_{s \in S} x_{s\varphi_o(s)}$. Let I_o be some semi-ideal for which $\sum I_o \geq a$. There exists $s_o \in S$ such that $I_o = A_{s_o}$. Now $x \leq x_{s_o\varphi_o(s_o)}$, so $x \in I_o$, and it follows that

$$\rho(a) \subseteq \bigcap \{I : I \text{ a semi-ideal}, \textstyle\sum I \geq a\}.$$

Conversely, suppose

$$x \in \bigcap \{I : I \text{ a semi-ideal}, \textstyle\sum I \geq a\}.$$

Note that for each $s \in S$ the semi-ideal I generated by A_s has the property that $\sum I \geq a$. It follows that there exists $\varphi \in \times_{s \in S} T_s$ such that $x \leq x_{s\varphi(s)}$ for each $s \in S$ and so $x \leq \prod_{s \in S} x_{s\varphi(s)}$ or $x \in \rho(a)$.

Exercise 15. (cf. Exercise 5) Prove that if L is a complete ring of subsets of a set X, then for every $A \in L$, $\rho(a)$ is the semi-ideal of L generated by the completely join irreducible elements of L contained in A. It follows that $\sum \rho(A) = \bigcup_{B \in \rho(A)} B = A$ and $\rho(A) = \bigcup_{B \in \rho(A)} \rho(B)$.

The following theorem generalizes Exercise 15.

Theorem 16. *Let L be a complete lattice. Then*

(i) $\rho: L \to \mathfrak{S}(L)$ *preserves arbitrary sums.*

(ii) *L is completely distributive if and only if* $\sum \rho(a) = a$ *for each* $a \in L$.

(iii) *If L is completely distributive, then* $\rho(a) = \cup_{x \in \rho(a)} \rho(x)$.

Proof. It is obvious from Theorem 14 that ρ preserves order. Let $A \subseteq L$. To show $\bigcup_{a\in A} \rho(a) = \rho(\sum_{a\in A} a)$. We only need to show $\rho(\sum_{a\in A} a) \subseteq \bigcup_{a\in A} \rho(a)$. Suppose $x \notin \bigcup_{a\in A} \rho(a)$. Thus $x \notin \rho(a)$ for each $a \in A$. Let $a \in A$ and let $\{A_s : s \in S_a\}$ be the set of subsets of L for which $\sum A_s \geq a$ for each $s \in S_a$. There exists, for each $a \in A$, an $s_a \in S_a$ such that $x \nleq y$ for each $y \in A_{s_a}$. Let B be the set-theoretic union of the sets $\{A_{s_a} : a \in A\}$, then $\sum B \geq \sum_{a\in A} a$. But for each $a \in A$, $x \nleq y$ for each $y \in A_{s_a}$, hence $x \nleq y$ for each $y \in B$. Therefore x is not in the semi-ideal generated by B and it follows from Theorem 14 that $x \notin \rho(\sum_{a\in A} a)$ which proves (i). (ii) First suppose L is completely distributive. Pick an element $a \in L$. Let $(A_s = (x_{st})_{s\in T_s})_{s\in S}$ be the sets, as in Definition 12; then $a \leq \sum_{t\in T_s} x_{st}$ for each $s \in S$ and thus

$$a \leq \prod_{s\in S} \sum_{t\in T_s} x_{st} = \sum_{\varphi \in \times_{s\in S} T_s} \prod_{s\in S} x_{s\varphi(s)} = \sum \rho(a).$$

It follows from Lemma 13 that $a = \sum \rho(a)$. Conversely, suppose that $a = \sum \rho(a)$ for each $a \in L$. By Theorem 10 it suffices to show that $f: \mathfrak{S}(L) \to L$ is a complete homomorphism. It follows from the first part of the proof of Theorem 10 that f preserves sums, regardless of whether L is completely distributive. Hence we only need to prove that f preserves products. Let $(I_s)_{s\in S}$ be a set of semi-ideals; then since f is order preserving, it suffices to show that $\prod_{s\in S} f(I_s) \subseteq f(\bigcap_{s\in S} I_s)$ or $\prod_{s\in S} (\sum I_s) \subseteq \sum (\bigcap_{s\in S} I_s)$. Now for each $s_o \in S$: $\rho(\prod_{s\in S} (\sum I_s)) \leq \rho(\sum I_{s_o})$. Also by (i), $\bigcup_{x\in I_s} \rho(x) = \rho(\sum I_s)$ for each $s \in S$ and thus $\rho(\prod_{s\in S} (\sum I_s)) \subseteq \bigcup_{x\in I_{s_o}} \rho(x)$ for each $s_o \in S$. Now let $t \in \rho(\prod_{s\in S} (\sum I_s))$ then $t \in \bigcup_{x\in I_s} \rho(x)$ for each $s \in S$. Thus for each $s \in S$, there exists an $x_s \in I_s$ such that $t \in \rho(x_s)$. But then $t \leq \sum \rho(x_s)$ for each $s \in S$. Also by Lemma 13, $\sum \rho(x_s) \leq x_s$ and thus $t \leq x_s$ for each $s \in S$ so that $t \in \bigcap_{s\in S} I_s$. It follows that $\rho(\prod_{s\in S} (\sum I_s)) \subseteq \bigcap_{s\in S} I_s$ and thus $\sum (\rho(\prod_{s\in S} (\sum I_s))) \leq \sum (\bigcap_{s\in S} I_s)$. But by hypothesis, the left side of this inequality is $\prod_{s\in S} (\sum I_s)$ which yields the desired result.

(iii) We have by (i) and (ii), $\rho(a) = \rho(\sum \rho(a)) = \rho(\sum_{x\in\rho(a)} x) = \bigcup_{x\in\rho(a)} \rho(x)$.

We have seen in Theorem 6 that a complete lattice is isomorphic to a complete ring of sets if and only if every element is the sum of completely join irreducible elements. A weaker result is the following, which is a corollary of Theorem 16.

Corollary 17. *Let L be a complete, completely distributive lattice. Then the map* $\rho: L \to \mathfrak{S}(L)$ *is an order embedding which preserves all sums. Hence L is isomorphic to a* $\sum$*-ring of sets (A* $\sum$*-ring of sets is a ring of sets which is closed under set-theoretic unions.). Dually, L is isomorphic to a* $\prod$*-ring of sets.*

Proof. First recall that ρ preserves order. That ρ preserves arbitrary sums follow from Theorem 16(i). If $\rho(a) \subseteq \rho(b)$, then it follows from Theorem 16(ii) that $a \leq b$. Hence L is isomorphic to a $\sum$-ring of sets. For the dual, consider the dual lattice $\check{L}$. Then $\check{L}$ is also completely distributive and hence $\check{L}$ is isomorphic to a $\sum$-ring R of subsets of a set X. For the sake of convenience we identify $\check{L}$ with R. Let $R' = \{X \sim A : A \in R\}$. Obviously, R' is a $\prod$-ring of subsets of X. Let $g: R \to R'$ be defined by $g(A) = X \sim A$. Let $h: L \to \check{L}$ be the map defined by $h(a) = a$, then $g \circ h: L \to R'$ is an isomorphism and it follows that L is isomorphic to a $\prod$-ring of sets.

Corollary 18. (S. Papert [1], also cf. G. Bruns [1]) *If L is a complete, completely distributive lattice, then L is isomorphic with the ring of open sets of a topological space and with the closed ring of closed sets of a topological space.*

Proof. The $\sum$-ring of subsets of Corollary 17 is a topology (of open sets) for the set L.

Exercise 19. (S. Papert [1]) Let L be a complete distributive lattice. Prove that the following are equivalent:

(i) L is isomorphic with the ring of closed sets of a topological space.

(ii) L is isomorphic to a $\prod$-ring of sets.

(iii) Every element of L is the sum of join irreducible elements.

We will now prove Raney's representation theorem for complete, completely distributive lattices, but before doing so we introduce the following notation. If X is a set and σ is a binary relation on X, then $\sigma \circ \sigma$ is the relation on X defined by $x\sigma \circ \sigma y \Leftrightarrow$ there exists $z \in X$ with $x\sigma z\sigma y$. (Recall that if $x\sigma x$ for each $x \in X$ and $\sigma \circ \sigma = \sigma$ then σ is called a *quasi-ordering*.)

Theorem 20. (G. N. Raney [2]) *Let L be a complete lattice. Then L is completely distributive if and only if L is a regular subdirect product of complete chains.*

Proof. By Theorem 1.16 we only need to prove the "only if" part. Thus, suppose L is completely distributive. Define a binary relation σ on L by $x\sigma y \Leftrightarrow x \in \rho(y)$. If $x \in \rho(y)$, then by Theorem 16(iii) there exists $z \in \rho(y)$ with $x \in \rho(z)$ thus $x\sigma y \Rightarrow x\sigma \circ \sigma y$. Conversely, by the same theorem, if $x\sigma z\sigma y$ then $x \in \rho(y)$ and thus $x\sigma y$. It follows that $\sigma = \sigma \circ \sigma$. Let $(M_s)_{s \in S}$ be the family of subsets of L such that for each $s \in S$,

(1) $x = y$ or $x\sigma y$ or $y\sigma x$ for each pair $x, y \in M_s$, and M_s is maximal subject to this condition.

It easily follows from Zorn's Lemma that if $\varnothing \neq A \subseteq L$ and for $x, y \in A$, either $x = y$ or $x\sigma y$ or $y\sigma x$, then A is contained in some M_s. In particular every element of L is contained in some M_s. Also note that $0 \in M_s$ for each $s \in S$ since $0 \in \rho(x)$ for each $x \in L$. We first show:

(2) For each $s \in S$ and for each $x, y \in M_s$, such that $x\sigma y$, there exists $z \in M_s$ with $x\sigma z\sigma y$.

Suppose that for some $s \in S$ and some $x, y \in M_s$, $x\sigma y$, but $x\sigma u\sigma y$ does not hold for each $u \in M_s$. But $\sigma \circ \sigma = \sigma$, so there exists $z \in L$ with $x\sigma z\sigma y$. Let $u \in M_s$, $u \neq x, y$. Then either $u\sigma x$ or $x\sigma u$. In the first case we have $u\sigma x\sigma z$ so $u\sigma z$. In the latter case we have $y\sigma u$. Indeed, otherwise $u\sigma y$, but then $x\sigma u\sigma y$ which is impossible. So $y\sigma u$. But then $z\sigma y\sigma u$ so $z\sigma u$. Thus we have that for each $u \in M_s$, $u \neq x, y$ either $u\sigma z$ or $z\sigma u$ and also $x\sigma z$ and $z\sigma y$. It follows that M_s is not maximal, a contradiction. We now define, for each $s \in S$, a map f_s from L into the power set of M_s by

$$f_s(a) = \{x \in M_s : x\sigma y\sigma a \text{ for some } y \in M_s\}$$

and let $C_s = f_s[L]$. We claim that C_s is a complete chain, partially ordered by set inclusion and that $f_s : L \to C_s$ is a complete homomorphism for each $s \in S$.

We show first that for each $s \in S$, C_s is a chain. Let $a, b \in L$ and suppose $f_s(a) \nsubseteq f_s(b)$. Then there exists $x \in f_s(a) \sim f_s(b)$, so there exists $y \in M_s$ with $x\sigma y\sigma a$ and $x\sigma u\sigma b$ for no $u \in M_s$. Let $z \in f_s(b)$, then there exists $u \in M_s$ with $z\sigma u\sigma b$. If $x\sigma z$, then $x\sigma z\sigma u\sigma b$ thus $x\sigma u\sigma b$, a contradiction. If $x = z$, then $x \in f_s(b)$ again a contradiction. It follows that $z\sigma x$, ($x, z \in M_s$!) and thus $z\sigma x\sigma y\sigma a$ which implies $z\sigma y\sigma a$. It follows that $z \in f_s(a)$ and thus $f_s(b) \subseteq f_s(a)$. Hence C_s is a chain for each $s \in S$.

Next we show that for each $s \in S$ and for each $T \subseteq L$,

$$(3) f_s(\sum T) = \bigcup f_s[T].$$

Suppose $x \in f_s(\sum T)$. Then there exists $y \in M_s$ with $x\sigma y\sigma(\sum T)$ or $x \in \rho(y)$ and $y \in \rho(\sum T)$. But by Theorem 16(i), $\rho(\sum T) = \bigcup \rho[T]$ so there exists $t_o \in T$ with $y \in \rho(t_o)$. Thus $x\sigma y\sigma t_o$ and since $y \in M_s$, we have that $x \in f_s(t_o)$ and thus $x \in \bigcup f_s[T]$. The converse is also straightforward.

It follows from (3) that sums in C_s exist, and that f_s preserves sums and thus also order. Each C_s also has a 0. In fact, for $x \in L$, $y \in M_s$, $x\sigma y\sigma 0$ implies that $x = 0$ and therefore $f_s(0) = \{0\}$ for each $s \in S$. It follows that each C_s is a complete chain.

We show next that for each $s \in S$ and $T \subseteq L$

$$(4) f_s(\prod T) = \prod f_s[T].$$

It suffices to show that the right side is included in the left side. Suppose $x \in \prod f_s[T]$. Now $\prod f_s[T] \in C_s$, so there exists $b \in L$ such that $f_s(b) = \prod f_s[T]$. Thus there exists $y \in M_s$ with $x\sigma y\sigma b$. Also $x \in M_s$ so by applying (2) twice, we have that there exist $u, v \in M_s$ such that $x\sigma u\sigma v\sigma y\sigma b$, so $v\sigma y\sigma b$. It follows that $v \in f_s(b) = \prod f_s[T] \subseteq \cap f_s[T]$ and so $v \in f_s(t)$ for each $t \in T$. Therefore, for each $t \in T$, there exists $w_t \in M_s$ with $v\sigma w_t\sigma t$, that is, $v \in \rho(w_t)$ and $w_t \in \rho(t)$ and thus by Theorem 16(iii) we have for each $t \in P$, $v \in \rho(t)$. It follows from Theorem 16(ii) that $v \leq t$ for each $t \in T$ and thus $v \leq \prod T$. Therefore $\rho(v) \subseteq \rho(\prod T)$. Now $u\sigma v \Rightarrow u \in \rho(v) \Rightarrow u \in \rho(\prod T) \Rightarrow u\sigma(\prod T) \Rightarrow x\sigma u\sigma(\prod T)$. Also $u \in M_s$ and so $x \in f_s(\prod T)$. We have now shown that for each $s \in S$, $f_s : L \to C_s$ is a complete homomorphism which is onto. The proof of the theorem will now be completed by showing that for $a, b \in L$, $a \neq b$, there exists an $s \in S$ with $f_s(a) \neq f_s(b)$. Suppose not, then $\bigcup_{s \in S} f_s(a)$

$= \bigcup_{s \in S} f_s(b)$. But $\rho(a) = \bigcup_{s \in S} f_s(a)$. Indeed if $x \in \rho(a)$, then $x\sigma a$ and thus there exists an $s_0 \in S$ such that $\{x, a\} \subseteq M_{s_0}$. Hence by (2) there exists $y \in M_{s_0}$ with $x\sigma y\sigma a$. It follows that $x \in f_{s_0}(a)$ and so $x \in \bigcup_{s \in S} f_s(a)$. Conversely, suppose $x \in \bigcup_{s \in S} f_s(a)$ then $x \in f_{s_0}(a)$ for some $s_0 \in S$. Then there exists $y \in M_{s_0}$ with $x\sigma y\sigma a$ and thus $x\sigma a$ or $x \in \rho(a)$. Similarly, $\rho(b) = \bigcup_{s \in S} f_s(b)$. It follows that $\rho(a) = \rho(b)$ and we infer from Corollary 17 that $a = b$ which completes the proof of the theorem.

Remark. Theorem 20 implies that every complete, completely distributive Boolean algebra is isomorphic to a complete field of sets (Theorem 2). Indeed, every C_s has at most two elements.

G. Bruns [1] has considered complete distributive lattices which satisfy conditions of complete distributivity for certain subsets, and he has generalized Raney's theorem to this situation (also cf. S. Papert [1]).

Exercise 21. (G. Raney [2], G. Birkhoff [4]) Let σ be a binary relation on a set X such that $\sigma \circ \sigma = \sigma$. For $A \subseteq X$, let $\Phi(A) = \{x \in A : x\sigma y \text{ for some } y \in A\}$. Let $L(\sigma) = \{\sigma(A) : A \subseteq X\}$ be ordered by set inclusion. Prove that $L(\sigma)$ is a complete, completely distributive lattice and that if $x\sigma x$ for each $x \in X$, then $L(\sigma)$ is a complete ring of sets.

5. Representation of distributive, α-complete lattices

In previous sections we have investigated when a complete distributive lattice L is isomorphic to a complete ring of sets, and when L is a complete homomorphic image of a complete ring of sets. We have seen that complete distributivity played an important role in our discussions. Indeed, recall that in order for L to be a complete homomorphic image of a complete ring of sets, complete distributivity is both necessary and sufficient (Theorem 4.10). On the other hand, this condition is not sufficient—although of course necessary—for L to be a complete ring of sets (Theorem 1.16 and Exercise 4.7). However, we have seen (Theorem 4.2) that if L is a Boolean algebra then this condition is sufficient.

In this section we will consider similar questions for distributive α-lattices. Thus we ask when a distributive α-lattice L is isomorphic to an α-ring of sets. We will, in particular, be interested in a second question: when is a lattice L an α-homomorphic image of an α-ring of sets? If L is an α-homomorphic image of an α-ring of sets, then L is called *α-representable*. Although a detailed treatment of Boolean algebras is beyond the scope of this book, it will be desirable to consider these questions in conjunction with the same questions for α-complete Boolean algebras (with the difference that in this case α-fields take the place of α-rings). In particular, the problem of α-representability is intimately related to the same problem for α-complete Boolean algebras. We make the convention that if we talk about an α-complete Boolean algebra L which is *α-representable*, then we mean that L is an α-homomorphic image of an *α-field of sets*. Again, we can expect that higher degrees of distributivity will play a role. It should be mentioned here that the study of α-representable, α-complete Boolean algebras was, in particular, stimulated and inspired by the classical theorem of Loomis-Sikorski, which states that every $\aleph_0$-complete Boolean algebra is $\aleph_0$-representable (R. Sikorski [2], L. H. Loomis [1]). It should also be pointed out here that there is an essential difference between Boolean algebras and distributive lattices in general, as far as the notion of α-representability is concerned. Indeed, suppose L is an α-complete Boolean algebra which is α-representable, then $L \cong F/I$ where F is an α-field and where I is an ideal which—as can be easily seen—is necessarily α-complete. But on the other hand, if L is an α-lattice which is α-representable, that is, L is an α-homomorphic image of an α-ring R, then L is not necessarily isomorphic to R/I, where I is an ideal in R. However, if R is an α-ring and I is an α-ideal in R, then the homomorphism h_I is necessarily an α-homomorphism (Lemma 4) and therefore R/I is α-representable. But not every α-representable α-lattice is isomorphic to an α-ring modulo an α-ideal. In fact, we will see that the class of α-representable, α-lattices is a proper subclass of the class of α-lattices which are α-homomorphic images of α-rings modulo α-ideals (Corollary 19). We close this introduction by mentioning that the notion of α-representability can also be defined for lattices which are not necessarily α-complete. However, we will restrict ourselves here to α-lattices (See, for example, the discussion of α-representability of Boolean algebras in general by R. Sikorski [8]).

Theorem 1. (C. C. Chang and A. Horn [2]) *Suppose L is a distributive α-lattice. Then L is isomorphic to an α-ring of sets if and only if for each $x, y \in L$, $x \nleq y$, there exists a prime ideal I of L such that $x \notin I$, $y \in I$ and such that I and $L \sim I$ are α-complete.*

The proof is left to the reader (cf. similar proof for ring representation, Section III.5).

Exercise 2. Show that [0, 1] (the real unit interval) is not isomorphic to a $2^{\aleph_0}$-ring of sets.

For the sake of convenience we will state here, again, the definition of α-representability.

Definition 3. Let L be a distributive α-lattice (α-complete Boolean algebra). L is *α-representable* if L is the α-homomorphic image of an α-ring (α-field) of sets.

Note that we could of course, have omitted the word "distributive" in the above definition.

We have already stated that we will also consider classes of lattices which are isomorphic to α-rings of sets modulo α-ideals. For practical reasons, we will consider a somewhat larger class. We first state the following lemma.

Lemma 4. *Let R be an α-ring and let I be an α-ideal in R. Then R/I is an α-lattice and $\nu_I\colon R \to R/I$ is an α-homomorphism.*

Proof. It suffices to show that if $(A_s)_{s\in S}$ and $(B_s)_{s\in S}$, $|S| \leq \alpha$, are subsets of R such that $(A_s, B_s) \in \boldsymbol{\theta}(I)$ for each $s \in S$, then $(\bigcup_{s\in S} A_s, \bigcup_{s\in S} B_s) \in \boldsymbol{\theta}(I)$ and

$(\bigcap_{s\in S} A_s, \bigcap_{s\in S} B_s) \in \boldsymbol{\theta}(I)$. There exists, for each $s \in S$, $U_s \in I$ such that $A_s \cup U_s = B_s \cup U_s$. So

$$\bigcup_{s\in S} A_s \cup \bigcup_{s\in S} U_s = \bigcup_{s\in S} B_s \cup \bigcup_{s\in S} U_s$$

but

$$\bigcup_{s\in S} U_s \in I, \text{ so } \left(\bigcup_{s\in S} A_s, \bigcup_{s\in S} B_s\right) \in \boldsymbol{\theta}(I).$$

$$\bigcap_{s\in S} A_s \cup \bigcup_{S_1 \subset S} \left(\bigcap_{s\in S_1} A_s \cap \bigcup_{s\in S\sim S_1} U_s\right) = \bigcap_{s\in S} (A_s \cup U_s) = \bigcap_{s\in S} (B_s \cup U_s)$$

$$= \bigcap_{s\in S} B_s \cup \bigcup_{S_1 \subset S} \left(\bigcap_{s\in S_1} B_s \cap \bigcap_{s\in S\sim S_1} U_s\right).$$

But obviously

$$\bigcup_{S_1\subset S} \left(\bigcap_{s\in S_1} A_s \cap \bigcap_{s\in S\sim S_1} U_s\right) \in I$$

and

$$\bigcup_{S_1\subset S} \left(\bigcap_{s\in S_1} B_s \cap \bigcap_{s\in S\sim S_1} U_s\right) \in I,$$

so

$$\left(\bigcap_{s\in S} A_s, \bigcap_{s\in S} B_s\right) \in \boldsymbol{\theta}(I).$$

Definition 5. $\mathbf{R}_\alpha$ is the class of α-lattices which are isomorphic to an α-sublattice of an α-ring of sets modulo an α-ideal.

(The reader should be warned that in C. C. Chang and A. Horn [2] a distributive α-lattice is called "α-representable" if it is isomorphic to an α-ring of sets modulo an α-ideal.)

Theorem 6. (C. C. Chang and A. Horn [2]) *Let L be an α-lattice. $L \in \mathbf{R}_\alpha$ if and only if L is isomorphic to an α-sublattice of an α-representable α-complete Boolean algebra.*

Proof. Sufficiency is obvious. For necessity, suppose L is an α-sublattice of R/I where R is an α-ring and I and α-ideal of R. Let F be the α-field generated by R (i.e. F is the smallest α-field containing R). Let J be the ideal in F generated by I. Then J is an α-ideal. Indeed, $J = \{t \in F: t \leq x, x \in I\}$. Thus if $\varnothing \neq T \subseteq J$, $|T| \leq \alpha$, then $\sum T \leq \sum X$ for some $X \subseteq I$, $|X| \leq \alpha$, but $\sum X \in I$, so $\sum T \in I$. It is easy to see that for $x, y \in R$, $\nu_I(x) = \nu_I(y) \Leftrightarrow \nu_J(x) = \nu_J(y)$.

Thus there exists $f \in [R/I, F/J]_{\mathscr{D}}$ which is one-one and satisfies $f \circ \nu_I = \nu_J$. It easily follows from Lemma 4 that f is an α-homomorphism. The map $f|L: L \to f[L]$ is the desired isomorphism.

Under certain conditions $L \in \mathbf{R}_\alpha$ implies that L is isomorphic to an α-ring modulo an α-ideal.

Theorem 7. *Suppose L is an α-lattice and an α-sublattice of an α-ring of sets R, modulo an α-ideal I, and suppose that L has a zero element which is also a zero element of R/I. Then L is isomorphic to an α-ring modulo an α-ideal.*

Proof. Let $R' = \nu_I^{-1}[L]$. Then R' is an α-ring and an α-sublattice of R. Indeed, suppose $(x_s)_{s\in S} \subseteq R'$, $|S| \leq \alpha$ then $\nu_I(x_s) \in L$ for each $s \in S$. So

$$\sum_{s\in S}^{L} \nu_I(x_s) = \sum_{s\in S}^{R/I} \nu_I(x_s) = \nu_I\left(\sum_{s\in S}^{R} x_s\right) \in L.$$

So $\sum_{s\in S}^{R} x_s \in R'$ and similarly for products. It follows that R' is an α-ring and an α-sublattice of R. We also have that $I \subseteq R'$. Indeed if $x \in I$, then $\nu_I(x) = 0_R/I = 0_L \in L$ so $x \in R'$. Thus I is also an α-ideal in R' and it follows that L is isomorphic to R' modulo I.

Theorem 8. (A. Horn [2]) *Suppose L is an α-lattice which is $(\alpha, \mathbf{2}^\alpha)$-meet distributive. Then L is α-representable.*

Proof. (cf. proof of Theorem 4.10) Let L^* be the set of semi-ideals A of L which are generated by subsets of L of cardinality at most $\mathbf{2}^\alpha$ and such that $\sum A$ exists. Since each principal ideal of L belongs to L^*, L^* is not void. We first show that L^* is an α-ring of subsets of L. Let $(A_s)_{s\in S} \subseteq L^*$, $|S| \leq \alpha$ and where for each $s \in S$, A_s is generated by $(a_{st})_{t\in T} \subseteq L$, where $|T| \leq \mathbf{2}^\alpha$. Then $\bigcup_{s\in S} A_s$ is a semi-ideal which is generated by the set $(a_{st})_{s\in S, t\in T}$ thus by a set of cardinality of at most $\mathbf{2}^\alpha$. Also

$$\sum_{\substack{s\in S\\ t\in T}} a_{st} = \sum_{s\in S}\sum_{t\in T} a_{st}$$

exists. Hence $\bigcup_{s\in S} A_s \in L^*$. Again,

$$\bigcap_{s\in S} A_s = \bigcap_{s\in S}\bigcup_{t\in T} (a_{st}]$$

$$= \bigcup_{\varphi\in T^S}\bigcap_{s\in S} (a_{s\varphi(s)}] = \bigcup_{\varphi\in T^S}\left(\prod_{s\in S} a_{s\varphi(s)}\right].$$

But $|T^S| \leq (\mathbf{2}^\alpha)^\alpha = \mathbf{2}^\alpha$. Hence $\bigcap_{s\in S} A_s$ is a semi-ideal which is generated by a set of cardinality of at most $\mathbf{2}^\alpha$. Now L is $(\alpha, \mathbf{2}^\alpha)$-meet distributive, hence

$$\prod_{s\in S}\sum_{t\in T} a_{st} = \sum_{\varphi\in T^S}\prod_{s\in S} a_{s\varphi(s)}.$$

Hence $\sum(\bigcap_{s\in S} A_s)$ exists and $\bigcap_{s\in S} A_s \in L^*$ and thus L is an α-ring. If we now define $h: L^* \to L$ by $h(A) = \sum A$, then it is easy to check that h is an α-homomorphism which is onto.

Corollary 9. *Every α-complete chain is α-representable.*

Proof. Apply Theorem 1.16.

We will now consider the class $\mathbf{R}_\alpha$. It is clear from Theorem 6 that it is useful to consider the class of α-complete, α-representable Boolean algebras first.

We start with a definition and a lemma.

Definition 10. Let L be a distributive α-lattice. Let I be a prime ideal of L and let $(a_s)_{s\in S} \subseteq L$, $|S| \leq \alpha$. *I preserves* $\sum_{s\in S} a_s$ if $(a_s)_{s\in S} \subseteq I$ implies $\sum_{s\in S} a_s \in I$. Likewise, *$L \sim I$ preserves* $\prod_{s\in S} a_s$ if $(a_s)_{s\in S} \subseteq L \sim I$ implies $\prod_{s\in S} a_s \in L \sim I$.

Lemma 11. *Let L be an α-complete Boolean algebra and let I be a prime ideal in L. Suppose $(a_{st})_{s\in S, t\in T} \subseteq L$, $|S| \leq \alpha$, $|T| \leq \alpha$. Then the following are equivalent.*

(i) *I preserves $\sum_{t\in T} a_{st}$ for each $s \in S$.*

(ii) $I \notin \bigcup_{s\in S} (\widehat{\sum_{t\in T} a_{st}} \sim \bigcup_{t\in T} \widehat{a_{st}})$.

The proof is straightforward and is left to the reader.

Exercise 12. Prove Lemma 11.

We now prove a basic theorem which characterizes this class.

Theorem 13. (C. C. Chang [1], R. S. Pierce [2], also cf. R. Sikorski [3]) *Let L be an α-complete Boolean algebra. L is α-representable if and only if, one of the following equivalent conditions hold*:

(i) *If $(a_{st})_{s\in S, t\in T} \subseteq L$, $|S| \leq \alpha$, $|T| \leq \alpha$ then for each $x \in L$, $x \neq 0$, there exists a prime ideal I in L, such that $x \notin I$ and such that I preserves $\sum_{t\in T} a_{st}$ for each $s \in S$.*

(ii) *If $(a_{st})_{s\in S, t\in T} \subseteq L$, $|S| \leq \alpha$, $|T| \leq \alpha$ then for each $x \in L$ $\hat{x} \subseteq \bigcup_{s\in S} (\widehat{\sum_{t\in T} a_{st}} \sim \bigcup_{t\in T} \widehat{a_{st}})$ implies $x = 0$.*

Proof. The equivalence of (i) and (ii) follows immediately from Lemma 11.

We first prove necessity. Thus suppose L is α-representable. We show that (ii) holds. There exists $h \in [L_1, L]_{\mathscr{B}}$ where h is an onto α-homomorphism and where L_1 is an α-field. Suppose

$$\hat{x} \subseteq \bigcup_{s\in S} \Big(\widehat{\sum_{t\in T} a_{st}} \sim \bigcup_{t\in T} \widehat{a_{st}}\Big),$$

$x \in L$, $(a_{st})_{t\in T} \subseteq L$ for each $s \in S$, $|S|$ and $|T| \leq \alpha$. For each $s \in S$, let $a_s = \sum_{t\in T} a_{st}$. Then

$$\bigcup_{s\in S} \Big(\widehat{\sum_{t\in T} a_{st}} \sim \bigcup_{t\in T} \widehat{a_{st}}\Big) = \bigcap_{\varphi\in T^S} \bigcup_{s\in S} \widehat{a_s \bar{a}_{s\varphi(s)}}.$$

Hence $\hat{x} \subseteq \bigcup_{s\in S} \widehat{a_s \bar{a}_{s\varphi(s)}}$ for each $\varphi \in T^S$. By the compactness of $\hat{x}$ there exists, for each $\varphi \in T^S$, a finite subset S_φ of S such that

$$\hat{x} \subseteq \bigcup_{s\in S_\varphi} \widehat{a_s \bar{a}_{s\varphi(s)}}$$

and thus

$$x \leq \sum_{s\in S_\varphi} a_s \bar{a}_{s\varphi(s)}.$$

Let $y \in L_1$ be such that $h(y) = x$. By Exercise II.6.18, we can pick, for each $\varphi \in T^S$ and for each $s \in S_\varphi$ an element $y_{s\varphi(s)} \in L_1$ such that $h(y_{s\varphi(s)}) = a_s\bar{a}_{s\varphi(s)}$ and $y \subseteq \bigcup_{s\in S_\varphi} y_{s\varphi(s)}$. (Recall that L_1 is an α-field, so sums and products in L_1 of at most α elements are unions and intersections respectively.) We now pick, for each $\varphi \in T^S$ and for each $s \in S \sim S_\varphi$, an element $y_{s\varphi(s)} \in L_1$ such that $h(y_{s\varphi(s)}) = a_s\bar{a}_{s\varphi(s)}$. It follows that for each $\varphi \in T^S$ we have $y \subseteq \bigcup_{s\in S} y_{s\varphi(s)}$ and thus

$$y \subseteq \bigcap_{\varphi\in T^S} \bigcup_{s\in S} y_{s\varphi(s)} = \bigcup_{s\in S} \bigcap_{t\in T} y_{st}.$$

Now, since h is an α-homomorphism and $|S|$ and $|T| \leq \alpha$, we obtain, after applying h: $x \leq \sum_{s\in S} \prod_{t\in T} a_s\bar{a}_{st}$. But it is easy to see that for each $s \in S$, $\prod_{t\in T} a_s\bar{a}_{st} = 0$ and it follows that $x = 0$. For sufficiency, let L_1 be the smallest α-field of subsets of $\mathscr{S}(L)$ containing $\hat{L}$. We will prove that $\hat{L}$ is an α-homomorphic image of L_1. Let I be the smallest α-ideal of L_1 containing all sets of the type $\widehat{\sum_{s\in S} a_s} \sim \bigcup_{s\in S} \widehat{a_s}$, $(a_s)_{s\in S} \subseteq L$, $|S| \leq \alpha$. It is easy to see that I consists of all subsets of $\mathscr{S}(L)$ which are contained in sets of the type $\bigcup_{s\in S} (\widehat{\sum_{t\in T} a_{st}} \sim \bigcup_{t\in T} \widehat{a_{st}})$, $(a_{st})_{s\in S, t\in T} \subseteq L$, $|S| \leq \alpha$, $|T| \leq \alpha$. Then by Lemma 4, L_1/I is α-complete and the canonical map $h: L_1 \to L_1/I$ is α-complete. Let $h': \hat{L} \to L_1/I$ be defined by $h' = h|\hat{L}$. Then h' is one-one. Indeed, suppose $h'(\hat{x}) = 0$ for some $x \in L$. But then

$$\hat{x} \subseteq \bigcup_{s\in S} \Big(\widehat{\sum_{t\in T} a_{st}} \sim \bigcup_{t\in T} \widehat{a_{st}}\Big)$$

for some $(a_{st})_{s\in S, t\in T} \subseteq L$, $|S| \leq \alpha$, $|T| \leq \alpha$. It follows from (ii) that $\hat{x} = \varnothing$. Next, we show that h' is α-complete. Suppose $(a_s)_{s\in S} \subseteq L$, $|S| \leq \alpha$ then we have

$$\sum_{s\in S}^{\hat{L}} \widehat{a_s} \supseteq \bigcup_{s\in S} \widehat{a_s},$$

so

$$h'\Big(\sum_{s\in S}^{\hat{L}} \widehat{a_s}\Big) = h\Big(\sum_{s\in S}^{\hat{L}} \widehat{a_s}\Big) = h\Big(\bigcup_{s\in S} \widehat{a_s} \cup \Big(\sum_{s\in S}^{\hat{L}} \widehat{a_s} \sim \bigcup_{s\in S} \widehat{a_s}\Big)\Big)$$

$$= h\Big(\bigcup_{s\in S} \widehat{a_s}\Big) \cup h\Big(\sum_{s\in S}^{\hat{L}} \widehat{a_s} \sim \bigcup_{s\in S} \widehat{a_s}\Big).$$

But

$$\sum_{s\in S}^{\hat{L}} \widehat{a_s} \sim \bigcup_{s\in S} \widehat{a_s} \in I$$

since

$$\sum_{s\in S}^{\hat{L}} \widehat{a_s} = \widehat{\sum_{s\in S} a_s}$$

so

$$h'\Big(\sum_{s\in S}^{\hat{L}} \widehat{a_s}\Big) = h\Big(\bigcup_{s\in S} \widehat{a_s}\Big) = \sum_{s\in S} h(\widehat{a_s}) = \sum_{s\in S} h'(\widehat{a_s}).$$

Thus h' preserves sums of at most α elements, and since h' is a homomorphism between Boolean algebras, h' preserves products of at most α elements. Therefore

h' is α-complete and $h'[\hat{L}]$ is α-regular in L_1/I. But L_1 is the smallest α-field containing $\hat{L}$, and it follows that $h'[\hat{L}] = L_1/I$ and so h' is onto. We conclude that h' is an isomorphism. The map $h'^{-1} \circ h$ is the desired α-complete homomorphism which maps the α-field L_1 onto $\hat{L}$.

An interesting application of Theorem 13 is the following theorem (cf. Theorem 8).

Theorem 14. (R. S. Pierce [1]) *If L is an α-complete, (α, α)-meet distributive Boolean algebra, then L is α-representable.*

Proof. Suppose

$$\hat{x} \subseteq \bigcup_{s \in S} \left(\widehat{\sum_{s \in S} a_{st}} \sim \bigcup_{t \in T} \widehat{a_{st}}\right)$$

for $x \in L$, $(a_{st})_{s \in S, t \in T} \subseteq L$, $|S| \leq \alpha$, $|T| \leq \alpha$. For each $s \in S$, let $a_s = \sum_{t \in T} a_{st}$, we have

$$\begin{aligned}\hat{x} \subseteq \bigcup_{s \in S} \left(\widehat{a_s} \sim \bigcup_{t \in T} \widehat{a_{st}}\right) &= \bigcup_{s \in S} \bigcap_{t \in T} \widehat{a_s \bar{a}_{st}} \\ &= \bigcap_{\varphi \in T^S} \bigcup_{s \in S} \widehat{a_s \bar{a}_{s\varphi(s)}} \subseteq \bigcap_{\varphi \in T^S} \widehat{\sum_{s \in S} a_s \bar{a}_{s\varphi(s)}} \\ &= \mathscr{S}(L) \sim \bigcup_{\varphi \in T^S} \widehat{\prod_{s \in S} (\bar{a}_s + a_{s\varphi(s)})}.\end{aligned}$$

But

$$1 = \prod_{s \in S} \sum_{t \in T} (\bar{a}_s + a_{st}) = \sum_{\varphi \in T^S} \prod_{s \in S} (\bar{a}_s + a_{s\varphi(s)}).$$

So by Exercise IV.3.15,

$$\hat{1} = \mathscr{S}(L) = \mathrm{Cl}\left(\bigcup_{\varphi \in T^S} \widehat{\prod_{s \in S} (\bar{a}_s + a_{s\varphi(s)})}\right).$$

It follows that $\hat{x} = \varnothing$.

We now return to the class $\mathbf{R}_\alpha$.

Lemma 15. (cf. Lemma 11) *Let L be a distributive α-lattice. Let I be a prime ideal of L and let $(a_{st})_{s \in S, t \in T} \subseteq L$ and $(b_{st})_{s \in S, t \in T} \subseteq L$, $|S| \leq \alpha$, $|T| \leq \alpha$. Then the following are equivalent*:

(i) *I preserves $\sum_{t \in T} a_{st}$ and $L \sim I$ preserves $\prod_{t \in T} b_{st}$ for each $s \in S$.*

(ii) $I \notin \bigcup_{s \in S} (\widehat{\sum_{t \in T} a_{st}} \sim \bigcup_{t \in T} \widehat{a_{st}}) \cup \bigcup_{s \in S} (\bigcap_{t \in T} \widehat{b_{st}} \sim \widehat{\prod_{t \in T} b_{st}})$.

The proof which is straightforward is left to the reader.

Exercise 16. Prove Lemma 15.

Theorem 17. (C. C. Chang and A. Horn [2]) *Let L be a distributive α-lattice. $L \in \mathbf{R}_\alpha$ if and only if one of the following equivalent conditions hold*:

(i) *If* $(a_{st})_{s\in S,t\in T} \subseteq L$ *and* $(b_{st})_{s\in S,t\in T} \subseteq L$, $|S| \leq \alpha$, $|T| \leq \alpha$, *then for every* $x, y \in L$, $x \nleq y$ *there exists a prime ideal I in L such that* $x \notin I$, $y \in I$ *and such that I preserves* $\sum_{t\in T} a_{st}$ *for each* $s \in S$ *and such that* $L \sim I$ *preserves* $\prod_{t\in T} b_{st}$ *for each* $s \in S$.

(ii) *If* $(a_{st})_{s\in S,t\in T} \subseteq L$, $(b_{st})_{s\in S,t\in T} \subseteq L$, $|S| \leq \alpha$, $|T| \leq \alpha$, *then for any* $x, y \in L$,

$$\hat{x} \sim \hat{y} \subseteq \bigcup_{s\in S} \left(\widehat{\sum_{t\in T} a_{st}} \sim \bigcup_{t\in T} \widehat{a_{st}}\right) \cup \bigcup_{s\in S} \left(\bigcap_{t\in T} \widehat{b_{st}} \sim \widehat{\prod_{t\in T} b_{st}}\right)$$

implies $x \leq y$.

Proof. The equivalence of (i) and (ii) follows immediately from Lemma 15.

We first prove necessity. Thus, suppose $L \in \mathbf{R}_\alpha$. We show that (i) holds. By Theorem 6 we may assume that L is an α-sublattice of an α-representable α-complete Boolean algebra L_1. Suppose $x \nleq y$ thus $x\bar{y} \neq 0$ so by Theorem 13 there exists a prime ideal I in L_1 such that $x\bar{y} \notin I$ and such that I preserves $\sum_{t\in T} a_{st}$ and $\sum_{t\in T} \bar{b}_{st}$ for each $s \in S$. Let $I' = I \cap L$. Since $x\bar{y} \notin I$ we have $x \notin I$ and $\bar{y} \notin I$ so $I' \neq \varnothing$, $I' \neq L$ and thus I' is a prime ideal in L such that $x \notin I'$ and $y \in I'$. Furthermore if, for $s \in S$, $(a_{st})_{t\in T} \subseteq I'$, then $(a_{st})_{t\in T} \subseteq I$ and thus

$$\sum_{t\in T}^{L_1} a_{st} \in I.$$

But L is an α-sublattice of L_1, thus

$$\sum_{t\in T}^{L} a_{st} = \sum_{t\in T}^{L_1} a_{st} \in I.$$

Likewise, if $(b_{st})_{t\in T} \subseteq L \sim I'$, for $s \in S$, then $(b_{st})_{s\in T} \subseteq L \sim I$ so $(\bar{b}_{st})_{t\in T} \subseteq I$ and thus

$$\sum_{t\in T}^{L_1} \bar{b}_{st} = \left(\overline{\prod_{t\in T}^{L_1} b_{st}}\right) \in I$$

or

$$\prod_{t\in T}^{L_1} b_{st} \notin I$$

and it follows again that

$$\prod_{t\in T}^{L} b_{st} \in L \sim I'.$$

Hence I' is the desired prime ideal in L.

The sufficiency part is similar to that of the proof of Theorem 13: Let L_1 be the smallest α-field of subsets of $\mathscr{S}(L)$ containing $\hat{L}$. We will prove that $\hat{L}$ is an α-sublattice of an α-homomorphic image of L_1. Let I be the smallest α-ideal of L_1 containing all sets of the type $\widehat{\sum_{s\in S} a_s} \sim \bigcup_{s\in S} \widehat{a_s}$ and $\bigcap_{s\in S} \widehat{b_s} \sim \widehat{\prod_{s\in S} b_s}$ for $(a_s)_{s\in S} \subseteq L$. $(b_s)_{s\in S} \subseteq L$, $|S| \leq \alpha$. Again it is easy to see that I consists of all subsets of $\mathscr{S}(L)$ which are contained in sets which are the unions of at most α sets of the above type. By Lemma 4, L_1/I is α-complete and $h_1 : L_1 \to L_1/I$ is α-complete. Let $h' : \hat{L} \to L_1/I$ be defined by $h' = h|\hat{L}$. We prove h' is one-one. Suppose $h'(\hat{x})$

$= h'(\hat{y})$, then $h'(\hat{x}) \leq h'(\hat{y})$ thus $h(\hat{x}) \leq h(\hat{y})$, and there exists $z \in L$, such that $\hat{z} \in I$ and such that $\hat{x} \leq \hat{y} \cup \hat{z}$. Hence

(1) $\hat{z} \subseteq \bigcup_{s\in S} (\widehat{\sum_{t\in T} a_{st}} \sim \bigcup_{t\in T} \widehat{a_{st}}) \cup \bigcup_{s\in S} (\bigcap_{t\in T} \widehat{b_{st}} \sim \widehat{\prod_{t\in T} b_{st}})$ for some $(a_{st})_{s\in S, t\in T}$ and $(b_{st})_{s\in S, t\in T} \subseteq L$, $|S| \leq \alpha$, $|T| \leq \alpha$. Now suppose $\hat{x} \nleq \hat{y}$, then by hypothesis (ii), $\hat{x} \sim \hat{y}$ is not contained in the right side of (1). Hence there exists $J \in \mathscr{S}(L)$ such that $J \in \hat{x} \sim \hat{y}$ but J does not belong to the right side of (1). Hence $J \notin \hat{z}$ and thus $J \notin \hat{y} \cup \hat{z}$. But $J \in \hat{x}$ and thus $\hat{x} \nsubseteq \hat{y} \cup \hat{z}$, a contradiction. It follows that h' is one-one. That h' is α-complete follows from an argument similar to that of the proof of the α-completeness of h' in Theorem 13 (except that one must now also prove that h' preserves products, but that goes in the same way as the proof for sums). It follows that $h'[\hat{L}]$ (which is an isomorphic copy of $\hat{L}$) is an α-sublattice of L_1/I completing the proof of this part of the theorem.

As an application of Theorem 17 we will prove that the class $\mathbf{R}_\alpha$ is a proper subclass of the class of α-representable distributive α-lattices. First, we will characterize the class of chains which are in $\mathbf{R}_\alpha$ and then we will exhibit an α-complete chain which is not in $\mathbf{R}_\alpha$ but which, by virtue of Corollary 9, is α-representable.

Theorem 18. (C. C. Chang and A. Horn [2]) *Let C be an α-complete chain. C belongs to $\mathbf{R}_\alpha$ if and only if every interval $[y, x]$, $y < x$ which is dense in itself (i.e. if $y \leq u < v \leq x$ then there exists z such that $u < z < v$) has cardinality $> \alpha$.*

Proof. Suppose first that $C \in \mathbf{R}_\alpha$ and that $[y, x]$, $y < x$, is dense in itself but has cardinality $\leq \alpha$. For each $z \in (y, x]$, let $(a_{zt})_{t\in T}$ be the set of elements of C which belong to $[y, z)$. Then $|T| \leq \alpha$. Thus $\sum_{t\in T} a_{zt}$ exists for each $z \in (y, x]$ but since $[y, x]$ is dense in itself we have $z = \sum_{t\in T} a_{zt}$ for each $z \in (y, x]$. Similarly, for each $z \in [y, x)$, let $(b_{zt})_{t\in T}$ be the set of elements of C which belong to $(z, x]$. Again $|T| \leq \alpha$ and $z = \prod_{t\in T} b_{zt}$ for each $z \in [y, x)$. (There is of course no objection to taking the same index set T for each z in both cases.) But $C \in \mathbf{R}_\alpha$ and thus by virtue of Theorem 17(i) there exists a prime ideal I in C such that $x \notin I$, $y \in I$, I preserves $\sum_{t\in T} a_{zt}$ for each $z \in (y, x]$ and $C \sim I$ preserves $\prod_{t\in T} b_{zt}$ for each $z \in [y, x)$. Let $I' = I \cap [y, x]$, then $I' \neq \varnothing$ since $y \in I'$ and $|I'| \leq \alpha$. Thus $u = \sum I'$ exists and $u \leq x$. If $u \in I'$, then $u \neq x$ since $x \notin I'$ thus $u < x$. Now $b_{ut} \in (u, x]$ for each $t \in T$. Hence $u < b_{ut}$ and thus $b_{ut} \notin I'$ for each $t \in T$. But then $b_{ut} \notin I$ for each $t \in T$, implying $\prod_{t\in T} b_{ut} \notin I$. It follows that $\prod_{t\in T} b_{ut} > u$. By density, there exists $t_o \in T$, such that $u < b_{ut_o} < \prod_{t\in T} b_{ut}$ which yields a contradiction. Next, suppose $u \notin I'$, then $u \neq y$. Now $a_{ut} \in [y, u)$ for each $t \in T$, thus $a_{ut} \in I'$ and so $a_{ut} \in I$ implying $\sum_{t\in T} a_{ut} \in I$. It follows that $\sum_{t\in T} a_{ut} < u$ which again yields a contradiction.

For the converse, we will show that (i) of Theorem 17 is satisfied. Thus let $(a_{st})_{s\in S, t\in T} \subseteq C$ and $(b_{st})_{s\in S, t\in T} \subseteq C$, $|S| \leq \alpha$, $|T| \leq \alpha$ and let $y, x \in C$ and $y < x$. We have two cases:

(1) $[y, x]$ is not dense in itself. In this case there exist $u, v \in [y, x]$ such that v covers u. Let $I = (u]$. It is easy to see that I satisfies condition (i) of Theorem 17;

(2) $[y, x]$ is dense in itself and the cardinality of $[y, x] > \alpha$. In this case, since the set $\{\prod_{t\in T} b_{st} : s \in S\}$ has cardinality $\leq \alpha$, there exists $z \in [y, x)$ such that

$z \notin \{\prod_{t\in T} b_{st} : s \in S\}$. Let $I = (z]$ then $y \in I$ and $x \notin I$. Also I is α-complete so if $(a_{st})_{t\in T} \subseteq I$ for some $s \in S$, then $\sum_{t\in T} a_{st} \in I$. Finally, if $(b_{st})_{t\in T} \subseteq C \sim I$ for some $s \in S$, then $\prod_{t\in T} b_{st} \neq z$ and so $\prod_{t\in T} b_{st} \in C \sim I$.

We have seen in Exercise 2 that the real unit interval [0, 1] is not isomorphic to a $\mathbf{2}^{\aleph_0}$-ring of sets. On the other hand, it follows from Corollary 5.9 that [0, 1] is α-representable for each α and finally, we infer from Theorem 17 that [0, 1] is not in $\mathbf{R}_\alpha$ for $\alpha \geq \mathbf{2}^{\aleph_0}$. So we have:

Corollary 19. *The real unit interval* [0, 1] *is α-representable for each α, but is not in $\mathbf{R}_\alpha$ for $\alpha \geq \mathbf{2}^{\aleph_0}$.*

We will close this section with an application of our results to the case $\alpha = \aleph_0$. It will be convenient for this purpose to restate part of Theorem 13 in a more topological form. Recall (Exercise IV.3.15) that if L is an α-complete Boolean algebra and $(a_s)_{s\in S} \subseteq L$, $|S| \leq \alpha$, then $\widehat{\sum_{s\in S} a_s} = \mathrm{Cl}(\bigcup_{s\in S} \widehat{a_s})$. For a topological space we will as usual define the *boundary* $\mathrm{Bd}A$ of a subset A by $\mathrm{Bd}A = \mathrm{Cl}\, A \sim \mathrm{Int}\, A$. It is then obvious that condition (i) of Theorem 13 can be reformulated:

Theorem 20. *Let L be an α-complete Boolean algebra. L is α-representable if and only if the following condition holds: If $(A_s)_{s\in S}$ is a family of open subsets of $\mathscr{S}(L)$, $|S| \leq \alpha$, such that for each $s \in S$, A_s is the union of at most α open and closed sets, then $\bigcup_{s\in S} \mathrm{Bd}A_s$ has an empty interior.*

For the case $\alpha = \aleph_0$ we now prove the following classical theorem.

Theorem 21. (L. H. Loomis [1], R. Sikorski [2]) *Every $\aleph_0$-complete Boolean algebra is $\aleph_0$-representable.*

Proof. By Theorem 20, it suffices to prove that if $(U_i)_{i=1,2,\ldots}$ is a countable set of open subsets of $\mathscr{S}(L)$, then

$$\hat{x} \subseteq \bigcup_{i=1}^{\infty} \mathrm{Bd}U_i$$

implies $\hat{x} = \varnothing$ for $x \in L$. Suppose $\hat{x} \neq \varnothing$. But $\mathrm{Bd}\, U_1$ is closed and has an empty interior, so $\hat{x} \sim \mathrm{Bd}\, U_1 \neq \varnothing$ and since $\hat{x} \sim \mathrm{Bd}\, U_1$ is also open, there exists $y_1 \in L$, $y_1 \neq 0$ such that $\widehat{y_1} \subseteq \hat{x} \sim \mathrm{Bd}\, U_1$. Applying the same argument to y_1 and $\widehat{y_1} \sim \mathrm{Bd}\, U_2$, we see that there exists $y_2 \in L$, $y_2 \neq 0$ such that $\widehat{y_2} \subseteq \widehat{y_1} \sim \mathrm{Bd}\, U_2$. Proceeding in this way, we construct a sequence of elements $y_1, y_2, \cdots$ of L, $y_i \neq 0$ for $i = 1, 2, \cdots$ such that $\hat{x} \supseteq \widehat{y_1} \supseteq \widehat{y_2} \supseteq \cdots$ and $\widehat{y_{i+1}} \subseteq \widehat{y_i} \sim \mathrm{Bd}\, U_{i+1}$ for $i = 1, 2, \cdots$. Since each $\widehat{y_i} \neq \varnothing$, we have by the compactness of $\mathscr{S}(L)$ that $\bigcap_{i=1} \widehat{y_i} \neq \varnothing$. Thus there exists a point $p \in \mathscr{S}(L)$ such that $p \in \widehat{y_i}$ for each i but then $p \notin \mathrm{Bd}\, U_i$ for each $i = 1, 2, \cdots$ so

$$p \notin \bigcup_{i=1}^{\infty} \mathrm{Bd}\, U_i$$

and thus $p \notin \hat{x}$. On the other hand, $p \in \widehat{y_1} \subseteq \hat{x}$, a contradiction.

We can now apply Theorem 21 to distributive $\aleph_0$-lattices. First we note that every α-ring of sets is of course $(2, \alpha)$-meet and join distributive and therefore if $L \in \mathbf{R}_\alpha$ then L is $(2, \alpha)$-meet and join distributive. We infer from this and from Theorem 21 and Corollary 2.16:

Corollary 22. (C. C. Chang, A. Horn [2]) *Suppose L is an $\aleph_0$-lattice. Then $L \in \mathbf{R}_{\aleph_0}$ if and only if L is $(2, \aleph_0)$-meet and join distributive.*

Bibliography

Abbott, J. C.
[1] *Sets, lattices and Boolean algebras*, Allyn and Bacon, Boston, 1969.
[2] *Trends in Lattice theory*, Van Nostrand Reinhold, New York, London, 1970.

Abian, S. and Brown, A.
[1] *A theorem on partially ordered sets with applications to fixed point theorems*, Canad. J. Math. **13** (1961), 78–82.

Adams, D.
[1] *Prime and maximal ideals in lattices*, Publ. Math. Debrecen. **17** (1970), 57–59.

Adams, M. E.
[1] *A problem of Monteiro* (to appear).
[2] *The Frattini sublattice of a distributive lattice*, Algebra Universalis **3** (1973), 216–228.
[3] *Maximal chains and antichains in countable distributive lattices* (manuscript).

Ajtai, M.
[1] *On congruence relations and idempotent endomorphisms of distributive lattices*, Ann. Univ. Sci. Budapest. Eötvös Sect. Math. **13** (1970), 93–99.

Amer. Math. Soc.
[1] *Lattice theory*, Proceedings of symposium in pure math. **2**, Amer. Math. Soc., Providence, 1961.

Anderson, F. W. and Blair, R. L.
[1] *Representations of distributive lattices as lattices of functions*, Math. Ann. **143** (1961), 187–211.

Areshkin, G. J.
[1] *On the congruence relations in the distributive structures with* 0 *element* (in Russian), Akad. Nauk SSR **90** (1953), 485–486.

Ash, C. J.
[1] *On countable n-valued Post algebras*, Algebra Universalis **2** (1972), 339–345.

Bacsich, P. D.
[1] *Extension of Boolean homomorphisms with bounding semimorphisms*, J. Reine Angew. Math. **253** (1972), 24–27.

Baker, K. A.
[1] *Equational axioms for classes of Heyting algebras* (manuscript).

Balachandran, V. K.
[1] *The Chinese remainder theorem for the distributive lattice*, J. Indian Math. Soc. **13** (1949), 76–80.
[2] *A characterization of $\Sigma\Delta$-rings of subsets*, Fund. Math. **41** (1954), 38–41.

Balbes, R.
[1] *Projective and injective distributive lattices*, Pacific J. Math. **21** (1967), 405–420.
[2] *A representation theory for prime and implicative semi-lattices*, Trans. Amer. Math. Soc. **136** (1969), 261–267.
[3] *On the partially ordered set of prime ideals of a distributive lattice*, Canad. J. Math. **23** (1971), 866–874.
[4] *The center of the free product of distributive lattices*, Proc. Amer. Math. Soc. **29** (1971), 434–436.
[5] *Solution to a problem concerning the intersection of maximal filters and maximal ideals in a distributive lattice*, Algebra Universalis **2** (1972), 389–392.

Balbes, R. and Dwinger, Ph.
[1] *A note on subdirect products of finite chains*, Algebra Universalis **1** (1971), 125–127.
[2] *Uniqueness of representations of a distributive lattice as a free product of a Boolean algebra and a chain*, Colloq. Math. **24** (1971), 27–35.
[3] *Coproducts of Boolean algebras and chains with applications to Post algebras*, Colloq. Math. **24** (1971), 15–25.

Balbes, R. and Grätzer, G.
[1] *Injective and projective Stone algebras*, Duke Math. J. **38** (1971), 339–347.

Balbes, R. and Horn, A.
[1] *Order sums of distributive lattices*, Pacific J. Math. **21** (1967), 421–435.
[2] *Projective distributive lattices*, Pacific J. Math. **33** (1970), 273–279.
[3] *Stone lattices*, Duke Math. J. **38** (1970), 537–545.
[4] *Injective and projective Heyting algebras*, Trans. Amer. Math. Soc. **148** (1970), 549–559.

Banaschewski, B.
[1] *Injectivity and essential extensions in equational classes of algebras*, Proceedings of the conference on Universal algebra, Queens University, Kingston, Ontario (1969), 131–147.

Banaschewski, B. and Bruns, G.
[1] *Categorical characterization of the MacNeille completion*, Arch. Math. **18** (1967), 369–377.
[2] *Injective hulls in the category of distributive lattices*, J. Reine Angew. Math. **232** (1968), 102–109.

Bell, J. L. and Fremlin, D. H.
[1] *The maximal ideal theorem for lattices of sets*, Bull. London Math. Soc. **4** (1972), 1–2.

Belnap, N. D. Jr. and Spencer, J. H.
[1] *Intensionally complemented distributive lattices*, Portugal. Math. **25** (1966), 99–104.

Berman, J.
[1] *On the congruence relations of pseudocomplemented lattices*, Algebra Universalis (to appear).
[2] *Homogeneous lattices and lattice ordered groups*, Colloq. Math. (to appear).
[3] *Algebras with modular lattice reducts and simple subdirectly irreducibles* (manuscript).

Berman, J. and Dwinger, Ph.
[1] *De Morgan algebras: free products and free algebras* (manuscript).
[2] *Finitely generated pseudocomplemented distributive lattices*, J. Austral. Math. Soc. (to appear).

Bernau, S. J.
[1] *The Boolean ring generated by a distributive lattice*, Proc. Amer. Math. Soc. **33** (1972), 423–424.

Bialynicki-Birula, A.
[1] *Remarks on quasi-Boolean algebras*, Bull. Acad. Polon. Sci. Sér. Sci. Math. Astronom. Phys. **5** (1957), 615–619.

Bialynicki-Birula, A. and Rasiowa, H.
[1] *On the representation of quasi-Boolean algebras*, Bull. Acad. Polon. Sci. Sér. Sci. Math. Astronom. Phys. **5** (1957), 259–261.

Birkhoff, G.
[1] *On the combination of subalgebras*, Proc. Cambridge Philos. Soc. **29** (1933), 441–464.
[2] *Applications of lattice algebras*, Proc. Cambridge Philos. Soc. **30** (1934), 115–122.
[3] *On the structure of abstract algebras*, Proc. Cambridge Philos. Soc. **31** (1935), 433–454.
[4] *On rings of sets*, Duke Math. J. **3** (1937), 443–454.
[5] *Subdirect unions in universal algebra*, Bull. Amer. Math. Soc. **50** (1944), 764–768.
[6] *Lattice Theory*, Amer. Math. Soc. Colloq. Publ., Vol. 25, second edition 1948; third edition, 1967, Providence.

Birkhoff, G. and Frink, O.
[1] *Representation of lattices by sets*, Trans. Amer. Math. Soc. **64** (1948), 299–316.

Blair, R. L.
See F. W. Anderson and R. L. Blair.

Daigneault, A.
[1] *Products of polyadic algebras and of their representations*, Ph.D. Thesis, Princeton University, (1959).

Davey, B.
[1] $\mathfrak{M}$-*Stone lattices*, Canad. J. Math. **24** (1972), 1027–1032.
[2] *A note on representable posets* (manuscript).
[3] *Some annihilator conditions on distributive lattices* (manuscript).

Davies, R., Hayes, A. and Rousseau, G.
[1] *Complete lattices and the generalized Cantor theorem*, Proc. Amer. Math. Soc. **27** (1971), 253–258.

Davis, A.
[1] *A characterization of complete lattices*, Pacific J. Math. **5** (1955), 311–319.

Day, A.
[1] *Injectives in non-distributive equational classes of lattices are trivial*, Arch. Math. **21** (1970), 113–115.
[2] *Injectivity in equational classes of algebras*, Canad. J. Math. **24** (1972), 209–220.
[3] *Varieties of Heyting algebras I* (manuscript).
[4] *Varieties of Heyting algebras II* (*amalgamation and injectivity*) (manuscript).

Day, G. W.
[1] *Maximal chains in atomic Boolean algebras*, Fund. Math, **67** (1970), 293–296

Dedekind, R.
[1] *Über Zerlegungen von Zahlen durch ihre grössten gemeinsamen Teiler*, Festschrift Techn. Hoch. Braunschweig (1897) and Ges. Werke, Vol. 2, 103–148.

Dilworth, R. P.
[1] *Lattices with unique complements*, Trans. Amer. Math. Soc. **57** (1945), 123–154.
[2] *The normal completion of the lattice of continuous functions*, Trans. Amer. Math. Soc. **68** (1950), 427–438.

Dilworth R. P. and McLaughlin, J. E.
[1] *Distributivity in lattices*, Duke Math. J. **19** (1952), 683–694
See also P. Crawley and R. P. Dilworth; M. Ward and R. P. Dilworth.

Dubreil-Jacotin, M. L., Lesieur, L., and Croisot, R.
[1] *Leçons sur la théorie des treillis des structures algébriques ordonnées et des treillis géométriques*, Gauthiers-Villars **21**, Paris, 1953.

Dunford, N. and Stone, M. H.
[1] *On the representation theorem for Boolean algebras*, Rev. Ci. (Lima) **43** (1941), 447–453.

Dwinger, Ph.
[1] *On the axiom of Baer in distributive complete lattices*, Indag. Math. **19** (1957), 220–226.
[2] *Introduction to Boolean algebras*, Physica. Verlag, Würzburg, second edition, 1971.
[3] *Notes on Post algebras I and II*, Indag. Math. **28** (1966), 462–478.
[4] *Notes on Boolean algebras* (manuscript).
[5] *Generalized Post algebras*, Bull Acad. Polon. Sci. Sér. Sci. Math. Astronom. Phys. **6** (1968), 560–565.
[6] *Ideals in generalized Post algebras*, Bull. Acad. Polon. Sci. Sér. Sci. Math. Astronom. Phys. **17** (1969), 483–486.
[7] *Free Post algebras and coproducts of post algebras*, Bull. Acad. Polon, Sci. Sér. Sci. Math. Astronom. Phys. **20** (1972), 535–537.
[8] *Subdirect products of chains*, Colloq. Math. **29** (1974) 201–207.

Dwinger, Ph. and Yaqub, F.
[1] *Generalized free products of Boolean algebras*, Indag. Math. **25** (1963), 225–232.
See also R. Balbes and Ph. Dwinger; J. Berman and Ph. Dwinger; S. D. Comer and Ph. Dwinger.

Engelking, R. and Pelczynsky, A.
[1] *Remarks on dyadic spaces*, Colloq. Math. **11** (1963), 55–63.

Epstein, G.
[1] *The lattice theory of Post algebras*, Trans. Amer. Math. Soc. **95** (1960), 300–317.
[2] *Multiple valued signal processing with limiting*, Symposium on multiple valued logic design, Buffalo, New York, 1972.

Epstein, G. and Horn, A.
[1] *P-algebras, an abstraction from Post algebras* (manuscript).
[2] *Chain based lattices* (manuscript).

Feinstein, D. C.
[1] *Categorical notions of distributive lattices*, Ph.D. Thesis, University of Illinois at Chicago Circle (1972).

Figa-Talamanca, A. and Franklin, S.
[1] *Multipliers of distributive lattices*, Indian J. Math. **12** (1970), 154–161.

Foster, A. L.
[1] *On the finiteness of free (universal) algebras*, Proc. Amer. Math. Soc. **7** (1956), 1011–1013.

Franklin, S.
See A. Figa-Talamanca and S. Franklin.

Fremlin, D. H.
See J. L. Bell and D. H. Fremlin.

Freyd, P.
[1] *Abelian categories*, Harper and Row, New York, 1964.

Frink, O.
[1] *Pseudocomplements in semi-lattices*, Duke Math. J. **29** (1962), 505–514.

Frink, O. and Smith, R.
[1] *On the distributivity of the lattice of filters of a groupoid*, (manuscript).
See also G. Birkhoff and O. Frink.

Fuchs, L.
[1] *Partially ordered algebraic systems*, Pergamon Press, 1963.

Funayama, N.
[1] *On the completion by cuts of distributive lattices*, Proc. of the Imp. Acad. Tokyo **20** (1944), 1–2.
[2] *Imbedding infinitely distributive lattices completely isomorphically into Boolean algebras*, Nagoya Math. J. **15** (1959), 71–81.

Funayama, N. and Nakayama, T.
[1] *On the distributivity of a lattice of lattice congruences*, Proc. of the Imp. Acad. Tokyo **18** (1942), 553–554.

Galvin, F. and Jónsson, B.
[1] *Distributive sublattices of a free lattice*, Canad. J. Math. **13** (1961), 265–272.

Gaskill, H. S.
[1] *On the relation of a distributive lattice to its lattice of ideals*, Bull. Austral. Math. Soc. **7** (1972), 377–388.

Georgescu, G.
[1] *Caractérisation des épimorphismes des algèbres de Lukasiewicz θ-valentes*, C. R. Acad. Sci. Paris Sér. A–B **271** (1970), A708–A710.
[2] *La dualité des algèbres de Post θ-valentes*, C. R. Acad. Sci. Paris Sér. A–B **271** (1970), A766–A768.

Georgescu, G. and Vraciu, C.
[1] *Le spectre maximal d'une algèbre de Lukasiewicz*, C. R. Acad. Sci. Paris Sér. A–B **268** (1969), A300–A317.
[2] *N-valued centered Lukasiewicz algebras*, Roum. Math. **6** (1969), 712–723.
[3] *On the characterization of centered Lukasiewicz algebras*, J. Algebra **16** (1970), 486–495.
[4] *La dualité des algèbres de Post θ-valentes*, J. Algebra **21** (1972), 74–86.

Glivenko, V.
[1] *Sur quelques points de la logique de M. Brouwer*, Bull. Acad. des Sci. de Belgique. **15** (1929), 183–188.

Goetz, A.
[1] *On various Boolean structures on a given Boolean algebra*, Publ. Math. Debrecen. **18** (1971), 103–107.

Grätzer, G.
[1] *A generalization of Stone's representation theorem for Boolean algebras*, Duke Math. J. **30** (1963), 469–474.
[2] *Universal algebra*, Van Nostrand, Princeton, 1968.
[3] *Stone algebras form an equational class (Notes on lattice theory III)*, J. Austral. Math. Soc. **9** (1969), 308–309.
[4] *Lattice theory: First concepts and distributive lattices*, W. H. Freeman Co., San Francisco, 1971.

Grätzer, G. and Lakser, H.
[1] *Chain conditions in the distributive free product of lattices*, Trans. Amer. Math. Soc. **44** (1969), 301–311.
[2] *The structure of pseudocomplemented distributive lattices II: Congruence extension and amalgamation*, Trans. Amer. Math. Soc. **156** (1971), 343–348.
[3] *The structure of pseudocomplemented distributive lattices III: Injective and absolute subretracts*, Trans. Amer. Math. Soc. **169** (1972), 475–487.

Grätzer, G. and Schmidt, E. T.
[1] *On a problem of M. H. Stone*, Acta. Math. Acad. Sci. Hungar. **8** (1957), 455–460.
[2] *Characterizations of relatively complemented distributive lattices*, Publ. Math. Debrecen. **5** (1958), 275–287.
[3] *On the generalized Boolean algebra generated by a distributive lattice*, Indag. Math. **20** (1958), 547–554.
[4] *Ideals and congruence relations in lattices*, Acta Math. Acad. Sci. Hungar. **9** (1958), 137–175.
[5] *On the lattice of all join endomorphisms of a lattice*, Proc. Amer. Math. Soc. **9** (1958), 722–726.
[6] *On congruence relations of lattices*, Acta Math. Acad. Sci. Hungar. **13** (1962), 179–185.
[7] *Standard ideals in lattices*, Acta Math. Acad. Sci. Hungar. **12** (1961), 17–86.

Grätzer, G. and Wolk, B.
[1] *Finite projective distributive lattices*, Canad. Math. Bull. **13** (1970), 139–140.
See also R. Balbes and G. Grätzer; C. C. Chen and G. Grätzer.

Halmos, P. R.
[1] *Injective and Projective Boolean algebras*, Lattice Theory, Proceedings of symposium in pure math. **2** Amer. Math. Soc., Providence, 1961.
[2] *Algebraic logic*, Chelsea Pub. Co., New York, 1962.
[3] *Lectures on Boolean algebras*, Van Nostrand, Princeton, 1963.

Halpern, J. D.
[1] *The independence of the axiom of choice from the Boolean prime ideal theorem*, Fund. Math. **55** (1964), 57–66.

[2] *On a question of Tarski and a maximal theorem of Kurepa*, Pacific J. Math. **41** (1972), 111–121.

Halpern, J. D. and Levy, A.
[1] *The Boolean prime ideal theorem does not imply the axiom of choice*, Proc. Symp. Pure Math. **18** part 1, Amer. Math. Soc. (1970), 83–134.

Hashimoto, J.
[1] *Ideal theory for lattices*, Math. Japon. **2** (1952), 149–186.

Hayes, A.
See R. Davies, A. Hayes and G. Rousseau.

Hecht, T. and Katrinák, T.
[1] *Equational classes of relative Stone algebras*, Notre Dame J. Formal Logic **13** (1972), 248–254.

Henkin, L., Monk, J. D., Tarski, A.
[1] *Cylindric algebras*, Part I, North Holland Pub. Co., 1971.

Heyting, A.
[1] *Die formalen Regeln der intuitionistischen Logik*, Sitzungsberichte der Preussischen Akadamie der Wissenschaften, Phys. mathem. Klasse (1930), 42–56.

Ho, N. C.
[1] *Generalized algebras of Post and their applications to many valued logics with infinitely long formulas*, Polish Acad. of Sciences Bull. of the section on Logic **1** (1972).

Hofmann, K. and Keimel, K.
[1] *A general character theory for partially ordered sets and lattices*, Mem. Amer. Math. Soc. **122** (1972).

Höft, M.
[1] *The order-sum in classes of partially ordered algebras*, Doctoral dissertation, University of Houston.

Holland, C.
[1] *The lattice ordered group of automorphisms of an ordered set*, Michigan Math. J. **10** (1963), 399–408.

Horn, A.
[1] *Normal completion of a subset of a complete lattice*, Pacific J. Math. **3** (1953), 137–152.
[2] *On α-homomorphic images of α-rings of sets*, Fund. Math. **61** (1962), 259–266.
[3] *The separation theorem of intuitionist propositional calculus*, J. Symbolic Logic **27** (1962), 391–399.
[4] Notes from lectures (1965).
[5] *A property of free Boolean algebras*, Proc. Amer. Math. Soc. **19** (1968), 142–143.

[6] *Logic with truth values in a linearly ordered Heyting algebra*, J. Symbolic Logic **34** (1969), 395–408.
[7] *Free L-algebras*, J. Symbolic Logic **34** (1969), 475–480.

Horn, A. and Kimura, N.
[1] *The category of semilattices*, Algebra Universalis **1** (1971), 26–38.

Horn, A. and Tarski, A.
[1] *Measures on Boolean algebras*, Trans. Amer. Math. Soc. **64** (1948), 467–497.
See also R. Balbes and A. Horn; C. C. Chang and A. Horn; G. Epstein and A. Horn.

Iturrioz, L. and Makinson, D.
[1] *Sur les filters premiers d'un treillis distributif et ses sous-treillis*, C. R. Acad. Sci. Paris **270** (1970), 575–577.

Jakubik, J.
[1] *Higher degrees of distributivity in lattices and lattice ordered groups*, Czechoslovak Math. J. **18** (93) (1968), 356–376.

Jónsson, B.
[1] *Universal relational systems*, Math. Scand. **4** (1956), 193–208.
[2] *Algebras whose congruence lattices are distributive*, Math. Scand. **21** (1967), 110–121.
See also F. Galvin and B. Jónsson.

Kalman, J. A.
[1] *Lattices with involution*, Trans. Amer. Math. Soc. **87** (1958), 485–491.

Kaplansky, I.
[1] *Lattices of continuous functions*, Bull. Amer. Math. Soc. 53 (1947), 617–622.

Karp, C. K.
[1] *A note on the representation of α-complete Boolean algebras*, Proc. Amer. Math. Soc. **14** (1963), 705–707.

Katrinák, T.
[1] *Pseudokomplementäre Halbverbände*, Mat. Casopis Sloven. Akad. Vied. **18** (1968), 121–143.
[2] *Charakterisierung der Verallgemeinerten Stoneschen Halbverbände*, Mat. Casopis Sloven. Akad. Vied. **19** (1969), 235–248.
[3] *Remarks on the W. C. Nemitz's paper "Semi-Boolean lattices"*, Notre Dame J. Formal Logic **11** (1970), 425–436.
[4] *Die Kennzeichnung der distributiven pseudokomplementären Halbverbände*, J. Reine Angew. Math. **241** (1970), 160–179.
[5] *Die Kennzeichnung der beschränkten Brouwerschen Verbände*, Czechoslovak Math. J. **22** (1972).
[6] *Notes on Stone lattices I* (Russian), Mat. Casopis Sloven. Akad. Vied. **16** (1966), 128–142.

[7] *Notes on Stone lattices II* (Russian), Mat. Casopis Sloven. Akad. Vied. **17** (1967), 20–37.
[8] *A new proof of the construction theorem for Stone algebras*, Proc. Amer. Math. Soc. **40** (1973), 75–78.
[9] *Die freien Stoneschen Verbände und Ihre Tripelcharakterisierung*, Acta Math. Acad. Sci. Hungar. **23** (1972), 315–326.
[10] *Über eine Konstruction der distributiven pseudokomplementären Verbände*, Math. Nachr. **53** (1971), 85–99.

Katrinák, T. and Mitschke, A.
[1] *Stonesche Verbände der Ordnung n und Postalgebren*, Math. Ann. **199** (1972), 13–20.
See also T. Hecht and T. Katrinák.

Keimel, K.
See K. Hofmann and K. Keimel.

Kimura, N.
See A. Horn and N. Kimura.

Kleene, S. C.
[1] *On notation for ordinal numbers*, J. Symbolic Logic **3** (1938), 150–155.
[2] *Introduction to metamathematics*, North Holland Pub. Co., Amsterdam, 1952.

Kleitman, D.
[1] *On Dedekind's problem: The number of monotone Boolean functions*, Proc. Amer. Math. Soc. **3** (1969), 677–682.

Klimovsky, G.
[1] *El teorema de Zorn y la existencia de filtros e ideales maximales en los reticulados distributivos*, Rev. Un. Mat. Argentina **18** (1958), 160–164.

Köhler, P.
[1] *Freie endlich erzeugte Heyting-Algebren, Diplomarbeit*, Mathematisches Institut Gießen (1973).

Lakser, H.
[1] *Injective hulls of Stone algebras*, Proc. Amer. Math. Soc. **24** (1970), 524–529.
[2] *The structure of pseudocomplemented distributive lattices I*, Trans. Amer. Math. Soc. (1971), 335–342.
[3] *Principal congruences of pseudocomplemented distributive lattices*, Proc. Amer. Math. Soc. **37** (1973), 32–38.
See also G. Bruns and H. Lakser; G. Grätzer and H. Lakser.

Lambek, J.
[1] *Lectures on rings and modules*, Blaisdell Pub. Co., Waltham, Toronto, London, 1966.

Lee, K. B.
[1] *Equational classes of distributive pseudocomplemented lattices*, Canad. J. Math. **22** (1970), 881–891.

Lesieur, L.
See M. L. Dubreil-Jacotin, L. Lesieur and R. Croisot.

Levy, A.
See J. D. Halpern and A. Levy.

Loomis, L. H.
[1] *On the representation of σ-complete Boolean algebras*, Bull. Amer. Math. Soc. **53** (1947), 757–760.

Loś, J. and Ryll-Nardzewski, C.
[1] *Effectiveness of the representation theory for Boolean algebras*, Fund. Math. **41** (1954), 49–56.

Lowig, H.
[1] *Note on self duality of the unrestricted distributive law in complete lattices*, Israel J. Math. **2** (1964), 170–172.

Lukasiewicz, J.
[1] *O logike trojwartosciowej*, Ruch Filozoficzny **5** (1920).
[2] *Elementy logiki matematyezny*, Warszawa, 1921.

Luxemburg, W. A. J.
[1] *A remark on Sikorski's extension theorem for homomorphisms in the theory of Boolean algebras*, Fund. Math. **55** (1964), 239–247.
[2] *Extensions of prime ideals and the existence of projections in Riesz spaces*, Indag. Math. **35** (1973), 263–279.

Luxemburg, W. A. J. and Zaanen, A. C.
[1] *Riesz Spaces, I*, North Holland Mathematical Library, Amsterdam, 1972.

MacLane, S.
[1] *Categories for the working mathematician*, Springer-Verlag, Berlin, Göttingen, Heidelberg, New York, 1971.

MacNeille, H. M.
[1] *Partially ordered sets*, Trans. Amer. Math. Soc. **42** (1937), 416–460.

Maeda, S.
[1] *Infinite distributivity in complete lattices*, Mem. Ehime Univ. Sect. 11 ser. A. **3** (1966), 11–13.

Makinson, D.
[1] *On the number of ultrafilters of an infinite Boolean algebra*. Z. Math. Logik Grundlagen Math. **15** (1969), 121–122.
See also L. Itturioz and D. Makinson.

Mandelker, M.
[1] *Relative annihilators in lattices*, Duke Math. J. **40** (1970), 377–386.

Marek, W. and Traczyk, T.
[1] *Generalized Lukasiewicz algebras*, Bull. Acad. Polon. Sci. Sér. Sci. Math. Astronom. Phys. Vol. 17, (1969), 789–792.

Markov, A. A.
[1] *Constructive Logic* (in Russian), Uspehi Mat. Nauk. **5** (1950), 187–188.

Mayer, R. D. and Pierce, R. S.
[1] *Boolean algebras with ordered bases*, Pacific J. Math. **10** (1960), 925–942.

McCoy, N. H. and Montgomery, D.
[1] *A representation of generalized Boolean rings*, Duke Math. J. **3** (1937), 455–459.

McKinsey, J. C. C. and Tarski, A.
[1] *The algebra of topology*, Ann. of Math. **45** (1944), 141–191.
[2] *On closed elements in closure algebras*, Ann. of Math. **47** (1946), 122–162.
[3] *Some theorems about the sentential calculi of Lewis and Heyting*, J. Symbolic Logic **13** (1948), 1–15.

McLaughlin, J. E.
See R. P. Dilworth and J. E. McLaughlin.

Mitchell, B.
[1] *Theory of categories*, Academic Press, New York, 1965.

Mitschke, A.
See T. Katrinák and A. Mitschke.

Moisil, G. C.
[1] *Recherches sur l'algèbre de la logique*, Annales scientifiques de l'université de Jassy **22** (1935), 1–117.
[2] *Recherches sur les logiques non-chrysippiennes*, Annales scientifiques de l'université de Jassy **26** (1940), 431–466.
[3] *Notes sur les logiques non-chrysippiennes*, Annales scientifiques de l'université de Jassy **27** (1941), 86–98.

Monk, J. D.
See L. Henkin, J. D. Monk, and A. Tarski.

Monteiro, A.
[1] *Sur l'arithmétique des filters premiers*, C. R. Acad. Sci. Paris **225** (1947), 846–848.
[2] *Matrices de Morgan caractéristiques pour le calcul propositional classique*, An. Acad. Brasil Ci. **52** (1960), 1–7.
[3] *Algèbres de Lukasiewicz trivalentes*, Lectures Univ. Nac. del Sur Bahía Blanca, 1963.

[4] *Construction des algèbres de Nelson finies*, Bull. Acad. Polon Sci. Sér. Sci. Math. Astronom. Phys. **11** (1963), 359–362.
[5] *Sur la définition des algèbres de Łukasiewicz trivalentes*, Bull. Acad. Polon Sci. Sér. Sci. Math. Astronom. Phys. **21** (1964).
[6] *Généralisation d'une théorème de R. Sikorski sur les algèbres de Boole*, Bull. Sci. Math. (2) **89** (1965), 65–74.
[7] *Construction des algèbres de Lukasiewicz trivalentes dans les algèbres de Boole monadiques I*, Math. Japon. **12** (1967), 1–23.
See also D. Brignole and A. Monteiro; A. Chateaubriand and A. Monteiro.

Monteiro, L.
[1] *Axiomes independents pour des algèbres de Lukasiewicz trivalentes*, Notas de Lógica Matemática, Instituto de Matemática Universidad del Sur Bahía Blanca **22** (1964).
[2] *Sur les algèbres de Heyting trivalents*, Notas de Lógica Matemática, Instituto de Matemática Universidad del Sur Bahía Blanca **19** (1964).
[3] *Sur les algèbres de Lukasiewicz injectives*, Proc. Japan Acad. **41** (1965), 578–581.

Monteiro, L. and Coppola, L.
[1] *Sur une construction des algèbres de Lukasiewicz trivalentes*, Notas de Lógica Matemática, Instituto de Matemática Universidad del Sur Bahía Blanca **7** (1964).

Montgomery, D.
See N. N. McCoy and D. Montgomery.

Mostowski, A. and Tarski, A.
[1] *Boolesche Ringe mit geordneter Basis*, Fund. Math. **32** (1939), 69–86.

Nachbin, L.
[1] *Une propriété caractéristique des algèbres Booliénnes*, Portugal. Math. **6** (1947), 115–118.
[2] *On a characterization of the lattice of ideals of a Boolean ring*, Fund. Math. **36** (1949), 137–142.

Nakayama, T.
See N. Funayama and T. Nakayama.

Nanzetta, Ph.
[1] *A representation theorem for relatively complemented distributive lattices*, Canad. J. Math. **20** (1968), 756–758.

Nelson, D.
[1] *Constructible falsity*, J. Symbolic Logic **14** (1959), 16–26.

Nemitz, W. C.
[1] *Implicative semi-lattices*, Trans. Amer. Math. Soc. **117** (1965), 128–142.
[2] *Semi-Boolean lattices*, Notre Dame J. Formal Logic **10** (1969), 235–238.

Nemitz, W. C. and Whaley, T.
[1] *Varieties of implicative semilattices*, Pacific J. Math. **37** (1971), 759–769.

Nerode, A.
[1] *Some Stone spaces and recursion theory*, Duke Math. J. **26** (1959), 397–406.

Neumann, J. von and Stone, M. H.
[1] *The determination of representative elements in the residual classes of a Boolean algebra*, Fund. Math. **25** (1935), 353–378.

Nishimura, I.
[1] *On formulas of one variable in intuitionistic propositional calculus*, J. Symbolic Logic **25** (1960), 327–331.

Nöbeling, G.
[1] *Grundlagen der analytischen Topologie*, Berlin, Göttingen, Heidelberg, 1954.

Ogasawara, T.
[1] *Relation between intuitionistic logic and lattice theory*, J. Sci. Hiroshima Univ. Ser. A-1 Math. **9** (1939), 157–164.

Ore, O.
[1] *Structures and group theory II*, Duke Math. J. **4** (1938), 247–267.

Papert, S.
[1] *Which distributive lattices are lattices of closed sets?* Proc. Cambridge Philos. Soc. **55** (1959), 172–176.

Pareigis, B.
[1] *Categories and functors*, Academic Press, New York, 1970.

Peirce, C. S.
[1] *On the algebra of logic*, Amer. J. Math. **3** (1880), 15–57.

Pelczynsky, A.
See R. Engelking and A. Pelczynsky.

Peremans, W.
[1] *Embedding a distributive lattice into a Boolean algebra*, Indag. Math. **60** (1957), 73–81.

Petrescu, I.
[1] *Injective objects in the category of Morgan algebras*, Rev. Roumaine Math. Pures. Appl. **16** (1971), 921–926.

Pierce, R. S.
[1] *Distributivity of Boolean algebras*, Pacific J. Math. **7** (1957), 983–992.
[2] *Representation theorems for certain Boolean algebras*, Proc. Amer. Math. Soc. **10** (1959), 42–50.

[3] *Introduction to the theory of abstract algebras*, Holt, Rinehart & Winston, New York, 1968.
[4] *Distributivity and normal completion of Boolean algebras*, Pacific J. Math. **8** (1958), 133–140.
See also R. D. Mayer and R. S. Pierce.

Pincus, D.
[1] *Independence of the prime ideal theorem from the Hahn-Banach theorem*, Bull. Amer. Math. Soc. **78** (1972), 766–770.

Plonka, J.
[1] *On distributive quasi-lattices*, Fund. Math. **60** (1967), 191–200.

Post, E. L.
[1] *Introduction to a general theory of elementary propositions*, Amer. J. Math. **43** (1921), 163–185.

Priestley, H.
[1] *Representation of distributive lattices by means of ordered Stone spaces*, Bull. London Math. Soc, **2** (1970), 186–190.
[2] *Ordered topological spaces and the representation of distributive lattices*, Proc. London Math. Soc. **3** (24) (1972), 507–530.

Quackenbush, R. W.
[1] *Free products of bounded distributive lattices*, Algebra Universalis **2** (1972), 393–394.

Rabinovic, M. G.
[1] *Remarks on infinitely distributive lattices* (Russian), Mat. Zametki **8** (1970), 95–103.

Raney, G. N.
[1] *Completely distributive lattices*, Proc. Amer. Math. Soc. **3** (1952), 677–680.
[2] *A subdirect-union representation for completely distributive complete lattices*, Proc. Amer. Math. Soc. **4** (1953), 518–522.
[3] *Tight Galois connections and complete distributivity*, Trans. Amer. Math. Soc. **97** (1960), 418–426.

Rasiowa, H.
[1] *N-lattices and constructive logic with strong negation*, Fund. Math. **46** (1958), 61–80.
[2] *Algebraische Charakterisation intuitionstichen Logik mit starker Negation. Constructivity in mathematics* (Proceedings of the Colloquium held at Amsterdam, 1957, edited by A. Heyting) Studies in logic and the foundations of mathematics, Amsterdam, 1959.
[3] *An algebraic approach to non-classical logics*, North Holland Publ. Co., Amsterdam, London, PWN-Polish Scientific Publishers, Warsaw, 1974.

Rasiowa, H. and Sikorski, R.
[1] *The Mathematics of Metamathematics*, Polska Akademia Nauk Monographic Mathematyczne **41**, Warszawa, 1963.
[2] *Algebraic treatment of the notion of satisfiability*, Fund. Math. **37** (1953), 62–95.
See also A. Bialynicki-Birula and H. Rasiowa.

Reznikoff, I.
[1] *Chaines and formules*, C. R. Acad. Sci. Paris **256** (1963), 5021–5023.

Ribenboim, P.
[1] *Characterization of the sup-complement in a distributive lattice with last element*, Summa Brasil. Math. **2** (1949), 43–49.

Rieger, L.
[1] *A note on topological representations of distributive lattices*, Casopis Pest. Mat. **74** (1949), 51–61.
[2] *On the lattice theory of Brouwerian propositional logic*, Acta facultatis rerum naturalium Univ. Carolonae **189** (1949), 1–40.

Rival, I.
[1] *Maximal sublattices of finite distributive lattices*, Proc. Amer. Math. Soc. **37** (1973).

Riviere, N. M.
[1] *Recursive formulas on free distributive lattices*, J. Combinatorial Theory **5** (1968), 229–234.

Rosenbloom, P. C.
[1] *Post algebras. I Postulates and general theory*, Amer. J. Math. **64** (1942), 167–183.

Rosser, J. B.
[1] *Simplified independence proofs*, Academic Press, New York, London, 1969.

Rotman, B.
[1] *Boolean algebras with ordered bases*, Fund. Math. **75** (1972), 187–197.

Rousseau, G.
[1] *Post algebras and pseudo-Post algebras*, Fund. Math. **67** (1970), 133–145.
See also R. Davies, A. Hayes and G. Rousseau.

Rubin, H. and Rubin, J.
[1] *Equivalents of the axiom of choice*, Studies in logic and the foundations of mathematics, North Holland Publishing Co., Amsterdam, 1963.

Rubin, H. and Scott, D.
[1] *Some topological theorems equivalent to the Boolean prime ideal theorem*, Bull. Amer. Math. Soc. (1954), 389.

Rubin, J.
See H. Rubin and J. Rubin.

Ryll-Nardzewski, C.
See J. Loś and C. Ryll-Nardzewski.

Sanin, N.
[1] *O proizvedenü topologiclskih protanstv*, Trudy Mat. Inst. Steklov. **24** (1948).

Sawicka, H.
[1] *On some properties of generalized Post algebras*, Bull. Acad. Polon. Sci. Sér. Sci. Math. Astronom. Phys. **19** (1971), 267–269.
[2] *On some properties of Post algebras with countable chains of constants*, Colloq. Math. **25** (1972), 201–209.

Schmidt, E. T.
[1] *Über die Kongruenzverbände der Verbände*, Publ. Math. Debrecen. **9** (1962), 243–256.
See also G. Grätzer and E. T. Schmidt.

Schröder, E.
[1] *Algebra der logik*, Teubner-Verlag, Leipzig, 1890.

Scott, D.
[1] *Prime ideal theorem for rings, lattices and Boolean algebras*, Bull. Amer. Math. Soc. (1954), 390.
[2] *The independence of certain distributive laws in Boolean algebras*, Trans. Amer. Math. Soc. **84** (1957), 258–261.

Scott, D. and Trotter, H. F.
[1] Proceedings of Symposium, Providence, R.I., 1961.
See also H. Rubin and D. Scott.

Sholander, M.
[1] *Postulates for distributive lattices*, Canad. J. Math. **3** (1951), 28–30.

Sicoe, C. O.
[1] *Sur les ideaux des algèbres Lukasiewicziennes polivalentes*, Rev. Roumaine Math. Pures. Appl. **12** (1967), 391–401.
[2] *Note supra algebrelor Lukasiewicziene polivalente*, Stud. Si. Cerc: Mat. **19** (1967), 1203–1207.
[3] *On many-valued Lukasiewicz algebras*, Proc. Japan Acad. **43** (1967), 725–728.
[4] *Sur la définition des algèbres Lukasiewicziennes polivalentes*, Rev. Roumaine Math. Pures. Appl. **13** (1968), 1027–1030.

Sikorski, R.
[1] *A theorem on extensions of homomorphisms*, Ann. Soc. Pol. Math. **21** (1948), 332–335.

Trotter, H. F.
See D. Scott and H. F. Trotter.

Urquhart, A.
[1] *Free Heyting algebras*, Algebra Universalis **3** (1973), 94–97.
[2] *Free distributive pseudocomplemented lattices*, Algebra Universalis **3** (1973), 13–15.

Varlet, J.
[1] *Contribution à l'étude des treillis pseudo-complémentés et des treillis de Stone*, Mém. Soc. Roy. Sci. Liège Coll. **8** (1963).
[2] *On the characterization of Stone lattices*, Acta Sci. Math. (Szeged) **27** (1966), 81–84.
[3] *Algèbres de Lukasiewicz trivalentes*, Bull. Soc. Roy. Sci. Liège. **37** (1968), 399–408.
[4] *A regular variety of type* (2, 2, 1, 1, 0, 0), Algebra Universalis **2** (1972), 218–223.

Venkatanarasimhan, P. V.
[1] *Pseudo-complements in posets*, Proc. Amer. Math. Soc. **28** (1971), 9–15.

Vorobev, N. N.
[1] *A constructive propositional calculus with strong negation* (Russian), Akad. Nauk SSR **85** (1952), 465–468.

Vraciu, C.
See G. Georgescu and C. Vraciu.

Wade, L. I.
[1] *Post algebras and rings*, Duke Math. J. **12** (1945), 389–395.

Wallman, H.
[1] *Lattices and topological spaces*, Ann. of Math. **39** (1938), 112–126.

Ward, M.
[1] *Note on the order of the free distributive lattice*, Bull. Amer. Math. Soc. (1946) Abstract 52–5–135.

Ward, M. and Dilworth, R. P.
[1] *Residuated lattices*, Trans. Amer. Math. Soc. **45** (1939), 335–354.

Weglorz, B.
[1] *A representation theorem for Post-like algebras*, Colloq. Math. **22** (1970), 35–39.

Whaley, T.
See W. Nemitz and T. Whaley.

Wilker, P.
[1] *f-closure algebras*, Fund. Math. **64** (1969), 105–119.

Wolk, B.
See G. Grätzer and B. Wolk.

Wolk, E. S.
[1] *Dedekind completeness and a fixed point theorem*, Canad. J. Math. **9** (1957), 400–405.

Yamamoto, K.
[1] *Logarithmic order of free distributive lattice*, Math. Soc. Japan **6** (1954), 343–353.

Yaqub, F. M.
See Ph. Dwinger and F. Yaqub.

Zaanen, A. C.
See W. A. J. Luxemburg and A. C. Zaanen.

Index of Authors and Terms

Index of Symbols